Animal Reproduction and Physiology

Animal Reproduction and Physiology

Edited by **Dominic Fasso**

SYRAWOOD
PUBLISHING HOUSE

New York

Published by Syrawood Publishing House,
750 Third Avenue, 9th Floor,
New York, NY 10017, USA
www.syrawoodpublishinghouse.com

Animal Reproduction and Physiology
Edited by Dominic Fasso

International Standard Book Number: 978-1-68286-185-1 (Hardback)

Contents

Preface

The aim of this book is to present researches that have transformed the study of animal physiology and aided its advancement. With a special emphasis on animal reproduction, this book on animal physiology covers an extensive range of topics such as breeding, genetics, animal health and nutrition, etc. It contains a multitude of innovative topics which will prove to be very insightful for students and research scholars of zoology, veterinary sciences and allied disciplines.

This book is a result of research of several months to collate the most relevant data in the field.

When I was approached with the idea of this book and the proposal to edit it, I was overwhelmed. It gave me an opportunity to reach out to all those who share a common interest with me in this field. I had 3 main parameters for editing this text:

1. Accuracy – The data and information provided in this book should be up-to-date and valuable to the readers.
2. Structure – The data must be presented in a structured format for easy understanding and better grasping of the readers.
3. Universal Approach – This book not only targets students but also experts and innovators in the field, thus my aim was to present topics which are of use to all.

Thus, it took me a couple of months to finish the editing of this book.

I would like to make a special mention of my publisher who considered me worthy of this opportunity and also supported me throughout the editing process. I would also like to thank the editing team at the back-end who extended their help whenever required.

Editor

The immune modifying effects of amino acids on gut-associated lymphoid tissue

Megan R Ruth and Catherine J Field[*]

Abstract

The intestine and the gut-associated lymphoid tissue (GALT) are essential components of whole body immune defense, protecting the body from foreign antigens and pathogens, while allowing tolerance to commensal bacteria and dietary antigens. The requirement for protein to support the immune system is well established. Less is known regarding the immune modifying properties of individual amino acids, particularly on the GALT. Both oral and parenteral feeding studies have established convincing evidence that not only the total protein intake, but the availability of specific dietary amino acids (in particular glutamine, glutamate, and arginine, and perhaps methionine, cysteine and threonine) are essential to optimizing the immune functions of the intestine and the proximal resident immune cells. These amino acids each have unique properties that include, maintaining the integrity, growth and function of the intestine, as well as normalizing inflammatory cytokine secretion and improving T-lymphocyte numbers, specific T cell functions, and the secretion of IgA by lamina propria cells. Our understanding of this area has come from studies that have supplemented single amino acids to a mixed protein diet and measuring the effect on specific immune parameters. Future studies should be designed using amino acid mixtures that target a number of specific functions of GALT in order to optimize immune function in domestic animals and humans during critical periods of development and various disease states.

Keywords: Amino acids, Arginine, Epithelium, Glutamate, Glutamine, Gut-associated lymphoid tissue, Intestine, Mucosa

Introduction

It is well established that protein deficiency suppresses the immune response and increases susceptibility to infection. In fact, protein energy malnutrition is hypothesized to be the leading contributor to immune deficiency globally [1]. Although the requirement for protein to support immunity is well defined and part of current recommendations, only recently have investigators begun to explore the potential use of individual dietary amino acids to optimize immune function. Early evidence suggested that amino acids are important energy substrates for immune cells [2-5] and for antioxidant defense mechanisms [6]. There are also critical health states (i.e. burns, trauma, infection, total parenteral (TPN) feeding) or periods of development (i.e. weaning, pregnancy) where it is now accepted that some dietary non-essential amino acids become conditionally essential.

These include arginine, glutamine, glutamate, glycine, proline, taurine and cysteine [7]. This change in need for these amino acids in the diet may be due in part because of their effects on immune function.

The intestine serves not only as the main site of nutrient absorption and amino acid metabolism, but is also the largest immune organ in the body. The intestinal epithelium, while facilitating nutrient absorption, also has a major role in protecting the host from oral pathogens, inducing oral tolerance and maintaining a healthy interaction with commensal bacteria. Indeed both protein and single amino acid deficiencies have been shown to impair the physical integrity and growth of the intestinal epithelium, as well as alter the immune response [8]. This manuscript will review our current understanding of Gut- Associated Lymphoid Tissue (GALT) and examine the immunomodulatory effects of specific amino acids on immunity that occurs or originates in the intestine.

* Correspondence: cjfield@ualberta.ca
Department of Agricultural, Food and Nutritional Science, 4-126A Li Ka Shing Health Research Innovation Centre, University of Alberta, Edmonton, AB T6G 2E1, Canada

The intestinal barrier and the gut associated immune system
GALT, the largest immune organ in the body of humans and domestic animals, contains a variety of immune cell types from the innate and acquired immune systems (as reviewed by [9]). Because of the proximity to the microbiome and the immediate contact with food, it is continually exposed to both 'normal' and potentially dangerous antigens. Accordingly, GALT develops in a manner that allows non-pathogenic substances, such as commensal bacteria, to survive and enables tolerance to food antigens, while protecting the host from pathogenic organisms and other potentially toxic substances [9]. GALT is considered a component of the mucosal immune system and is composed of aggregated tissue including Peyer's patches (PPs) and solitary lymphoid follicles, and non-aggregated cells in the lamina propria, intestinal epithelial cells (IECs), intraepithelial lymphocytes (IELs), as well as mesenteric lymph nodes (MLNs) [9]. Collectively, GALT plays a critical role in the development of the systemic immune response. As a primary site of antigen exposure it primes naïve T- and B-lymphocytes that develop into effector cells which migrate from the intestine to other sites of the body to protect against immune challenges, such as invading pathogens (Figure 1).

GALT has an important role in first line mucosal defenses. The epithelium is protected from large pathogens or particles by a layer of mucin, a glycoprotein secreted from the specialized goblet cell within the endothelium [10]. The IELs are dispersed among the IECs that line intestinal villi and both cell types play a role in gut immune function (Figure 1). Tight junction proteins, such as claudin, occludin and ZO-1, determine the mucosal permeability and regulate the flow of solutes between the IECs [10]. IECs are involved in the intestinal immune response and some consider them an integral part of GALT. They can activate or suppress IELs via secretion of antimicrobial peptides, cytokines and chemokines or through the processing and presentation of antigen in the context of MHC Class I and MHC class II molecules to the IELS [11]. IELs are primarily T-cells but have functions distinct from peripheral T-cells [12]. The types of T-cells present vary widely by species and disease states [13], but the majority are CD8+, CD45RO + (antigen mature), and express adhesion molecules that are thought to be homing signals [12]. In mice and cows/calves, but not humans, the majority of T-cells are $\gamma\delta$ T-cell Receptor + (TCR+) and the remainder are $\alpha\beta$TCR + [13-15]. The primary role of $\gamma\delta$TCR + cells is to induce tolerance and the primary role of the $\alpha\beta$TCR + cells is to induce IgA production [13]. The difference between species may be related to the degree of exposure to the microbiota and different dietary exposure and requirements.

The PPs are lymphoid aggregates that line the intestine and colon and are the primary inductive sites of the mucosal humoral immune response (Figure 1) [16]. The follicle associated epithelium (FAE) layer of the PP contains highly specialized cells called microfold or M cells that continually sample the intestinal contents bringing them in contact with the resident immune cells (primarily B-cells and small numbers of macrophages, dendritic

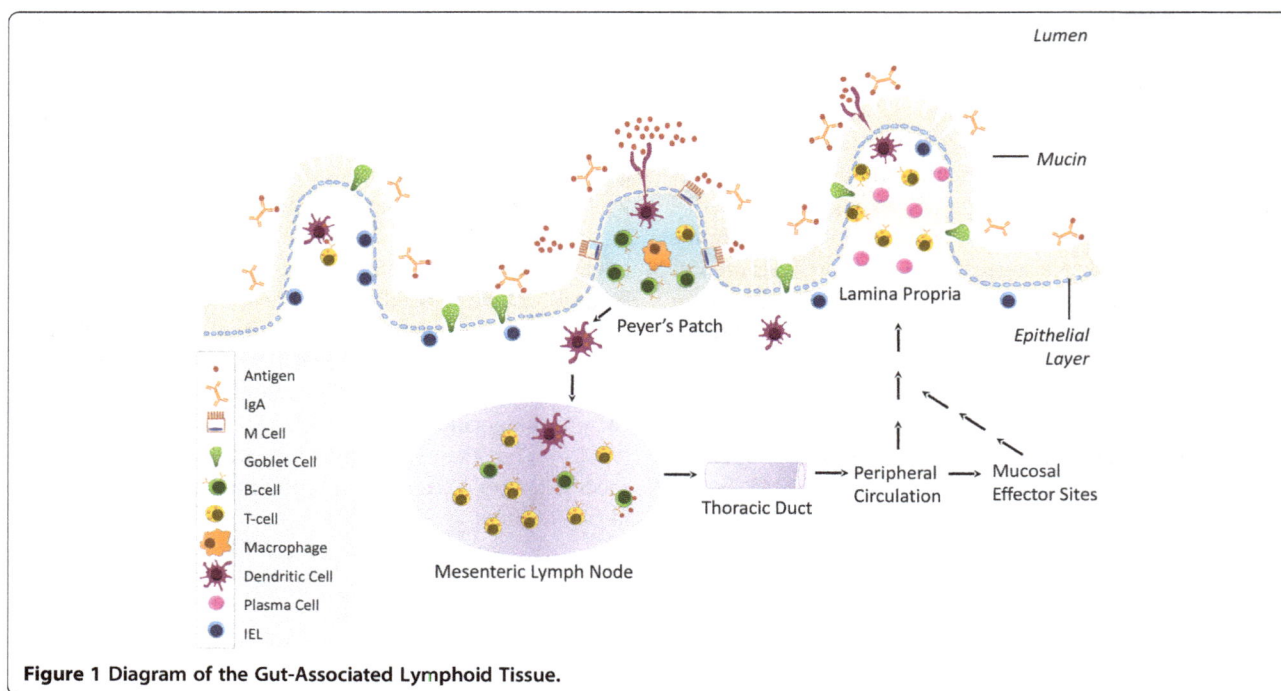

Figure 1 Diagram of the Gut-Associated Lymphoid Tissue.

cells and T-cells) [17]. Dendritic cells can also extend through the intestinal epithelial cells to directly sample antigen [18]. Antigen presenting cells, particularly dendritic cells, migrate from the PP or epithelium to the MLN where they educate naïve T-cells [19]. The MLNs act as the interphase between the peripheral immune system and the gut and it is believed that they are the primary sites of oral tolerance induction [17]. Oral tolerance is mechanistically defined as the process by which dendritic cells present peptides to CD4+ T-cells and through a series of signals (cell surface and secreted) induce regulatory T-cells and subsequently the tolerance to the antigen/peptide. In rats, MLN are composed primarily of T-helper cells (55%), but also contain cytotoxic T-cells (15%), B-cells (25%) and dendritic cells (5%) [20]. Pigs have slightly different phenotypes, with approximately 12% CD4+ CD8+, 25-28% CD4+ (single positive), 27-32% CD8+ (single positive) [21] and the rest composed of B-cells and other antigen presenting cells [22].

After exposure to antigen in the PPs and MLNs, immune cells circulate in the periphery and migrate to other mucosal effector sites and home back to the lamina propria (Figure 1) [23]. This is the major effector component of GALT as these cells are antigen mature and primed to respond to foreign antigens. The lamina propria is comprised primarily of IgA secreting plasma cells and effector T-cells (50% T-helper and 30% cytotoxic T-cells) [24]. Secretory IgA (sIgA) is the most abundant immunoglobulin in the mammalian intestine and acts by binding pathogens and facilitating the entrapment in mucous and removal from the intestinal track [25]. Indeed a deficiency or inability to produce IgA results in frequent intestinal infections [26].

Amino acids and the immune system

Although it has long been established that adequate nutrition is essential to the development and maintenance of the immune system, there is a rapidly growing body of literature that demonstrates the immune benefits of supplementation with specific nutrients, particularly during critical stages of development or disease states, when animals may have a higher demand for essential and non-essential nutrients. Such states include weaning, infectious diseases or chronic inflammatory conditions. The importance of individual amino acids to gut function and immunity has become apparent in recent years due to studies that have supplemented amino acids to animals/humans fed: 1) intravenously (total parenteral nutrition(TPN)), which demonstrates not only the importance of GALT but also the importance to immune functions beyond the intestine; 2) during weaning, which demonstrates the importance of these amino acids to the normal growth and development of the intestine and GALT; and 3) during infection or

chronic inflammation, which has demonstrated the role in regulating inflammation and infectious challenges.

Glutamine

Glutamine has been the most extensively studied amino acid with regards to its effects on GALT and the intestine. In health, glutamine is categorized as a non-essential amino acid and represents the amino acid in highest proportions in the body. However, during periods of stress and during critical stages of development the essentiality of exogenous sources of glutamine is now well-established to support growth [7] and health in young animals [27].

Role as a precursor and energy substrate for immune and epithelial cells

Glutamine is an important energy substrate and precursor for other amino acids and derivatives in immune cells and enterocytes (Table 1) [2-4,28]. In fact, both cell types cannot function without at least some exogenous glutamine [29]. In immune cells, particularly lymphocytes, neutrophils and macrophages, glutamine is used rapidly and metabolized to glutamate, aspartate, lactate and CO_2. Wu et al. [30] demonstrated the main metabolic fates of glutamine in enterocytes from weaning piglets are ammonia, glutamate, alanine, aspartate and CO_2. As a precursor for glutamate, glutamine facilitates the production of glutathione (GSH), an important regulator of redox in enterocytes and lymphocytes [31]. It also provides nitrogen for the synthesis of nucleic acids and proteins that are needed for lymphocytes to proliferate and produce signals such as cytokines [32].

Effects on intestinal function

In addition to its role as an energy substrate, glutamine is important for intestinal development and function, including maintaining the integrity of the gut barrier, the structure of the intestinal mucosa and redox homeostasis (Table 1).

Experimental evidence suggests that glutamine supplementation to weaning animals is beneficial to intestinal health. Wu et al. [28] first demonstrated that oral supplementation with 1% w/w glutamine prevented the decrease in jejunal villus height that occurs following weaning. Oral supplementation of glutamine (0.5-1.0% w/w) to healthy weaning piglets improves measures of intestinal health, including increasing villus height and crypt depth, reducing oxidative stress, lowering the proportion of apoptotic IECs and increasing the proliferative rates of IECs [33,34]. Glutamine supplementation has been demonstrated to reduce the adverse effects of TPN on intestinal function in healthy animals. A TPN solution containing 2% w/v glutamine has improved villus length, crypt depth, tight junction protein expression (occludin, JAM1 and ZO-1), and epithelial permeability [35-37].

Table 1 Summary of the role of amino acids in GALT and the intestine

Amino acid	Functions
Glutamine	• Oxidative substrate for immune cells and IECs
	• Precursor for glutamate/GSH
	• Intestinal growth, structure and function (young animals and disease states)
	• Supports proliferative rates and reduces apoptosis of IECs
	• Protects against E.coli/LPS-induced damage to intestinal structure and barrier function
	• Lowers inflammatory and increases immunoregulatory cytokine production
	• Improves the proliferative responses of IELs and MLN cells
	• Intestinal IgA levels
	• Increases lymphocyte numbers in PP, lamina propria and IELs
Glutamate	• Oxidative substrate for immune cells and IECs
	• Precursor for GSH and other amino acids (i.e. arginine)
	• Intestinal growth, structure and function
	• Acts as Immunotransmitter between dendritic cells and T-cells*
	• Facilitates T-cell proliferation and Th1 and proinflammatory cytokine production
Arginine	• Precursor for NO and glutamate in IECs and immune cells
	• Intestinal growth, structure and function
	• Supports microvasculature of intestinal mucosa
	• Increases expression of HSP70 to protect intestinal mucosa
	• Protects against E.coli/LPS-induced damage to intestinal structure and barrier function
	• Facilitates neutrophil and macrophage killing through iNOS-mediated NO production
	• Increases intestinal IgA levels
	• Lowers inflammatory cytokine levels in intestine
	• Increases T-lymphocytes in lamina propria, PPs, intraepithelial spaces
Methionine & Cysteine	• Precursor for GSH, taurine and cysteine
	• Reduces intestinal oxidative stress
	• Intestinal structure
	• Increases goblet cells and proliferating crypt cells
	• Protects against DSS-induced intestinal damage (colitis model) by lowering inflammation, crypt damage and intestinal permeability.
Threonine	• Mucin synthesis
	• Intestinal structure and function
	• Intestinal IgA levels

*No direct evidence of effects on immune cells in GALT.

In addition to the beneficial effects in healthy animals, we and others have demonstrated that glutamine supplementation may be protective to intestinal health during *E. coli* infection in animals at weaning (Table 1). We previously demonstrated that supplementing the weaning diet of piglets with glutamine (at 4.4% w/w) improved intestinal barrier function (decreased ion movement across mucosa), and maintained tight junction (claudin-1 and occludin) protein expression after an *E. coli* challenge [38]. Similarly, Yi *et al.* [39] reported that weaned piglets fed 2% w/w glutamine for 12 days prior to oral *E. coli* challenge maintained villus height, area and volume similar to uninfected piglets. Suckling piglets supplemented with oral glutamine (3.42 mmol/kg body weight) were

protected against LPS-induced damage to the intestine [40]. Glutamine supplementation (5% w/w) was also reported to improve gut barrier function in a rat model of colitis [41].

Effects on GALT

In vitro and *in vivo* studies have demonstrated the importance of glutamine to B- and T-lymphocyte, neutrophil and macrophage functions (as reviewed by [42]). *In vitro*, glutamine supports the proliferative response of T-cells, plasma cell generation, macrophage inflammatory cytokine production and phagocytosis of neutrophils and macrophages [42]. We and others have shown that glutamine supplementation lowers inflammatory cytokine levels, improves intestinal cytokine mRNA expression, increases immunoregulatory cytokine concentrations and increases the proliferative responses of MLN cells to a B- and T-cell mitogen (pokeweed mitogen) in healthy weaning piglets [22,34,38,43]. We also reported a lower proportion of IgA + cells in the MLN of weaning piglets fed glutamine relative to the control group, suggestive of less intestinal permeability and subsequently lower MLN lymphocyte activation with supplementation (Table 1) [22].

In addition to healthy weaning animals, there is support for a protective effect of glutamine in models of sepsis suggesting a therapeutic role for this amino acid in the infected animal (Table 1). Oral glutamine supplementation (1.1-2% w/v) prior to the induction of sepsis or endotoxemia increased the number of lymphocytes in PPs and lamina propria and normalized intestinal IgA levels of control animals [44-46]. Interestingly, even a single IV bolus of glutamine given immediately following the induction of sepsis appears to be protective. Previous studies have demonstrated that a 0.75 g/kg bolus of glutamine normalized systemic and intestinal inflammatory cytokine levels, increased the number of CD8αα + TCRαβ + and TCRγδ+/CD8αα + IELs , lowered the expression of inflammatory mediators in IELs and reduced IEL apoptosis [47,48]. Glutamine supplementation (4% w/w) also increased the proportion of IgA + cells in the lamina propria in rat models of short bowel syndrome [49].

The importance of glutamine to the intestine is also evident when provided systemically. In healthy mice, a TPN solution containing 2% w/v glutamine was reported to restore intestinal IgA levels, the total number of lymphocytes in PPs, IEL and lamina propria, and improved intestinal levels of regulatory cytokines, IL-10 and IL-4 [37,50,51].

Summary

Overall, animal studies have demonstrated that dietary supplementation with glutamine (0.5%–5% w/w) is required to maintain a healthy intestinal mucosa and support several GALT functions during weaning (lymphocyte counts and proliferative responses, decreased inflammatory cytokine production and increased immunoregulatory cytokines), infection (increased lymphocytes and sIgA levels, decreased inflammatory cytokine levels and IEL apoptosis, intestinal barrier function and structure and IEL proliferation and decreased oxidative stress) and other intestinal inflammatory states (increased sIgA levels). Providing glutamine systemically (TPN studies) have established the importance of glutamine to the health of the intestinal barrier (maintained intestinal structure and function) and for some GALT and other mucosal immune responses (maintained sIgA, lymphocyte and regulatory cytokine levels).

Glutamate

Glutamate is one of the most abundant dietary amino acids, but is found in very low concentrations in plasma [6,52]. This is likely the result of glutamate being a major energy substrate for intestinal epithelial cells [6]. It also serves as a precursor for other amino acids (L-alanine, L-aspartate, L-ornithine and L-proline) and for GSH in the intestine [53]. GSH is essential to maintaining the thiol redox state, which is vital to adequate functioning of enterocytes and immune cells (Table 1) [6].

Effects on intestinal function

Glutamate has a very low capacity to cross biological membranes, and enterocytes contain glutamate transporters in the plasma membrane [54] making them one of the few cells that can rapidly transport and metabolize exogenous glutamate [55]. This contributes to glutamate's recognition as the single most important oxidative substrate for IECs [55]. Dietary glutamate, as both a carbon and nitrogen donor, is the precursor of the conditionally essential amino acid, arginine [55]. Maintaining endogenous arginine synthesis in piglet enterocytes has been demonstrated to be essential for optimal growth [31].

In vitro and in vivo studies have reported that providing glutamate can modulate the intestinal epithelium (Table 1). In an in vitro model of intestinal hyperpermeability (Caco2 cells), glutamate treatment reduced hyperpermeabilty up to 30% [56]. Wu et al. [57] reported that weanling piglets fed 1% w/w dietary glutamate for 20 days had increased jejunal villus height, mucosal thickness and intestinal epithelial cell proliferation. Although the immune functions of the intestine were not specifically measured in these studies, these changes would be consistent with improved intestinal immune function. However, Tsuchioka et al. [58] reported that rats given TPN supplemented with glutamate (6.3% w/v) for 5 days had lower mucosal thickness and villous height in the small intestine relative to control TPN, suggesting a negative effect on the intestinal epithelium when glutamate is provided systemically.

Effects on immune function and GALT

Although immune cells produce considerable amounts of glutamate when provided glutamine [4], investigations into the effects of glutamate on immune cells are limited. It has been recently reported that T-cells, B-cells, dendritic cells and macrophages express glutamate receptors [59,60], suggesting that glutamate likely has an important role in immune cell function. In support, Sturgill et al. [60] reported that purified B-cells and peripheral mononuclear cells produced more IgG and IgE when cultured with glutamate in vitro. In T-cells, glutamate may function as an immunotransmitter, akin to its role as a neurotransmitter, as extracellular concentrations of glutamate have been shown to regulate T-cell responses (Table 1). Pacheco et al.

[61] demonstrated that dendritic cells release glutamate during antigen presentation to T-cells and this released glutamate influences T-cell proliferation and cytokine production. During the early stages of dendritic cell-T-cell interaction, glutamate binds to the constitutively expressed mGlu5R on T-cells to inhibit proliferation and cytokine production; however, later in the interaction glutamate binds to mGlu1R to induce T-cell proliferation and Th1 and proinflammatory cytokine production [61]. This study demonstrates that glutamate plays an essential role in regulating antigen-specific T-cell activation and suggests that the high concentrations of glutamate in the intestine may play an important role in T-cell regulation in the gut.

Despite glutamate being present in high concentrations in the intestinal lumen and immune cells having unique glutamate receptors, there have not been dietary studies that have directly assessed the effect on GALT. Due to the high oxidation rate of glutamate by enterocytes and immune cells, and its role as a precursor for GSH and other amino acids [62] it is reasonable to postulate that changes in the availability of glutamate modulates aspects of GALT (Table 1). We recently reviewed the evidence and presented a hypothesis for a novel role of glutamate receptors on immune cells as the means by which changes in glutamate availability modulates specific immune functions [6]. In that review, we proposed that due to its immunosuppressive effects at concentrations above plasma levels, glutamate may have a key role in the development and maintenance of oral tolerance [6], a unique aspect of immunity in the intestine.

Despite the lack of investigation into the immune modulating properties of glutamate on GALT, it is likely that it has an essential role. To date, the effects of glutamate on GALT have not been examined *in vivo*. However, it is likely that glutamate has an essential role as an oxidative substrate to both enterocytes and immune cells. It is also a precursor for the synthesis of GSH, which is required to protect the intestinal mucosa and optimize immune cell function. And, finally, glutamate is a precursor for arginine, the substrate for the synthesis of NO. A high rate of NO synthesis by neutrophils is required during the innate immune response to infection. This is an important role of the immune system in the intestine.

Summary

Dietary glutamate appears essential for intestinal barrier function and likely other immune functions of the IEC, primarily as a precursor for GSH and as an oxidative substrate for enterocytes. Based on the available data, we can only hypothesize that the availability of glutamate to the cells in GALT has an immunoregulatory role. Studies conducted in systemic immune cells suggest that glutamate is essential for T-cell activation and B-cell immunoglobulin

production and we postulate from indirect evidence that glutamate has a role in the induction of oral tolerance (that originates in GALT) and protection from enteric infections.

Arginine

In most adult mammals, arginine is considered a dietary non-essential amino acid as it can be synthesized from glutamine, glutamate and proline, but becomes conditionally essential during periods of stress [63,64]. Moreover, the absence of arginine in the diet has been shown to have adverse effects in adults, including reproductive, metabolic and neurological derangements [29]. Arginine is classified as an essential amino acid in young mammals as endogenous synthesis cannot meet demands [29]. Several studies have demonstrated that arginine supplementation, either to the piglet's diet or to the lactating sow, improves growth performance in piglets [65-68]. The immune system is particularly sensitive to changes in arginine availability during early development and various disease states.

Metabolism

Arginine is the most plentiful nitrogen carrier in animals and is a precursor for urea, polyamines, proline, creatinine, agmatine, glutamate and protein [64]. Perhaps most importantly, for the immune system, arginine is the only precursor for nitric oxide synthase (all isoforms) for the synthesis of nitric oxide (NO). In both the intestine and immune system, NO is essential for optimal functioning, including regulating the inflammatory response, facilitating killing of microbes by neutrophils and macrophages, and facilitating lymphocyte functions [63].

Effects on intestinal function

The structure and function of the intestine is sensitive to the amount of arginine in the diet during critical periods of development and disease states (Table 1). Studies have shown that arginine supplementation supports the growth and the development of the intestine and mucosal barrier in weanling piglets [65,69,70]. Dietary L-arginine supplementation ranging from 0.6% to 1.0% w/w increased intestinal growth, mucosa microvasculature (0.7% but not 1.2% w/w), villus height, crypt depth, and goblet cell counts in the piglets [65,69,70]. A proposed mechanism is that feeding arginine (0.6% w/w) increases expression of heat shock protein 70 (HSP70) which prevents protein denaturation and associated cellular stress [65].

In addition to supporting normal growth and development, supplementation with arginine has also been reported to reduce intestinal damage induced by *E. coli* derived LPS (Table 1). Sukhotnik *et al.* [71] demonstrated that arginine (2% w/v in drinking water) ameliorated the adverse effects of LPS on the rat intestine, including improving intestinal weight, villous height, epithelial cell proliferation and mucosal DNA and protein. In addition, arginine (0.5 or 1.0% w/w) supplemented to weaned piglets abolished the villous atrophy and morphological changes induced by LPS infection [72]. Arginine supplementation (1% v/v in water) lowered serum concentrations of endotoxin suggestive of improved gut permeability in a rat model of acute pancreatitis [73]. In support of this finding, other researchers have reported that arginine supplementation reduces bacterial counts in mesenteric lymph nodes (4% w/w arginine) [74] and improves gut barrier function (0.33 g/d arginine) [75].

Effects on GALT

The immunomodulatory properties of L-arginine are well established and have been reviewed elsewhere [63,76,77]. Arginine has a fundamental role in both the innate and adaptive immune responses. One of the primary functions of arginine in leukocytes is as a substrate for inducible nitric oxide synthase (iNOS) to produce NO. Macrophages and neutrophils utilize NO to kill a variety of pathogens and malignant cells [63,76]. NO also appears to be important for B-cell development and T-cell receptor function [63]. The effects of arginine on GALT have been studied in both healthy and disease states and the available evidence suggest a beneficial effect on immune function.

Feeding arginine has been shown to be beneficial to GALT in inflammatory and trauma animal models, as well as healthy animals (Table 1). Rats fed diets containing 1% w/w arginine orally prior to the induction of acute pancreatitis had a higher proportion of T-helper cells and an increased ratio of CD4+:CD8+ cells in the intestinal lamina propria, as well as a greater concentration of fecal sIgA [73]. Similarly, Fan *et al.* [78] reported that supplementing arginine (1 g/kg) to severely burned mice for 7 days increased the number of lymphocytes isolated from PPs and intestinal IgA concentrations. Arginine supplemented mice (1 g/kg) also had intestinal cytokine profiles favouring a less inflammatory state (increased IL-4 and IL-10 and lower IFN-γ and IL-2) [78]. In chickens, feeding diets containing 2% w/w arginine improved intraepithelial cytotoxicity to viral infection and improved the antibody response to vaccine, suggesting effects on both cell types of the acquired immune system [79].

Animal models of TPN in both health and disease states have demonstrated that arginine supplementation can reverse the negative effects that TPN (not providing nutrients to the intestine) has on GALT. Mice supplemented with arginine (2 g/kg), prior to (oral diet) and following (TPN), had greater numbers of PPs and lymphocytes isolated from PPs, higher intestinal IgA levels and greater PHA-stimulated IL-10 production (splenocytes) relative to mice given no arginine prior to induction of sepsis [80]. This study suggests that dietary arginine may be essential to maintaining the intestinal immune system during acute infection. Despite these improvements in immune parameters, arginine supplementation in this model of sepsis did not significantly improve survival [80]. However, arginine supplementation to healthy animals fed by TPN also seems to have a similar beneficial effect on GALT. TPN supplemented with 1% w/v arginine given to healthy mice increased the proportion of αβTCR + T-cells and CD4+ T-cells in PPs and intraepithelial spaces compared to mice supplemented with 0.3% w/v arginine [81]. These studies strongly support an essential role for a systemic supply of arginine to maintaining GALT, particularly when the intestine is not receiving nutrients directly from the diet.

Summary

There is considerable support that in health and stressed conditions oral ingestion of arginine (0.6% to 2% w/w) has a beneficial effect on GALT, with particular improvements in aspects of the acquired immune response. Arginine also supports the growth, development and maintenance of a healthy intestinal mucosa during critical periods of development (weaning) and under certain health conditions. These effects on the intestinal mucosa and GALT may be partly explained by arginine's role as an essential precursor for NO.

Other amino acids
Methionine and cysteine–sulfur containing amino acids
The dietary essentiality of methionine and conditional essentiality of cysteine to humans and animals has been well established [82,83]. Currently, there is little direct evidence demonstrating that these sulfur-containing amino acids alter immune function. However, indirectly their efficacy is supported by evidence that their metabolites (taurine, GSH and homocysteine) have immunomodulatory properties *in vitro* [82]. GSH (also see glutamate section) functions as a free radical scavenger and may support proper immune cell function through a role in T-cell proliferation, and inflammatory cytokine regulation [6,82,83]. GSH also has a crucial role in protecting the intestinal epithelium from electrophile and fatty acid hydroperoxide damage [29]. There is evidence that taurine and homocysteine have immunodulatory properties. Taurine is an end product of cysteine metabolism and diets devoid of taurine in cats resulted in reduced lymphocyte numbers, and mononuclear cells with impaired respiratory burst capacity [82]. *In vitro* evidence suggests that taurine chloramine can suppress NF-kappaB activation and pro-inflammatory cytokine (IL-6 and TNF-α) production and in stimulated macrophages [82]. In an *in vitro* model, homocysteine promoted monocyte activation and increased their adhesion to endothelial cells [84]. At present there are no feeding studies to provide direct support for the effect of homocysteine or taurine on immune function in GALT.

There is some evidence that dietary methionine and cysteine are important to ensure the health of the intestine and immune function during development and in inflammatory states (Table 1). For example, Bauchart-Thevret et al. [85] demonstrated that relative to healthy neonatal piglets fed a deficient diet, piglets supplemented with cysteine (0.25 g/kg) and methionine (25 g/kg) had less intestinal oxidative stress, improved villus height and area and crypt depth, higher number of goblet cells and Ki-67+ proliferative crypt cells. Cysteine also appears to be therapeutic in stressed inflammatory states, through improving intestinal inflammation and permeability. An infusion of L-cysteine (0.144 g/kg) given to pigs after DSS-induced colitis lowered mRNA expression of IL-8, MCP-1, MIP-1α, and MIP-2, and normalized IL-6, TNF- α, IFN-γ, IL-12, IL-1β and IL-10 in colon tissue [86]. In addition, less inflammatory cell infiltration, crypt damage and lower intestinal permeability were observed in the pigs supplemented with L-cysteine (Table 1) [86]. While these studies demonstrate the importance of sulfur containing amino acids to gut health in healthy and stressed animals, there is no direct evidence of the effects on lymphocyte or macrophage cell function in GALT.

Threonine

Threonine is a dietary essential amino acid that has been shown to have a particularly high retention rate in the intestine, which suggests an important function in the gut [55,87]. Threonine has a major role in mucin synthesis, a glycoprotein that is required to protect the intestinal epithelium (Table 1) [88]. Mucin production is reduced in diets low or deficient in threonine in healthy rats and piglets [88-91]. Feeding a diet low in threonine (0.37% w/w) was found to adversely affect tight junction ultrastructure in the intestinal epithelium and induce villus atrophy in pigs [91,92], supporting the importance of a dietary supply of threonine in maintaining gut barrier function. Consistent with this, threonine-deficient piglets were found to have higher paracellular permeability which would increase the risk of infectious organisms or their products coming in contact with the body [92]. To date, there are no studies examining the effect of feeding threonine on the function of immune cells in GALT. However, Hamard et al. [92] reported that pigs fed a 30%-reduced threonine diet for two weeks had increased expression of genes involved in inflammation and immunity in the ileum, including MHC Class I antigen (HLA-B), T-cell differentiation antigen CD6, and chemokine receptors. Chickens fed 0.4% w/w threonine in the diet for 8 weeks had higher IgA concentrations in the ileum than chickens fed 0%, 0.1% or 0.2% threonine [93], suggesting an effect on B cell function in the lamina propria (Table 1).

Conclusion

The intestine and the GALT are essential components of immune defense, protecting the animal/human from foreign antigens and pathogens, while allowing the absorption and tolerance of dietary nutrients. Feeding trials, primarily conducted in pigs and rodents, have established convincing evidence that not only the total protein intake but the availability of specific dietary amino acids, in particular glutamine, glutamate, and arginine, and perhaps methionine, cysteine and threonine, are essential to optimizing the immune functions of the intestine and specific immune cells located in GALT. These amino acids modulate their effects by maintaining the integrity, growth and immune functions of the epithelial cells in the intestine, as well as improve T-cell numbers and function, the secretion of IgA, and regulate inflammatory cytokine secretion. The studies conducted using feeding regimes (TPN) that bypass the oral route suggest that amino acids delivered in the blood from other parts of the body are important for maintaining GALT.

To date the majority of the studies have focussed on modulating single amino acids in a diet that contains many different proteins (combinations of amino acids) and determined function by measuring selective (often single parameters) functions. Evidence for some of these immunoactive amino acids comes primarily from in vitro studies or cells isolated from the systemic immune system (blood). Future studies should be designed using amino acid mixtures based on the existing knowledge to optimize immune function and growth in domestic animals and humans during critical periods of intestinal and GALT development in order to optimize health.

Abbreviations

FAE: Follicle associated epithelium; HSP70: Heat shock protein 70; IEC: Intestinal epithelial cell; IEL: Intraepithelial lymphocyte; IgA: Immunoglobulin A; IL: Interleukin; iNOS: Inducible nitric oxide; GALT: Gut-associated lymphoid tissue; GSH: Glutathione; LPS: Lipopolysaccharide; MLN: Mesenteric lymph node; NO: Nitric oxide; PP: Peyer's patches; sIgA: Secretory IgA; TCR: T-cell receptor; Th1: T-helper 1; TNF-α: Tumour necrosis factor-alpha; TPN: Total parenteral nutrition.

Competing interests

MRR and CJF do not have any competing interests to disclose.

Authors' contributions

CJF conceived of the manuscript's purpose and design and critically revised the manuscript. MRR wrote and revised the manuscript according to CJF's suggestions. Both authors read and approved the final manuscript submitted.

Acknowledgements

We thank Ms. Amanda Leong and Ms. Xiaoming Jia for their assistance in developing the figure for this manuscript. This work was supported by CJ Field's funding from the Natural Sciences and Engineering Council of Canada (NSERC).

References

1. Field CJ, Johnson IR, Schley PD: **Nutrients and their role in host resistance to infection.** *J Leukoc Biol* 2002, **71**:16–32.
2. Wu GY, Field CJ, Marliss EB: **Glutamine and glucose metabolism in thymocytes from normal and spontaneously diabetic BB rats.** *Biochem Cell Biol* 1991, **69**:801–808.
3. Wu GY, Field CJ, Marliss EB: **Elevated glutamine metabolism in splenocytes from spontaneously diabetic BB rats.** *Biochem J* 1991, **274**(Pt 1):49–54.
4. Wu GY, Field CJ, Marliss EB: **Glutamine and glucose metabolism in rat splenocytes and mesenteric lymph node lymphocytes.** *Am J Physiol* 1991, **260**:E141–E147.
5. Field CJ, Wu G, Marliss EB: **Enhanced metabolism of glucose and glutamine in mesenteric lymph node lymphocytes from spontaneously diabetic BB rats.** *Can J Physiol Pharmacol* 1994, **72**:827–832.
6. Xue H, Field CJ: **New role of glutamate as an immunoregulator via glutamate receptors and transporters.** *Front Biosci (Schol Ed)* 2011, **3**:1007–1020.
7. Wu G: **Functional amino acids in nutrition and health.** *Amino Acids* 2013 [Epub ahead of print].
8. Ziegler TR, Evans ME, Fernandez-Estivariz C, Jones DP: **Trophic and cytoprotective nutrition for intestinal adaptation, mucosal repair, and barrier function.** *Annu Rev Nutr* 2003, **23**:229–261.
9. Wershil BK, Furuta GT: **4. Gastrointestinal mucosal immunity.** *J Allergy Clin Immunol* 2008, **121**:S380–S383.
10. Turner JR: **Intestinal mucosal barrier function in health and disease.** *Nat Rev Immunol* 2009, **9**:799–809.
11. Artis D: **Epithelial-cell recognition of commensal bacteria and maintenance of immune homeostasis in the gut.** *Nat Rev Immunol* 2008, **8**:411–420.
12. Kunisawa J, Takahashi I, Kiyono H: **Intraepithelial lymphocytes: their shared and divergent immunological behaviors in the small and large intestine.** *Immunol Rev* 2007, **215**:136–153.
13. McGhee J: **Mucosa-Associated Lymphoid Tissue (MALT).** In *Encyclopedia of Immunology.* 2nd edition. Edited by Roitt I, Delves P. United Kingdom: Elsevier Ltd, Academic Press Inc.; 1998:1774–1780.
14. Wyatt CR, Brackett EJ, Perryman LE, Davis WC: **Identification of gamma delta T lymphocyte subsets that populate calf ileal mucosa after birth.** *Vet Immunol Immunopathol* 1996, **52**:91–103.
15. Asai K, Komine Y, Kozutsumi T, Yamaguchi T, Komine K, Kumagai K: **Predominant subpopulations of T lymphocytes in the mammary gland secretions during lactation and intraepithelial T lymphocytes in the intestine of dairy cows.** *Vet Immunol Immunopathol* 2000, **73**:233–240.
16. Bilsborough J, Viney JL: **Getting to the guts of immune regulation.** *Immunology* 2002, **106**:139–143.
17. Newberry RD, Lorenz RG: **Organizing a mucosal defense.** *Immunol Rev* 2005, **206**:6–21.
18. Rescigno M, Rotta G, Valzasina B, Ricciardi-Castagnoli P: **Dendritic cells shuttle microbes across gut epithelial monolayers.** *Immunobiology* 2001, **204**:572–581.
19. Brandtzaeg P: **The gut as communicator between environment and host: immunological consequences.** *Eur J Pharmacol* 2011, **668**(Suppl 1):S16–S32.
20. Ruth MR, Proctor SD, Field CJ: **Effects of feeding fish oil on mesenteric lymph node cytokine responses in obese leptin receptor-deficient JCR: LA-cp rats.** *Int J Obes (Lond)* 2009, **33**:96–103.
21. Zuckermann FA, Gaskins HR: **Distribution of porcine CD4/CD8 double-positive T lymphocytes in mucosa-associated lymphoid tissues.** *Immunology* 1996, **87**:493–499.
22. Johnson IR, Ball RO, Baracos VE, Field CJ: **Glutamine supplementation influences immune development in the newly weaned piglet.** *Dev Comp Immunol* 2006, **30**:1191–1202.
23. Brandtzaeg P: **Mucosal immunity: induction, dissemination, and effector functions.** *Scand J Immunol* 2009, **70**:505–515.
24. McGhee JR, Mestecky J, Dertzbaugh MT, Eldridge JH, Hirasawa M, Kiyono H: **The mucosal immune system: from fundamental concepts to vaccine development.** *Vaccine* 1992, **10**:75–88.
25. Mantis NJ, Forbes SJ: **Secretory IgA: arresting microbial pathogens at epithelial borders.** *Immunol Invest* 2010, **39**:383–406.
26. Agarwal S, Mayer L: **Pathogenesis and treatment of gastrointestinal disease in antibody deficiency syndromes.** *J Allergy Clin Immunol* 2009, **124**:658–664.
27. Wu QY, Li F, Wang XY: **Evidence that amino-acid residues are responsible for substrate synergism of locust arginine kinase.** *Insect Biochem Mol Biol* 2008, **38**:59–65.
28. Wu G, Meier SA, Knabe DA: **Dietary glutamine supplementation prevents jejunal atrophy in weaned pigs.** *J Nutr* 1996, **126**:2578–2584.
29. Wu G: **Amino acids: metabolism, functions, and nutrition.** *Amino Acids* 2009, **37**:1–17.
30. Wu G, Knabe DA, Yan W, Flynn NE: **Glutamine and glucose metabolism in enterocytes of the neonatal pig.** *Am J Physiol* 1995, **268**:R334–R342.
31. Wu G, Fang YZ, Yang S, Lupton JR, Turner ND: **Glutathione metabolism and its implications for health.** *J Nutr* 2004, **134**:489–492.
32. Calder PC, Yaqoob P: **Glutamine and the immune system.** *Amino Acids* 1999, **17**:227–241.
33. Domeneghini C, Di GA, Bosi G, Arrighi S: **Can nutraceuticals affect the structure of intestinal mucosa? Qualitative and quantitative microanatomy in L-glutamine diet-supplemented weaning piglets.** *Vet Res Commun* 2006, **30**:331–342.
34. Wang J, Chen L, Li P, Li X, Zhou H, Wang F, *et al*: **Gene expression is altered in piglet small intestine by weaning and dietary glutamine supplementation.** *J Nutr* 2008, **138**:1025–1032.
35. O'Dwyer ST, Smith RJ, Hwang TL, Wilmore DW: **Maintenance of small bowel mucosa with glutamine-enriched parenteral nutrition.** *JPEN J Parenter Enteral Nutr* 1989, **13**:579–585.
36. Li J, Langkamp-Henken B, Suzuki K, Stahlgren LH: **Glutamine prevents parenteral nutrition-induced increases in intestinal permeability.** *JPEN J Parenter Enteral Nutr* 1994, **18**:303–307.
37. Nose K, Yang H, Sun X, Nose S, Koga H, Feng Y, *et al*: **Glutamine prevents total parenteral nutrition-associated changes to intraepithelial lymphocyte phenotype and function: a potential mechanism for the preservation of epithelial barrier function.** *J Interferon Cytokine Res* 2010, **30**:67–80.
38. Ewaschuk JB, Murdoch GK, Johnson IR, Madsen KL, Field CJ: **Glutamine supplementation improves intestinal barrier function in a weaned piglet model of Escherichia coli infection.** *Br J Nutr* 2011, **106**:870–877.
39. Yi GF, Carroll JA, Allee GL, Gaines AM, Kendall DC, Usry JL, *et al*: **Effect of glutamine and spray-dried plasma on growth performance, small intestinal morphology, and immune responses of Escherichia coli K88 + –challenged weaned pigs.** *J Anim Sci* 2005, **83**:634–643.
40. Haynes TE, Li P, Li X, Shimotori K, Sato H, Flynn NE, *et al*: **L-Glutamine or L-alanyl-L-glutamine prevents oxidant- or endotoxin-induced death of neonatal enterocytes.** *Amino Acids* 2009, **37**:131–142.
41. Vicario M, Amat C, Rivero M, Moreto M, Pelegri C: **Dietary glutamine affects mucosal functions in rats with mild DSS-induced colitis.** *J Nutr* 2007, **137**:1931–1937.
42. Newsholme P: **Why is L-glutamine metabolism important to cells of the immune system in health, postinjury, surgery or infection?** *J Nutr* 2001, **131**:2515S–2522S.
43. Zhong X, Li W, Huang X, Wang Y, Zhang L, Zhou Y, *et al*: **Effects of glutamine supplementation on the immune status in weaning piglets with intrauterine growth retardation.** *Arch Anim Nutr* 2012, **66**:347–356.
44. Alverdy JA, Aoys E, Weiss-Carrington P, Burke DA: **The effect of glutamine-enriched TPN on gut immune cellularity.** *J Surg Res* 1992, **52**:34–38.
45. Lai YN, Yeh SL, Lin MT, Shang HF, Yeh CL, Chen WJ: **Glutamine supplementation enhances mucosal immunity in rats with Gut-Derived sepsis.** *Nutrition* 2004, **20**:286–291.
46. Manhart N, Vierlinger K, Spittler A, Bergmeister H, Sautner T, Roth E: **Oral feeding with glutamine prevents lymphocyte and glutathione depletion of Peyer's patches in endotoxemic mice.** *Ann Surg* 2001, **234**:92–97.
47. Lee WY, Hu YM, Ko TL, Yeh SL, Yeh CL: **Glutamine modulates sepsis-induced changes to intestinal intraepithelial gammadeltaT lymphocyte expression in mice.** *Shock* 2012, **38**:288–293.
48. Tung JN, Lee WY, Pai MH, Chen WJ, Yeh CL, Yeh SL: **Glutamine modulates CD8alphaalpha TCRalphabeta intestinal intraepithelial lymphocyte expression in mice with polymicrobial sepsis.** *Nutrition* 2013, **29**:911–917.
49. Tian J, Hao L, Chandra P, Jones DP, Willams IR, Gewirtz AT, *et al*: **Dietary glutamine and oral antibiotics each improve indexes of gut barrier function in rat short bowel syndrome.** *Am J Physiol Gastrointest Liver Physiol* 2009, **296**:G348–G355.
50. Li J, Kudsk KA, Janu P, Renegar KB: **Effect of glutamine-enriched total parenteral nutrition on small intestinal gut-associated lymphoid tissue and upper respiratory tract immunity.** *Surgery* 1997, **121**:542–549.

51. Kudsk KA, Wu Y, Fukatsu K, Zarzaur BL, Johnson CD, Wang R, et al: Glutamine-enriched total parenteral nutrition maintains intestinal interleukin-4 and mucosal immunoglobulin A levels. JPEN J Parenter Enteral Nutr 2000, 24:270–274.

52. Blachier F, Boutry C, Bos C, Tome D: Metabolism and functions of L-glutamate in the epithelial cells of the small and large intestines. Am J Clin Nutr 2009, 90:814S–821S.

53. Reeds PJ, Burrin DG, Stoll B, Jahoor F, Wykes L, Henry J, et al: Enteral glutamate is the preferential source for mucosal glutathione synthesis in fed piglets. Am J Physiol 1997, 273:E408–E415.

54. Berger UV, Hediger MA: Distribution of the glutamate transporters GLT-1 (SLC1A2) and GLAST (SLC1A3) in peripheral organs. Anat Embryol (Berl) 2006, 211:595–606.

55. Reeds PJ: Dispensable and indispensable amino acids for humans. J Nutr 2000, 130:1835S–1840S.

56. Vermeulen MA, De Jong J, Vaessen MJ, van Leeuwen PA, Houdijk AP: Glutamate reduces experimental intestinal hyperpermeability and facilitates glutamine support of gut integrity. World J Gastroenterol 2011, 17:1569–1573.

57. Wu X, Zhang Y, Liu Z, Li TJ, Yin YL: Effects of oral supplementation with glutamate or combination of glutamate and N-carbamylglutamate on intestinal mucosa morphology and epithelium cell proliferation in weanling piglets. J Anim Sci 2012, 90(Suppl 4):337–339.

58. Tsuchioka T, Fujiwara T, Sunagawa M: Effects of glutamic acid and taurine on total parenteral nutrition. J Pediatr Surg 2006, 41:1566–1572.

59. Rezzani R, Corsetti G, Rodella L, Angoscini P, Lonati C, Bianchi R: Cyclosporine-A treatment inhibits the expression of metabotropic glutamate receptors in rat thymus. Acta Histochem 2003, 105:81–87.

60. Sturgill JL, Mathews J, Scherle P, Conrad DH: Glutamate signaling through the kainate receptor enhances human immunoglobulin production. J Neuroimmunol 2011, 233:80–89.

61. Pacheco R, Oliva H, Martinez-Navio JM, Climent N, Ciruela F, Gatell JM, et al: Glutamate released by dendritic cells as a novel modulator of T cell activation. J Immunol 2006, 177:6695–6704.

62. Nakamura H, Kawamata Y, Kuwahara T, Torii K, Sakai R: Nitrogen in dietary glutamate is utilized exclusively for the synthesis of amino acids in the rat intestine. Am J Physiol Endocrinol Metab 2013, 304:E100–E108.

63. Li P, Yin YL, Li D, Kim SW, Wu G: Amino acids and immune function. Br J Nutr 2007, 98:237–252.

64. Wu G, Meininger CJ, Knabe DA, Bazer FW, Rhoads JM: Arginine nutrition in development, health and disease. Curr Opin Clin Nutr Metab Care 2000, 3:59–66.

65. Wu X, Ruan Z, Gao Y, Yin Y, Zhou X, Wang L, et al: Dietary supplementation with L-arginine or N-carbamylglutamate enhances intestinal growth and heat shock protein-70 expression in weanling pigs fed a corn- and soybean meal-based diet. Amino Acids 2010, 39:831–839.

66. Tan B, Li XG, Kong X, Huang R, Ruan Z, Yao K, et al: Dietary L-arginine supplementation enhances the immune status in early-weaned piglets. Amino Acids 2009, 37:323–331.

67. Mateo RD, Wu G, Bazer FW, Park JC, Shinzato I, Kim SW: Dietary L-arginine supplementation enhances the reproductive performance of gilts. J Nutr 2007, 137:652–656.

68. Kim SW, Wu G: Dietary arginine supplementation enhances the growth of milk-fed young pigs. J Nutr 2004, 134:625–630.

69. Yao K, Guan S, Li T, Huang R, Wu G, Ruan Z, et al: Dietary L-arginine supplementation enhances intestinal development and expression of vascular endothelial growth factor in weanling piglets. Br J Nutr 2011, 105:703–709.

70. Zhan Z, Ou D, Piao X, Kim SW, Liu Y, Wang J: Dietary arginine supplementation affects microvascular development in the small intestine of early-weaned pigs. J Nutr 2008, 138:1304–1309.

71. Sukhotnik I, Mogilner J, Krausz MM, Lurie M, Hirsh M, Coran AG, et al: Oral arginine reduces gut mucosal injury caused by lipopolysaccharide endotoxemia in rat. J Surg Res 2004, 122:256–262.

72. Zhu H, Liu Y, Xie X, Huang J, Hou Y: Effect of L-arginine on intestinal mucosal immune barrier function in weaned pigs after Escherichia coli LPS challenge. Innate Immun 2013, 19:242–252.

73. Qiao SF, Lu TJ, Sun JB, Li F: Alterations of intestinal immune function and regulatory effects of L-arginine in experimental severe acute pancreatitis rats. World J Gastroenterol 2005, 11:6216–6218.

74. Gurbuz AT, Kunzelman J, Ratzer EE: Supplemental dietary arginine accelerates intestinal mucosal regeneration and enhances bacterial clearance following radiation enteritis in rats. J Surg Res 1998, 74:149–154.

75. Ersin S, Tuncyurek P, Esassolak M, Alkanat M, Buke C, Yilmaz M, et al: The prophylactic and therapeutic effects of glutamine- and arginine-enriched diets on radiation-induced enteritis in rats. J Surg Res 2000, 89:121–125.

76. Field CJ, Johnson I, Pratt VC: Glutamine and arginine: immunonutrients for improved health. Med Sci Sports Exerc 2000, 32:S377–S388.

77. Popovic PJ, Zeh HJ III, Ochoa JB: Arginine and immunity. J Nutr 2007, 137:1681S–1686S.

78. Fan J, Meng Q, Guo G, Xie Y, Li X, Xiu Y, et al: Effects of early enteral nutrition supplemented with arginine on intestinal mucosal immunity in severely burned mice. Clin Nutr 2010, 29:124–130.

79. Tayade C, Koti M, Mishra SC: L-Arginine stimulates intestinal intraepithelial lymphocyte functions and immune response in chickens orally immunized with live intermediate plus strain of infectious bursal disease vaccine. Vaccine 2006, 24:5473–5480.

80. Shang HF, Wang YY, Lai YN, Chiu WC, Yeh SL: Effects of arginine supplementation on mucosal immunity in rats with septic peritonitis. Clin Nutr 2004, 23:561–569.

81. Fukatsu K, Ueno C, Maeshima Y, Hara E, Nagayoshi H, Omata J, et al: L-arginine-enriched parenteral nutrition affects lymphocyte phenotypes of gut-associated lymphoid tissue. JPEN J Parenter Enteral Nutr 2004, 28:246–250.

82. Grimble RF: The effects of sulfur amino acid intake on immune function in humans. J Nutr 2006, 136:1660S–1665S.

83. Shoveller AK, Stoll B, Ball RO, Burrin DG: Nutritional and functional importance of intestinal sulfur amino acid metabolism. J Nutr 2005, 135:1609–1612.

84. Koga T, Claycombe K, Meydani M: Homocysteine increases monocyte and T-cell adhesion to human aortic endothelial cells. Atherosclerosis 2002, 161:365–374.

85. Bauchart-Thevret C, Stoll B, Chacko S, Burrin DG: Sulfur amino acid deficiency upregulates intestinal methionine cycle activity and suppresses epithelial growth in neonatal pigs. Am J Physiol Endocrinol Metab 2009, 296:E1239–E1250.

86. Kim CJ, Kovacs-Nolan J, Yang C, Archbold T, Fan MZ, Mine Y: L-cysteine supplementation attenuates local inflammation and restores gut homeostasis in a porcine model of colitis. Biochim Biophys Acta 2009, 1790:1161–1169.

87. Stoll B, Henry J, Reeds PJ, Yu H, Jahoor F, Burrin DG: Catabolism dominates the first-pass intestinal metabolism of dietary essential amino acids in milk protein-fed piglets. J Nutr 1998, 128:606–614.

88. Law GK, Bertolo RF, Adjiri-Awere A, Pencharz PB, Ball RO: Adequate oral threonine is critical for mucin production and gut function in neonatal piglets. Am J Physiol Gastrointest Liver Physiol 2007, 292:G1293–G1301.

89. Faure M, Moennoz D, Montigon F, Mettraux C, Breuille D, Ballevre O: Dietary threonine restriction specifically reduces intestinal mucin synthesis in rats. J Nutr 2005, 135:486–491.

90. Nichols NL, Bertolo RF: Luminal threonine concentration acutely affects intestinal mucosal protein and mucin synthesis in piglets. J Nutr 2008, 138:1298–1303.

91. Wang W, Zeng X, Mao X, Wu G, Qiao S: Optimal dietary true ileal digestible threonine for supporting the mucosal barrier in small intestine of weanling pigs. J Nutr 2010, 140:981–986.

92. Hamard A, Mazurais D, Boudry G, Le Huerou-Luron I, Seve B, Le FN: A moderate threonine deficiency affects gene expression profile, paracellular permeability and glucose absorption capacity in the ileum of piglets. J Nutr Biochem 2010, 21:914–921.

93. Azzam MM, Zou XT, Dong XY, Xie P: Effect of supplemental L-threonine on mucin 2 gene expression and intestine mucosal immune and digestive enzymes activities of laying hens in environments with high temperature and humidity. Poult Sci 2011, 90:2251–2256.

Evaluation of steroidogenic capacity after follicle stimulating hormone stimulation in bovine granulosa cells of Revalor 200® implanted heifers

Andrea D Stapp[1], Craig A Gifford[1], Dennis M Hallford[2] and Jennifer A Hernandez Gifford[1,3]*

Abstract

Background: Heifers not used as breeding stock are often implanted with steroids to increase growth efficiency thereby altering hormone profiles and potentially changing the environment in which ovarian follicles develop. Because bovine granulosa cell culture is a commonly used technique and often bovine ovaries are collected from abattoirs with no record of implant status, the objective of this study was to determine if the presence of an implant during bovine granulosa cell development impacts follicle stimulating hormone-regulated steroidogenic enzyme expression. Paired ovaries were collected from 16 feedlot heifers subjected to 1 of 3 treatments: non-implanted (n = 5), Revalor 200 for 28 d (n = 5), or Revalor 200 for 84 d (n = 6). Small follicle (1 to 5 mm) granulosa cells were isolated from each pair and incubated with phosphate buffered saline (n = 16) or 100 ng/mL follicle stimulating hormone (n = 16) for 24 h.

Results: Granulosa cells of implanted heifers treated with follicle stimulating hormone produced medium concentrations of progesterone similar ($P = 0.22$) to non-implanted heifers, while medium estradiol concentrations were increased ($P < 0.10$) at 28 and 84 d compared to non-implanted heifers indicating efficacy of treatment. Additionally, real-time PCR analysis in response to follicle stimulating hormone treatment demonstrated a decrease in steroidogenic acute regulatory protein ($P = 0.05$) mRNA expression in heifers implanted for 84 d and an increase in P450 side chain cleavage mRNA in granulosa cells of heifers implanted for 28 ($P < 0.10$) or 84 d ($P < 0.05$) compared to non-implanted females. However, no difference in expression of 3-beta-hydroxysteroid dehydrogenase ($P = 0.57$) and aromatase ($P = 0.23$) were demonstrated in implanted or non-implanted heifers.

Conclusions: These results indicate follicles which develop in the presence of high concentrations of androgenic and estrogenic steroids via an implant tend to demonstrate an altered capacity to respond to follicle stimulating hormone stimulation. Thus, efforts should be made to avoid the use of implanted heifers to study steroidogenesis in small follicle granulosa cell culture systems.

Keywords: Bovine, Follicle stimulating hormone, Granulosa cells, Implant, Steroidogenesis

Background

Combination trenbolone acetate (TBA) and estradiol-17β (E$_2$) implants are commonly used in feedlot cattle to increase feed efficiency and muscle mass [1]. However, exposure to exogenous hormones also influences other physiological functions. SJ Jones, RD Johnson, CR Calkins and ME Dikeman [2] demonstrated that implanted bulls had reduced cortisol and testosterone serum concentrations

and smaller testicular size compared to non-implanted bulls. These data indicate that combination implants alter adrenal and gonadal steroid production and normal gonadal development.

In females, elevated concentrations of hormones, including estradiol, can alter ovarian function and steroid hormone synthesis. Anabolic agents used to enhance growth inhibit pituitary release of gonadotropins [3] as a result of the androgenic activity as exhibited by TBA [4] or the estrogenic activity [5]. Consequently, implanting developing heifers has marked impacts on reproductive function. Heifers receiving a TBA and E$_2$ implant at 84 d of age had

* Correspondence: jah.hernandez_gifford@okstate.edu
[1]Department of Animal Science, Oklahoma State University, Stillwater, OK 74078, USA
Full list of author information is available at the end of the article

delayed puberty and retardation in reproductive tract development [6]. Ewes prenatally treated with testosterone during mid-gestation did not display a delay in onset of puberty but demonstrated absent or disrupted progestogenic cycles, and had larger follicles with prolonged presence [7]. Heifers receiving a TBA implant at estrus or d 13 of the estrous cycle were anestrus for a period of time during growth promotant release, thereafter d 13 implanted heifers remained anestrus due to follicle or luteal cysts [8].

Though steroid implants are not intended for use in breeding females, bovine ovaries are often harvested from abattoirs for GC culture to investigate mechanisms regulating follicle maturation and differentiation. To our knowledge there is no study demonstrating the impact of elevated levels of androgens and estrogens on the developing follicle. Therefore, the objective of this study was to determine if the presence of anabolic and estrogenic steroids impacts follicle stimulating hormone (FSH)-regulated steroidogenic enzyme expression.

Methods
Animals
All procedures involving animals were approved by the Oklahoma State University Institutional Animal Care and Use Committee (AG-12-4). Sixteen predominantly Angus heifers (361 kg) were randomly assigned to one of three implant groups: non-implanted (n = 5), implanted for 28 d with a combination implant (200 mg TBA + 20 mg E_2; Revalor 200®; Intervet, Inc., Millsboro, DE, USA; 28 d; n = 5), and 84 d with Revalor 200® (84 d; n = 6). Assigned heifers were implanted on d 0 (group implanted for 84 d) or d 56 (group implanted for 28 d) and were not re-implanted. Heifers were harvested on d 84 and 85 and paired ovaries were harvested (Robert M. Kerr Food and Agriculture Products Center, Oklahoma State University, Stillwater, OK) from each heifer for GC collection and culture.

Granulosa cell culture
Small follicle (1 to 5 mm) GC were isolated from ovaries of each animal and each animal's GC were cultured separately using methods previously described [9]. Follicle size selection was based on the intent of investigating FSH signaling cascades in implanted and non-implanted heifers. Previous observations indicate that 1) recruitment of bovine follicles able to respond to FSH occurs at a diameter of 1 to 3 mm [10]; 2) GC acquire FSH receptors prior to follicular recruitment [11,12]; and 3) GC of recruited follicles express steroidogenic enzyme mRNAs before LH receptor mRNA is detected [13]. Briefly, GC were resupended and washed twice in short-term media (1:1 mixture of Dulbecco's Modified Eagle Medium (DMEM) and Ham's F12 containing 0.12 mmol/L gentamycin and 38.5 mmol/L sodium bicarbonate) obtained from Sigma-Aldrich (St. Louis, MO,

USA). After the final wash, cells were re-suspended in 0.5 to 2 mL of resuspension medium (serum-free medium with 2.5 mg/mL collagenase and 1 mg/mL DNase) (Sigma-Aldrich) to prevent cell clumping prior to plating. Cell number and viability were determined via hemocytometer using trypan blue dye exclusion. Granulosa cells from each animal were seeded in two-60-mm culture dishes at a density of 5.2×10^5 cells in DMEM complete medium (1:1 DMEM and Ham's F-12 containing 10% fetal bovine serum, 0.12 mmol/L gentamycin, 2.0 mmol/L glutamine, and 38.5 mmol/L sodium bicarbonate). Incubation of cells occurred at 38.5°C and 5% CO_2 and medium was changed every 24 h until cell confluency reached 70-75%. Once confluency was reached, medium and unattached cells were removed. To test how each animal's GC responded to FSH treatment, one culture dish of GC from each animal were incubated with phosphate buffered saline (PBS; Con; n = 16) in serum free media supplemented with 10^{-7} mol/L testosterone propionate (Sigma-Aldrich) for 24 h, allowing each animal to serve as its own control. The second culture dish was treated with 100 ng/mL purified human FSH (S1AFP-B-3; National Hormone and Peptide Program, National Institute of Diabetes and Digestive and Kidney Diseases, National Institutes of Health, Bethesda, MD, USA; n = 16) in serum free media supplemented with 10^{-7} mol/L testosterone propionate (Sigma-Aldrich) for 24 h. Treatment medium was collected and frozen at −80°C until analysis. Treatments were terminated by removing medium and rinsing cells once with ice cold PBS. Cells were scraped into 1 mL TRIzol (Invitrogen, Grand Island, NY, USA) reagent and stored at −80°C until isolation of RNA.

RNA extraction and quantitative real-time PCR
RNA was isolated from cultured GC using TRIzol reagent according to the manufacturer's protocol and stored at −80°C. Integrity of RNA was assessed by visualization of 18S and 28S ribosomal RNA resolved by agarose gel electrophoresis. RNA purity and quantity was determined using a NanoDrop, ND 1000 Spectophometer (Thermo Fisher Scientific, Wilmignton, DE, USA). Purity was determined by 260/280 nm absorbance ratios, absorbance ratios above 1.8 were considered acceptable. Total RNA (1 µg) was treated with 1 µL DNase I Amplification Grade (Invitrogen) to remove genomic DNA contamination following manufacturer's instructions. First strand cDNA was synthesized from total RNA using oligo(dT) primers and 1 µL Superscript II Reverse Transcriptase (Invitrogen) for each FSH-treated and non-treated samples. Samples were stored at −20°C until analysis. All gene specific primers were designed using Primer3 [14] and synthesized by Integrated DNA Technologies (Coralville, IA, USA). Forward and reverse primer

sequences are listed in Table 1. Primers were validated at a concentration of 300 nmol/L using a 7-log dilution curve as previously reported [15].

A working solution of cDNA was prepared by diluting 1:10 with DEPC-treated water. Five microliters of cDNA working solution was added to 20 μL master mix containing 13 μL SYBR green and fluorescein mix (Bioline, Taunton, MA, USA) and 0.75 μL of each forward primer (10 μmol/L) and reverse primer (10 μmol/L). Real-time PCR analysis for each sample was carried out in duplicate using a CFX real-time PCR detection system (Bio-Rad Laboratories, Hercules, CA, USA). Standard thermocycler conditions were as follows: 95°C for 10 min, followed by 40 cycles of 95°C for 15 s, 60°C for 30 s, and 72°C for 30 s. Relative fold change in target mRNAs was quantified using the ΔΔCq method where the FSH ΔΔCq for each animal was determined by subtracting each animals Con ΔCq from their FSH ΔCq [16]. All reverse-transcribed cDNA samples were assayed in duplicate for each gene, and melt curve analyses were performed to ensure specificity of amplification. Melt curve analysis was carried out for 81 cycles with 0.5°C temperature increase from 55°C to 95°C.

To determine the appropriate reference gene to normalize cDNA variability between samples, a panel of three reference genes was analyzed including, glyceraldehyde-3-phosphate dehydrogenase (GAPDH), cyclophilin A (PPIA), and mitochondrial ribosomal protein L19 (MRPL19). The raw Cq values were obtained for each gene in all samples and analyzed using GeNorm (Biogazell qbasePLUS2, Zwijnaarde, Belgium) to determine the most stable normalization factor. The most stable housekeeping gene for target gene normalization was determined to be GAPDH and was used as the reference gene [17].

Radioimmunoassay

Granulosa cell culture medium was analyzed for E_2 and progesterone (P_4) by solid-phase radioimmunoassay using components of Siemens Medical Diagnostics Corp (Los Angeles, CA, USA) commercial kits as previously described [9]. The E_2 concentration in samples of cell culture medium was determined in 200 μL of medium and the specific binding was 62.5%. Detection limit (95% of maximum binding) of the assay was 2 pg/mL. Intra-assay CV for E_2 was 6.5% for cell culture medium. The P_4 concentration in samples of GC medium was assayed at 10 μL. The specific binding was 58.8%. Detection limit (95% of maximum binding) of the assay was 0.1 ng/mL. Intra-assay CV for P_4 was 4.1% for cell culture medium.

Statistical analysis

Experiments were analyzed by analysis of variance for a completely randomized design in which three treatments were included; non-implanted (n = 5), Revalor 200® for 28 d (n = 5), and Revalor 200® for 84 d (n = 6). Relative fold changes in gene expression for steroidogenic acute regulatory protein (STAR), 3β-hydroxysteroid dehydrogenase (3β-HSD), P450 side chain cleavage (CYP11A1), and aromatase (CYP19A1) mRNA, and medium hormonal concentration of P_4 and E_2 are presented as the least square means ± standard error of the mean. For all culture experiments, GC from each animal were kept separate and each animal's GC were subjected to either control treatment or FSH treatment. Thus, fold change values are each animal's FSH response relative to that animal's non-treated controls. A value in CYP11A1 mRNA expression of a non-implanted heifer at least three standard deviations from the mean and a missing fold change for CYP19A1 in the 84 d treatment group were excluded from statistical analysis. Quantitative real-time PCR data and hormone concentrations were

Table 1 Primer sequences used in real-time PCR

| | | Sequences of primers (5'-3') | |
Gene	Accession no.	Forward	Reverse
[1]3β-HSD	NM_174343	CCACACCAAAGCTACGATGA	GCAAGCCAGTACTGCAGAGA
[2]CYP11A1	NM_176644	AAGTTTGACCCAACCAGGTG	GTGGATGAGGAAGAGGGTCA
[3]CYP19A1	NM_174305	CAACAGCAGAGAAGCTGGAAGACA	CACCCACAACAGTCTGGATTTCCCT
[4]GAPDH	NM_001034034	GGGTCATCATCTCTGCACCT	GGTCATAAGTCCCTCCACGA
[5]MRPL19	NM_001046068	GGAAAGCAGGTTCTTGAGTCC	TGGCATATGGGTCAGCAGTA
[6]PPIA	XM_002690515	GGTACTGGTGGCAAGTCCAT	GCCATCCAACCACTCAGTCT
[7]STAR	BC110213	CCCATGGAGAGGCTTTATGA	CGTGAGTGATGACCGTGTCT

[1]3β-HSD = 3β-hydroxysteroid dehydrogenase.
[2]CYP11A1 = P450 side chain cleavage.
[3]CYP19A1 = aromatase.
[4]GAPDH = glyceraldehyde-3-phosphate dehydrogenase.
[5]MRPL19 = mitochondrial ribosomal protein L19.
[6]PPIA = cyclophilin A.
[7]STAR = steroidogenic acute regulatory protein.

analyzed using the GLM procedures of SAS (SAS Institute, Cary, NC, USA). Data were tested for homogeneity of variance using Hartley's F max test and *STAR* and E_2 were corrected by log transformation (log + 3 and log + 1, respectively). When a significant treatment effect was observed, means were separated using the least significant test computed by the predicted difference option of SAS. Statistical significance was set at $P < 0.10$.

Results and discussion

The anabolic effects of implants are likely a consequence of altering the endogenous hormonal milieu. This concept is supported by the demonstrated increase in plasma GH concentrations in response to E_2 [18] or TBA and E_2 [19] implants. High levels of anabolic hormones can also modulate reproduction as demonstrated by TBA induced anestrus in cows [8,20] and delayed puberty and decreased fertility in TBA plus E_2 implanted heifers compared to non-implanted controls [6,21].

To determine the steroidogenic capacity of small follicle GC from an environment of elevated anabolic steroids, we first evaluated steroid accumulation in cell culture medium. Hormonal output was analyzed via radioimmunoassay. The ability of cultured GC to synthesize P_4 was not affected by heifer implant status ($P = 0.22$; Figure 1A). These results are similar to previous studies in which bovine GC were cultured with 200 ng/mL FSH and also did not show an effect on production of P_4 in cell culture medium compared to controls [22]. As expected, FSH treated GC had increased ($P < 0.01$) concentrations of E_2 compared to PBS-treated controls (Figure 1B) demonstrating successful induction of the FSH signaling pathway. Additionally, GC from heifers implanted for 28 d or 84 d produced greater concentrations of E_2 (78 and 80 ± 21 pg/mL, respectively) compared to non-implanted heifers (26 ± 21 pg/mL; $P < 0.10$; Figure 2B) in response to FSH.

Based on the apparent change in estrogen production as a result of implant status and that estrogen production by the follicle is determined by FSH regulation of genes encoding key steroidogenic enzymes, we next evaluated gene expression of the steroidogenic enzymes of non-implanted and implanted heifers in response to FSH. Analysis of steroidogenic enzyme mRNAs of pubertal heifers implanted with TBA and E_2 in the presence or absence of FSH demonstrated differences in expression as compared to non-implanted heifers. The first rate limiting step in steroid synthesis is the delivery of cholesterol to the inner mitochondrial membrane which is mediated by steroid acute regulatory protein (*STAR*). Expression of *STAR* was reduced ($P < 0.05$) in response to FSH in cells from heifers exposed to TBA and E_2 for 84 d when compared to non-implanted heifers and heifers implanted for 28 d (Figure 2A). *STAR*

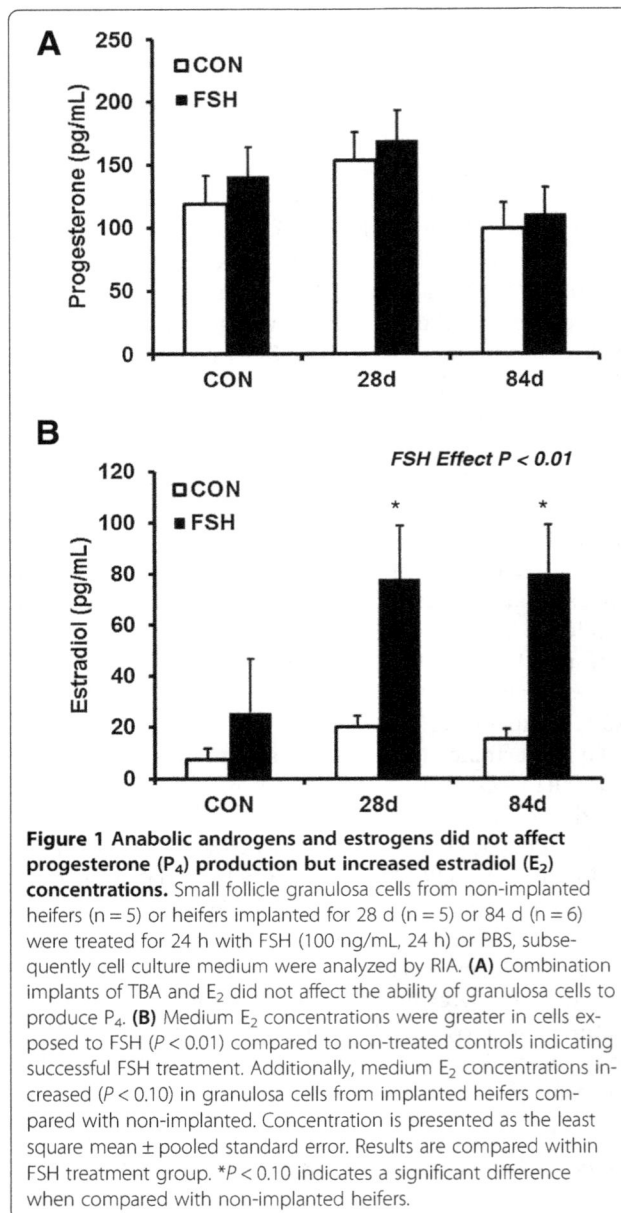

Figure 1 Anabolic androgens and estrogens did not affect progesterone (P_4) production but increased estradiol (E_2) concentrations. Small follicle granulosa cells from non-implanted heifers (n = 5) or heifers implanted for 28 d (n = 5) or 84 d (n = 6) were treated for 24 h with FSH (100 ng/mL, 24 h) or PBS, subsequently cell culture medium were analyzed by RIA. **(A)** Combination implants of TBA and E_2 did not affect the ability of granulosa cells to produce P_4. **(B)** Medium E_2 concentrations were greater in cells exposed to FSH ($P < 0.01$) compared to non-treated controls indicating successful FSH treatment. Additionally, medium E_2 concentrations increased ($P < 0.10$) in granulosa cells from implanted heifers compared with non-implanted. Concentration is presented as the least square mean ± pooled standard error. Results are compared within FSH treatment group. *$P < 0.10$ indicates a significant difference when compared with non-implanted heifers.

is fundamental to the biosynthesis of steroid hormones as it provides cholesterol to the cytochrome P450 side-chain cleavage enzyme (*CYP11A1*). Mitochondrial *CYP11A1* catalyzes the cleavage of the cholesterol side chain to form pregnenolone and this reaction represents the first committed step in steroidogenesis. Follicle stimulating hormone increased mRNA expression of *CYP11A1*, in GC from 28 d ($P < 0.10$) and 84 d ($P < 0.05$) implanted heifers as compared to non-implanted (Figure 2B). Studies indicate that both delivery of cholesterol to the enzyme system and the expression of *CYP11A1* are important factors controlling the rate of steroid hormone synthesis [23] and may contribute to

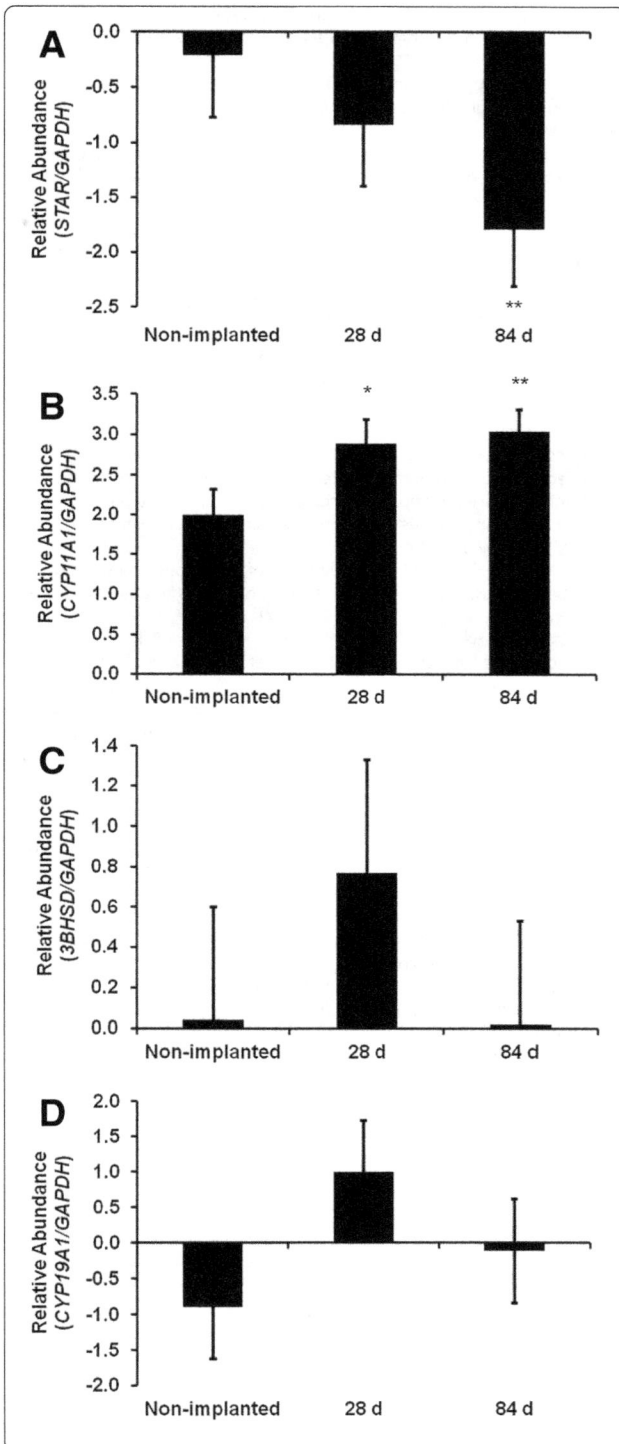

Figure 2 Expression of key steroidogenic enzymes is altered in FSH stimulated granulosa cells from implanted heifers. Bovine small follicle granulosa cells were treated as described in Figure 1. Gene expression data are represented as fold change for each group compared to its respective control to demonstrate individual group response to FSH treatment. Fold change of **(A)** STAR, **(B)** CYP11A1, **(C)** 3βHSD, and **(D)** CYP19A1 mRNA expression was analyzed by real time-PCR. Statistical significance is presented as the least square mean ± pooled standard error. $*P < 0.10$ and $**P < 0.05$ indicates a significant difference when compared with non-implanted heifers.

the increase in medium estrogen detected on d 28 and 84. In the female, ovarian androgens and estrogens regulate release of LH and FSH by feedback mechanisms on the hypothalamus and pituitary and it is not unexpected that exposure to anabolic steroids may disrupt this delicate balance. Elevated concentrations of anabolic and estrogenic steroids from the implants did not have a marked effect on mRNA expression of 3β-HSD ($P = 0.57$; Figure 2C) the enzyme responsible for converting pregnenolone to progesterone. Next, we evaluated CYP19A1, the enzyme in granulosa cells responsible for converting androgens to estrogen. However, no change in gene expression was detected for CYP19A1 in control versus heifers implanted for 28 or 84 d ($P = 0.22$; Figure 1D). This may be explained in part by the relatively short 3 h half-life of CYP19A1 in FSH-stimulated bovine granulosa cells compared to the more stable 14 h half-life demonstrated for CYP11A1 [24].

Additionally, although implant status did not result in significant changes in CYP19A1 mRNA expression, E_2 concentrations however, were elevated in both FSH treated implanted groups. This is consistent with previous work in cattle showing that minimal regulation in CYP19A1 gene expression can contribute to measurable differences in E_2 synthesis [9,25]. Normal follicular development relies on increasing concentrations of E_2 corresponding with follicle maturation and resultant proliferation and differentiation of GC [26]. Additionally, elevated levels of E_2 in culture medium improves growth of oocytes from early antral follicles [27] supporting an important role of E_2 in folliculogenesis. However, elevated exogenous concentration of E_2 can increase the chance of developing cystic follicles and decreases fertility in heifers [6].

In conclusion, these results indicate that follicles which develop in the presence of high concentrations of androgenic and estrogenic steroids via an implant have an altered ability to respond to FSH stimulation as demonstrated by varied steroidogenic enzyme expression and elevated estradiol production. Thus, efforts should be made to avoid the use of implanted heifers to study steroidogenesis in small follicle GC culture systems.

Competing interests
The authors declare they have no competing interest.

Authors' contribution
ADS carried out granulosa cell isolation and cell culture including treatment and collection of cells and media. ADS conducted real-time PCR gene expression analysis. DMH carried out the radioimmunoassays. CAG and JAHG were involved in experimental design, interpretation of data and writing of this manuscript. All authors read and approved the final manuscript.

Acknowledgements
Research supported by the Oklahoma Agric. Exp. Sta, Stillwater (OKL02789). The authors would like to thank the NIDDK's National Hormone & Peptide Program and A.F. Parlow for supplying the FSH reagent. Additionally, authors appreciate the Willard Sparks Beef Research Center personnel for overseeing care of experimental animals and lab members for assistance in granulosa cell collection.

Author details
[1]Department of Animal Science, Oklahoma State University, Stillwater, OK 74078, USA. [2]Department of Animal and Range Science, New Mexico State University, Las Cruces, NM 88003, USA. [3]Department of Animal Science, 114B Animal Science Building, Oklahoma State University, Stillwater, OK 74078, USA.

References
1. Johnson BJ, Anderson PT, Meiske JC, Dayton WR: **Effect of a combined trenbolone acetate and estradiol implant on feedlot performance, carcass characteristics, and carcass composition of feedlot steers.** J Anim Sci 1996, **74**(2):363–371.
2. Jones SJ, Johnson RD, Calkins CR, Dikeman ME: **Effects of trenbolone acetate on carcass characteristics and serum testosterone and cortisol concentrations in bulls and steers on different management and implant Schemes.** J Anim Sci 1991, **69**(4):1363–1369.
3. Cooper RA: **Some aspects of the use of the growth promoter zeranol in ewe lambs retained for breeding: III. effect on plasma LH levels.** Br Vet J 1985, **141**(4):424–426.
4. Neumann F: **Pharmacological and endocrinological studies on anabolic agents.** Environ Qual Saf Suppl 1976, **5**:253–264.
5. Katzenellenbogen BS, Katzenellenbogen JA, Mordecai D: **Zearalenones: characterization of the estrogenic potencies and receptor interactions of a series of fungal beta-resorcylic acid lactones.** Endocrinology 1979, **105**(1):33–40.
6. Moran C, Prendiville DJ, Quirke JF, Roche JF: **Effects of oestradiol, zeranol or trenbolone acetate implants on puberty, reproduction and fertility in heifers.** J Reprod Fertil 1990, **89**(2):527–536.
7. Manikkam M, Steckler TL, Welch KB, Inskeep EK, Padmanabhan V: **Fetal programming: prenatal testosterone treatment leads to follicular persistence/luteal defects; partial restoration of ovarian function by cyclic progesterone treatment.** Endocrinology 2006, **147**(4):1997–2007.
8. Reynolds IP, Harrison LP, Mallinson CB, Harwood DJ, Heitzman RJ: **The effect of trenbolone acetate on the bovine estrous-cycle.** Anim Reprod Sci 1981, **4**(2):107–116.
9. Castañon BI, Stapp AD, Gifford CA, Spicer LJ, Hallford DM, Gifford JAH: **Follicle-stimulating hormone regulation of estradiol production: possible involvement of WNT2 and β-catenin in bovine granulosa cells.** J Anim Sci 2012, **90**:3789–3797.
10. Jaiswal RS, Singh J, Adams GP: **Developmental pattern of small antral follicles in the bovine ovary.** Biol Reprod 2004, **71**(4):1244–1251.
11. Xu Z, Garverick HA, Smith GW, Smith MF, Hamilton SA, Youngquist RS: **Expression of follicle-stimulating hormone and luteinizing hormone receptor messenger ribonucleic acids in bovine follicles during the first follicular wave.** Biol Reprod 1995, **53**(4):951–957.
12. Evans AC, Fortune JE: **Selection of the dominant follicle in cattle occurs in the absence of differences in the expression of messenger ribonucleic acid for gonadotropin receptors.** Endocrinology 1997, **138**(7):2963–2971.
13. Bao B, Garverick HA, Smith GW, Smith MF, Salfen BE, Youngquist RS: **Changes in messenger ribonucleic acid encoding luteinizing hormone receptor, cytochrome P450-side chain cleavage, and aromatase are associated with recruitment and selection of bovine ovarian follicles.** Biol Reprod 1997, **56**(5):1158–1168.
14. Rozen S, Skaletsky H: **Primer3 on the WWW for general users and for biologist programmers.** Methods Mol Biol 2000, **132**:365–386.
15. Gifford CA, Racicot K, Clark DS, Austin KJ, Hansen TR, Lucy MC, Davies CJ, Ott TL: **Regulation of interferon-stimulated genes in peripheral blood leukocytes in pregnant and bred, nonpregnant dairy cows.** J Dairy Sci 2007, **90**(1):274–280.
16. Kubista M, Andrade JM, Bengtsson M, Forootan A, Jonák J, Lind K, Sindelka R, Sjöback R, Sjögreen B, Strömbom L, et al: **The real-time polymerase chain reaction.** Mol Aspects Med 2006, **27**(2–3):95–125.
17. Vandesompele J, De Preter K, Pattyn F, Poppe B, Van Roy N, De Paepe A, Speleman F: **Accurate normalization of real-time quantitative RT-PCR data by geometric averaging of multiple internal control genes.** Genome Biol 2002, **3**(7). research0034.0031 - research0034.0011.
18. Rumsey TS, Elsasser TH, Kahl S: **Roasted soybeans and an estrogenic growth promoter affect growth hormone status and performance of beef steers.** J Nutr 1996, **126**(11):2880–2887.
19. Hongerholt DD, Crooker BA, Wheaton JE, Carlson KM, Jorgenson DM: **Effects of a growth hormone-releasing factor analogue and an estradiol-trenbolone acetate implant on somatotropin, insulin-like growth factor I, and metabolite profiles in growing Hereford steers.** J Anim Sci 1992, **70**(5):1439–1448.
20. Heitzman RJ, Harwood DJ: **Residue levels of trenbolone and oestradiol-17beta in plasma and tissues of steers implanted with anabolic steroid preparations.** Br Vet J 1977, **133**(6):564–571.
21. Heitzman RJ, Harwood DJ, Kay RM, Little W, Mallinson CB, Reynolds IP: **Effects of implanting prepuberal dairy heifers with anabolic steroids on hormonal status, puberty and parturition.** J Anim Sci 1979, **48**(4):859–866.
22. Langhout DJ, Spicer LJ, Geisert RD: **Development of a culture system for bovine granulosa cells: effects of growth hormone, estradiol, and gonadotropins on cell proliferation, steroidogenesis, and protein synthesis.** J Anim Sci 1991, **69**(8):3321–3334.
23. Miller WL: **Molecular biology of steroid hormone synthesis.** Endocr Rev 1988, **9**(3):295–318.
24. Sahmi M, Nicola ES, Price CA: **Hormonal regulation of cytochrome P450 aromatase mRNA stability in non-luteinizing bovine granulosa cells in vitro.** J Endocrinol 2006, **190**(1):107–115.
25. Luo W, Gumen A, Haughian JM, Wiltbank MC: **The role of luteinizing hormone in regulating gene expression during selection of a dominant follicle in cattle.** Biol Reprod 2011, **84**(2):369–378.
26. Richards JS, Midgley AR Jr: **Protein hormone action: a key to understanding ovarian follicular and luteal cell development.** Biol Reprod 1976, **14**(1):82–94.
27. Endo M, Kimura K, Kuwayama T, Monji Y, Iwata H: **Effect of estradiol during culture of bovine oocyte–granulosa cell complexes on the mitochondrial DNA copies of oocytes and telomere length of granulosa cells.** Zygote 2012:1–9.

3

Calcium potentiates the effect of estradiol on PGF2α production in the bovine endometrium

Claudia Maria Bertan Membrive[1], Pauline Martins da Cunha[2], Flávio Vieira Meirelles[3] and Mario Binelli[2*]

Abstract

Background: Estradiol (E_2) is required for luteolysis in cows and its injection stimulates prostaglandin F2α (PGF2α) release. The main goal of our study was to investigate the ability of endometrial explants and cells treated with E_2 and the calcium ionophore (CI) A23187 to synthesize PGF2α.

Results: Treatment with E_2 *in vivo* resulted in a 48.4% increase of PGF2α production by endometrial explants treated *in vitro* with A23187. Production of PGF2α was better stimulated with A23187 at concentrations of 10^{-6} and 10^{-5} mol/L compared with other concentrations used. The concentration of PGF2α for untreated bovine endometrial cell cultures was 33.1 pg/mL, while for cultures treated with E_2, A23187, or a combination of E_2 and A23187, the PGF2α concentration was 32.5, 92.4 and 145.6 pg/mL, respectively.

Conclusions: Treatment with A23187 tended to stimulate PGF2α production. In the presence of E2, A23187 significantly stimulated PGF2α synthesis. It appears that A23187 potentiates the effects of E_2 with respect to synthesis of endometrial PGF2α in cattle.

Keywords: Animal reproduction, Cattle, Estradiol, Luteolysis, PGF2α synthesis, Reproductive physiology

Background

In cattle, administration of 17β-estradiol (E2; 3 mg) as early as day 13 of the estrous cycle causes an increase in the plasma concentration of 13,14-dihydro-15-keto-prostaglandin F2α (PGFM), the main metabolite of PGF2α [1,2]. Consistently, ablation of ovarian follicles, which causes a decrease in plasma concentrations of E_2, delays luteolysis in ruminants [3-5]. E_2 is capable of inducing endometrial synthesis of PGF2α [6]; furthermore, E_2 can induce luteolysis [7].

The endometrial synthesis of PGF2α results from a complex cascade of highly coordinated events. Arachidonic acid (AA), stored in the phospholipid membranes, is the primary precursor of prostaglandins [8]. Oxytocin (OT), acting through its receptor on the endometrial cell membrane, activates a guanosine nucleotide-binding protein (G protein), which promotes the activation of phospholipase C (PLC) [9]. PLC then cleaves phosphatidylinositol triphosphate into inositol triphosphate (IP_3) and diacylglycerol (DAG). IP_3 binds to receptors on the endoplasmic reticulum,

promoting an increase in cytoplasmic calcium concentration. DAG activates protein kinase C (PKC), a serine/threonine kinase that is dependent on calcium for activation. Activated PKC phosphorylates phospholipase A_2 (PLA_2). The IP_3-induced increase in cytosolic calcium stimulates calcium-dependent PLA_2 activity [10]. PLA_2 preferentially cleaves the sn-2 position of phosphatidylcholine, releasing AA [11]. The free AA is then converted into prostaglandin H_2 (PGH_2) by prostaglandin endoperoxide synthase 2 (PTGS2). Finally, PGH_2 is converted into PGF2α by PGF synthases, such as Aldo-keto reductase family 1 member C1 (AKR1C1). Calcium is a known cofactor of PKC and PLA_2, which are enzymes involved in PGF2α production. Although the proteins involved in the synthesis of PGF2α have been identified, the role of E_2 in this process remains unknown.

The main goal of our study was to determine the ability of endometrial explants and cells to synthesize PGF2α. Specifically, we sought to evaluate the capacity of E_2, when administered *in vivo*, to stimulate PGF2α synthesis in endometrial explants incubated with a calcium ionophore (CI), melittin or oxytocin. We also sought to determine the CI dose that was capable of stimulating the synthesis of PGF2α in cultured endometrial cells, and to evaluate

---
* Correspondence: binelli@usp.br
[2]Department of Animal Reproduction, School of Veterinary Medicine and Animal Science, University of São Paulo, São Paulo, Brazil
Full list of author information is available at the end of the article

the capacity of bovine endometrial (BEND) cells to synthesize PGF2α following treatment with E_2 and/or a CI.

Methods

Experiment 1

We used 13 cyclic cross-bred beef heifers (*B. taurus* × *B. indicus*) in our study. Animals procedures were approved by Ethics and Animal Handling Committee of the Universidade de São Paulo. Animals were fed pasture (*Brachiaria decumbens* var. *marandu*) supplemented with minerals and had access to water *ad libitum*. Animals were implanted with a device containing 1 g of P4 (CIDR®, Pfizer, USA) along with an intramuscular injection of gonadorelin (100 μg; Fertagil®, Intervet, The Netherlands) on the first day of the synchronization protocol. Devices were removed 7 days later and cows were administered D-cloprostenol (150 μg; Preloban®, Intervet) intramuscularly and were marked on the tailhead using an All-Weather Paintstik® (LA-CO Industries Inc., USA). Estrous behavior was observed twice daily for 48–120 h after PGF2α was injected. The start of standing estrus was considered day 0 of the estrous cycle (day 0). On day 6, an ultrasonographic examination was performed (Aloka, SSD-500, linear probe 7.50 MHz), and females with a dominant follicle ≥7.5 mm received an intramuscular injection of gonadorelin (100 μg; Fertagil®, Intervet, The Netherlands) to induce ovulation of the dominant follicle and promote the emergence of a new follicular wave [12]. Ultrasonographic examination was performed to verify the presence of an accessory corpus luteum on day 16 of the cycle. Heifers were paired according to the day of standing estrus and randomly chosen for intravenous treatment with 0 (*n* = 6) or 3 mg of E_2 (*n* = 7) on day 17. At 2 h post-treatment, animals were stunned by cerebral concussion from a pneumatic pistol and euthanized by jugular exsanguination. Genital tracts were transported to our laboratory on ice immediately after animals were euthanized. The endometrium of the ipsilateral horn to the original corpus luteum was dissected, and fragments of the intercaruncular region weighing 80–100 mg were conditioned in 12 mm × 75 mm borosilicate tubes containing 0.5 mL Krebs-Hensleit bicarbonate medium (KHB; 118 mmol/L NaCl, 4.7 mmol/L KCl, 2.56 mmol/L $CaCl_2$, 1.13 mmol/L $MgCl_2$, 25 mmol/L $NaHCO_3$, 1.15 mmol/L NaH_2PO_4, 5.55 mmol/L glucose, 20 mmol/L Hepes and 0.013 mmol/L phenol red, pH 7.4). Cultures were maintained according to procedures described by Burns et al. [13]. Tubes containing explants were kept at 37°C in a shaking waterbath (40 rpm) for 1 h. Culture media was discarded and explants were washed twice with 0.5 mL of KHB. Explants were incubated for 1 h in KHB, washed, and then treated *in vitro* with 1 mL of either: KHB medium (control); KHB supplemented with 10^{-5} mol/L A23817, a CI (C-7522, Sigma Chemicals, USA);

KHB supplemented with 10^{-5} mol/L melittin (M-2272, Sigma Chemicals, USA); or KHB supplemented with 10^{-6} mol/L OT (O-6379, Sigma Chemicals, USA). Explants from each cow received all treatments in triplicate. The concentrations of drugs we used were based on those previously reported [13-15]. Samples comprising 100 μL of culture medium were removed immediately after the administration of treatments (time 0) and 60 min later. Samples were stored at –20°C until required. The concentration of PGF2α (pg/mL/mg of endometrial tissue) in the culture medium was measured using radio immuno assays as described by Danet-Desnoyers et al. [14]. Intra-assay variation coefficients were 23.7 and 14.1%, and the inter-assay coefficients were 23.4 and 13.1% for standards containing 250 and 1,000 pg/mL PGF2α, respectively.

Experiment 2

We obtained BEND cells [16] from the American Type Culture Collection (ATCC CRL-2398; USA). Cells were suspended in 4 mL of complete culture medium [40% HAM F-12 (N6760, Sigma Chemicals, USA); 40% minimal essential medium (MEM; M0643, Sigma Chemicals, USA); 200 IU/L insulin (I5500, Sigma Chemicals, USA); 10% (v/v) fetal bovine serum (FBS; 10270-106, Gibco Life, USA); 10% (v/v) equine serum (Nutricel, Brazil); and 1% (v/v) antibiotic and antimycotic solution (A7292, Sigma Chemicals, USA)] and seeded in 6-well tissue culture plates (Corning Incorporated, USA). Cells were cultured at 38.5°C/5% CO_2 until 90% confluent. Subsequently, cells were maintained in serum-free medium for 24 h, washed twice in the same medium, and incubated for 12 h in 4 mL of serum-free medium supplemented with 0, 10^{-7}, 10^{-6} or 10^{-5} mol/L A23817 (C-7522, Sigma Chemicals, USA) in triplicate wells. Samples of media (500 μL) were collected immediately after the administration of treatments (time 0) and 12 h later. After sample collection at time 0, the same volume of medium containing the specific treatment was replaced in culture wells. Samples were stored at –20°C until required. Experiments were repeated three times; intra-assay variation coefficients were 13.2 and 18.8%, and inter-assay variation coefficients were 10.3 and 22.0% for standards containing 250 and 1,000 pg/mL PGF2α, respectively.

Experiment 3

We seeded BEND cells in 24-well plates (4×10^4 cells/well; Corning Incorporated) with 1.5 mL of complete culture medium. Cultures were incubated at 38.5°C/5% CO_2 until cells were 90% confluent. Cells were then cultured in serum-free medium for 24 h, washed twice with the same medium, and incubated for 12 h in serum-free medium supplemented with 10^{-13} mol/L E_2 (E-8875, Sigma Chemicals, USA) and 10^{-6} mol/L A23817 (Sigma Chemicals, USA). Control wells received serum-free medium without supplements; all experiments were conducted in triplicate. Samples of media (300 μL) were collected immediately after the administration of drugs (time 0) and 12 h later. After sample collection at time 0, the same volume of medium containing the specific treatment was replaced in wells. Samples were stored at −20°C until subsequent measurement of PGF2α concentration. Experiments were conducted three times; intra-assay variation coefficients were 16.0 and 20.6%, and the inter-assay variation coefficients were 4.0 and 18.9% for standards containing 250 and 1,000 pg/mL PGF2α, respectively.

Statistical analysis

Data that did not meet the assumptions of normality of residues (Shapiro-Wilk Test, $P \leq 0.01$) or homogeneity of variances (F Test, $P \leq 0.01$) were transformed by square roots and reanalyzed. Data were analyzed by ANOVA using the GLM procedure with the SAS program and are presented as untransformed least squares means ± SEM. Treatment means were compared by orthogonal contrasts. In Experiment 1, the dependent variable was DIF60 (the difference between PGF2α concentrations at 0 and 60 min), and the independent variable was treatment. For Experiments 2 and 3 the dependent variable was DIF12 (the difference between PGF2α concentrations at 0 and 12 h) and the independent variables were experiment, treatment and the interaction between experiment and treatment. Statistical significance was considered at $P < 0.05$.

Results

Experiment 1

The DIF60 for the various treatments are shown in Table 1. Endometrial explants that remained untreated *in vitro* produced similar concentrations of PGF2α whether they were obtained from cows injected with E$_2$ (3 mg) (20.7 ± 4.2 pg/mL/mg of tissue) or not (24.7 ± 4.6 pg/mL/mg of tissue; Table 1). Regarding control animals, PGF2α synthesis was similar among untreated explants (24.7 ± 4.6 pg/mL/mg of tissue) and explants treated with OT (28.2 ± 4.6 pg/mL/mg of tissue), melittin (26.9 ± 5.0 pg/mL/mg of tissue) or A23187 (29.0 ± 4.6 pg/mL/mg

Table 1 Concentration of PGF2α (pg/mL/mg of tissue) produced by endometrial explants from cross-bred heifers on Day 17 of the estrous cycle treated *in vivo* with 17β-estradiol and *in vitro* with A23817, melittin or OT

Variable	Treatment (mean ± SEM)			
	Control	E$_2$	Stimulator	E$_2$/Stimulator
DIF60 −A23817	24.7 ± 4.6[b]	20.7 ± 4.2[b]	29.0 ± 4.6[b]	43.0 ± 4.2[a]
DIF60 - melittin	24.7 ± 4.6	20.7 ± 4.2	26.9 ± 5.0	31.7 ± 4.6
DIF60 - OT	24.7 ± 4.6	20.7 ± 4.2	28.2 ± 4.6	33.2 ± 4.2

[a,b]significantly different, $P \leq 0.01$.

of tissue; Table 1). In contrast, synthesis of PGF2α from explants originating from animals treated with E$_2$ was 48.4% greater following exposure to A23817 (43.0 ± 4.2 pg/mL/mg of tissue) compared with explants from control animals (29.0 ± 4.6 pg/mL/mg of tissue; $P \leq 0.01$). Treatment with E$_2$ *in vivo* did not affect the responses to melittin (26.9 ± 5.0 *vs.* 31.7 ± 4.6 pg/mL/mg of tissue) or OT (28.2 ± 4.6 *vs.* 33.2 ± 4.2 pg/mL/mg of tissue).

Experiment 2

We used A23187 at 10^{-7}, 10^{-6} and 10^{-5} mol/L, and these concentrations stimulated dose-dependent increase in PGF2α production (Figure 1). A23187 stimulated the synthesis of PGF2α in all treatment groups compared with that seen in the control group ($P \leq 0.01$). The level of stimulation was greater when A23187 was used at 10^{-6} and 10^{-5} mol/L, in comparison with A23817 at 10^{-7} mol/L ($P \leq 0.01$). We did not observe a significant difference in stimulation levels between 10^{-6} and 10^{-5} mol/L A23817.

Experiment 3

Mean PGF2α concentrations were 33.1, 32.5, 92.4 and 145.6 pg/mL (SEM: 21.8 pg/mL) for untreated cells, cells treated with E$_2$, cells treated with A23817, and cells treated with E$_2$ and A23817, respectively (Figure 2). Production of PGF2α was similar between cells treated with E$_2$ and untreated cells. The cells treated with A23817 tended ($P \leq 0.08$) to produce higher quantities of PGF2α compared with cells treated with E$_2$, or cells that were left untreated . The combination of E$_2$ and A23817 stimulated greater synthesis of PGF2α compared with the other treatments ($P < 0.01$). A23817 was responsible for a 179% increase in PGF2α production relative to the control group. When A23817 was combined with E$_2$, the production of PGF2α increased by 340% compared with that in the control group.

Discussion

In Experiment 1, PGF2α synthesis was not stimulated in the endometrial explants from cows treated with E$_2$ *in vivo* and received no treatment *in vitro* (Table 1). These results oppose those reported by Mann [17]. In contrast, Asselin et al. [18] and Xiao et al. [19] reported that the addition of E$_2$ to the culture medium was not able to stimulate the release of PGF2α. Because of the inconsistencies in the previously reported data, it was considered appropriate to administer E$_2$ *in vivo* in this experiment. Previous reports indicated that the administration of E$_2$ could stimulate the production of PGFM in cows on days 13 [2,20,21], 17 [20-22], 18 [1,20,23] and 19 [21] of the estrous cycle. Bertan et al. [21] verified that E$_2$ promoted an increase in the plasma concentration of PGFM 4 h after injection; its concentration peaked within 6.5 h and returned to basal levels 9 h after

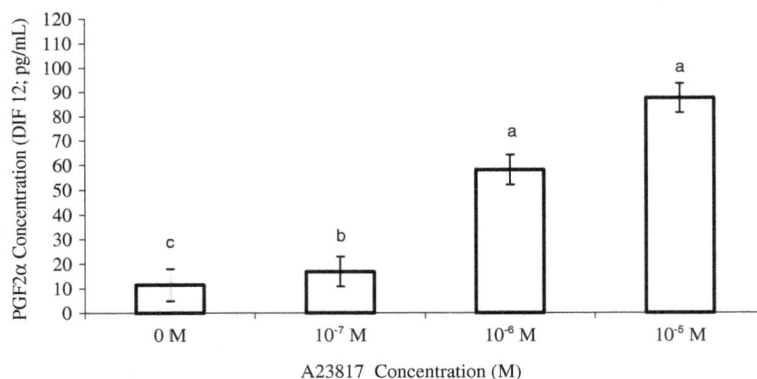

Figure 1 PGF2α production after 12 h in the culture medium of bovine endometrial (BEND) cells treated with A23817 (DIF12). [a,b,c]significantly different, $P < 0.01$.

treatment. Given this time frame, it was considered appropriate to administer E_2 *in vivo* 2 h prior to euthanizing cows. This procedure allowed the explants to be incubated with different stimulants approximately 6.5 h after the E_2 injection.

It was expected that E_2 would stimulate the synthesis of enzymes involved in the synthesis of PGF2α. Thus, E_2 would increase the concentrations of the corresponding proteins in cellular compartments. It was also expected that the activity of these proteins during PGF2α production would be amplified following specific stimulation. The CI A23817 promotes an increase in intracellular calcium concentration that is responsible for the activation of PKC and PLA2. The stimulant melittin specifically activates PLA2. Melittin and A23817 increase the synthesis of PGF2α in ovine [13,24] and bovine endometrial explants [13,14,17,25,26]; however, this stimulation was not observed

in our studies when A23817, melittin or OT was used on their own.

In our Experiment 1, synthesis of PGF2α was observed only when A23817 was used in combination with E_2. Other researchers observed that the effects of E_2 were frequently associated with an increase in the concentrations of free calcium in the cytosol [27-29]. We suggest that PKC and PLA2 are present in greater concentrations in the explants from animals treated in vivo with E_2. Increased production of PGF2α would result in response to this greater concentration of enzymes stimulated by A23817. However, it is possible that higher concentrations of PLA2 in the endometrial explants that were previously exposed to E_2 could amplify the synthesis of PGF2α in the presence of melittin. This could be explained by the fact that calcium is likely to be the major limiting factor in the activation of PLA2. In

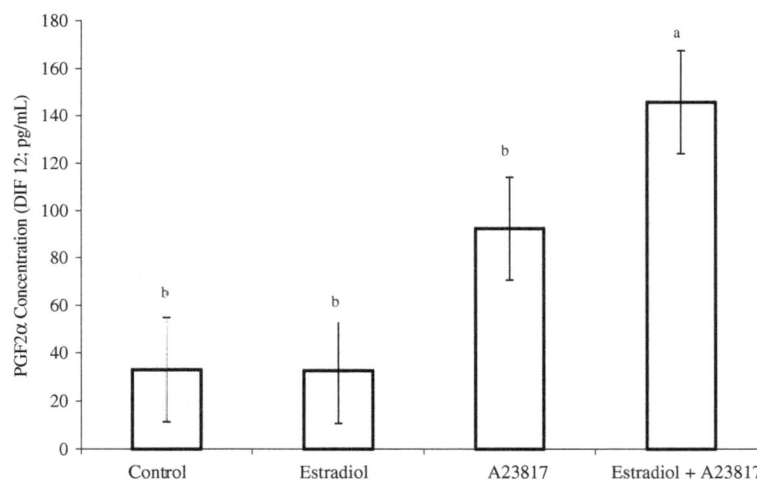

Figure 2 PGF2α production after 12 h in the culture medium of BEND cells treated with estradiol and/or A23817 (DIF12). [a,b]significantly different, $P < 0.01$.

summary, we suggest that higher concentrations of PKC and PLA2 were present in the explants treated with E_2, and that these enzymes promoted an increase in the synthesis of PGF2α in the presence of A23817.

In Experiment 2 (Figure 1), A23817 stimulated the synthesis of PGF2α for all treatment groups. When A23817 was administered at doses of 10^{-6} and 10^{-5} mol/L, stimulation was greater than that when the CI was used at 10^{-7} mol/L. The capacity of BEND cells to synthesize PGF2α [30-32] and the cellular model for the synthesis of PGF2α described by Burns et al. [13] indicate that calcium is responsible for the activation of enzymes such as PKC and PLA2. These enzymes are essential for the synthesis of PGF2α by endometrial cells. Treatment with a CI has stimulated PGF2α synthesis in endometrial explants from sows [33], guinea pigs [34,35], ewes [36-38] and cows [14,15,39].

In our study, CI stimulated PGF2α synthesis in BEND cells in a dose-dependent manner. Other studies have verified the dose-dependent effects of CI on the synthesis of PGF2α in endometrial explants. Lafrance and Goff [40] observed that a CI at 2.6 µg/mL was sufficient to stimulate the synthesis of PGF2α in the endometrial explants of heifers on days 19 or 20 of the estrous cycle. Danet-Desnoyers et al. [14] reported that the biosynthesis of PGF2α stimulated by a CI in the endometrium of cows on day 17 of the estrous cycle exhibited a dose-dependent response. Synthesis was increased by 0, 67 and 107% when doses of 2, 4 and 10 µg/mL were administered, respectively. In the present experiment, doses of 0.052 (10^{-7} mol/L), 0.52 (10^{-6} mol/L), and 5.2 µg/mL (10^{-5} mol/L) increased the production of PGF2α by 46, 105 and 661%, respectively. We propose that lower doses of A23817 stimulate the synthesis of PGF2α in BEND cells, with a significantly amplified response compared with that in endometrial explants. The increase in intracellular calcium concentration is associated with PLA2 activity [41,42] and in the increased availability of AA for the synthesis of PGF2α [41]. Arnold et al. [15] verified that the synthesis of PGF2α was stimulated in bovine endometrium cultured with a CI, regardless of whether it was supplemented with PLA2. Synthesis was greater when the endometrium was treated with a CI and PLA2 compared with explants treated with PLA2 alone. We propose that calcium has an additive effect on the stimulation of PLA2-induced synthesis of PGF2α.

Our hypothesis for Experiment 3 was that E_2 increases the sensitivity of endometrial cells to calcium. We verified that A23817 promoted an increase of 179% in the production of PGF2α compared with untreated cells. However, when A23817 was used in combination with E_2, this increase was approximately 340% (Figure 2), similar to our results in Experiment 1. Thus, E_2 increased the sensitivity of endometrial cells to calcium, supporting the hypothesis

of our experiment. Several researchers reported that the synthesis of PGF2α by endometrial explants was not increased by supplementing cultures with E_2 [26,43,44]. Consistently, in Experiment 3, synthesis of PGF2α was not stimulated in BEND cells treated with E_2 only. The genomic action of E_2 is promoted by the E_2-receptor complex, which activates transcription factors that are bound to DNA, ultimately resulting in new protein synthesis [29]. *In vivo*, E_2 most likely activates the synthesis of PGF2α by stimulating the transcription and translation of proteins involved in the production of PGF2α. However, the activity of E_2-induced proteins might need further stimulation *in vitro* because of limited access to calcium under culture conditions. In our study, the addition of A23817 to BEND cell cultures increased intracellular calcium concentrations, to enhance intracellular mechanisms that depend on this ion. Therefore, PKC and PLA2, which are involved in the synthesis of PGF2α, might have been activated, to increase synthesis of PGF2α.

In summary, from our experiments, E_2 did not stimulate the synthesis of PGF2α by endometrial cells and explants cultured *in vitro*. Many other *in vitro* studies have shown that the administration of E_2 alone in endometrial explants does not stimulate PGF2α synthesis [26,43,44]. We suggest that the participation of E_2 in the synthesis of PGF2α involves other endocrine and paracrine factors that are absent in cell and explant culture systems. Bertan et al. [21] clearly showed that E_2 did not immediately affect the cows that were treated on day 17 of the estrous cycle. The interval between the injection of E_2 and the increase in PGFM serum concentrations suggests that E_2 acts on the synthesis of PGF2α through a genomic pathway. We speculate that E_2 directly stimulates the synthesis of proteins involved in PGF2α production; these proteins might include PKC and PLA2, as they contain calcium-dependent activation domains. Further studies are required to test this proposition.

Competing interests
The authors declare that they have no competing interests.

Authors' contributions
CMBM participated in the design of the study, in all experiments and drafted the manuscript. PMC participated in all experiments, FVM participated in *in vitro* experiments, and MB designed and coordinated the study. All authors read and approved the final manuscript.

Acknowledgments
This study was funded by Fundação de Amparo à Pesquisa do Estado de São Paulo (FAPESP). We are grateful to Professor William Thatcher from the University of Florida for antibodies against PGF2α.

Author details
[1]São Paulo State University, Rod. Comandante João Ribeiro de Barros (SP 294) Km 651, Dracena, SP 17900-000, Brazil. [2]Department of Animal Reproduction, School of Veterinary Medicine and Animal Science, University of São Paulo, São Paulo, Brazil. [3]Department of Veterinary Medicine, School of Animal Sciences and Food Engineering, University of São Paulo, Pirassununga, Brazil.

References

1. Knickerbocker JJ, Thatcher WW, Foster DB, Wolfwrson D, Bartol FF, Caton D: Uterine prostaglandin and blood flow responses to estradiol-17β in cyclic cattle. *Prostaglandins* 1986, **31**:757-770.

2. Thatcher WW, Terqui M, Thimonier J, Mauleon P: Effects of estradiol-17β on peripheral plasma concentration of 15-keto-13, 14-dihydro PGF2α and luteolysis in cyclic cattle. *Prostaglandins* 1986, **31**(4):745-756.

3. Hughes TL, Villa-Godoy A, Kesner JS, Fogwell RL: Destruction of bovine ovarian follicles: effects on the pulsatile release of luteinizing hormone and prostaglandin F2α-induced luteal regression. *Biol Reprod* 1987, **36**:523-529.

4. Karsch FJ, Noveroske JW, Roche JF, Norton HW, Nalbandov AV: Maintenance of ovine corpora lutea in the absence of ovarian follicle. *Endocrinology* 1970, **87**:1228-1236.

5. Villa-Goddoy A, Ireland JJ, Wortman JA, Ames NK, Hughes TL, Fogwell RL: Effect of ovarian follicles on luteal regression in heifers. *J Anim Sci* 1985, **60**(2):519-527.

6. Hansel W, Alila HW, Dowd JP: Control of steroidogenesis in small and large bovine luteal cells. *Aust J Biol Sci* 1987, **40**(3):331-347.

7. Auletta JF, Flint APF: Mechanisms controlling corpus luteum function in sheep, cows, non human primates and women, especially in relation to the time of luteolysis. *Endocr Rev* 1988, **9**:88-105.

8. McCracken JA: Luteolysis a neuroendocrine-mediated event. *Physiol Rev* 1999, **79**:263-323.

9. Burns PD, Hayes SH, Silvia WJ: Cellular mechanisms by which oxytocin mediates uterine prostaglandin F2 alpha synthesis in bovine endometrium: roles of calcium. *Domest Anim Endocrinol* 1998, **5**:477-487.

10. Clark JD, Lin LL, Kriz RW, Ramesha CS, Sultzman LA, Lin AY, Milona N, Knopf JL: A novel arachidonic acid-selective cytosolic PLA2 contains a Ca(2+)-dependent translocation domain with homology to PKC and GAP. *Cell* 1991, **65**:1043-1051.

11. Gijón MA, Leslie CC: Regulation of arachidonic acid release and cytosolic phospholipase A2 activation. *J Leukoc Biol* 1999, **65**:330-336.

12. Diaz FJ, Anderson LE, Wu YL, Rabot A, Tsai SJ, Wiltbank MC: Regulation of progesterone and prostaglandin F2α production in the CL. *Mol Cell Endocrinol* 2002, **191**:65-68.

13. Burns PD, Graf GA, Hayes SH, Silvia WJ: Cellular mechanisms by which oxytocin stimulates uterine PGF2 alpha synthesis in bovine endometrium: roles of phospholipases C and A2. *Domest Anim Endocrinol* 1997, **14**(3):181-191.

14. Danet-Desnoyers G, Meyer MD, Gross TS, Johnsor JW, Thatcher WW: Regulation of endometrial prostaglandin synthesis during early pregnancy in cattle: effects of phospholipase and calcium in vitro. *Prostaglandins* 1995, **50**:313-330.

15. Arnold DR, Binelli M, Vonk J, Alexenko AP, Drost M, Wilcox CJ, Thatcher WW: Intracellular regulation of endometrial PGF2α and PGE2 production in dairy cows during early pregnancy and following treatment with recombinant interferon-τ. *Domest Anim Endocrinol* 2000, **18**:199-216.

16. Staggs KL, Austin KJ, Johnson GA, Teixeira MG, Talbot CT, Dooley VA, Hansen TR: Complex induction of bovine uterine proteins by interferon-τ. *Biol Reprod* 1998, **59**:277-283.

17. Mann GE: Hormone control of prostaglandin F2α production and oxytocin receptor concentrations in bovine endometrium in explant culture. *Domest Anim Endocrinol* 2001, **20**:217-225.

18. Asselin E, Goff AK, Bergeron H, Fortier MA: Influence of sex steroids on the production of prostaglandin F2α and E2 and response to oxytocin in cultured epithelial and stromal cells of bovine endometrium. *Biol Reprod* 1996, **54**:371-379.

19. Xiao CW, Goff AK: Hormonal regulation of estrogen and progesterone receptors in cultured bovine endometrial cells. *J Reprod Fertil* 1999, **115**:101-109.

20. Bertholazzi A, Bertan JP, Paula LA C e, Da Cunha PM, Loureiro JGP, Teixeira AB, Barros CM, Binelli M: Efeitos do benzoato de estradiol (EB) e cipionato de estradiol (ECP) na liberação de prostaglandina F2α (PGF2α) embovinos. In *Proceedings of the Congresso de Integração em Biologia da Reprodução: 12-15 April 2003; Ribeirão Preto, Brazil*. Edited by Alzira Amélia Rosa e Sllva: Arte Ciência Vilipress; 2003:307.

21. Bertan CM: *Mecanismos endócrinos e moleculares pelos quais o estradiol estimula a síntese de prostaglandina F2a no endométrio de fêmeas bovinas*. São Paulo: Tese (Doutorado em Reprodução Animal) – Faculdade de MedicinaVeterinária e Zootecnia, Universidade de São Paulo; 2004:180.

22. Jorge P, Bertan CM, Bertholazzi A, Paula LA C e, Da Cunha PM, Goissis M, Bressan FF, Madureira EH, Binelli M: Efeitos Da Associação do 17β-estradiol (E2) com o hormônioliberador de gonadotrofinas (GnRH) e com gonadotrofina coriônica humana (hCG) na liberação de prostaglandina F2α (PGF2α) embovinos. In *Congresso de Integração em Biologia da Reprodução: 12-15 April 2003; Ribeirão Preto, Brazil*. Edited by Alzira Amélia Rosa e Sllva: Arte Ciência Vilipress; 2003:305.

23. Rico LW, Thatcher WW, Drost M, Wolfeenson D, Terqui M: Plasma PGFM responses to estradiol injection in pregnant and cycling cows. *J Anim Sci* 1981, **53**(Suppl 1):363.

24. Lee JS, Silvia WJ: Cellular mechanisms mediating the stimulation of ovine endometrial prostaglandin F2a in response to oxytocin: role of phospholipase A2. *J Endocrinol* 1994, **141**:491-496.

25. Burns PD, Hayes SH, Silvia WJ: Cellular mechanisms by which oxytocin mediates uterine prostaglandin F2α synthesis in bovine endometrium: role of calcium. *Domest Anim Endocrinol* 1998, **15**(6):477-487.

26. Skarzynski D, Bogacki M, Kotwica J: Involvement of ovarian steroids in basal and oxytocin-stimulated prostaglandin F2α secretion from bovine endometrium in vitro. *Theriogenology* 1999, **52**:385-397.

27. Pietras RJ, Szego CM: Endometrial cell calcium and oestrogen action. *Nature* 1975, **253**:357-359.

28. Stefano GB, Cadet P, Breton C, Goumon Y, Prevot V, Dessaint JP, Beauvillain JC, Roumier AS, Welters I, Salzet M: Estradiol-stimulated nitric oxide release in human granulocytes is dependent on intracellular calcium transients: evidence of a cell surface estrogen receptor. *Blood* 2000, **95**:3951-3958.

29. Acconcia F, Marino M: Synergism between genomic and non genomic estrogen action mechanism. *IUBMB Life* 2003, **55**(3):145-150.

30. Binelli M, Guzeloglu A, Badinga L, Arnold DR, Sirois J, Hansen TR, Thatcher WW: Interferon-τ modulates phorbol Ester-induced production of prostaglandin and expression of cyclooxygenase-2 and phospholipase-2 from bovine endometrial cells. *Biol Reprod* 2000, **63**:417-424.

31. Guzeloglu A, Binelli M, Badinga L, Hansen TR, Thatcher WW: Inhibition of phorbol ester-induced PGF2α secretion by IFN-τ is not through regulation of protein kinase C. *Prostaglandins* 2004, **74**:87-99.

32. Guzeloglu A, Michel A, Michel F, Thatcher WW: Differential effects of interferon-τ on the prostaglandin synthetic pathway in bovine endometrial cells treated with phorbol ester. *J Dairy Sci* 2004, **87**:2032-2041.

33. Basha SMM, Bazer FW, Roberts RM: Effect of the conceptus on quantitative and qualitative aspects of uterine secretion in pigs. *J Reprod Fertil* 1980, **60**:41.

34. Poyser NL: Effects of various factors on prostaglandin synthesis by the guinea-pig uterus. *J Reprod Fertil* 1987, **141**:491-496.

35. Riley SC, Poyser NL: Prostaglandin production by the guinea-pig endometrium: is calcium necessary? *J Endocrinol* 1987, **113**:463-472.

36. Silvia WJ, Homanics GE: Role of phospholipase C in mediating oxytocin-induced release of prostaglandin F2 alpha from ovine endometrial tissue. *Prostaglandins* 1988, **35**:535-548.

37. Raw RE, Silvia WJ, Curry TE: Effects of progesterone and estradiol on prostaglandin endoperoxide sythase in ovine endometrial tissue. *Anim Reprod Sci* 1995, **40**:17-30.

38. Silvia WJ, Lewis GS, Mccracken JA, Thatcher WW, Jr W: L:Hormonal regulation of uterine secretion of PGF2α during luteolysis in ruminants. *Biol Reprod* 1991, **45**:655-663.

39. Tysseling KA, Thaycher WW, Bazer FW, Hansen PJ, Mirando MA: Mechanisms regulating prostaglandin F2α secretion from the bovine endometrium. *J Dairy Sci* 1998, **81**:382-389.

40. Lafrance M, Goff AK: Effects of P4 and estradiol 17-β on oxytocin induced releasing of prostaglandin F2α in heifers. *J Reprod Fertil* 1988, **82**:429-436.

41. Van den Bosh H: Intracellular phospholipases. *Biochem Biophys* 1980, **604**:191-199.

42. Ho AK, Klein DC: Activation of alpha 1-adrenoreceptors, protein kinase C, or treatment with intracellular free Ca^{+2} elevating agents increases pineal phospholipase A2 activity. *J Biol Chem* 1987, **262**:11764-11769.

43. Paula LA C e, Loreiro JGP, Membrive CMB, Da Cunha PM, Bertholazi A, Jorge P, Madureira EH, Binelli M: **Effects of estradiol-17β on prostagandin F2α (PGF) secretion in cattle: in vitro and in vivo responses.** *Biol Reprod* 2002, **66**(Suppl 01):324.

44. Raw RE, Silvia WJ: **Activity of phospholipase C and release of prostaglandin F2α by endometrial tissue from ovariectomized ewes receiving progesterone and estradiol.** *Biol Reprod* 1991, **44**:404–412.

4

Escherichia coli challenge and one type of smectite alter intestinal barrier of pigs

Juliana Abranches Soares Almeida[1], Yanhong Liu[1], Minho Song[1,2], Jeong Jae Lee[1], H Rex Gaskins[1], Carol Wolfgang Maddox[3], Orlando Osuna[4] and James Eugene Pettigrew[1*]

Abstract

An experiment was conducted to determine how an *E. coli* challenge and dietary clays affect the intestinal barrier of pigs. Two groups of 32 pigs (initial BW: 6.9 ± 1.0 kg) were distributed in a 2×4 factorial arrangement of a randomized complete block design (2 challenge treatments: sham or *E. coli*, and 4 dietary treatments: control, 0.3% smectite A, 0.3% smectite B and 0.3% zeolite), with 8 replicates total. Diarrhea score, growth performance, goblet cell size and number, bacterial translocation from intestinal lumen to lymph nodes, intestinal morphology, and relative amounts of sulfo and sialo mucins were measured. The *E. coli* challenge reduced performance, increased goblet cell size and number in the ileum, increased bacterial translocation from the intestinal lumen to the lymph nodes, and increased ileal crypt depth. One of the clays (smectite A) tended to increase goblet cell size in ileum, which may indicate enhanced protection. In conclusion, *E. coli* infection degrades intestinal barrier integrity but smectite A may enhance it.

Keywords: Barrier function, *E. coli*, Pigs, Smectite, Zeolite

Background

Weaning is a stressful period for piglets due to environmental, social and nutritional changes. During this period, pigs are also vulnerable because of their immature immune and digestive systems [1]. The stress may result in depressed feed intake which may lead to poor performance and changes in the intestinal structure and microbiota, thus increasing the susceptibility of pigs to enteric diseases [2]. Post-weaning diarrhea caused by *Escherichia coli* is a common enteric disease in weaned pigs; it causes economic losses due to mortality, morbidity, decreased growth performance and cost of medication [3]. Diarrhea also impairs nutrient absorption, increases permeability in the intestine, decreases tight junction integrity, increases paracellular movements of molecules and increases infection [4]. Among a large number of potential mechanisms are mucosal injury, villous atrophy, increased mast cell number, and reduction in numbers of lymphocytes subsets (CD8[+] T and CD4[+] T) in jejunum and ileum [4,5].

Antibiotics suppress growth of certain microorganisms and are widely used as growth promoters in the swine industry [6]. However, concern over their potential contribution to antibiotic resistance in bacteria infecting humans has led to tightening restrictions on antibiotic use in animals, including cessation of their use as growth promoters in Denmark in May 1995 [7] and elsewhere more recently. The resulting reduction of growth performance and increase in the morbidity in nursery pigs in Denmark indicate the need for prophylaxis [7]. Therefore, it is important to find other reliable strategies to maintain pig health. Among several alternatives, clays have shown promise [8].

Clays have been used in human medicine to ameliorate diarrhea [9], and they are also used in the pig industry with some success [8,10,11]. In the livestock industry, clays are used mainly as mycotoxin binders and as additives that contribute to improve the flow of the feed in bins and feeders, reducing problems with caking of feed. Clays have not been shown to consistently alter growth performance [12-14]. Several types of clays are available and they appear to have different applications and modes of action. Clays with both the 1:1 layer structure (e.g. kaolinite) and the 2:1 layer structure

* Correspondence: jepettig@illinois.edu
[1]Department of Animal Sciences, University of Illinois, Urbana 61801, USA
Full list of author information is available at the end of the article

(e.g. smectite) have positive effects on gastrointestinal health of the animals [15,16]. Song et al. reported [8] that, when pigs were challenged with a pathogenic *E. coli*, feeding dietary clays including smectite, zeolite, kaolinite or combinations of them at 0.3% of the diet reduced diarrhea. Thus, the effect of clays on gastrointestinal health seems more consistent and beneficial than the effect of clays on performance.

Knowledge of the mechanisms through which clays specifically improve gastrointestinal health is lacking, but there are indications [15,16] that clays may strengthen the mucus layer of the intestinal barrier. Moreover, the effects of a challenge with a pathogenic *E. coli* on bacterial translocation from intestinal lumen to mesenteric lymph nodes and goblet cell size and number in weaned pigs has not yet been reported. Our objectives were to determine the effects of a pathogenic *E. coli* challenge and of dietary clays on the intestinal barrier of pigs.

Materials and methods

The Institute of Animal Care and Use Committee of the University of Illinois reviewed and approved the animal care procedures for this experiment.

Animals, experimental design and diets

Two groups of 32 weanling pigs each (about 21 d old; initial BW: 6.9 ± 1.0 kg) were obtained from the Swine Research Center of the University of Illinois. Pigs were housed in disease-containment chambers of the Edward R. Madigan Laboratory building at the University of Illinois at Urbana-Champaign from weaning to about 35 d of age. Pigs had 6 d of adaptation period before challenge. There were a total of 32 individual pens, 4 in each of 8 chambers in each suite. There were 2 suites that were used for either challenged or unchallenged pigs and in each suite, 4 chambers in each suite were used. The treatments were arranged in a 2×4 factorial design [without or with *E. coli* challenge and 4 dietary treatments: control, and 0.3% of 3 different clays added to the control diet: smectite A (**SMA**), smectite B (**SMB**) and zeolite (**ZEO**)]. The enterotoxigenic (**ETEC**) *E. coli* used for the challenge was isolated from a field disease outbreak, (isolate number UI-VDL 05–27242). It is an F-18 fimbria + *E. coli* strain that produces the heat-labile toxin, heat-stable toxin b, and Shiga-like toxin-2 [17]. The pigs were orally inoculated with *E. coli* (10^{10} cfu per 3 mL dose) in PBS daily for 3 d continuously to cause mild diarrhea [17]. The unchallenged treatment (sham) received a 3 mL dose of PBS daily for 3 d. Both inoculations were given orally beginning 6 d after weaning (d 0). Personnel conducting the experiment were blind to the dietary treatments.

The complex nursery basal diet [8] was formulated to meet or exceed NRC [18] estimates of requirements of weanling pigs (Table 1). All the other experimental diets were made from the basal and the addition of 0.3% of each dietary clay. It did not include spray-dried plasma, antibiotics, or zinc oxide to avoid their antibacterial or physiological effects. The experimental diets were introduced at weaning (d –6).

Feeding and sample collection

Pigs and feeders were weighed on the d of weaning (d –6), the d of the first inoculation (d 0), and d 5, for calculation of average daily gain (**ADG**), average daily feed intake (**ADFI**), and gain to feed ratio (**G:F**). Diarrhea score was assessed visually with a score from 1 to 5 (1 = normal feces, 2 = moist feces, 3 = mild diarrhea, 4 = severe diarrhea, and 5 = watery diarrhea) daily from d 0 by 1 scorer who was blind to the dietary treatments. Frequency of

Table 1 Ingredient composition of experimental control diet (as-fed basis)

Ingredient, %	Control diet
Corn, ground	40.93
Dried whey	20.00
Soybean meal, 47%	10.00
Fishmeal	10.00
Lactose	7.22
Soy protein concentrate	5.00
Poultry byproduct meal	3.22
Soybean oil	2.92
Mineral premix[1]	0.35
Vitamin premix[2]	0.20
L-Lys HCl	0.06
DL-Met	0.05
L-Thr	0.03
L-Trp	0.02
Calculated energy and nutrient levels	
ME, kcal/kg	3,480
CP, %	22.53
Fat, %	6.48
Ca, %	0.80
P, %	0.73
Available P, %	0.51
Lys, %	1.50
Lactose, %	21.00

[1]Provided as milligrams per kilogram of diet: 3,000 of NaCl; 100 of Zn from zinc oxide; 90 of Fe from iron sulfate; 20 of Mn from manganese oxide; 8 of Cu from copper sulfate; 0.35 of I from calcium iodide; 0.30 of Se from sodium selenite.
[2]Provided per kilogram of diet: 2,273 μg of retinyl acetate; 17 μg of cholecalciferol; 88 mg of DL-α-tocopheryl acetate; 4 mg of menadione from menadione sodium bisulfite complex; 33 mg of niacin; 24 mg of D-Ca-pantothenate; 9 mg of riboflavin; 35 μg of vitamin B_{12}; 324 mg of choline chloride.

diarrhea was calculated by counting pig d with diarrhea score of 3 or higher.

The standard *E. coli* vaccine was withheld from the dams of the pigs used in this experiment, as were all routine treatments of the piglets with antibiotics. Prior to weaning, fecal samples of the sows from which we obtained the piglets for this experiment were collected to verify if they were negative for β-hemolytic coliforms by plating on blood and McConkey agars. Plates were incubated at 37°C and 5% CO_2 for 24 h before reading. Populations of both total coliforms and β-hemolytic coliforms on blood agar were assessed visually. In the present study β-hemolytic coliforms were detected in the sow feces but they were not the pathogenic *E. coli* we used.

One-half of the pigs (16 from the challenged group (4 from each dietary treatment) and 16 from the sham group (4 from each dietary treatment)) were euthanized on d 5 post inoculation (**PI**) and the remainder on d 6 PI. Prior to euthanasia, pigs were anesthetized by intramuscular injection of a 1-mL combination of telazol, ketamine, and xylazine (2:1:1) per 23 kg of body weight. The final mixture contained 100 mg telazol, 50 mg ketamine, and 50 mg xylazine in 1 mL (Fort Dodge Animal Health, For Dodge, IA). After anesthesia, pigs were euthanized by intracardiac injection of 78 mg sodium pentobarbital per 1 kg of BW (Fort Dodge Animal Health, For Dodge, IA).

Mesenteric lymph nodes were aseptically collected then pooled within pig, ground, diluted and plated on brain heart infusion agar for measurement of total bacteria and the results were expressed as CFU per g of lymph node [19].

Three-cm samples of ileum and colon were collected and cut with scissors longitudinally in the mesenteric border. Tissues were gently washed in buffered saline then fixed in Carnoy's solution for 2–3 h. Subsequently tissue samples were placed in 100% ethanol, 95% ethanol, and 70% ethanol for 30 min each and maintained in 70% ethanol until the staining process. The fixed intestinal tissues were embedded in paraffin, sectioned at 5 μm and stained with high iron diamine (**HID**) and alcian blue (**AB**), pH 2.5, as previously described [20].

Sample processing and analysis

After staining, the slides were scanned by NanoZoomer Digital Pathology System (Hamamatsu Co., Bridgewater, NJ), and the measurements were conducted in NanoZoomer Digital Pathology Image Program (Hamamatsu Co., Bridgewater, NJ). Measurements included villus height, crypt depth, and the cross-sectional area of sulfo- (stained brown) and sialomucin (stained blue). The measurements for villus height and crypt depth were performed on 10 well-oriented villi [21] scanned at 40× resolution.

The total number of goblet cells per villus was counted and NDP.view software was used to measure the cross-sectional area (μm^2) of individual goblet cells. The measurements were performed in 3 well-oriented villi scanned at 40x resolution.

Statistical analysis

Data were subjected to an analysis of variance using the Proc Mixed procedure (SAS Inst. Inc., Cary, NC). Pig was the experimental unit. The statistical model included effects of *E. coli* challenge, diet, and their interaction as fixed effects and group as a random effect. Specific contrasts were used to test comparisons between the control and the clay treatments collectively within each challenge group. In addition, differences among the clay treatments within each challenge group were tested by pair-wise comparisons when the overall main effect or the diet x challenge interaction was significant. The χ^2 test was used for the frequency of diarrhea. The α levels of 0.05 and between 0.05 and 0.10 were used for determination of significance and tendency, respectively, among means.

Results and discussion

After the challenge, fecal samples were collected from pigs from sham and *E.coli*-challenged groups and it was observed that both groups of pigs carried β-hemolytic *E. coli*. Subsequent PCR analysis [22] showed that the sham-challenged pigs carried *E. coli* that produced cytotoxic necrotizing factor. This minor background infection with a wild strain of *E. coli* occurred in some of the sham-challenged and *E.coli*-challenged pigs in this experiment indicating that the sham-challenged pigs had pathogenic organisms and that the *E.coli*-challenged pigs could have other pathogenic organisms besides the challenge one, so the model represents a multiple infection rather than an uncomplicated single-pathogen challenge.

Cytotoxic necrotizing factor is produced by 40% of pathogenic *E. coli* strains involved in urinary tract infections and 5-30% of those involved in diarrheic infections [23]; it increases adherence of the pathogen to epithelial cells. The impact of infection with this wild strain on the response to the challenge strain is unclear, but if clays provide protection from diarrhea by strengthening the mucus barrier, they should provide similar protection from both of these strains of *E. coli*.

Diarrhea score and growth performance

The *E. coli* challenge was successful as it increased diarrhea score moderately from d 3 to 5 (Table 2) and reduced ADG from d 0 to 5 PI (Table 3), consistent with previous results [8]. The diarrhea scores were low during the first d after challenge, apparently reflecting a lag period after the inoculation before the clinical signs appeared (Table 2).

Table 2 Effect of clays on diarrhea score of pigs experimentally infected with a pathogenic E. coli[1]

| | Treatment[2] | | | | | | | | | P-value | | | | |
| | Sham | | | | E. coli | | | | | Main effect[3] | | | CON vs. Clays[4] | |
Item	CON	SMA	SMB	ZEO	CON	SMA	SMB	ZEO	SEM	E. coli	Diet	E x D	Sham	E. coli
d 0 to 2[5]	2.02	2.33	2.23	2.21	1.50	1.94	2.00	1.60	0.19	0.03	0.46	0.91	0.27	0.45
d 3 to 5	2.37	1.98	2.52	1.87	2.64	3.04	2.94	2.50	0.24	0.01	0.45	0.66	0.64	0.52
d 0 to 5	2.20	2.16	2.37	2.04	2.07	2.49	2.47	2.05	0.17	0.65	0.39	0.81	0.34	0.98
Pig d[6]	48	48	48	48	48	48	48	48	-	-	-	-	-	-
Diarrhea d[7]	3	4	4	4	3	8	8	5	-	-	-	-	-	-
Frequency, %[8]	6.25	8.33	8.33	8.33	6.25	16.67	16.67	10.42	-	0.13	0.14	0.08	0.64	0.13

[1]n = 8 pigs/treatment.
[2]Sham = unchallenged; E. coli = E. coli challenged; CON = control diet; SMA = 0.3% smectite A; SMB = 0.3% smectite B; ZEO = 0.3% zeolite.
[3]E. coli = E. coli challenge effect; Diet = diet effect; E x D = interaction between E. coli and diet effects.
[4]Contrast between CON and all clay treatments within challenge treatments.
[5]Diarrhea score = 1, normal feces, 2, moist feces, 3, mild diarrhea, 4, severe diarrhea, 5, watery diarrhea.
[6]Pig d = number of pigs x the number of d of diarrhea scoring.
[7]Diarrhea d = number of pig days with diarrhea score ≥ 3. Statistical analysis was conducted by chi-square test.
[8]Frequency (frequency of diarrhea during the entire experimental period) = diarrhea days*100/pig days.

During this period the E.coli-challenged pigs actually had lower diarrhea scores ($P < 0.05$) than did the sham-challenged ones. During the active disease, from d 3 to 5 PI, the E.coli-challenged pigs had a higher diarrhea score than the sham-challenged pigs ($P < 0.05$), as expected (Table 3).

There were no dietary effects on either diarrhea scores (Table 2) or growth performance (Table 3), in contrast to the beneficial effects of clays on diarrhea score is in our earlier results [8]. Our earlier experiments [8] continued for 12 d after inoculation, well into the recovery phase. The pigs in the present experiment were euthanized at around the peak of disease (d 5 and 6 PI) in order to measure physiological effects of the E. coli challenge and the clays at that crucial time. Therefore, diarrhea was assessed for only a short time, with the critical period being d 3–5 PI. It is not clear if we would have observed the same effects on diarrhea score as we did earlier [8] if the experiment had been carried out until

the recovery phase. In one of our earlier experiments clays reduced diarrhea during d 3–6 PI; whereas in the other there was only a trend during d 3–6 PI but clearer effects later [8]. The benefits of clays in reducing diarrhea that we reported [8] are supported by research in humans, as a meta-analysis of 9 studies showed that children with acute gastroenteritis consistently had lower duration of diarrhea when treated with smectite along with re-hydration compared with a placebo group without smectite [24].

Goblet cell number and size

Goblet cells in the intestine produce mucins, the proteins that comprise the bulk of the mucus layer which acts as the first line of defense against enteric infections [25]. The present results show that the E. coli challenge increased both the number and size of goblet cells in the ileum (Table 4), consistent with an increase in mucin secretion in response to pathogenic bacteria or intestinal

Table 3 Effect of clays on growth performance of pigs experimentally infected with a pathogenic E. coli[1]

| | Treatment[2] | | | | | | | | | P-value | | | | |
| | Sham | | | | E. coli | | | | | Main effect[3] | | | CON vs. SM[4] | |
Item	CON	SMA	SMB	ZEO	CON	SMA	SMB	ZEO	SEM	E. coli	Diet	E x D	Sham	E. coli
d −6 to 0														
ADG, g	6.25	29.17	−2.08	−25.00	12.50	−25.00	33.33	8.33	42.4	0.80	0.86	0.38	0.87	0.84
ADFI, g	394	442	319	367	421	421	329	406	212	0.74	0.31	0.96	0.79	0.60
d 0 to 5														
ADG, g	237	180	157	187	137	132	122	85	63.71	< 0.01	0.52	0.73	0.16	0.58
ADFI, g	715	715	557	632	632	627	455	517	193	0.11	0.15	1.00	0.42	0.32
G:F[5]	0.34	0.26	0.33	0.32	0.23	0.24	0.24	0.24	0.048	0.11	0.95	0.92	0.64	0.95

[1]n = 8 pigs/treatment.
[2]Sham = unchallenged; E. coli = E. coli challenged; CON = control diet; SMA = 0.3% smectite A; SMB = 0.3% smectite B; ZEO = 0.3% zeolite.
[3]E. coli = E. coli challenge effect; Diet = diet effect; E x D = interaction between E. coli and diet effects.
[4]Contrast between CON and all clay treatments within challenge treatments.
[5]G:F was not reported for period −6 to 0 because of the negative values for ADG.

Table 4 Effect of clays on goblet cell number and size in ileum and colon of pigs experimentally infected with a pathogenic E. coli[1]

Item	Treatment[2] Sham				E. coli				SEM	P-value Main effect[3]			CON vs. SM[4]	
	CON	SMA	SMB	ZEO	CON	SMA	SMB	ZEO		E. coli	Diet	E x D	Sham	E. coli
Ileum														
Number[5]	25.54	23.58	23.67	25.62	27.42	25.00	32.42	26.87	3.08	< 0.01	0.16	0.06	0.49	0.71
Size[6,7], μm^2	29.47^b	29.58^b	31.88^{a,b}	30.72^b	31.00^{a,b}	35.96^a	30.65^b	31.89^{a,b}	0.764	0.01	0.18	0.01	0.32	0.15
Colon														
Number	28.21	24.75	27.54	25.83	27.71	28.85	26.67	24.18	12.71	0.84	0.43	0.39	0.29	0.60
Size, μm^2	26.26	27.33	27.60	30.37	25.56	28.31	25.82	27.69	0.801	0.20	0.04	0.41	0.09	0.21

^{a,b}Means with different superscripts in the same row differ.
[1]n = 8 pigs/treatment.
[2]Sham = unchallenged; E. coli = E. coli challenged; CON = control diet; SMA = 0.3% smectite A; SMB = 0.3% smectite B; ZEO = 0.3% zeolite.
[3]E. coli = E. coli challenge effect; Diet = diet effect; E x D = interaction between E. coli and diet effects.
[4]Contrast between CON and all clay treatments within challenge treatments.
[5]Goblet cell number; total number of goblet cells per villus, average of 3 villi.
[6]Goblet cell size, cross-sectional area.
[7]Con vs. SMA (Tukey adjustment) P = 0.07.

microbes that has been previously reported [21,26,27]. Perhaps the increased mucin production is a protective response. One of the clays (SMA) tended to increase goblet cell size in the ileum (P = 0.07) when compared to BAS in the E.coli-challenged group. There was a trend (P = 0.06) for an interaction between diet and challenge on ileal goblet cell number in which one clay (SMB) increased the number of goblet cells in challenged pigs only. There was a diet effect on goblet cell size in the colon (Table 4) in which the clays generally increased goblet cell size, mostly in the sham group. These modest increases in goblet cell size and number during the acute phase of the infection when clays were fed may reflect enhanced protection and may at least partially explain the reduction in diarrhea observed previously in pigs [8] and children [24].

Bacterial translocation
The E. coli challenge clearly increased bacterial translocation from the lumen to the lymph nodes but the dietary treatments did not detectably alter it (Table 5). To our knowledge, bacterial translocation from the intestinal lumen to the mesenteric lymph nodes has not been reported for pigs challenged with a pathogenic E. coli strain. Chicks infected with Eimeria acervulina, E.

maxima, and Clostridium perfringes exhibited increased bacterial translocation from intestinal lumen to the spleen when compared with control birds [26] indicating that enteric infections reduce the integrity of the intestinal barrier. The increased total bacterial translocation caused by E. coli in the present study (Table 5) indicates that the infection reduced the effectiveness of the intestinal barrier, which was expected.

Intestinal morphology
Weaning triggers a reduction in villus height and in the villus height:crypt depth ratio, caused at least partially by interruption of voluntary feed intake [28], and restoration of villus height may be important for health and growth performance of the pig. In the present study, the challenge increased crypt depth and tended to reduce the villus height:crypt depth ratio (VH:CD; Table 6) as shown previously [17]. These effects of disease may exacerbate the detrimental impact of weaning on pig health and growth. The response to E. coli is inconsistent across experiments. Our observed values for the sham group are similar to previously reported in some cases [29] but smaller than those previously reported [17,30] in others. We did not detect any effect of clays or challenge on intestinal morphology (Table 6) except

Table 5 Effects of clays on bacteria in lymph nodes of pigs experimentally infected with a pathogenic E. coli[1]

Item	Treatment[2] Sham				E. coli				SEM	P-value Main effect[3]			CON vs. Clays[4]	
	CON	SMA	SMB	ZEO	CON	SMA	SMB	ZEO		E. coli	Diet	E x D	Sham	E. coli
Log$_{10}$ CFU[5]	1.05	0.74	0.65	0.60	1.87	2.12	2.03	1.69	0.30	0.01	0.88	0.90	0.44	0.87

[1]n = 64 (8 pigs/treatment).
[2]Sham = unchallenged; E. coli = E. coli challenged; CON = control diet; SMA = 0.3% smectite A; SMB = 0.3% smectite B; ZEO = 0.3% zeolite.
[3]E. coli = E. coli challenge effect; Diet = diet effect; E x D = interaction between E. coli and diet effects.
[4]Contrast between CON and all clay treatments within challenge treatments.
[5]Log$_{10}$ CFU/g of lymph node.

Table 6 Effect of clays on intestinal morphology of pigs experimentally infected with a pathogenic *E. coli*[1]

	Treatment[2]									P-value				
	Sham				*E. coli*					Main effect[3]			CON vs. Clays[4]	
Item	CON	SMA	SMB	ZEO	CON	SMA	SMB	ZEO	SEM	*E. coli*	Diet	E x D	Sham	*E. coli*
Duodenum														
VH[5]	384.6	374.6	380.6	359.1	356.6	393.8	382.5	365.0	21.06	0.99	0.78	0.80	0.64	0.40
CD[6]	264.9	276.3	263.8	262.0	273.5	257.1	259.9	275.45	39.15	0.98	0.95	0.63	0.87	0.55
VH:CD[7]	1.55	1.48	1.53	1.47	1.39	1.84	1.61	1.45	0.26	0.42	0.25	0.15	0.70	0.07
Ileum														
VH	299.0	310.7	288.7	305.8	305.4	289.8	282.1	309.9	9.15	0.64	0.36	0.72	0.85	0.45
CD	208.4	214.1	212.8	222.4	228.5	232.3	230.8	217.6	8.37	0.05	0.96	0.48	0.45	0.88
VH:CD	1.44	1.49	1.37	1.38	1.34	1.25	1.25	1.45	0.08	0.10	0.61	0.35	0.76	0.81
Colon														
CD	236.0	229.0	247.7	227.5	228.8	244.2	223.1	216.4	73.94	0.28	0.36	0.18	0.90	0.93

[1]n = 8 pigs/treatment.
[2]Sham = unchallenged; *E. coli* = *E. coli* challenged; CON = control diet; SMA = 0.3% smectite A; SMB = 0.3% smectite B; ZEO = 0.3% zeolite.
[3]*E. coli* = *E. coli* challenge effect; Diet = diet effect; E x D = interaction between *E. coli* and diet effects.
[4]Contrast between CON and all clay treatments within challenge treatments.
[5]Villus height, μm.
[6]Crypt depth, μm.
[7]Villus height:crypt depth ratio.

for a tendency ($P = 0.07$) for the effects of clays in increasing VH:CD in the *E.coli* challenged pigs. Beneficial effects of small amounts of dietary clays have been reported previously. For example, montmorillonite increased villus height and villus height: crypt depth ratio in jejunum when fed to weanling pigs at 0.15% of the diet [13]. Similar results were obtained in broiler chickens. Previous authors [14,30,31] reported that feeding 0.1%, or 0.2% montmorillonite increased villus height and reduced crypt depth in the duodenum and jejunum.

Sulfo- and sialomucin
Mucins can be acidic or neutral. Acidic mucins are comprised of sulfo- and sialomucins. The body often reacts to infection by increasing the secretion of sulfomucins [32] as a protective mechanism; the present data do not show that response (Table 7). The present results do not show effects of either infection or dietary clays on the relative amount of sulfo- and sialomucins within goblet cells (Table 7).

Conclusions
The present results provide novel information regarding the physiological responses in the intestinal barrier of pigs to a challenge with a pathogenic *E. coli* strain. To our knowledge, it is the first time that bacterial translocation from intestinal lumen to mesenteric lymph nodes and goblet cell size and number in weaned pigs

Table 7 Effect of clays on relative amounts of sulfo- and sialomucin area of pigs experimentally infected with a pathogenic *E. coli*[1]

	Treatment[2]									P-value				
	Sham				*E. coli*					Main effect[3]			CON vs. Clays[4]	
Item	CON	SMA	SMB	ZEO	CON	SMA	SMB	ZEO	SEM	*E. coli*	Diet	E x D	Sham	*E. coli*
Ileum														
Sulfo[5]	44.31	37.59	32.86	37.95	31.28	32.31	37.38	37.10	5.80	0.52	0.97	0.73	0.37	0.65
Sialo[6]	55.69	62.41	67.14	62.05	68.62	67.69	62.62	62.90	5.80	0.52	0.97	0.73	0.37	0.65
Colon														
Sulfo	92.39	92.74	95.09	94.96	93.37	90.49	87.60	94.29	1.57	0.14	0.43	0.26	0.45	0.33
Sialo	7.61	7.26	4.91	5.03	6.63	9.51	12.40	5.71	1.57	0.14	0.43	0.26	0.45	0.33

[1]n = 8 pigs/treatment.
[2]Sham = unchallenged; *E. coli* = *E. coli* challenged; CON = control diet; SMA = 0.3% smectite A; SMB = 0.3% smectite B; ZEO = 0.3% zeolite.
[3]*E. coli* = *E. coli* challenge effect; Diet = diet effect; E x D = interaction between *E. coli* and diet effects.
[4]Contrast between CON and all clay treatments within challenge treatments.
[5]Sulfo = % of total sulfo- and sialomucin area that is sulfamucin.
[6]Sialo = % of total sulfo- and sialomucin area that is sialomucin.

challenged with a pathogenic *E. coli* is reported. Both the infection and SMA altered goblet cell size and number. The clinical benefits of clays in the face of enteric infections that we observed in previous experiments with pigs, such as the reduction in diarrhea score, did not occur in this shorter experiment, but it is unclear whether they may have appeared if the experiment had been longer. However, it was important to explore the potential beneficial of the clays during the acute phase of an enteric infection.

Abbreviations
PI: Post inoculation; SMA: Smectite A; SMB: Smectite B; ZEO: Zeolite; ETEC: Enterotoxigenic; HID: High iron diamine; AB: Alcian blue.

Competing interests
Dr. Orlando Osuna, is employed by Milwhite, a company that manufactures and markets clays.

Authors' contributions
JASA carried out the animal work, processed the samples, participated in the design of the study, performed the statistical analysis, and drafted the manuscript. YL carried out the lymph node assay and participated in the design of the study. MS participated in the design of the study and performed the training for diarrhea score assessment. JJL helped with the animal work, and carried out the goblet cell size and number quantification. HRG participated in the design of the experiment. CWM provided the *E. coli* for the challenge, and participated in the design of the experiment. OO participated in the design of the experiment. JEP participated in the design of the experiment and helped to draft the manuscript. All authors read and approved the final manuscript.

Acknowledgements
Financial support from Milwhite, Inc., Brownsville, TX, is appreciated.

Author details
[1]Department of Animal Sciences, University of Illinois, Urbana 61801, USA. [2]Current address: Department of Animal Science and Biotechnology, Chungnam National University, Daejeon, South Korea. [3]Department of Pathobiology, University of Illinois, Urbana 61801, USA. [4]Milwhite, Inc., Brownsville, TX, USA.

References
1. Pluske JR, Pethick RDW, Hopwood DE, Hampson DJ: **Nutritional influences on some major enteric bacterial diseases of pigs.** *Nutr Res Rev* 2002, **15**:333–371.
2. Pluske JR, Thompson MJ, Williams IH: **Factors influencing the structure and function of the small intestine in the weaned pig: a review.** *Livest Prod Sci* 1997, **51**:215–236.
3. Fairbrother JM, Nadeau É, Gyles CL: *Escherichia coli* **in postweaning diarrhea in pigs: an uptake on bacterial types, pathogenesis, and prevention strategies.** *Anim Health Res Rev* 2005, **6**:17–39.
4. Zhu HL, Liu YL, Xie XL, Huang J, Hou Y: **Effect of L-arginine on intestinal mucosal immune barrier function in weaned pigs after** *Escherichia coli* **LPS challenge.** *Innate Immun*. 2013, **19**:242–252.
5. Dean P, Kenny B: **Intestinal barrier dysfunction by enteropathogenic** *Escherichia coli* **is mediated by two effector molecules and a bacterial surface protein.** *Mol. Microbiol*. 2004, **54**:665–675.
6. Cromwell GL: **Why and how antibiotics are used in swine production.** *Anim Biotechnol* 2002, **13**:7–27.
7. World Health Organization (WHO): [http://www.who.int/gfn/en/Expertsreportgrowthpromoterdenmark.pdf]
8. Song M, Liu Y, Soares JA, Che TM, Osuna O, Maddox CW, Pettigrew JE: **Dietary clays alleviate diarrhea of weaned pigs.** *J Anim Sci* 2012, **90**:345–360.
9. Carretero MI: **Clay minerals and their beneficial effects upon human health.** *A review. Appl Clay Sci* 2002, **21**:155–163.
10. Schell TC, Lindemann MD, Kornegay ET, Blodgett DJ, Doerr JA: **Effectiveness of different types of clay for reducing the detrimental effects of aflatoxin-contaminated diets on performance and serum profiles of weanling pigs.** *J Anim Sci* 1993, **71**:1226–1231.
11. Trckova M, Vondruskova H, Zraly Z, Alexa P, Hamrik J, Kummer V, Maskova J, Mrlik V, Krizova K, Slana I, Leva L, Pavlik I: **The effect of kaolin feeding on efficiency, health status and course of diarrhoeal infections caused by enterotoxigenic** *Escherichia coli* **strains in weaned pigs.** *Vet Med* 2009, **54**:47–63.
12. Shurson GC, Ku PK, Miller ER, Yokoyama MT: **Effects of zeolite a or clinoptilolite in diets of growing swine.** *J Anim Sci* 1984, **59**:1536–1545.
13. Xia MS, Hu H, Xu ZR: **Effects of copper bearing montmorillonite on the growth performance, intestinal microflora and morphology of weanling pigs.** *Anim Feed Sci Technol* 2004, **118**:307–317.
14. Xia MS, Hu H, Xu ZR: **Effects of copper-bearing montmorillonite on growth performance, digestive enzyme activities, and intestinal microflora and morphology of male broilers.** *Poult Sci* 2004, **83**:1868–1875.
15. Droy-Lefaix MT: **Effects of treatment with smectite on gastric and intestinal glycoproteins in the rat: a histochemical study.** *Histochem J* 1987, **19**:665–670.
16. Gonzales RF, Medina S, Martinez-Augustin O, Nieto A, Galvez J, Risco S, Zarzuelo A: **Anti-inflammatory effect of diosmectite in hapten-induced colitis in the rat.** *Br J Pharmacol* 2004, **141**:951–960.
17. Perez-Mendoza V: **Effects of distillers dried grains with solubles and dietary fiber on the intestinal health of young pigs and chicks.** In *PhD thesis.* University of Illinois, Animal Sciences Department; 2010.
18. NRC: *Nutrient requirements of swine.* Washington, DC: Natl Acad Press; 1998:10.
19. Swildens B, Stockhofe-Zurwieden N, der Meulen JV, Wisselink HJ, Nielen M, Niewold TA: **Intestinal translocation of** *Streptococcus suis* **type 2 EF + in pigs.** *Vet Microbiol* 2004, **103**:29–33.
20. Deplancke B, Gaskins HR: **Microbial modulation of innate defense: goblet cells and the intestinal mucus layer.** *Am J Clin Nutr* 2001, **73**:1131S–1141S.
21. Fasina YO, Hoerr FJ, McKee SR: **Influence of** *Salmonella enterica* **serovar** *Typhimurium* **infection on intestinal goblet cells and villous morphology in broiler chicks.** *Avian Dis* 2010, **54**:841–847.
22. DebRoy C, Maddox CW: **Assessing virulence of gastroenteric** *Escherichia coli* **isolates of veterinary significance.** *Anim Health Res Rev* 2001, **2**:129–140.
23. Hofman P, Le Negrate G, Mograbi B, Hofman V, Brest P, Alliana-Schmid A, Flatau G, Bouquet P, Rossi B: *Escherichia coli* **cytotoxic necrotizing factor-1 (CNF-1) increases the adherence to epithelia and the oxidative burst of human polymorphonuclear leukocytes but decreases bacteria phagocytosis.** *J Leukoc Biol* 2000, **68**:522–528.
24. Szajewska HL, Dziechciarz P, Mrukowicz J: **Meta-analysis: smectite in the treatment of acute infectious diarrhea in children.** *Aliment Pharmacol and Ther* 2006, **23**:217–227.
25. Forder RE, Howarth GS, Tivey DR, Hughes RJ: **Bacterial Modulation of small intestinal goblet cells and mucin composition during early posthatch development of poultry.** *Poult Sci* 2007, **86**:2396–2403.
26. Collier CT, Hofacre CL, Payne AM, Anderson DB, Kaiser P, Mackie RI, Gaskins HR: **Coccidia-induced mucogenesis promotes the onset of necrotic enteritis by supporting** *Clostridium perfringens* **growth.** *Vet Immunol Immunopathol* 2008, **122**:104–115.
27. Deplancke B, Hristova KR, Oakley HA, McCracken VJ, Aminov R, Mackie RI, Gaskins HR: **Molecular ecological analysis of the succession and diversity of sulfate-reducing bacteria in the mouse gastrointestinal tract.** *Appl Environ Microbiol* 2000, **66**:2166–2174.
28. Pluske JR, Thompson MJ, Atwood CS, Bird PH, Williams IH, Hartmann PE: **Maintenance of villus height and crypt depth, and enhancement of disaccharide digestion and monosaccharide absorption, in piglets fed on cows' whole milk after weaning.** *Brit J Nutr* 1996, **76**:409–422.
29. Liu Y, Song M, Che TM, Almeida JAS, Lee JJ, Bravo D, Maddox CW, Pettigrew JE: **Dietary plant extracts alleviate diarrhea and alter immune responses of weaned pigs experimentally infected with a pathogenic** *Escherichia coli.* *J Anim Sci* 2013, **91**:5294–5306.
30. Owusu-Asiedu A, Nyachoti CM, Baidoo SK, Marquardt RR, Yang X: **Response of early-weaned pigs to an enterotoxigenic** *Escherichia coli* **(K88) challenge when fed diets containing spray-dried porcine plasma or pea protein isolate plus egg yolk with antibody.** *J Anim Sci* 2003, **81**:1781–1789.

31. Xu ZR, Hu H, Xia MS, Zhan XA, Wang MQ: **Effects of dietary fructooligosaccharide on digestive enzyme activities, intestinal microflora and morphology of male broilers.** *Poult Sci* 2003, **82**:648–654.
32. Ma YL, Guo T: **Intestinal morphology, brush border and digesta enzyme activities of broilers fed on a diet containing Cu^{2+}-loaded montmorillonite.** *Br Poult Sci* 2008, **49**:65–73.

5

Preslaughter diet management in sheep and goats: effects on physiological responses and microbial loads on skin and carcass

Govind Kannan, Venkat R Gutta, Jung Hoon Lee[*], Brou Kouakou, Will R Getz and George W McCommon

Abstract

Sixteen crossbred buck goats (Kiko x Spanish; BW = 32.8 kg) and wether sheep (Dorset x Suffolk; BW = 39.9 kg) were used to determine the effect of preslaughter diet and feed deprivation time (FDT) on physiological responses and microbial loads on skin and carcasses. Experimental animals were fed either a concentrate (CD) or a hay diet (HD) for 4 d and then deprived of feed for either 12-h or 24-h before slaughter. Blood samples were collected for plasma cortisol and blood metabolite analyses. *Longisimus* muscle (LM) pH was measured. Skin and carcass swabs were obtained to assess microbial loads. Plasma creatine kinase activity (863.9 and 571.7 ± 95.21 IU) and non-esterified fatty acid concentrations (1,056.1 and 589.8 ± 105.01 mEq/L) were different ($P < 0.05$) between sheep and goats. Species and diet treatments had significant effects on the ultimate pH of LM. Pre-holding total coliform (TCC) and aerobic plate counts (APC) of skin were significantly different between species. Goats had lower ($P < 0.05$) TCC (2.1 vs. 3.0 $\log_{10}$ CFU/cm^2) and APC (8.2 vs. 8.5 $\log_{10}$ CFU/cm^2) counts in the skin compared to sheep. Preslaughter skin *E. coli* counts and TCC were different ($P < 0.05$) between species. Goats had lower ($P < 0.05$) counts of *E. coli* (2.2 vs. 2.9 $\log_{10}$ CFU/cm^2) and TCC (2.3 vs. 3.0 $\log_{10}$ CFU/cm^2) in the skin compared with those in sheep. Diet, species, and FDT had no effect ($P > 0.05$) on *E. coli* and TCC in carcass swab samples. The APC of carcass swab samples were only affected ($P < 0.05$) by the FDT. The results indicated that preslaughter dietary management had no significant changes on hormone and blood metabolite concentrations and sheep might be more prone for fecal contamination than goats in the holding pens at abattoir.

Keywords: *E. coli* contamination diet, Goats, Physiology, Sheep

Background

The hide and viscera of animals entering the abattoir are potential sources of contamination of carcasses with pathogenic bacteria [1]. The hide of the live animal becomes contaminated with pathogenic and non-pathogenic microorganisms from a wide range of sources such as feces, soil, water, and vegetation [2]. Animals can spread the contaminants to other animals during preslaughter transport and holding, directly via physical contact with one another or with the contaminated floor [3]. Fecal shedding of bacteria can be controlled by manipulating the preslaughter diet [4] and feed deprivation time [5] in ruminants.

Preslaughter dietary manipulation may not only affect the micro flora in gastrointestinal tracts in ruminants,

but may also influence the variables related to meat quality and animal welfare [6,7]. Feeding grain diets can change the rumen and intestinal microbial populations [8]. Overfeeding cattle with grain has been shown to cause a 2 log scale increase in total coliform counts [9].

Stress and dehydration resulting from preslaughter management methods can adversely affect production variables such as live and carcass weights as well as meat quality [10]. A switch to hay feeding from a concentrate diet is likely to influence carcass weights, although Stanton and Schultz [11] indicated that such a diet change did not have a dramatic impact on carcass characteristics and final body weights in cattle. However, Kannan et al. [12] reported that 18 h of feed deprivation resulted in a 10% live weight shrinkage in goats. Earlier studies also showed that fasting sheep for 24 h resulted in about 7% live weight loss due to reduction in gut contents [13,14]. Feed deprivation is one

* Correspondence: leej@fvsu.edu
Agricultural Research Station, Fort Valley State University, 1005 State University Drive, Fort Valley, GA 31030, USA

of the preslaughter stress factors that may be responsible for depletion of muscle glycogen prior to slaughter [15]. Preslaughter depletion of muscle glycogen may result in an abnormally high pH of meat, which may have adverse effects on meat quality such as dark cutters [16] and poor shelf life due to microbial spoilage [17].

Blood hormone and metabolites in ruminants are also influenced by feed deprivation. Plasma cortisol concentration, a good indicator of welfare status during the preslaughter period in food animals [18], increases in sheep [19] and goats [12] due to feed deprivation. Feed deprivation also alters plasma glucose [20,21], urea nitrogen [10,12], and non-esterified fatty acid [22,23]. Kannan et al. [24] reported an increase in creatine kinase activity in the circulation during preslaughter feed deprivation in goats.

The objectives of this study were, therefore, to estimate the efficacy of preslaughter diet (concentrate vs roughage) and feed deprivation time (12 vs. 24 h) on *E. coli* and other enteric bacterial population on skin and carcass, as well as to determine the effects on blood hormone and metabolites in sheep and goats.

Methods

Animal feeding and feed deprivation treatments

Experimental procedures involving animals were conducted with approval of the Fort Valley State University (FVSU) Institutional Animal Care and Use Committee. Animals were obtained from the Georgia Small Ruminant Research and Extension Center at (FVSU). Sixteen crossbred wether sheep (Dorset x Suffolk; BW = 39.9 ± 0.88 kg) and buck goats (Kiko x Spanish; BW = 32.8 ± 0.91 kg) grazed on winter pea and rye grass dominant forages were assigned in a completely randomized design to a feeding trial consisting of two dietary treatments: primarily corn based concentrate (Table 1) and Bermuda grass hay diets. Each treatment was replicated in two pens with either four sheep or goats per pen. Each pen of four experimental

Table 1 Ingredient composition of concentrate diet[1,2] fed to sheep and goats

Ingredient	Composition,%
Cottonseed hull	14.0
Ground corn	67.8
Soybean meal	13.6
Poultry fat	2.73
Trace minerals[3]	0.5
Vitamin premixed	0.5
Dicalcium phosphate	0.9

[1]Predicted digestible Energy (DE) = 4.0 Mcal/kg.
[2]Crude protein = 12.9%.
[3]Composition: NaCl, 45 to 50%; Ca 9.0 to 10.8%; P, >4.5%; Mg, >1.5%; K, >0.9%; S, >0.3%; Zn, >1.55%; and I, >180 ppm; Fe, >2,000 ppm; Mn, >4,000 ppm; Se, >60 ppm; vitamin A, >2,200,000 IU; vitamin D₃, >165,000 IU; and vitamin E, >6,600 IU/kg.

animals was fed twice a day either a concentrate (CD) or hay diet (HD) with ad libitum access to water for 4 days. At the end of the 4-d feeding trial, half of animals from each pen (n = 16) were randomly selected and transported to the university slaughter and processing facility. Each animal was weighed and then assigned to a pen in the holding area according to the original pen numbers in order to maintain the same social group. This group of animals was deprived of feed for a 24-h period with continuous access to water. Other half of animals (n = 8/pen) were assigned to deprive of feed for a 12 h period according to previously descried in the 24 h feed deprivation. Both feed deprivation time (FDT) groups were processed on the same day such that harvest occurred within the same time frame for both group.

Animal behavior

Behavior of each animal was monitored for a 90-min period before slaughter. The weather conditions were identical on the experimental days. Minimum temperatures ranged from 5 to 7°C and maximum temperatures ranged from 18 to 20°C. Standing, moving, agonistic (ramming, jumping, horning, and head butting) lying and drinking behaviors were recorded. Behavioral observations were made every minute using the scan sampling method in each pen (from pen 1 to 8) [25]. At each monitoring period, the number of animals performing each behavior was recorded. Animals were slaughtered in a predetermined order and rotated among pens to avoid confounding of effects.

Blood sampling and analysis

Blood samples were collected from each animal at the beginning of the feeding trial (pretrial) and prior to slaughter. Blood samples were collected by trained personnel via jugular venipuncture into 10 mL Vacutainer tubes containing 81 µL of 15% EDTA solution and immediately placed on ice. All efforts were made not to agitate the animals during sampling. Plasma was separated by centrifugation at 1,000 × g for 30 min in a Sorvall Superspeed model 5RC2-B automatic refrigerated centrifuge (Ivan Sorvall Inc., Newton, CT) and stored in a 10-mL vial at −20°C for determination of plasma cortisol, glucose, creatine kinase (CK), urea nitrogen (PUN), and non-esterified fatty acid (NEFA) concentrations.

Plasma cortisol concentrations were determined using a Coat-A-Count radioimmunoassay (RIA) kit (Diagnostic Product Corp., Los Angeles, CA) as described by Kannan et al. [12]. Blood glucose and PUN concentrations and CK activity were analyzed using an IDEXX VetTest® instrument (IDEXX Laboratories Inc., Westbrook, ME). The plasma sample was delivered into a pipette tip and dispensed onto each metabolite testing slide. As the sample was absorbed and filtered through the layers of the slides, color changes occurred due to biochemical reactions. The color and intensity were measured by an optical system. Plasma NEFA concentrations were analyzed using a commercially available kit (Wako Chemicals, Richmond, VA) as described by Kannan et al. [24]. The assay was performed using acetyl CoA synthetase/acetyl-CoA oxidase method (NEFA C Code No. 994–75409 E). The absorbance values were determined using a Shimadzu® (Model UV-2401 PC) UV–VIS spectrophotometer (Shimadzu Scientific Instruments, Inc., Columbia, MD).

Carcass yield and muscle pH

Animals were weighed prior to slaughter and then processed at the FVSU slaughter and meat processing facility. After final carcass wash, hot carcass weights were recorded. Dressed carcasses were stored at 2°C for 24 h before fabrication. After 24 h cooling, cold carcass weights were also recorded. Dressing percent of each carcass was reported as carcass yield. Muscle pH was recorded at 0- (immediately after skinning) and 24-h postmortem using a portable pH meter (Fisher Scientific, Pittsburgh, PA) with a penetrating probe (Pakton® Model OKPH1000N, Fisher Scientific). The probe was inserted directly into the *longissimus* muscle of each carcass to measure pH.

Microbial counts

Sterile sponges, hydrated with 10 mL of buffered peptone water (BioPro Enviro-Sponge Bags, International BioProducts, Redmond, WA) with disposable sterile paper templates (5 cm × 5 cm) were used for collection of skin and carcass swab samples. The swab sampling procedure for skin was adopted from Kannan et al. [26]. Samples were obtained from each animal at the beginning of the feeding trial and prior to slaughter by swabbing the hind leg within the 25 cm^2 template area with five vertical wipes and five horizontal wipes. Carcass swab sampling was followed by the USDA procedure used for genetic *E. coli* testing [27] as modified by Kannan et al. [26]. The modification was the smaller sampling area and fewer wipes to suit the smaller size of goat carcasses instead of the 10 cm × 10 cm template recommended by the USDA. Swab samples were collected from each carcass after skinning and evisceration, but before washing by swabbing three different anatomical locations (flank, brisket, leg) within the 25-cm^2 template area for a total sampling area of 75 cm^2. The swab samples were placed in sterilized sponge bags, transported on ice, and stored under refrigeration until analysis.

After swabbing, the sponges were transferred into sterilized stomacher bags and 90-mL of 0.1% sterile buffered peptone water (Difco Laboratories, Detroit, MI) was added to each bag. The contents of the bag were pummeled in a stomacher (Seward Model 400, Tekmar Co. Cincinnati, OH) for 1 min. Serial dilutions were prepared with 0.1% sterile buffered peptone water. The 3M™ Petrifilm

plate techniques were used to enumerate microbial loads on skin and carcass samples as recommended by the manufacturer [28]. Appropriate sample dilutions were inoculated on Petrifilm plates (3M™ Microbiology Products, St. Paul, MN) to determine *E. coli* and total coliform (3M™ Petrifilm™ *E. coli*/coliform Counts Plates) counts (TCC), and aerobic plate (3M™ Petrifilm™ aerobic Count Plates) counts (APC) as prescribed by the supplier. Colonies were counted after 24-h incubation in a Fisher Isotemp incubator (Fisher Scientific, Pittsburgh, PA) at 35°C for *E. coli* and total coliform counts, and after 48-h incubation for aerobic plate counts. Bacterial counts of skin and carcass samples were converted to log$_{10}$ CFU/cm^2 values.

Statistical analysis

The body weight (BW) data were analyzed as a Completely Randomized Design (CRD) with repeated measures using the PROC MIXED procedure of SAS (SAS institute Inc., Cary, NC), with individual animal as experimental unit. The effects of species, diet, FDT, and their interactions were considered to be fixed effects. Behavior data were analyzed as a CRD with 2 × 2 factorial treatment arrangement using the PROC MIXED procedures of SAS, with animal as a random effect and species, diet, and their interactions considered as fixed effects. Blood data were also analyzed as a CRD with 2 × 2 × 2 factorial treatment using the PROC MIXED procedures of SAS, with animal considered to be a random effect. The effects of species, diet, FDT, and their interactions were considered to be fixed, with pretrial concentrations as covariate. Carcass yield, muscle pH, and microbial data were also analyzed as a CRD with 2 × 2 × 2 factorial treatment using the PROC MIXED procedures of the SAS, with animals considered to be a random effect and species, diet, and feed deprivation considered to be fixed effects.

The following statistical models were used to analyze 1) body weight data; 2) behavior data (standing, moving, agonistic, lying and drinking); 3) blood data (plasma cortisol, glucose, creatine kinase, plasma urea nitrogen, and non-esterified fatty acids); and 4) carcass yield, muscle pH and microbial data (*E. coli*, *Enterobacteriaceae*, total coliform, and aerobic plate counts):

$$
\begin{aligned}
Y_{ijklm} = {} & \mu + S_i + D_j + F_k + T_l + SD_{ij} + SF_{ik} + ST_{il} \\
& + DF_{jk} + DT_{jl} + FT_{kl} + + SDF_{ijk} + SDT_{ijl} \quad (1) \\
& + SFT_{ikl} + DFT_{jkl} + SDFT_{ijkl} + e_{ijklm}
\end{aligned}
$$

$$
Y_{ijk} = \mu + S_i + D_j + SD_{ij} + e_{ijk} \tag{2}
$$

$$
\begin{aligned}
Y_{ijklm} = {} & \mu + S_i + D_j + SD_{ij} + DF_{jk} + SF_{ik} \\
& + SDF_{ijk} + B_l + e_{ijklm}
\end{aligned} \tag{3}
$$

$$
\begin{aligned}
Y_{ijkl} = {} & \mu + S_i + D_j + F_k + SD_{ij} + DF_{jk} + SF_{ik} \\
& + SDF_{ijk} + e_{ijkl}
\end{aligned} \tag{4}
$$

Where Y_{ijkl} or Y_{ijklm} = dependent variables, μ = overall means, S_i = species, D_j = diet, F_k = feed deprivation, T_l = body weights (prior to diet, feed deprivation, and slaughter) as repeated measures, B_l = pretrial concentrations of plasma cortisol, glucose, creatine kinase, plasma urea nitrogen, or non-esterified fatty acids, e_{ijkl} or e_{ijklm} = residuals.

From each analysis, least squares means were generated and when significant by ANOVA, separated using the PDIFF option of SAS for main or interaction effects. Pearson correlation analysis (SAS Institute Inc.) was performed to study the relationships among selected dependent variables [29]. Significance was determined at $P < 0.05$, but difference of $0.05 \leq P < 0.1$ was considered as trends.

Results

Body weight and animal behavior

The mean BW of sheep (39.6 ± 0.47 kg) was significantly higher than that of goats (32.1 ± 0.47 kg) in this experiment (Figure 1). Diet and FDT significantly influenced ($P < 0.05$) BW of experimental animals. The BW were 36.9 ± 0.47 and 34.9 ± 0.47 kg in CD and HD groups, respectively. Mean BW of animals in the 12- and 24-h FDT groups were 36.8 ± 0.47 and 35.0 ± 0.47 kg, respectively. Species × diet and diet × FDT interactions also had significant effects on BW of animals during the experimental period (Figure 1).

Frequencies of standing, moving, and agonistic behaviors were higher ($P < 0.05$) in goats than sheep (Table 2). Sheep spent more time lying down than goats ($P < 0.01$). Animals from the CD group had significantly higher frequencies of moving and agonistic behaviors than those from the HD group. The frequencies of standing and lying behaviors were higher in animals from the HD group than those from the CD group. Animals rarely drank water during the preslaugher holding period and thus the frequency of drinking behavior was not affected ($P > 0.05$) by either

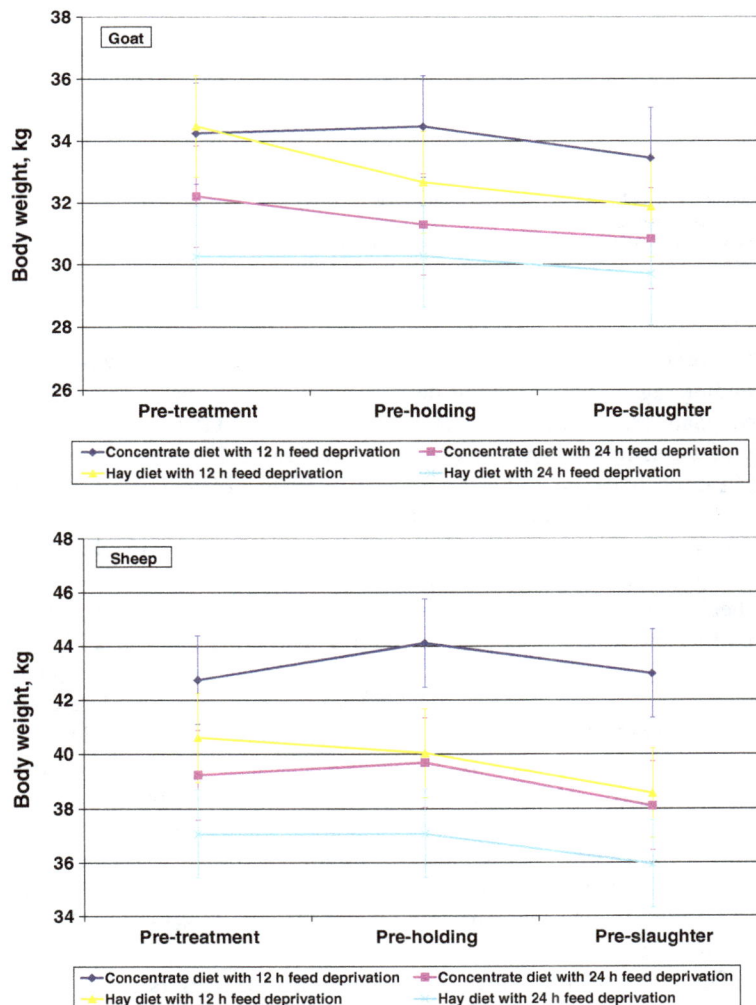

Figure 1 Body weights of sheep and goats measured prior to feeding (pre-treatment), holding (pre-holding), and slaughter (pre-slaughter).

Table 2 Effects of species, diet, and feed deprivation time (FDT) on blood hormone, metabolite, behavior, and muscle pH in sheep and goats

Response	Species			Diet			FDT			
	Goat	Sheep	P-value	HD	CD	P-value	12-hrs	24-hrs	P-value	SE
n	16	16		16	16		16	16		
Blood hormone										
Plasma cortisol, ng/mL	54.76	80.11	0.2027	62.79	72.08	0.6452	64.84	70.02	07911	13.612
Blood metabolite[1]										
Plasma glucose, mg/dL	130.39	128.54	0.8888	122.60	136.24	0.2159	132.37	126.47	0.5888	7.551
PUN, mg/dL	17.24	18.30	0.6820	17.24	18.30	0.6468	17.16	18.38	0.5944	1.603
Plasma CK activity, IU	571.74[b]	863.94[a]	0.0392	732.97	702.72	0.8265	747.15	688.54	0.6798	95.210
Plasma NEFA, mEq/L	589.78[b]	1056.05[a]	0.0054	953.56	692.27	0.1105	729.05	916.79	0.2108	105.01
Behavioral observations[2]										
Standing	3.49[a]	3.24[b]	0.0008	3.46[a]	3.26[b]	0.0084				
Moving	0.37[a]	0.28[b]	0.0440	0.19[b]	0.46[a]	0.0001				
Agonistic	0.13[a]	0.01[b]	0.0001	0.03[b]	0.11[a]	0.0024				
Lying	0.00[b]	0.48[a]	0.0001	0.32[a]	0.16[b]	0.0043				
Drink	0.00	0.003	0.3176	0.002	0.00	0.3176				
Dressing percent,%	43.02	42.84	0.7380	42.59	43.28	0.2049	42.79	43.07	0.5959	0.3752
Muscle pH[3]										
Initial	6.96	6.96	0.6000	6.97	6.91	0.2740	6.92	6.97	0.3688	0.039
Ultimate	6.02[a]	5.84[b]	0.0002	5.98[a]	5.87[b]	0.0141	5.92	5.93	0.8678	0.029

HD = hay diet; CD = concentrate diet.
[1]PUN = plasma urea nitrogen; CK = creatine kinase; NEFA = non-esterified fatty acids.
[2]Observed or 90 min before staring the slaughtering process.
[3]Inital = pH of *longissimus* muscle at immediately after skinning; Ultimate = pH of *longissimus* muscle at 24 h postmortem.
[a,b]Within a row, least squares means that do not have a common superscript letter differ ($P < 0.05$).

species or diet. Species × diet interaction effect was significant for frequencies of standing, agonistic, and lying behaviors (Figure 2). Hay-fed goats (3.7 ± 0.08/min) had a higher ($P < 0.05$) frequency of the standing behavior compared to other treatment groups (species × diet); and, concentrate-fed goats (0.2 ± 0.03/min) also had a higher ($P < 0.05$) frequency of agonistic behavior compared to other groups. However, sheep (0.6 ± 0.06/min) fed with the concentrate had a higher ($P < 0.05$) frequency of the lying behavior compared to other groups.

Blood hormone and metabolite concentrations
Plasma cortisol, glucose, and plasma urea nitrogen (PUN) concentrations were not influenced ($P > 0.05$) by any of the factors studied (Table 2). The interaction effects were also not significant for plasma cortisol or any of the metabolic concentrations (glucose, creatine kinase, plasma urea nitrogen, and non-esterified fatty acids). However, plasma creatine kinase (CK) activities and non-esterified fatty acids (NEFA) levels were different ($P < 0.05$) between species (Table 2). The HD animals tended to have higher ($P = 0.11$) plasma NEFA levels than CD animals (Table 2).

Carcass yield and muscle pH
Carcass yield ranged from 40 to 45% in the present experiment, but was not influenced by species, diet, or FDT (Table 2). However, goats (43.7 ± 0.53%) deprived of feed for 24-h had a higher mean carcass yield than those deprived for 12-h (42.3 ± 0.53%), while an opposite trend was noticed in sheep (species × FDT, $P < 0.05$, Figure 3).

The initial pH of LM was not affected ($P > 0.05$) by species, diet, or FDT (Table 2). However, species x FDT interaction effect was significant (Figure 4). Sheep subjected to 12-h feed deprivation (7.02 ± 0.055) had lower ($P < 0.05$) pH values than those subjected to 24-h feed deprivation (6.83 ± 0.055), while the initial pH of goat carcasses were not affected by FDT. The ultimate pH was higher in goats compared to sheep ($P < 0.05$) and higher in HD compared to CD animals ($P < 0.05$, Table 2). However, the interaction effects were not significant ($P > 0.05$) for the ultimate muscle pH values in the current study.

Skin bacterial counts
Pre-holding *E. coli* counts of skin samples were not influenced ($P < 0.05$) by species, diet, or diet × species interaction

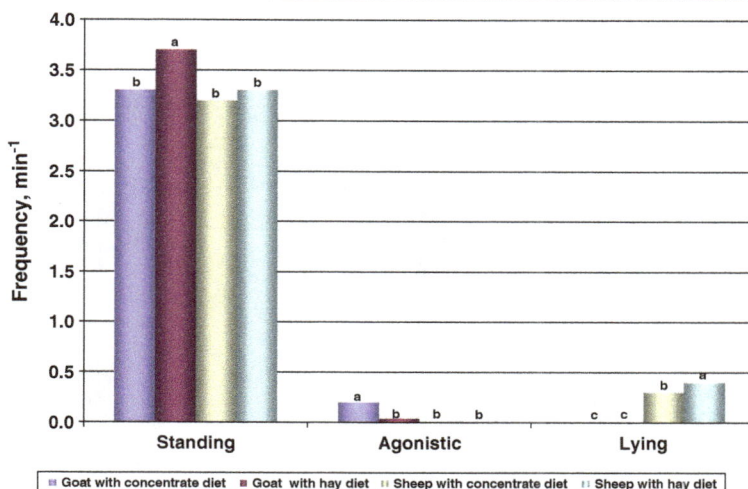

Figure 2 Effect of species and diet on standing (SE = 0.076), agonistic (SE = 0.026), and lying (SE = 0.056) behaviors in the holding pens during a 90-min period prior to slaughter. For any behavior, bars bearing different letters are different ($P < 0.05$).

effects (Table 3). However, TCC of skin were significantly different between the two species. Goats had lower ($P < 0.05$) coliform counts in the skin than sheep (Table 3). Diet treatments or diet x species interaction factors had no significant effects on TCC (Table 3). The TCC was influenced ($P < 0.05$) by the species × diet × FDT interaction (Figure 5), with concentrate-fed sheep having the highest TCC ($P < 0.05$, $3.73 \pm 0.353 \log_{10}$ CFU/cm^2) after 12 h feed deprivation compared to all other groups. Goats also had lower ($P < 0.05$) skin APC compared to sheep (Table 3). No significant effects of diet and diet x species interaction were detected in aerobic plate counts of skin.

Preslaughter skin *E. coli* counts and TCC were different ($P < 0.05$) between species (Table 3), and goats had lower ($P < 0.05$) counts of *E. coli* and TCC in skin swab samples

compared with sheep. Diet, FDT, or interaction effects were not significant ($P > 0.05$) for skin *E. coli* and total coliform counts (Table 3). However, skin swab samples of the 24-h feed deprivation group tended ($P = 0.12$) to have higher *E. coli* counts than the 12 h group. Total coliform counts tended ($P = 0.14$) to be higher in the 24-h group than 12-h feed deprivation group (Table 3). Aerobic plate counts of skin swab samples were not influenced ($P > 0.05$) by any of the main effects or interactions.

Carcass bacterial counts

Diet, species, FDT, and their interactions had no significant effects on *E. coli* and coliform counts (Table 3). Aerobic plate counts of carcass swab samples were also not influenced ($P > 0.05$) by any of the treatment factors or

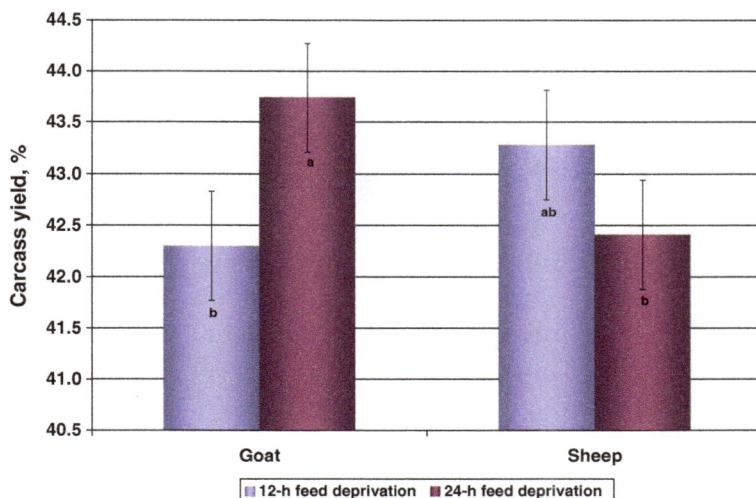

Figure 3 Effect of species and feed deprivation time on carcass yields (SE = 0.531). Bars bearing different letters are different ($P < 0.05$).

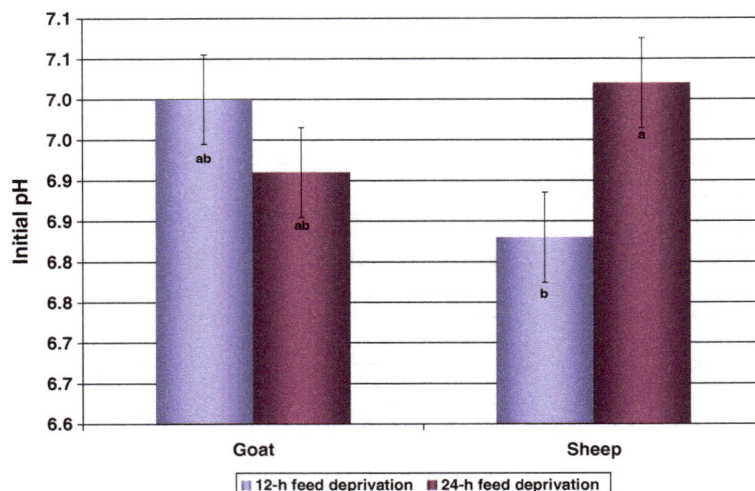

Figure 4 Effect of species and feed deprivation time on the initial pH of *longissimus* muscle (SE = 0.055). Bars bearing different letters are different ($P < 0.05$).

their interactions, except the feed deprivation (FD) time (Table 3). Carcasses from the 12 h feed deprivation group had higher ($P < 0.05$) APC than those from 24 h group. Carcass swab samples from sheep tended ($P = 0.07$) to have higher APC than those from goats (Table 3).

Discussion

In sheep, the body weights appeared to increase due to concentrate feeding, but did not change or decrease due to hay feeding. In goats, there was no clear pattern in body weight changes due to diet. However, body weights decreased due to feed deprivation in both sheep and goats. Live weight losses during the pre-slaughter period are of major concern in small ruminants. Live weight shrinkage can be about 10% in goats after 18 h feed deprivation, and about 7% in sheep after 24 h feed deprivation [12,13]. These live weight losses can be attributed to reductions in gut weights, since the

Table 3 Effects of species, diet, and feed deprivation time (FDT) on the microbial counts (log$_{10}$ CFU/cm^2) on skin and carcass of sheep and goats

Response	Species			Diet			FDT			
	Goat	Sheep	*P*-value	HD	CD	*P*-value	12-hrs	24-hrs	*P*-value	SE
n	16	16		16	16		16	16		
Skin, pre-holding[1]										
E. coli count	2.00	2.21	0.2129	2.06	2.16	0.5527				
Total coliform count	2.12[b]	3.04[a]	0.0011	2.43	2.73	0.2413				
Aerobic plate count	8.21[b]	8.53[a]	0.0305	8.42	8.31	0.4511				
Skin, pre-slaughtering[2]										
E. coli count	2.23[b]	2.93[a]	0.0118	2.56	2.60	0.8650	2.38	2.78	0.1238	0.180
Total coliform count	2.26[b]	3.01[a]	0.0135	2.60	2.67	0.8074	2.42	2.85	0.1352	0.197
Aerobic plate count	8.31	8.30	0.8017	8.33	8.28	0.3198	8.34	8.28	0.2163	0.035
Carcass, pre-washing[3]										
E. coli count	2.28	2.51	0.2212	2.34	2.44	0.6018	2.28	2.41	0.8959	0.134
Total coliform count	2.29	2.56	0.1999	2.35	2.50	0.4587	2.43	2.43	1.0000	0.141
Aerobic plate count	7.93	8.28	0.0718	8.16	8.04	0.5076	8.35[a]	7.85[b]	0.0128	0.131

HD = hay diet; CD = concentrate diet.
[1]Skin swabs from hind leg before depriving feed to experimental animals.
[2]Skin swabs from hind leg right before starting slaughtering.
[3]Carcass swabs from flank, brisket and leg regions right before washing after skinning and gut removed.
[a,b]Within a row, least squares means that do not have a common superscript letter differ ($P < 0.05$).

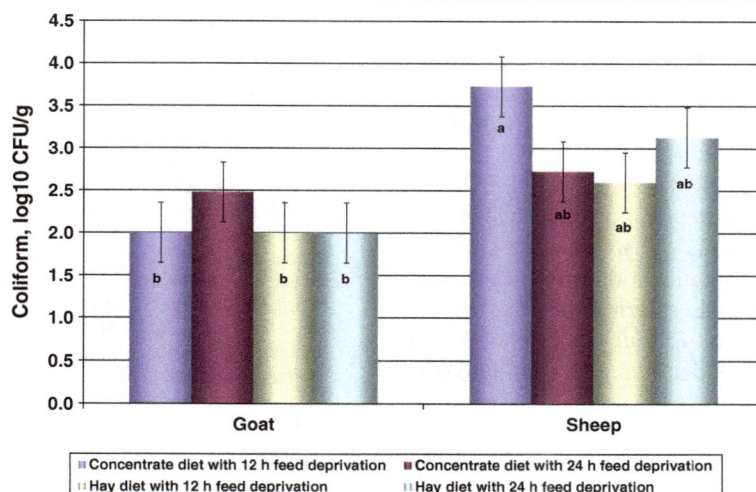

Figure 5 Effect of species, diet, and feed deprivation time on pre-holding total coliform counts of skin swab samples (SE = 0.353). Bar bearing different letters are different ($P < 0.05$).

gastrointestinal tract contributes to a major proportion of live weight in small ruminants [30].

Plasma cortisol and certain metabolite concentrations are good indicators of the physiological status of animals as influenced by preslaughter dietary treatment and FDT. The cortisol concentrations were not influenced by species, diet, or FDT in the present experiment. Feed deprivation combined with a 2.5-h transportation has been reported to elevate cortisol concentrations in goats [12]. Transporting animals from the experimental facility to the slaughter plant was completed within 10 min in the present study. It appears that feed deprivation alone for 12 or 24 h is not stressful enough to elevate circulating cortisol concentrations in sheep and goats. A similar effect was observed in a previous study when Spanish does were feed deprived for 7, 14, or 21 h [24]. However, feed deprivation has been reported to elevate cortisol concentrations in sheep [19]. Creatine kinase activities were higher in sheep compared to goats, although behavioral observations showed that goats were more active than sheep during preslaughter holding. Creatine kinase activity in blood increases due to muscle damage or increased muscular activity in animals [31]. Plasma NEFA concentrations were also higher in sheep than goats. The higher CK and NEFA levels may be attributed simply to a species difference. It is not clear if size of animals could have contributed to this effect, since the sheep used in this study were heavier than goats. Hay-fed animals tended to have higher NEFA concentrations than concentrate-fed animals. Furthermore, animals subjected to 24 h of feed deprivation tended to have higher NEFA concentrations than those subjected to 12 h of feed deprivation. Knowles et al. [22] reported that the plasma NEFA concentrations increased in sheep after 24 h feed deprivation. Kouakou et al. [23] found that feed restriction

elevated plasma NEFA concentration in goats because feed deprivation increases lipolysis in animals, which in turn increases free fatty acid levels in the blood [32].

In the present study, carcasses from CD group had lower ultimate LM pH values than HD animals (Table 2). Variation in glycogen content of muscles may be responsible for the differences in the ultimate pH. Glycogen content of muscles at the time of slaughter is the important factor that affects muscle ultimate pH and meat quality [33]. Forage-finished animals have been reported to produce lower quality meat than grain-finished animals [34,35]. Immonem et al. [36] reported variations in the ultimate pH of muscle due to energy levels in the diet of cattle, and they found that cattle fed a high energy diet had a lower ($P < 0.05$) ultimate muscle pH (5.69 ± 0.03) value compared to cattle fed a low energy diet (5.93 ± 0.03). In contrast, Diaz et al. [37] did not find any significant differences in meat quality and muscle pH (measured immediately after slaughter, after 45 min and after 24 h) from pasture-fed and concentrate-fed lambs.

The skin of a live animal becomes contaminated with microorganisms derived from a wide range of sources such as feces, soil, water and vegetation [2]. The APC, TCC, and *E. coli* counts of skin swab samples collected at two different times in the present study was not influenced by diet. Skin swab samples collected from sheep showed higher bacterial counts than goats. Sheep fleece may be responsible for picking up fecal material from the pen floor and retaining the contamination for longer time. Behavioral observations of the animals during holding period revealed that sheep tended to spend more time lying down in the pens than goats. It is possible that goats were not able to withstand the cold temperature of concrete floors, which would have prevented them from lying down. This experiment was conducted in the month of December. Skin swab samples collected from the 24-h feed deprivation group showed higher *E. coli* counts than the 12-h group. There is always more chance for skin contamination with fecal material if the animals spend more time in the holding pens.

Preslaughter diet and feed deprivation time had no effect on TCC and *E. coli* counts of carcass swab samples. Major sources of carcass contamination are unclean animal skin and viscera of animals entering the slaughter facility [1]. Carcass TCC and *E. coli* counts were not correlated with skin counts. Elder et al. [38] found no correlation between the prevalence of *E. coli* O157 contamination on cattle hides and that resulting on carcasses. In their study, the prevalence of *E. coli* O157 on the carcasses was higher than that on hides.

Conclusions
Preslaughter diet and FDT did not influence the physiological status of sheep and goats according to plasma cortisol, glucose, CK, PUN, and NEFA. Feed deprivation may significantly decrease body weights in sheep and goats. Sheep had higher skin contamination than goats, probably due to differences in their behavior during preslaughter holding. Sheep spent more time lying down than goats in holding pens. Diet and FDT did not influence skin contamination in sheep and goats. There was no relationship between skin contamination and carcass contamination in the present study. Preslaughter diet may have an effect on the energy reserves in muscles, as the LM ultimate pH was lower in the concentrate-fed group. The results indicate that diet can be manipulated without significant effects on physiological responses in sheep and goats.

Competing interests
The authors declare that they have no competing interests.

Authors' contributions
All authors made significant contributions to design and perform the research. Especially GK and VRG conducted all data analyses and drafted the initial manuscript. All authors read and agreed the final manuscript.

Acknowledgments
The authors would like to thank Dr. Isabel Blackman for advice with microbiological analysis and Krishna Gadiyaram, Shirely Wang, and Nealie Moye for technical assistance.

References
1. Grau FH: **Prevention of microbial contamination in the export beef abattoir.** In *Elimination of Pathogenic Organisms from Meat and Poultry*. Edited by Smulders FJM. Amsterdam: Elsevier; 1987:221–234.
2. McEvoy JM, Doherty AM, Finnerty M, Sheridan JJ, McGuire L, Blair IS, MacDowell DA, Harrington D: **The relationship between hide cleanliness and bacterial numbers on beef carcasses at a commercial abattoir.** *Lett Appl Microbiol* 2000, **30**:390.
3. Reid CA, Small A, Avery S, Bunic S: **Presence of food-borne pathogens on cattle hides.** *Food Control* 2002, **13**:411–415.
4. Kudva IT, Hunt CW, Williams CJ, Nance UM, Hovde CJ: **Evaluation of dietary influences on *Escherichia coli* O157:H7 shedding by sheep.** *Appl Environ Microbiol* 1997, **63**:3878–3886.
5. Reid CA, Avery SM, Warriss PD, Buncic S: **The effect of feed withdrawal on *Escherichia coli* shedding in beef cattle.** *Food Control* 2002, **13**:393–398.
6. Callaway TR, Anderson RC, Edrington TS, Elder RO, Genovese KJ, Bischoff KM, Poole TL, Jung YS, Harvey RB, Nisbe DJ: **Preslaughter intervention strategies to reduce food-borne pathogens in food animals.** *J Anim Sci* 2002, **81**(E. Suppl. 2):E17–E23.
7. Jacobson HL, Tanya AN, Gregory NG, Bell RG, Roux GL, Haines JM: **Effect of feeding pasture-finished cattle different conserved forages on *Escherichia coli* in the rumen and feces.** *Meat Sci* 2002, **2002**(62):93–106.
8. Byers FM, Schelling GT: **Microbiology of the rumen and intestine.** In *The Ruminant Animal Digestive Physiology and Nutrition*. Edited by Church DC. NJ: Prentice Hall; 1988:125–171.
9. Allison MJ, Robinson IM, Dougherty RW, Bucklin JA: **Grain overload in cattle and sheep: changes in microbial population in the cecum and rumen.** *Amer J Vet Res* 1975, **36**:181–185.
10. Gregory NG: *Animal Welfare and Meat Science*. NY, USA: CABI Pub; 1998.
11. Callaway TR, Anderson RC, Edrington TS, Elder RO, Genovese KJ, Bischoff KM, Poole TL, Jung YS, Harvey RB, Nisbet DJ: **Preslaughter intervention strategies to reduce food-borne pathogens in food animals.** *J Anim Sci* 2003, **81**(E. Suppl. 2):E17–E23.
12. Kannan G, Terrill TH, Kouakou B, Gazal OS, Gelaye S, Amoah EA, Samake S: **Transportation of goats: Effect on physiological stress response and live weight loss.** *J Anim Sci* 2000, **78**:1450–1457.
13. Kirton AH, Moss RA, Talylor AG: **Weight losses from milk and weaned lamb in mid Canterbury resulting from different lengths of starvation before slaughter.** *N Z J Agric Res* 1971, **14**:149–160.
14. Chillard Y, Doreau M, Bocquier F, Lobley GE: **Digestive and metabolic adaptations of ruminants to variations in food supply.** In *Recent Developments in the Nutrition of Herbivores*. Edited by Journet M, Grenet E, Farce MH, Theriez M, Demarquilly C. France: INRA Editions; 1995.
15. Silva JA, Patarata L, Martins C: **Influence of ultimate pH on bovine meat tenderness during ageing.** *Meat Sci* 1993, **52**:453–459.
16. Lawrie RA: **Metabolic stresses which affect muscle.** In *The Physiology and Biochemistry of Muscle as Food*. Edited by Cassens EJ, Briskey RG, Trautman JC. Madison: The University of Wisconsin Press; 1996:137–164.
17. Braggins TJ: **Effect of stress-related changes in sheep meat ultimate pH on cooked odor and flavor.** *J Agri Food Chem* 1996, **44**:2352–2360.
18. Warriss PD: **Antemorterm factors influencing the yield and quality of meat from farm animals.** In *Quality and Grading of Carcasses of Meat Animals*. Edited by Morgan Jones SD. New York: CRC Press; 1995:1–15.
19. Murayama S, Moriya K, Saaki Y: **Changing pattern of plasma cortisol level associated with feeding in sheep.** *Jpn J Zootech Sci* 1986, **57**:317–323.
20. Shorthose WR, Wythes JR: **Transport of sheep and cattle.** In *34th Int. Cong. Meat Sci. Technol., Part A*. Brisbane, Australia; 1988.
21. Schaefer AL, Jones SDM, Stanley RW: **The use of electrolyte solutions for reducing transport stress.** *J Anim Sci* 1997, **75**:258–265.
22. Knowles TG: **A review of the road transport of cattle.** *Vet Rec* 1999, **144**:197–201.
23. Kouakou B, Gazal OS, Terrill TH, Kannan G, Galaye S, Amoah EA: **Effects of plane of nutrition on blood metabolites and hormone concentration in goats [abstract].** *J Anim Sci* 1999, **77**(Suppl.1):267.
24. Kannan G, Terrill TH, Kouakou B, Gelaye S, Amoah EA: **Simulated preslaughter holding and isolation effects on stress responses and live weight shrinkage in meat goats.** *J Anim Sci* 2002, **80**:1771–1780.
25. Barroso FG, Alados CL, Boza J: **Social hierarchy in the domestic goat: effect on food habits and production.** *Appl Anim Behav Sci* 2000, **69**:35–53.
26. Kannan G, Jenkins AK, Eega KR, Kouakou B, McCommon GW: **Preslaugher spray-washing effects on physiological stress responses and skin and carcarss microbial counts in goat.** *Small Rumin Res* 2007, **67**:14–19.
27. FSIS-USDA: **Pathogen reduction: hazard analysis and critical control point (HACCP) systems, final rule.** *Fed Reg* 1996, **61**:38805–38989.
28. The 3M Products: *Manufacture's Interpretation Guide*. MN: The 3M Microbiology Products; 1999.
29. Steel RGD, Torrie JH, Dickey DA: *Principles and Procedures of Statistics: A Biometrical Approach*. 3rd edition. New York: McGraw-Hill Book Co.; 1996.
30. Romans JR, Costello WJ, Carlson CW, Greaser ML, Jones KW: *The Meat We Eat*. 13th edition. IL: Interstate publishers; 1994.

31. Wilson BW, Nieberg PS, Buhr RJ, Kelly BJ, Shultz FT: **Turkey muscle growth and focal myopathy.** *Poult Sci* 1990, **69:**1553–1562.

32. Warriss PD, Bevis EA, Brown SN, Ashby JG: **An examination of potential indices of fasting time in commercially slaughtered sheep.** *Br Vet J* 1989, **145:**242–248.

33. Bidner TD, Schupp AR, Montgomery RE, Carpenter JC Jr: **Acceptability of beef finished on all-forage, forage plus-grain or high energy diets.** *J Anim Sci* 1981, **53:**1181–1187.

34. Bowling RA, Smith GC, Carpenter ZL, Dutson TR, Oliver WM: **Comparison of forage-finished and grain-finished beef carcasses.** *J Anim Sci* 1977, **45:**209–215.

35. Melton SL, Amiri M, Davis GW, Backus WR: **Flavor and chemical characteristics of ground beef from grass-, forage-grain- and grain-finished steers.** *J Anim Sci* 1982, **55:**77–87.

36. Immonem K, Puolanne E, Hissa K, Ruusunen M: **Bovine muscle glycogen concentration in relation to finishing diet, slaughter and ultimate pH.** *Meat Sci* 2000, **55:**25–31.

37. Diaz MT, Velasco S, Caneque V, Lauzurica S, Ruiz de Huidoro F, Perez C, Gonzalez J, Manzanares C: **Use of concentrate or pasture for fattening lambs and its effect on carcass and meat quality.** *Small Rumin Res* 2002, **43:**257–268.

38. Elder RO, Keen JE, Siragusa GR, Barkocy-Gallagher GA, Koohmaraie M, Laegreid WW: **Correlation of enterohemorrhagic** *Escherichia coli* **O157 prevalence in feces, hides, and carcasses of beef cattle during processing.** *Proc Natl Acad Sci* 2000, **97:**2999–3003.

6

Evaluation of alfalfa inter-seeding effect on bahiagrass baleage fermentation and lactating Holstein performance

Michael E McCormick[1*], Kun Jun Han[2], Vinicius R Moreira[3] and David C Blouin[4]

Abstract

Background: Previous research indicates that bahiagrass may be successfully conserved as baleage, but nutritive value is typically low for lactating dairy cows. The purpose of this study was to determine the effect of adding modest amounts of alfalfa forage (22%), achieved by inter-seeding alfalfa into an existing bahiagrass pasture, on baleage nutritive value and lactation performance of Holstein cows. Forage treatments employed were monoculture bahiagrass baleage (MBB; negative control), bahiagrass-alfalfa mixture baleage (BAB) and conventional corn silage (CCS; positive control). Thirty six mid lactation Holstein cows [34.8 ± 5.8 kg 3.5% fat-corrected milk and 112 ± 19 d in milk (DIM)] were stratified according to milk yield and DIM and assigned randomly to 1 of 3 forage treatments. Cows were trained to Calan feeding gates and were offered a common CCS-based TMR in a 10-d covariance period followed by a 42-d treatment feeding period.

Results: The BAB contained more protein and less NDF than MBB (12.6 vs 10.3% CP and 71.8 vs 76.6% NDF). Diet DMI was similar for MBB and BAB (19.5 vs 21.6 kg/hd/d), but cows consumed more of the CCS diet (25.5 kg/hd/d) than either baleage-based diet. Cows offered BAB tended to produce more milk than cows offered MBB based TMR (28.4 vs 26.1 kg/hd/d), but both baleage diets generated less milk than CCS-based diets (33.1 kg/hd/d). Milk composition was similar across diets except for milk protein concentrations which were higher for CCS than either MBB or BAB diets; however, milk urea nitrogen (MUN) was lowest for cows fed CCS diets. Cow BW gain was higher for BAB than MBB implying that a portion of the higher energy contributed by the alfalfa was being used to replenish weight on these mid lactation cows.

Conclusions: Data from this study indicate that alfalfa inter-seeded in bahiagrass sod that produces BAB with as little as 22% alfalfa may improve nutritive value compared to monoculture bahiagrass baleage and marginally improve lactation performance of Holstein cows. However, the CCS diet was vastly superior to either MBB or BAB-based diets for milk production.

Keywords: Alfalfa, Bahiagrass, Baleage, Corn silage, Lactating dairy cows

Background

Bahiagrass (*Paspalum notatum*, L.) is a common perennial forage base for many dairies in southern Louisiana and Mississippi, USA [1]. When properly managed, this forage is high yielding and persistent, particularly given the often harsh environmental conditions of the area. However, nutritive content of conserved bahiagrass is often below requirements of young growing animals and lactating dairy cattle [2]. In spite of its low nutritive value and water soluble carbohydrate content, bahiagrass may be successfully conserved as baleage [3].

Interseeding legumes into perennial grass pastures often increases animal growth and milk production above that of animals grazing pure grass stands [4,5]. Most of the research on interseeding legumes into grass pastures has been conducted in the Northeastern and Midwestern USA where conditions are favorable for legume production.

* Correspondence: memccormick@agctr.lsu.edu
[1]Southeast Region LSU Agricultural Center, 21549 Old Covington Hwy, Hammond, LA 70403, USA
Full list of author information is available at the end of the article

Researchers in Louisiana attempted to inter-seed red clover (*Trifolium pretense*) into bermudagrass (*Cynodon dactylon*) and bahiagrass pastures for lactating dairy cattle [6]. They recorded a 10% (4 pounds FCM) increase in milk yield for cows grazing grass-clover pastures compared to those grazing grass alone. During the last decade, alfalfa varieties have been developed which are especially adapted for grazing in southern climates [7]. Research using these varieties has focused on interseeding (no-till drilled) in existing bermudagrass pastures [8,9]. Bermudagrass stand loss has been a concern with interseeding due to earlier spring emergence of alfalfa and shading of bermudagrass plants. Increasing alfalfa row spacing from 20 to 60 cm aided in maintaining the bermudagrass stand and reduced alfalfa seeding rate from 19 to 6.4 kg/ha. No information on overall forage quality was provided with these reports nor is information available on more traditional hay type (upright growing) alfalfa plants inter-seeded in summer perennial grasses. Research on inter-seeding alfalfa in existing bahiagrass stands and its effect on conserved forage nutritive value are not available. Though bahiagrass has a dense sod, it was speculated that a late fall interseeding of alfalfa would allow sufficient early spring alfalfa production to improve the nutritive value of the resulting bahiagrass-alfalfa bale silage crop. A lactation performance study was conducted to evaluate the nutritive value of subsequent bahiagrass-alfalfa baleage fed to lactating dairy cows. Monoculture bahiagrass served as a negative control and corn silage (*Zea Mays*, L.) was included as a positive control.

Materials and methods
Forage production, storage and sampling

All agronomic and animal procedures described in this report were conducted at the Louisiana State University Agricultural Center's Southeast Research Station located approximately 8 km west of Franklinton, LA (Lat 30° 47' 04" N and Lon 90 12' 19" W). A 4.9 ha field of 'Argentine' bahiagrass (*Paspalum notatum*, var. *latiflorum*) was top-dressed with 4.5 Mg/ha dolomitic lime on May 25, 2006. On December 19, 2006 the bahiagrass was inter-seeded with 7.4 kg live seed/ha of 'Amerigraze 702' alfalfa (*Medicago sativa*, L.; America's Alfalfa, Madison, WI) with a Bush Hog Model 9690 no-till drill (M & L Industries, Inc., Baton Rouge, LA). Alfalfa seeds were factory-coated with material containing a fungicide (metalaxyl-M) and rhizobia (*Sinorhizobium meliloti*) constituting 34% of planted seed weight. Alternating seed drops on the planter were closed to provide row widths of approximately 30 cm. In late February the inter-seeded bahiagrass-alfalfa was top-dressed with 140 kg K, 60 kg P, 28 kg S, 24 kg N and 4 kg B fertilizer per ha. Following an initial late-April hay harvest, the bahiagrass-alfalfa field was top-dressed with 125 kg K, 17 kg P, 15 kg N, 26 kg S and 3 kg B fertilizer per ha. A nearby (<300 m distant) 5.7 ha field of 'Argentine'

bahiagrass was used as a negative control i.e. monoculture bahiagrass. Fertilizing and harvest schedules were similar to that of the inter-seeded field except lime was applied at 2.25 Mg/ha and fertilizer consisted of 449 kg/ha of a blended fertilizer containing 24% nitrogen, 10% phosphorus and 17% potassium.

On June 11th, 2007 at approximately 1000 h, the bahiagrass-alfalfa field (second harvest; 43 d regrowth) was harvested with a Kuhn model FC-300C disc cutter equipped with a flail-type conditioner (Kuhn North American, Inc., Broadhead, WI). The regrowth contained bahiagrass with less than 10% seed heads and 1/10-bloom stage alfalfa. One meter-length windrow samples were collected from four random locations within the field and separated into bahiagrass, alfalfa and weed fractions. Forage was allowed to wilt in 1.2 m windrows for approximately 24 h after which high-moisture forage was baled into 1.5 m × 1.2 m round bales using a model 568 John Deere baler equipped with a high moisture kit (John Deere, Inc., Cary, NC). Prior to wrapping, six randomly selected bales were weighed on a digitized platform scale (Rice Lake Weighing Systems, Rice Lake, WI) and core-sampled to a depth of 45 cm in four locations and forage was composted and stored on ice until processed in the Southeast Research Station Forage Quality Laboratory. Within 1.5 h postbaling, bales were individually wrapped (Ag Wrap, Ag Nations Products, Inc., Canton, OH) in six layers of 25-μm-thick white stretch film stretched 50% (Silage Wrap Sun Film, AEP Industries, Inc., Chino, Ca, USA). After wrapping, bales were moved to a well-drained location and stored outside on the ground. Hereafter, wrapped high moisture bales from the bahiagrass field inter-seeded with alfalfa will be designated as bahiagrass-alfalfa baleage (BAB).

The adjacent 5.7 ha field of monoculture bahiagrass was cut on June 12, 2007. Forage was managed in a manner similar to that described for the BAB. Windrow samples were handled as above and evaluated for bahiagrass and weed content. Hereafter, wrapped high moisture bales from the monoculture bahiagrass field will be referred to as bahiagrass baleage (MBB).

Corn silage was produced from Pioneer '31R87RR' corn seed (Pioneer Hi-Bred International, Inc., Johnston, IA, USA) planted with a John Deere Maxi-Merge no-till planter (John Deere, Inc., Cary, NC, USA) in late March 2007 at a seeding rate of 64,200 plants ha. The non-irrigated corn was fertilized with 202 kg N/ha as urea, 31 kg/ha P and 88 kg/ha K at planting. Silage was harvested in mid-July at 1/3 milk line stage, chopped to 1.2 cm theoretical length and stored in a 2.7 m × 42.0 m silage bag (Ag Bag Model 9135) Ag Bag Systems, Inc., Astoria, OR, USA. Six green forage samples were collected every 7 m of bag length during silo filling and transported to the Southeast Forage Quality Laboratory for analysis. In future references within this report, this positive control forage will be referred to as conventional corn silage (CCS).

Animal management

Thirty six mid lactation Holstein cows (34.8 ± 5.8 kg 3.5% fat-corrected milk (FCM) and 112 ± 19 d in milk (DIM)) were stratified according to lactation number (18 multiparous and 18 primiparous), DIM and milk yield and allotted to forage treatments. Cows were housed in a free stall barn equipped with Calan Gates (American Calan, Norwood, NH, USA) for measuring individual intake. Cows were trained to Calan Gates for 2 wk, after which they were subjected to a 10-d covariance period followed by a 42-d treatment feeding period. During the standardization period, all cows were fed a corn silage-based total mixed ration (TMR) containing 17.1% crude protein (CP), 22.7% acid detergent fiber (ADF) and 37.8% neutral detergent fiber (NDF) formulated to meet energy, CP, rumen undegradable protein (RUP) and mineral requirements for a 613 kg Holstein cow producing 39 kg FCM daily [10]. All animal procedures employed in this study were approved by the Louisiana State University Animal Use and Care Committee.

Milk weights were automatically recorded for each cow at each a.m. and p.m. milking using the AFI-Milk 2000 Information System (Germania Dairy Automation, Waunakee, WI, USA). Morning and evening milk sample composites were collected weekly, preserved with 2-bromo, 2 nitropropane-1, 3-diol and analyzed for fat, protein, lactose and urea N via infrared spectroscopy and SCC was determined via flow cytometry (Louisiana DHIA, Baton Rouge, LA, USA). Body weights and body condition score (BCS; 1 = extremely thin and 5 = extremely obese) [11] were recorded on two consecutive days at the beginning and end of the trial. Body condition scores were assigned by three independent observers.

On the morning of d 42, 10 mL of rumen fluid was collected pre-feeding from each animal via rumenocentesis [12]. Rumen fluid pH was measured immediately with an Orion model 230A portable pH meter (Orion Research Inc., Boston, MA) and the sample was then placed on ice until transported to the Southeast Research Station Forage Quality Laboratory. Five mL of rumen fluid was sterilized with 1.0 mL 25% metaphosphoric acid and samples were centrifuged at 7,000 g in a refrigerated centrifuge (4°C), filtered through a 2.0 micron filter and stored at −20°C.

Diet preparation and Lab analyses

Forages conserved as baleage (BAB and MBB) were ground with a flail type hay grinder (Model 2554 HayBuster, McConnell Machinery Co., Lawrence KS) into a vertical auger TMR mixer (Jaylor Model 3650, Orton, Ontario, Canada). A fresh bale of each treatment baleage

was processed each morning between 0700 and 0830 h during the 42 d feeding period. Ground forage treatments were augured from the TMR mixer wagon onto a conveyor and into a Calan Data Ranger (American Calan, Inc., Norwood, NH, USA) whereupon concentrates were added to complete the experimental TMRs. Corn silage was handled in a manner similar to ground baleages. Cows were individually fed once per day between 0800 h and 0930 h to leave 10% or more orts.

Orts were collected and weighed prior to feeding each day. Ort and TMR grab samples were collected daily and stored at 4°C. These samples were composited weekly and used to determine TMR nutritive value and DM intake. Grab samples of ground baleage and corn silage were also taken daily at the conveyor, composited weekly and later analyzed for nutritive value in the Southeast Research Station Forage Quality Laboratory (Franklinton, LA, USA). A 1.0 kg sample of ground forage was collected weekly and manually evaluated for particle length (Table 1). During the second half of the study (wk 4–6), ambient and ort temperatures were recorded using a hand-held Raynger ST60 laser thermometer (Raytek, Inc., Santa Cruz, CA, USA). Ort temperatures were measured only for those cows leaving at least 5.0 kg fresh weight. Temperatures were taken from the approximate center of the ort TMR mass.

Weekly silage and TMR composite samples were weighed, dried at 55°C for 48 h and ground through a 1-mm screen. Ground samples were analyzed for DM and CP (micro Kjeldahl) according to AOAC [13] procedures. Sample ADF and NDF concentrations were determined using the detergent fiber methods of Van Soest et al. [14]. Sodium sulfite and amylase were added to NDF solutions before fiber analyses were conducted. *In vitro* digestibility (IVTD) of silage samples was determined by the two stage rumen fermentation-pepsin technique as described by Goering and Van Soest [15] and water soluble carbohydrate (WSC) concentrations were determined using the phenol-sulfuric acid procedure [16]. Volatile fatty acids (VFA) for silage and rumen fluid samples were measured via gas–liquid chromatography [17]. Rumen fluid lactic acid concentration was determined via the enzymatic conversion of L-lactate to pyruvate using the YSI 2700 Select Biochemistry Analyzer (YSI, Inc., Yellow Springs, OH).

Bahiagrass net energy was estimated by the equation: NE_L (Mcal/kg) = $2.2 \times [1.085 - (0.0124 \times ADF)]$ and corn silage by the equation: NE_L (Mcal/kg) = $2.2 \times [1.044 - (0.0124 \times ADF)]$ [3]. Concentrate mix NE_L was estimated based on NRC [10] estimates of NE_L for grain supplement ingredients [10]. Concentrations of Ca, K, Mg, Mn, Zn and Cu in silages and TMRs were determined by dry ashing, solubilizing in 20% HCl and analyzing via atomic absorption spectroscopy (Analyst 30, Perkin Elmer, Norwalk, CT)

Table 1 Chemical composition, fermentation characteristics and particle size of silage crops (% of DM)[1]

Item	Silage crop[2]		
	MBB	BAB	CCS
DM	55.77 ± 13.03	51.60 ± 9.70	29.23 ± 1.13
CP	10.32 ± 0.38	12.64 ± 0.89	9.53 ± 0.46
ADF	42.18 ± 1.19	41.50 ± 0.60	26.55 ± 0.83
NDF	76.56 ± 1.73	71.80 ± 1.76	45.70 ± 1.94
IVTD[3]	68.12 ± 5.97	71.34 ± 3.54	80.61 ± 2.50
NE_L, Mcal/kg	1.23 ± 0.04	1.32 ± 0.06	1.56 ± 0.04
WSC[3]	2.22 ± 0.96	3.77 ± 1.05	10.26 ± 4.20
pH	5.71 ± 0.55	5.39 ± 0.32	3.76 ± 0.09
Lactate	0.73 ± 0.29	0.58 ± 0.12	3.47 ± 0.59
Acetate	0.24 ± 0.12	0.27 ± 0.05	4.78 ± 1.57
Propionate	0.00	0.00	0.69 ± 0.29
Butyrate	0.00	0.03 ± 0.01	0.05 ± 0.02
Isobutyrate	0.02 ± 0.01	0.00	0.00
Valerate	0.04 ± 0.10	0.03 ± 0.06	0.00
Isovalerate	0.01 ± 0.01	0.00	0.00
Calcium	0.32 ± 0.03	0.57 ± 0.07	0.15 ± 0.05
Phosphorus	0.17 ± 0.01	0.27 ± 0.02	0.87 ± 0.21
Potassium	1.49 ± 0.29	1.88 ± 0.12	0.96 ± 0.23
Magnesium	0.25 ± 0.05	0.23 ± 0.02	0.13 ± 0.01
Copper, ppm	4.00 ± 1.45	3.61 ± 1.66	2.29 ± 2.01
Manganese, ppm	236.1 ± 121.1	147.0 ± 63.0	35.38 ± 8.79
Zinc, ppm	25.71 ± 4.18	22.2 ± 3.65	24.63 ± 1.58
Post-ensiling particle size[4]		% of total	
< 1.0 cm	24.9 ± 14.4	22.5 ± 12.1	37.5 ± 29.0
1 to 7.4 cm	13.9 ± 3.5	13.2 ± 1.7	13.2 ± 4.4
7.5 to 14.8 cm	17.4 ± 2.0	25.9 ± 1.6	16.7 ± 6.9
14.9 to 22.2 cm	16.2 ± 4.9	22.3 ± 4.8	17.2 ± 10.6
22.3 to 29.6 cm	13. 1 ± 5.3	10.5 ± 4.3	8.6 ± 9.1
> 29.7 cm	14.4 ± 10.5	5.6 ± 5.1	6.8 ± 5.9

[1]Determined from forage samples collected each day and composited weekly (n = 7).
[2]MBB = monoculture bahiagrass baleage, BAB = bahiagrass-alfalfa baleage and CCS = conventional corn silage.
[3]IVTD = *in vitro* true digestibility; WSC = water soluble carbohydrates.
[4]Based on fresh samples collected biweekly and manually measured for particle length.

[13]. Phosphorus was determined by the molybdovanadate colorimetric method [13].

Statistical analyses

Lactation performance data were statistically analyzed using the MIXED procedure of SAS Institute [18]. All daily intake and milk production data were reduced to cow-week means before statistical analysis. The model for the lactation performance data included forage treatment, week and forage treatment x week as fixed effects; cow nested within treatments as random effects; and cows by week as residual error effects. Performance data collected during the standardization period were used as covariables for the dependent variables. Treatment means were generated via Tukey-Kramer procedures and results are accompanied by the highest standard error. Statistical differences between means were declared at $P < 0.05$ and tendencies were reported at $P > 0.05$ and < 0.10.

Results and discussion

Forage agronomic and nutritive value assessment

Although alfalfa broke dormancy several weeks prior to bahiagrass, alfalfa constituted less than 10% of the forage mix in the initial late April harvest. However, at the second cutting on June 11, 2007 the alfalfa accounted for 22.3% of the forage mixture, bahiagrass accounted for 76.8% and weeds represented the remaining 1.5% of the forage mass (BAB treatment). The monoculture bahiagrass field contained 97.1% bahiagrass and 2.9% weeds at the second cutting (MBB treatment). The bahiagrass-alfalfa harvest used for the study generated forty five 1.5 m × 1.2 m round bales weighing an average of 659 kg fresh weight. Average DM yield for the second cutting was 3,235 kg/ha for BAB and 3,450 kg/ha for MBB treatment.

Chemical composition, fermentation characteristics and particle size of silage crops are presented in Table 1. The CP levels were low and NDF high for MBB (10.3 and 76.5%, respectively), but these concentrations are similar to values reported by Florida researchers for 5-wk bahiagrass regrowth [19]. Crude protein concentration in BAB was 22.5% higher and NDF was 7.3% lower than MBB, reflecting the higher nutritive value of alfalfa present in the BAB mixture [20]. *In vitro* digestibility and net energy for lactation of bahiagrass baleage were also improved by interseeded alfalfa. In general, research with tropical grass-legume mixtures indicates nutritive values similar to the ratios of the grass and legumes present in the mixture [21]. Corn silage *in vitro* digestibility and NE_L were considerably higher than those of either bale silage crop, though this was an exceptionally high-energy corn silage crop for dry land corn grown in southeastern Louisiana [2]. Poor bahiagrass baleage nutritive value i.e. low protein and high NDF, was also reported in previous baleage work at this unit [3], but NDF levels were higher in the present study likely due to the later stage of maturity of harvested bahiagrass.

Dry matter content in baleage crops MBB and BAB was similar and was well within the optimum range of 40-60% described by Muck [22]. Since DM differences between MBB and BAB were minimal, pH and fermentation characteristics were more likely influenced by forage composition than baleage moisture content. Residual WSC concentrations were much lower for baleage crops than corn silage and coupled with high pH indicate a restricted fermentation for MBB and BAB. A slight increase in silage WSC and decrease in pH was noted for BAB compared to MBB, but pH values for both baleage crops were above 5.0, the maximum pH threshold required for inhibition of undesirable clostridial fermentations [22]. In spite of this restricted fermentation, high levels of mold were not observed in bale interiors probably a result of relatively low baleage moisture concentrations [23] and airtight bale wrapping. Lactic acid was the predominant organic acid present in baleage extracts though concentrations were less than one fourth of that found (4% of DM) in high WSC-containing cool season grasses [24]. As McEniry et al. [23] noted, extent of baleage fermentation is often limited in low moisture forage crops. Similarly, Foster et al. [19] recorded bahiagrass baleage WSC of 3.72% post-ensiling and lactate concentrations equaled less than 1% of DM. However, Foster et al. [19] noted butyrate concentrations of 0.39%, whereas we were unable to detect any butyrate in our MBB silage. In fact, only nominal levels of VFAs were detected for either baleage crop with the exception of acetate whose average concentration was 0.24 and 0.27% for MBB and BAB, respectively. Overall, the pH, organic acid concentrations and nutritive values were quite similar between the two bahiagrass baleage crops with each data set indicating that bahiagrass is a forage crop low in WSC and overall nutritive value, which under the management conditions imposed, may none the less be successfully ensiled as baleage. Corn silage pH was typical of that recorded for well-preserved chopped whole plant corn (<4.0), but acetic acid concentrations were inordinately high (4.78% of DM) suggestive of hetero-fermenting bacteria dominant ensiling [24].

Post-grinding particle sizes for silage crops presented in Table 1 varied considerably between MBB, BAB and CCS (non-ground). More than 27% of the MBB forage exceeded 22 cm in length compared to 16.4 and 15.4% for BAB and corn silage, respectively. Lower levels of long particles in BAB than MBB suggest that alfalfa forage length was more efficiently reduced by grinding than bahiagrass forage as would be expected based on lower alfalfa NDF concentrations. In addition, nearly 50% of corn silage particles were less than 7.4 cm in length compared to only about 35% for the baleage crops. In spite of our best efforts to reduce baleage particle length via a hay grinder, baleage crop particle lengths remained substantially longer than those of corn silage during the feeding study.

Ingredient and chemical composition of experimental diets are presented in Table 2. Forages represented approximately 47.5% of DM for all experimental diets. Diet CP, ADF, NDF, and IVTD concentrations were similar for MBB and BAB. These data contradict the individual silage compositions in which CP and IVTD were higher and NDF was slightly lower for BAB than MBB (Table 1).

Table 2 Ingredient and chemical composition of experimental diets (% of DM)[1]

Item	Diets[2]		
	MBB	BAB	CCS
Ingredient composition			
Corn, grd	21.93	23.92	21.93
Soybean meal, 55%	17.74	15.73	17.75
Cottonseed	8.34	8.34	8.34
Mineral mix[3]	1.74	1.74	1.74
Sodium bicarbonate	0.89	0.89	0.89
Rumen inert fat[4]	0.85	0.85	0.85
SC[5]	0.63	0.63	0.63
Calcium carbonate	0.43	0.43	0.43
Forage	47.45	47.47	47.44
Chemical composition			
CP	16.38 ± 1.65	16.10 ± 0.93	17.07 ± 1.08
ADF	27.02 ± 1.65	26.90 ± 1.79	19.25 ± 2.62
NDF	50.30 ± 2.82	47.61 ± 3.21	32.11 ± 2.23
NE_L, Mcal/kg	1.41 ± 0.08	1.47 ± 0.10	1.62 ± 0.11
IVTD[6]	74.90 ± 2.16	74.30 ± 6.35	87.38 ± 3.57
Calcium	0.62 ± 0.08	0.67 ± 0.07	0.69 ± 0.13
Phosphorus	0.34 ± 0.03	0.36 ± 0.02	0.45 ± 0.02
Potassium	1.77 ± 0.18	1.77 ± 0.23	1.35 ± 0.16
Magnesium	0.30 ± 0.04	0.29 ± 0.03	0.29 ± 0.02
Copper, ppm	16.40 ± 3.75	14.29 ± 2.82	17.54 ± 5.51
Manganese, ppm	222.4 ± 89.9	145.9 ± 40.2	87.4 ± 28.8
Zinc, ppm	89.0 ± 11.3	82.3 ± 12.8	100.3 ± 6.6

[1]Chemical composition of total mixed rations determined from samples collected daily and composited weekly (n = 7).
[2]MBB = mono culture bahiagrass baleage-based diet, BAB = bahiagrass-alfalfa baleage-based diet and CCS = conventional corn silage-based diet.
[3]Mineral and vitamin mix containing 23% Ca, 3% P, 7% Mg, 6% K, 3.0% S, 15 mg Co/kg, 650 mg Cu/kg, 50 mg I/kg, 1200 mg Mn/kg, 2700 mg Zn/kg, 18 mg Se/kg, 300,000 IU Vitamin A/kg, 30,000 IU Vitamin D/kg, and 1,500 IU Vitamin E/kg.
[4]MegaLac rumen inert fat manufactured by Church and Dwight, Inc., Princeton, NJ.
[5]Saccharomyces cerevisiae yeast product sold by Diamond V Mills, Inc. Cedar Rapids, IA.
[6]IVTD = in vitro true digestibility.

These findings are in contrast to those of Titterton and Bareeba [21] who noted that nutritive value of tropical grass – legume mixtures was generally a reflection of individual nutrient composition and the ratio of grass to legume. In our study, windrow samples indicated about 22% alfalfa content in the mixed sward which should have led to higher BAB protein content than experienced based on forage and TMR analyses. This implies that actual legume contribution to BAB may have been lower than recorded or sampling procedures failed to generate representative samples. We did note considerable variation in field-wide alfalfa growth prior to forage harvest. This variation in percentage bahiagrass and alfalfa was also likely present within BAB bales. As expected, the corn silage-based TMR was lower in fiber fractions and higher in IVTD than either baleage-based TMR.

Lactation performance and rumen characteristics

Lactation performance data are presented in Table 3. Alfalfa inclusion in the bahiagrass baleage did not improve DM intake though cattle offered BAB showed numerically higher (10.5%) diet consumption. Cattle fed CCS-based TMR consumed 24.2% more ($P < 0.05$) DM than the average consumed by the cows fed baleage-based TMRs. Expressed as a percent of final body weight, DMI was 3.16, 3.35 and 3.74% of BW for MBB, BAB and CCS, respectively. Intake for the corn silage based diets was similar to that predicted by NRC [10], but cattle consumed the baleage-based diets at considerably lower levels than predicted. Differences in DMI between baleage diets and CCS

were likely related to the higher digestibility, and smaller particle size of corn silage compared to the baleage treatments (Table 2). Also, NDF intake as percent BW was 1.65% for baleage-based diets compared to only 1.25% for corn silage-based diets. Long particle length and high NDF concentrations have been shown to limit DMI for both tropical [25] and cool season grasses [26].

Secondary fermentation i.e. diet heating, may also have influenced MBB consumption. From wk 4 to 5 of the study, minimum ambient temperature increased from 7.3 to 15.2°C which increased TMR refusal temperature 8.1 degrees (from 10.0 to 18.1°C) for MBB compared to a 5.5 degree increase (from 9.7 to 15.2°C) for CCS (Figure 1). Higher pH and lower VFA concentrations, especially lower acetic acid in baleages compared to CCS may have allowed more rapid proliferation of molds and yeasts which led to more extensive TMR heating and lower palatability [27].

Milk yield tended to be higher ($P < 0.10$) for cows offered BAB than MBB (28.44 vs 26.11 kg/hd); however, when expressed on a 3.5% FCM basis milk yields between the two baleage crops were similar. Since the percentage improvement in milk yield with alfalfa introduction was about 9% for both actual and FCM, lack of statistical difference between the two baleage treatments for FCM was likely due to higher variability in milk fat concentration as evidenced by a SE for actual milk yield of 0.71 compared to 1.20 kg/hd for FCM. Cows fed CCS-based diets produced substantially more ($P < 0.05$) actual and FCM than cows fed either BAB or MBB-based diets as expected based on differences in diet NDF and NE_L concentrations (Table 2). Actual milk yields and NRC [10] predictions for energy allowable milk were similar among diets. Feed efficiency averaged 1.43 kg FCM/kg DMI and did not differ with dietary treatment. Research studying inclusion of cool season perennial legumes in warm-season perennial grass silage diets is scant, but substitution of red clover for bermudagrass pasture or ryegrass silage indicated a substantial improvement in DMI and FCM yield but not in feed efficiency [6,28].

Percentage milk fat concentration did not differ among dietary treatments. Since both baleage diets contained higher NDF concentrations and longer particle lengths, it was anticipated that baleage treatments would generate higher milk fat concentrations than CCS-based diets. The fact that they did not may partially be explained by the higher DMI for CCS, but may also be related to inherently low NDF digestibility for the bahiagrass-based diets [19]. However, rumen fluid analysis demonstrated that acetate, the major precursor for milk fat synthesis, was present in higher concentrations for MBB than the other two silages (Table 4). These inconsistencies may be related to time of rumen fluid sampling (4 h fast) that

Table 3 Lactation performance of Holstein cows fed total mixed diets containing monoculture bahiagrass baleage (MBB), bahiagrass-alfalfa baleage (BAB) or conventional corn silage (CCS)

Item	Silage crop			
	MBB	BAB	CCS	SE
Cows	12	12	12	
DM intake, kg	19.51[a]	21.56[a]	25.50[b]	0.75
CP intake, kg	2.86[a]	3.20[ab]	3.48[b]	0.24
NDF intake, kg	9.81[a]	10.26[a]	8.18[b]	0.85
Milk yield, kg[1]	26.11[a]	28.44[b]	33.12[c]	0.71
3.5% FCM, kg	27.58[a]	29.83[a]	35.26[b]	1.20
Fat,%	3.87[a]	3.95[a]	3.83[a]	0.12
Fat, kg	1.00[a]	1.09[a]	1.28[b]	0.06
Protein,%	3.08[a]	3.04[a]	3.24[b]	0.04
Protein, kg	0.89[a]	0.89[a]	0.95[b]	0.01
Lactose,%	4.64[a]	4.68[a]	4.78[b]	0.02
Lactose, kg	1.37[a]	1.38[a]	1.40[a]	0.01
MUN, mg/dL	16.80[a]	15.76[a,b]	14.31[b]	0.53
SCC × 1,000	304[a]	209[a]	221[a]	89
FE[2], kg fcm/kg DMI	1.49[a]	1.42[a]	1.38[a]	0.06
Initial BW, kg	611.17[a]	611.26[a]	657.12[a]	21.50
Final BW, kg	616.90[a]	643.38[a]	681.29[a]	21.60
BW gain, kg	5.73[a]	32.12[b]	24.17[ab]	4.71
Initial BCS	2.83[a]	2.63[a]	2.64[a]	0.15
Final BCS	2.78[a]	2.67[a]	3.03[a]	0.13
BCS change	−0.05[a]	0.04[a]	0.39[b]	0.11

a, b, c Means in a row with different superscripts differ significantly ($P < 0.05$).
[1] Milk yield for BAB tended to be higher than MBB ($P < 0.10$). Milk yield for CCS was higher than either MBB or BAB ($P < 0.05$).
[2] FE = feed efficiency.

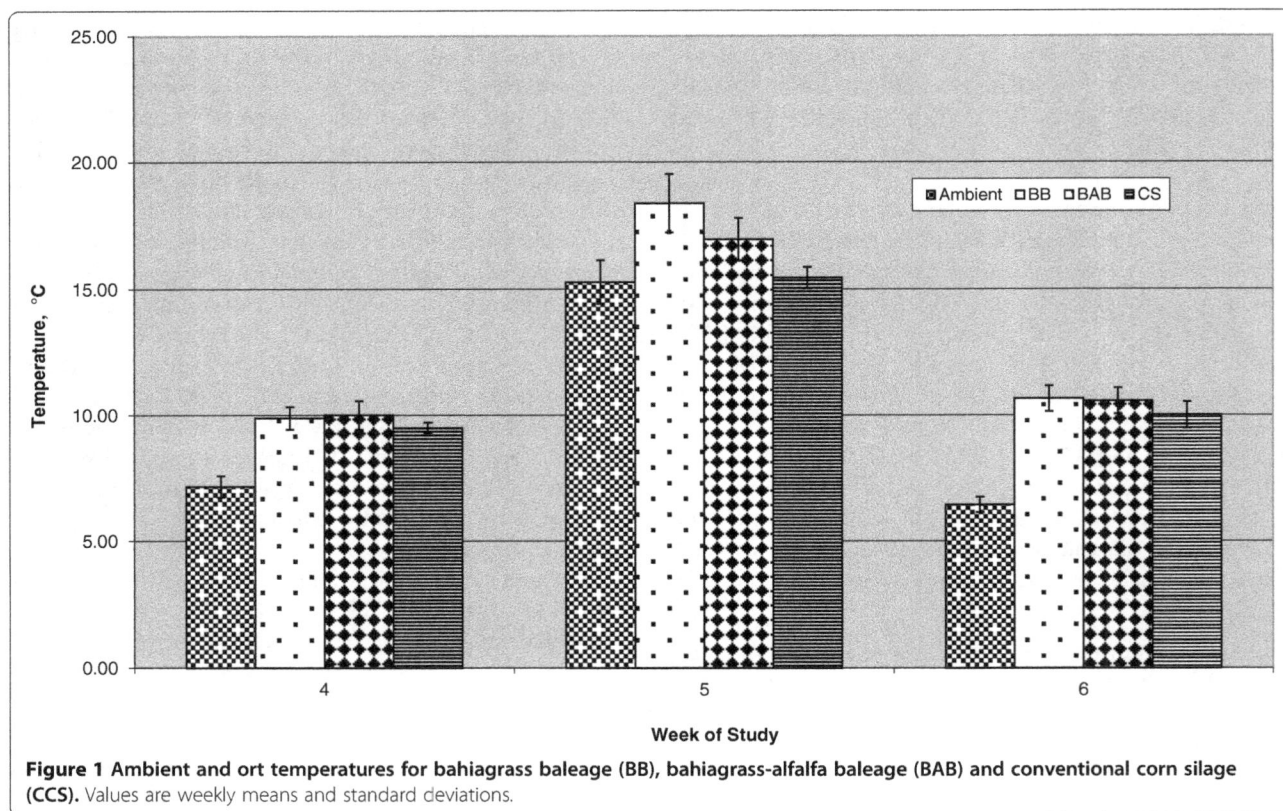

Figure 1 Ambient and ort temperatures for bahiagrass baleage (BB), bahiagrass-alfalfa baleage (BAB) and conventional corn silage (CCS). Values are weekly means and standard deviations.

favored larger particle sizes and slower fiber digestion rates for bahiagrass baleage-based TMRs vs corn silage-based TMRs.

Percent milk protein and lactose did not differ among cows fed BAB or MBB, but cows offered CCS generated higher ($P < 0.05$) levels of milk protein and lactose than either baleage crop. Higher milk protein and other component concentrations for CCS may be related to greater microbial protein synthesis and or enhanced dietary N utilization due to more of easily fermentable carbohydrate in the rumen than baleage-based diets. Milk urea nitrogen concentrations, though similar between BAB and MBB, were higher for baleage-based diets than CCS diets (16.3 vs 14.3 mg/dL) which may again suggest poorer N utilization by cows fed the baleage-based diets. Differences in MUN between bahiagrass and corn silage TMR were not likely due to differences in total dietary N intake as CCS diet was higher in CP intake than MBB and BAB (Table 3), but more likely differences were related to higher soluble N concentrations for the baleage crops compared to corn silage [29].

Body weight and condition score changes were generally positive, as expected for cows that averaged 162 DIM at study conclusion (Table 3). The BW gain was higher ($P < 0.05$) for cows on BAB than MBB (32.1 vs 5.7 kg/h/d) suggesting that a portion of the higher

energy and numerically higher DMI for BAB may have been partitioned toward BW gain rather than milk production. Cows consuming MBB had the lowest BW gain and, on the average, experienced slight losses in BCS. Differences in BW gain between BAB and CCS were

Table 4 Rumen fluid characteristics of Holstein cows fed total mixed diets containing monoculture bahiagrass baleage (MBB), bahiagrass-alfalfa baleage (BAB) and conventional corn silage (CCS)[1]

Item	Silage crop			SE
	MBB	BAB	CCS	
pH	6.40[a]	6.61[a]	6.55[a]	0.11
Lactate	0.02[a]	0.03[a]	0.05[a]	0.01
Acetate	55.23[a]	39.60[b]	42.30[b]	3.11
Propionate	13.69[a]	9.20[b]	11.83[ab]	0.99
Butyrate	0.36[a]	1.08[a]	0.18[a]	0.26
Isobutyrate	9.02[a]	6.04[a]	6.30[a]	1.13
Valerate	17.90[a]	10.15[a]	14.06[a]	2.54
Isovalerate	0.12[a]	0.12[a]	0.54[a]	0.25
Total VFA	96.20[a]	66.19[b]	75.21[b]	5.75
A:P[2]	4.08[ab]	4.34[a]	3.67[b]	0.13

[a, b, c]Means in a row with different superscripts differ significantly ($P < 0.05$).
[1]Rumen fluid collected via rumenocentesis on d-42 prior to a.m. feeding (4 h fast).
[2]Acetate to propionate ratio.

small, but BCS change was 0.35 and 0.44 units higher (*P* < 0.05) for CCS than either BAB or MBB, respectively. Again, higher DMI for CCS allowed cows to begin replenishing body condition earlier than experienced by cows receiving baleage-based diets.

Conclusion

In conclusion, modest levels of alfalfa (22%) in bahiagrass-alfalfa mixture fed in this study improved forage energy and protein concentrations and tended to improve actual milk yield compared to monoculture bahiagrass forage. Neither MBB nor BAB was consumed as readily as CCS-based TMR by mid-lactation Holstein cows and cows offered CCS diets produced 23.0% more FCM than the average of cows offered MBB or BAB-based diets. Never the less, inter-seeding alfalfa into bahiagrass pastures may serve as a means of improving bahiagrass pasture and stored forage nutritive value.

Abbreviations

MBB: Monoculture bahiagrass; BAB: Bahiagrass-alfalfa baleage; CCS: Conventional corn silage; DIM: Days in milk; BW: Body weight; FCM: Fat-corrected milk; TMR: Total mixed ration; SCC: Somatic cell count; MUN: Milk urea nitrogen; DMI: Dry matter intake; FE: Feed efficiency; CP: Crude protein; ADF: Acid detergent fiber; NDF: Neutral detergent fiber; NRC: National Research Council; WSC: Water soluble carbohydrates; IVTD: *In vitro* true digestibility; VFA: Volatile fatty acids; BCS: Body condition score.

Competing interests

The authors declare that they have no competing interests.

Authors' contributions

MM conceived the trial, generated forage treatments, conducted the feeding experiment and drafted the initial manuscript. KJH managed forage production and lab analyses; VM was instrumental in project development, animal training and animal data collection. DB participated in the design of the study and conducted all statistical analyses. All authors read the manuscript, participated in revisions and approved the final manuscript.

Acknowledgements

The authors wish to thank Shannon Forbes, Tara Martin, Doug McKean and Bill Barber of the Southeast Research Station (Franklinton, LA) for assisting with animal management and laboratory analyses during this study.

Author details

[1]Southeast Region LSU Agricultural Center, 21549 Old Covington Hwy, Hammond, LA 70403, USA. [2]Louisiana State University School of Plant, Soil and Environmental Sciences, Baton Rouge 70803, USA. [3]Louisiana State University Agricultural Center, Southeast Research Station, P.O. Drawer 569, Franklinton 70438, USA. [4]Department of Experimental Statistics, Louisiana State University, Baton Rouge 70803, USA.

References

1. Han KJ, McCormick ME, Walz R: Forage utilization trends as determined from eight years of forage quality analyses in Louisiana and Mississippi. *Southeast Res Stn Field Day Summ* 2007, 38–43. http://www.lsuagcenter.com/en/our_offices/research_stations/Southeast/Features/extension/2007+Southeast+Research+Station+Field+Day+Summaries.htm (accessed 02/15/14).
2. Han KJ, McCormick ME, Walz R: Forage quality results from Louisiana and Mississippi producer forage samples, 1999–2006. *Southeast Res Stn Field Day Summ* 2007, 35–37. http://www.lsuagcenter.com/en/our_offices/research_stations/Southeast/Features/extension/2007+Southeast+Research+Station+Field+Day+Summaries.htm (accessed 02/15/14).
3. McCormick ME, Han KJ, Moreira VR, Blouin DC: Forage conservation efficiency and lactation response to bahiagrass conserved as barn-stored hay, outdoor-stored hay or baleage. *J Dairy Sci* 2011, 24(5):2500–2507.
4. Barker JM, Buskirk DD, Ritchie HD, Rust SR, Leep RH, Barclay BJ: Intensive grazing management of smooth bromegrass with or without alfalfa or birdsfoot trefoil: heifer performance and sward characteristics. *Prof Anim Sci* 1999, 15:130–135.
5. Wu Z, Kanneganti VR, Massingill LJ, Wiltbank MW, Walgenbach RP, Satter LD: Milk production of fall calving dairy cows during summer grazing of grass or grass-clover pasture. *J Dairy Sci* 2001, 84:1166–1173.
6. Morgan EB, Nelson BD, Kilgore L, Mason L, Schilling PE, Montgomery CR: Evaluation of summer perennial grasses with and without inter-planted clover under grazing with lactating dairy animals. *Southeast Res Sta Ann Prog Rep* 1980, 127–133.
7. Bouton J: Breeding tall fescue, alfalfa and white clover in Georgia. In *Proceedings 57th Pasture and Forage Crop Improvement Conference*. Athens, GA: 2002. http://agrilife.org/spfcic/annual-proceedings/57th/breeding-tall-fescue, white- clover-in-georgia/ (accessed 2/4/2014).
8. Stringer WC, Khalilian A, Undersander DJ, Stapleton GS, Bridges JW Jr: Row spacing and nitrogen: Effect on alfalfa-bermudagrass yield and botanical composition. *Agron J* 1994, 86:72–76.
9. Haby VA, Davis JV, Leonard AT: Response of over-seeded alfalfa and bermudagrass to row spacing and nitrogen rate. *Agron J* 1999, 91:902–910.
10. National Research Council: *Nutrient Requirements of Dairy Cattle. 7th rev. ed.* Washington, D.C: Natl Acad Sci; 2001.
11. Edmonson AJ, Lean IJ, Weaver LD, Farver T, Webster G: A body condition scoring chart of Holstein dairy cows. *J Dairy Sci* 1989, 72:68–78.
12. Norlund KV, Garrett EF: Rumenocentesis: a technique for the diagnosis of subacute rumen acidosis in dairy herds. *Bovine Practitioners* 1994, 28:109–122.
13. *A O A C Official Methods of Analysis. 15th ed.* Arlington, VA: AOAC; 1990.
14. Van Soest PJ, Robertson JB, Lewis BA: Methods for dietary fiber, neutral detergent fiber and non-starch polysaccharides in relation to animal nutrition. *J Dairy Sci* 1971, 74:3583–3597.
15. Goering HK, Van Soest PJ: *Forage fiber analyses (apparatus, reagents, procedures, and some applications)*. Washington, DC: USDA-ARS, U.S. Gov. Print. Office; 1970.
16. Dubois MK, Giles M, Hamilton JK, Rebers PA, Smith F: Colorimetric method for determination of sugars and related substances. *Anal Chem* 1956, 28:350–357.
17. Bateman HG, Williams CC, Chung YC: Effects of supplemental zinc in high quality diets on rumen fermentation and degradation of urea in vitro and in vivo. *Prof Anim Sci* 2002, 18:363–367.
18. *SAS User's Guide: Statistics, Version 8.1, 1st ed.* Cary, NC: SAS Inst. Inc; 2001.
19. Foster JL, Carter JN, Sollenberger LE, Blount AR, Meyer RO, Maddox MK, Phatak SC, Adesogan AT: Nutritive value, fermentation characteristics and in situ disappearance kinetics of ensiled warm-season legumes and bahiagrass. *J Dairy Sci* 2011, 94:2042–2050.
20. McCormick ME, Doughty T, Walz R: Alfalfa baleage production, quality and persistence – three year summary. *LSU Agric Center, Southeast Res Stn Field Day Summ* 2007, 28–30. http://www.lsuagcenter.com/en/our_offices/research_stations/Southeast/Features/extension/2007+Southeast+Research+Station+Field+Day+Summaries.htm (accessed 02/15/12).
21. Titterton M, Bareeba FB: Grass and legume silages in the tropics. Paper 4.0. In *Proceedings of the FAO Electronic Conference on Tropical Silage*. FAO Plant Production and Protection; 2000:161. http:www.fao.org/docrep/005/x8486e/x8486e0c.htm. Accessed March 20, 2013.
22. Muck RE: Fermentation characteristics of round bale silages. In *Proceedings Southern Pasture & Forage Crop Improvement Conference*. Auburn, AL: Auburn University Press; 2006:1–11.
23. McEniry JP, O'Kiely NJ, Clipson W, Forristal PD, Doyle EM: The microbial and chemical composition of baled and precision-chop silage on a sample of farms in county Meath. *Irish Agric and Food Res* 2006, 45:73–83.
24. Woolford MK: Chap. 4, Heterolactic silage fermentations. In *The Silage Fermentation*. Edited by Allen I, Laskin Richard I, Mateles I. New York and Basel: Marcel Deckers; 1984.
25. Panditharatne S, Allen SG, Fontenot JP, Jayasuriya MC: Ensiling characteristics of tropical grasses as influenced by stage of growth, additives and chopping length. *J Anim Sci* 1986, 63:197–207.
26. Petit HV, Tremblay GF, Savoie P, Tremblay D, Wauthy JM: Milk yield, intake, and blood traits of lactating cows fed grass silage conserved under different harvesting methods. *J Dairy Sci* 1993, 76:1365–1374.

27. Kung L: **Aerobic stability of silages**. In *Proceeding on the Conference on Silage for Dairy Farms*. Harrisburg, PA: Pennsylvania State University Press; 2005:1–13.

28. Moorby JM, Lee MRF, Davies DR, Kim EJ, Nute JR, Ellis NM, Scolan ND: **Assessment of dietary ratios of red clover and grass silages on milk production and milk quality in dairy cows**. *J Dairy Sci* 2008, **92**:1148–1160.

29. Jonkers JS, Kohn RA, Erdman RA: **Using milk urea nitrogen to predict nitrogen excretion and utilization efficiency in lactating dairy cows**. *J Dairy Sci* 1998, **81**:2681–2692.

Changes in fetal mannose and other carbohydrates induced by a maternal insulin infusion in pregnant sheep

Laura D Brown[1,2*], Stephanie R Thorn[1], Alex Cheung[1], Jinny R Lavezzi[1], Frederick C Battaglia[1] and Paul J Rozance[1,2]

Abstract

Background: The importance of non-glucose carbohydrates, especially mannose and inositol, for normal development is increasingly recognized. Whether pregnancies complicated by abnormal glucose transfer to the fetus also affect the regulation of non-glucose carbohydrates is unknown. In pregnant sheep, maternal insulin infusions were used to reduce glucose supply to the fetus for both short (2-wk) and long (8-wk) durations to test the hypothesis that a maternal insulin infusion would suppress fetal mannose and inositol concentrations. We also used direct fetal insulin infusions (1-wk hyperinsulinemic-isoglycemic clamp) to determine the relative importance of fetal glucose and insulin for regulating non-glucose carbohydrates.

Results: A maternal insulin infusion resulted in lower maternal (50%, $P < 0.01$) and fetal (35-45%, $P < 0.01$) mannose concentrations, which were highly correlated ($r^2 = 0.69$, $P < 0.01$). A fetal insulin infusion resulted in a 50% reduction of fetal mannose ($P < 0.05$). Neither maternal nor fetal plasma inositol changed with exogenous insulin infusions. Additionally, maternal insulin infusion resulted in lower fetal sorbitol and fructose ($P < 0.01$).

Conclusions: Chronically decreased glucose supply to the fetus as well as fetal hyperinsulinemia both reduce fetal non-glucose carbohydrates. Given the role of these carbohydrates in protein glycosylation and lipid production, more research on their metabolism in pregnancies complicated by abnormal glucose metabolism is clearly warranted.

Keywords: Fructose, Glucose, Inositol, Insulin, Mannose, Pregnancy

Background

The importance of non glucose carbohydrates for normal fetal and neonatal development is becoming increasingly recognized, especially for mannose and inositol [1-4]. Carbohydrates not only serve as substrates for the glycolytic pathway, but also are critical for glycoprotein formation [5], phospholipid and glycerol production [2,6], and neural development [7,8]. Insights into fetal metabolism of mannose and inositol have been provided by recent reports which show that, like glucose, the fetus is dependent on its mannose supply from the mother. Conversely, the fetus and placenta endogenously produce inositol from glucose [9,10]. In the term neonate, the utilization rates of mannose and inositol are much higher than the rate which could be obtained from human milk, suggesting endogenous production of both substrates after birth to meet daily requirements [11,12].

Pathological conditions in pregnancy such maternal under nutrition, placental insufficiency, and diabetes have the potential to increase or decrease the delivery of carbohydrates to the fetus and adversely affect growth, body composition, and long term outcomes in human and livestock pregnancy [13,14]. For example, diabetic pregnancies are characterized by increased glucose transfer to the fetus with subsequent elevations in fetal insulin concentrations [15]. In adults, insulin regulates mannose concentrations independent of glucose by suppression of hepatic mannose production [16-19]. Thus, the fetus is potentially at risk for hypomannosemia, both from fetal hyperinsulinemia as well as from decreased mannose delivery from the mother if insulin is used to manage maternal diabetes. Placental insufficiency in

* Correspondence: Laura.Brown@ucdenver.edu
[1]Perinatal Research Center, Division of Neonatology, Department of Pediatrics, University of Colorado Denver School of Medicine, Aurora, CO, USA
[2]Center for Women's Health Research, University of Colorado Denver School of Medicine, Aurora, CO, USA

humans and sheep, on the other hand, restricts glucose delivery to the fetus leading to decreased fetal glucose and insulin concentrations [20,21]. Placental insufficiency in pregnant sheep also decreases fetal sorbitol and inositol concentrations [22]. Thus, a better understanding of the metabolism of non glucose carbohydrates during pregnancy is warranted, especially when carbohydrate supply to the fetus is compromised.

We experimentally decreased maternal glucose concentrations in sheep pregnancies using an insulin infusion both short term (2-wk) and long term (8-wk) to test the hypothesis that a maternal insulin infusion would suppress maternal and thus fetal mannose and inositol concentrations. We also used direct fetal insulin infusions in the form of a 1-wk hyperinsulinemic-isoglycemic clamp to determine the relative importance of fetal glucose and insulin concentrations for regulating fetal non glucose carbohydrates. Finally, we report for the first time the concentrations of several other carbohydrates and polyols under each of these conditions.

Methods
Surgical preparation
Three separate groups of Columbia-Rambouillet ewes with singleton pregnancies were used for the following experiments: 1) 2-wk maternal insulin infusion, 2) 8-wk maternal insulin infusion, and 3) 1-wk direct fetal insulin infusion. All maternal and fetal surgical preparations and post operative care have been previously described [23-25]. Maternal femoral venous and arterial catheters were placed through a left groin incision. Fetal infusion catheters were placed into fetal femoral veins via hind limb pedal veins and sampling catheters were placed into the fetal abdominal aorta via a hind limb pedal artery. Ewes were fed Premium Alfalfa Pellets (Standlee; Kimberly ID) and intake was not different between control and hypoglycemic groups (1.80 ± 0.10 kg/d control; 1.59 ± 0.10 kg/d hypoglycemic).

Care and use of animals
All animal procedures have followed established standards for the humane care and use of animals and were in compliance with guidelines of the United States Department of Agriculture, the National Institutes of Health, and the American Association for the Accreditation of Laboratory Animal Care. The animal care and use protocols were approved by the University of Colorado Institutional Animal Care and Use Committee.

Experimental design
Two-Wk maternal insulin infusion
For the 2-wk maternal insulin infusion group and their respective controls, maternal and fetal catheters were surgically placed at 122.8 ± 1.3 d of gestation (dGA;

term = 148 dGA) [23,26-28]. One randomly assigned group received a continuous maternal infusion of intravenous insulin for 2-wk. Maternal arterial plasma glucose was measured at least twice daily and the insulin infusion was adjusted to achieve a 40-50% reduction in glucose concentrations (2-wk HG; n = 8). Maternal insulin concentrations are approximately doubled by this experimental design [29]. The control group received a maternal saline infusion at rates matched to the insulin infusion rates (2-wk C; n = 13). Fetal arterial plasma was sampled at the end of the 2-wk maternal infusion period for insulin and carbohydrate measurements.

Eight-Wk maternal insulin infusion
For the 8-wk maternal insulin group and their respective controls, an initial surgery was performed at 70.0 ± 0.8 dGA to place maternal catheters [25]. One randomly assigned group (n = 9) received a continuous maternal infusion of intravenous insulin for 8-wk. Maternal arterial plasma glucose was measured at least twice daily and the insulin infusion was adjusted to achieve a 40-50% reduction in glucose concentrations. Maternal insulin concentrations are approximately doubled by this experimental design [30,31]. The control group received a maternal saline infusion at rates matched to the insulin infusion rates (8-wk C; n = 5). At 119.4 ± 0.5 dGA, a second surgery was performed to place fetal catheters. After the second surgery, 8-wk HG ewes were further randomly divided into 2 groups. Fetuses in one of these groups received a direct fetal insulin infusion for the final week of the study (8-wk HG+I; n = 4). The insulin infusion was kept constant at 100 mU/h (using necropsy weights = 38.9 mU/kg/h ± 2.8) and ran concurrently with a direct fetal infusion of 33% dextrose (wt/vol) to prevent a further fall in fetal glucose concentrations. Fetal arterial plasma glucose concentrations were measured at least twice daily and the dextrose infusion was adjusted accordingly. The other group received a direct fetal saline infusion matched at equal infusion rates to the combined insulin and dextrose infusion (8-wk HG; n = 5). Finally, fetuses in the 8-wk C group also received a direct fetal saline infusion at equal rates. Fetal arterial plasma was sampled at the end of the infusions for insulin and carbohydrate measurements.

One-Wk fetal insulin infusion
Six late gestation animals were used in this experiment. Fetal catheters were placed at 118.5 ± 0.6 dGA. All fetuses received an insulin infusion with a concurrent dextrose infusion into a fetal hind limb vein. The insulin infusion rate was progressively increased such that infusion rates ranged from 36.6 ± 8.4 mU/h on the first day to 121.8 ± 1.2 mU/h on the final d. The fetal dextrose infusion was adjusted to prevent a fall in glucose concentrations based

on measurement of fetal arterial plasma glucose once or twice daily. Baseline, 4d, and 7d fetal arterial plasma was sampled for insulin and carbohydrate measurements.

Biochemical analysis

Whole blood was collected in EDTA-coated syringes and immediately centrifuged (14,000 g) for 3 min at 4°C. Plasma was removed and the glucose concentration immediately determined using the YSI model 2700 select biochemistry analyzer (Yellow Springs Instruments, Yellow Springs, OH) [23]. The remainder of the plasma was stored at −70°C for insulin and carbohydrate measurements. Insulin (Alpco; inter-assay and intra-assay coefficients of variation: 2.9 and 5.6%) was measured by enzyme-linked immunosorbent assay [23]. Plasma was analyzed for mannose, inositol, fructose, mannitol, erythritol, arabinol, sorbitol and ribitol by HPLC as we have previously described [9].

Postmortem exam

Liver tissue from the right hepatic lobe was obtained in the 8-wk HG, 8-wk C, and 8-wk HG+I fetuses under conditions closely approximating *in vivo* study conditions as previously described for measurement of sorbitol and fructose [25].

Statistical analysis

Statistical analysis was performed with SAS v.9.2 (SAS Institute) and GraphPad Prism 4. Results are expressed as mean ± SEM. *P*-values less than 0.05 were considered significant. The 2-wk HG and C fetuses were compared with Student's *t* test (parametric data) or the Mann–Whitney test (non parametric data). The 8-wk HG, HG+I, and C fetuses were compared with a one-way ANOVA. For measurements taken at multiple time points within an animal a mixed models ANOVA was used with terms for experimental group, time, group by time interaction as indicated. Repeated measurements made within an animal were accounted for. Post-test comparisons were made using Fishers least squares difference if the overall ANOVA had a *P* < 0.05.

Results

Two-Wk maternal insulin infusion
Fetal insulin, carbohydrates, and polyols

As previously reported [23,26-28], gestational ages were similar, but fetal weight, fetal arterial plasma insulin, and glucose were 24%, 63%, and 49% lower, respectively in the 2 wk HG group (*P* < 0.01, Table 1). We found lower fetal arterial plasma concentrations of mannose (35%, *P* < 0.01), sorbitol (57%, *P* < 0.01), and fructose (60%, *P* < 0.01) in the 2-wk HG group (Table 1). Although the mean arterial plasma inositol concentration was nearly doubled in the 2-wk HG fetuses this did not reach statistical

Table 1 Two-Wk maternal insulin infusion

Measurement	Control	Hypoglycemic
Gestational age, d	138.5 ± 0.4	137.5 ± 0.8
Fetal weight, kg	4.37 ± 0.14	3.30 ± 0.18**
Fetal plasma arterial insulin, ng/mL	0.30 ± 0.03	0.11 ± 0.01**
Fetal plasma arterial carbohydrates, µmol/L		
Glucose	1,159 ± 64	594 ± 47**
Mannose	24.5 ± 1.2	15.9 ± 1.4**
Inositol	638.5 ± 138.6	1,129.0 ± 256.1
Sorbitol	123.8 ± 12.7	53.9 ± 10.5**
Fructose	4771 ± 450	1,909 ± 356**
Erythritol	399.6 ± 28.1	345.9 ± 20.8
Arabinol	245.0 ± 11.9	225.3 ± 15.6
Ribitol	128.1 ± 9.1	148.5 ± 18.8
Mannitol	72.8 ± 9.7	49.8 ± 10.1

Values are means ± SEM. ** refer to significant differences between Control (n = 13) and Hypoglycemic (n = 8); *P* < 0.01 by Students *t* test or the Mann–Whitney test.

significance. Other fetal plasma carbohydrate concentrations were similar between 2-wk HG and C groups (Table 1).

Eight-Wk maternal insulin infusion
Maternal plasma carbohydrates and polyols

Since 2-wk maternal insulin infusion experiments focused only on plasma carbohydrate changes in the fetal circulation, we included maternal carbohydrate analysis with the 8-wk insulin infusion studies. Consistent with study design, maternal arterial plasma glucose concentrations in the 8-wk HG group were approximately 40% lower compared to 8-wk C group throughout the insulin infusion period as previously reported (Table 2, *P* < 0.01) [25]. Maternal arterial plasma mannose concentrations also were 48% lower in the 8 wk HG group (Table 2, *P* < 0.01), but maternal arterial plasma inositol, sorbitol, erythritol, arabinol, and ribitol did not change (Table 2).

Fetal insulin, carbohydrates, and polyols

As previously reported, gestational age at the time of study was not different between the groups. Fetal weights were 40% lower in the 8-wk HG group compared to 8-wk C (Table 2, *P* < 0.01) [25]. Fetal arterial plasma insulin and glucose were lower in the 8-wk HG group (*P* < 0.05, Table 2) [25]. Fetal arterial plasma mannose was 44% lower (*P* < 0.01), as were sorbitol (71%, *P* < 0.01) and fructose (64%, *P* < 0.01, Table 2). Furthermore, the arterial plasma maternal - fetal mannose difference also was lower (*P* < 0.01, Figure 1A), and fetal and maternal arterial plasma mannose concentrations were highly correlated (Figure 1B). Fetal arterial plasma inositol, erythritol,

Table 2 Eight-Wk maternal insulin infusion

Measurement	Control	Hypoglycemic	HG+I
Gestational age, d	133.2 ± 1.1	133.0 ± 1.3	134.5 ± 0.9
Maternal plasma arterial carbohydrates, μmol/L			
Glucose	3,605 ± 203	1,715 ± 145**	1,817 ± 95**
Mannose	67.0 ± 4.3	34.7 ± 3.5**	31.4 ± 5.0**
Inositol	27.7 ± 4.4	34.9 ± 3.9	34.6 ± 3.1
Sorbitol	13.4 ± 3.4	18.8 ± 4.2	11.8 ± 1.3
Erythritol	52.8 ± 5.5	44.5 ± 3.1	39.6 ± 8.5
Arabinol	96.2 ± 13.3	125.6 ± 6.5	118.0 ± 23.3
Ribitol	25.6 ± 3.0	25.9 ± 3.2	16.1 ± 4.3
Fetal weight, kg	3.66 ± 0.12	2.20 ± 0.14**	2.61 ± 0.21**
Fetal plasma arterial insulin, ng/mL	0.46 ± 0.09	0.16 ± 0.03*	0.66 ± 0.17
Fetal plasma arterial carbohydrates, μmol/L			
Glucose	1,553 ± 282	587 ± 56**	511 ± 37**
Mannose	30.4 ± 3.8	17.0 ± 1.1**	7.9 ± 0.9**
Inositol	741.4 ± 240.8	910.1 ± 157.5	792.2 ± 199.5
Sorbitol	188.5 ± 31.9	55.2 ± 11.9**	41.3 ± 8.3**
Fructose	4,508 ± 981	1,631 ± 301**	800 ± 154**
Erythritol	394.6 ± 30.7	372.2 ± 9.3	311.6 ± 24.6
Arabinol	235.3 ± 17.1	237.5 ± 15.1	214.9 ± 54.3
Ribitol	124.3 ± 15.9	139.6 ± 21.7	156.8 ± 57.6
Mannitol	47.4 ± 8.5	31.4 ± 6.1	33.8 ± 8.7
Fetal hepatic sorbitol and fructose, nmol/g			
Sorbitol	589.0 ± 90.1	205.8 ± 42.1**	161.3 ± 56.0**
Fructose	1,679 ± 219	398 ± 119**	215 ± 95**

Values are means ± SEM. *, ** refer to significant differences from Control (n = 5) animals; $P < 0.05$, 0.01, respectively, by ANOVA.

arabinol, ribitol, and mannitol were similar between 8-wk HG and C groups (Table 2).

A direct fetal insulin infusion for the final wk of the 8-wk maternal insulin infusion with a concurrent direct fetal dextrose infusion to prevent a further fall in fetal arterial plasma glucose was used to determine the effect of fetal glucose and insulin concentrations on regulating fetal non glucose carbohydrates. The direct fetal insulin infusion with a concurrent dextrose infusion (HG+I) resulted in fetal arterial plasma insulin concentrations 3-fold higher than HG fetuses ($P < 0.05$) without a change in fetal arterial plasma glucose concentrations (Table 2), as previously reported [25]. HG+I fetuses demonstrated even lower arterial plasma mannose concentrations when compared to the HG group (53%, $P < 0.05$), and had mean fructose concentrations which were over 50% lower, though this failed to reach statistical significance ($P = 0.058$ by post hoc Student's t test). There was no effect on fetal arterial plasma inositol, sorbitol, erythritol, arabinol, ribitol, and mannitol (Table 2). Fetal hepatic sorbitol and fructose concentrations were lower in

the 8-wk HG group compared to C fetal livers ($P < 0.01$, Table 2). HG+I fetuses did not have changes in hepatic sorbitol or fructose compared to the 8-wk HG fetuses (Table 2).

One-Wk fetal insulin infusion

In order to determine the effect of fetal glucose and insulin concentrations for regulating non glucose carbohydrates in fetuses whose mothers did not receive a chronic insulin infusion, we infused 6 fetuses directly with insulin with a concurrent dextrose infusion to maintain fetal glucose concentrations for 1 wk beginning on 125.2 ± 0.7 dGA. Fetal plasma arterial glucose concentrations were stable throughout the infusion period, however the dextrose infusion was increased progressively during the insulin infusion period to maintain euglycemia, starting at 3.6 ± 1.1 mg/min and ending at 12.1 ± 2.5 mg/min. Fetal arterial plasma insulin, carbohydrate, and polyol concentrations were measured at baseline and on d 4 and d 7 of the insulin infusion (Table 3). Fetal arterial plasma insulin concentrations were increased

Changes in fetal mannose and other carbohydrates induced by a maternal insulin infusion...

55

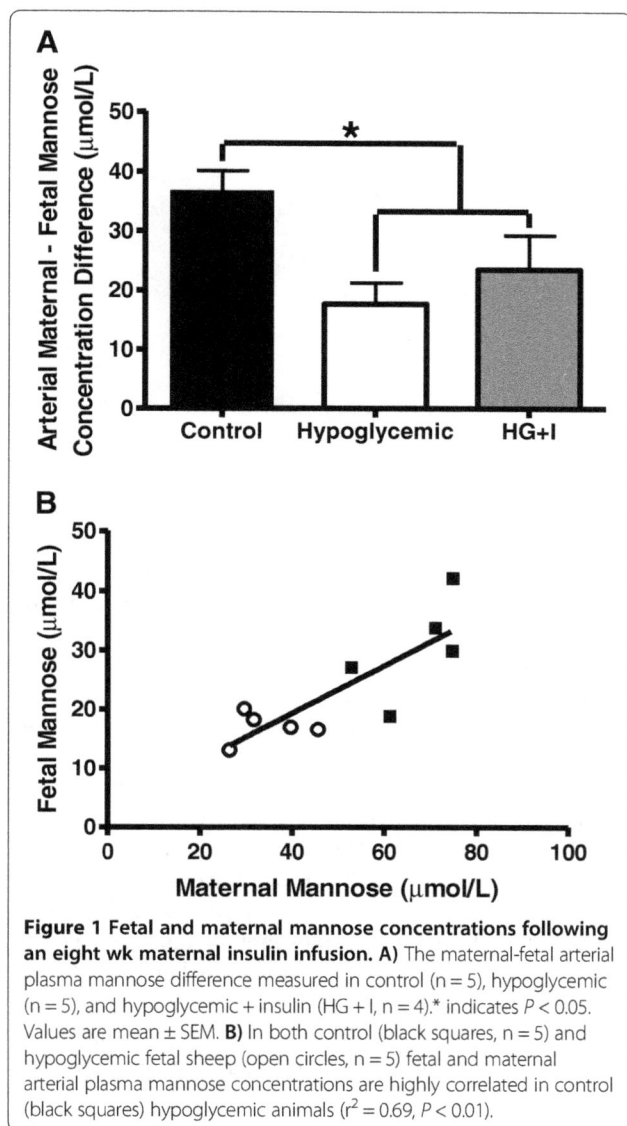

Figure 1 Fetal and maternal mannose concentrations following an eight wk maternal insulin infusion. A) The maternal-fetal arterial plasma mannose difference measured in control (n = 5), hypoglycemic (n = 5), and hypoglycemic + insulin (HG + I, n = 4).* indicates $P < 0.05$. Values are mean ± SEM. **B)** In both control (black squares, n = 5) and hypoglycemic fetal sheep (open circles, n = 5) fetal and maternal arterial plasma mannose concentrations are highly correlated in control (black squares) hypoglycemic animals ($r^2 = 0.69$, $P < 0.01$).

($P < 0.01$) and fetal arterial plasma mannose, fructose, and erythritol concentrations were decreased ($P < 0.05$).

Discussion

Non glucose carbohydrates are important for normal fetal development [1-4]. Pathological conditions in pregnancy that adversely affect glucose delivery to the fetus could affect the delivery of non glucose carbohydrates. Therefore, the primary goal of the current study was to determine how a maternal insulin infusion with an associated reduction in glucose supply to the fetus affected the plasma concentrations of these metabolites in both the maternal and fetal circulations. We found that a maternal insulin infusion which reduced glucose supply to the fetus of both short (2-wk) and long (8-wk) durations resulted in decreased maternal and fetal mannose

concentrations along with decreased fetal concentrations of sorbitol and fructose. As in human pregnancies, fetal and maternal arterial plasma mannose concentrations were highly correlated, suggesting that maternal mannose concentration determines fetal mannose concentrations. However, a physiological increase in fetal insulin after prolonged fetal hypoglycemia further reduced circulating fetal mannose concentrations, indicating that insulin also plays a key role in regulating mannose concentrations. Finally, we found that neither maternal nor fetal arterial plasma inositol concentrations changed with exogenous insulin infusions.

The most striking findings in this study relate to the regulation of fetal mannose concentrations. There has been emerging evidence in both normal human and sheep pregnancies that the fetus is dependent on placental delivery of mannose [9,10,32]. In fact, in human pregnancies, it appears that over 95% of fetal circulating mannose is derived from transplacental transfer from the mother [10]. Our findings also support transplacental transfer of mannose from the mother as the dominant source of circulating fetal mannose, as we found that maternal and fetal plasma arterial mannose concentrations were highly correlated when an experimental maternal insulin infusion decreased maternal mannose concentrations. Furthermore, we found a significant decrease in the maternal to fetal mannose concentration difference in 8-wk HG sheep. This finding suggests that decreased fetal mannose concentrations were due to decreased maternal mannose concentrations rather than a decrease in placental transport capacity. In fact, when fetal and maternal mannose concentrations were measured in a sheep model of chronic placental insufficiency there was a significant increase in the maternal to fetal mannose concentration gradient, indicating that placental insufficiency has the potential to disrupt maternal to fetal mannose delivery [22].

Our results also demonstrate a key role for insulin in the regulation of both maternal and fetal plasma mannose concentrations. We report for the first time that long term (8-wk) maternal insulin infusion during pregnancy results in a 50% reduction in maternal mannose concentrations. As these maternal ewes were also chronically hypoglycemic, we cannot determine whether chronic hypoglycemia or chronic hyperinsulinemia was directly responsible for reductions in mannose concentrations. However, previous work in rodents and adults has shown that insulin is independently involved in lowering plasma mannose by suppressing hepatic glycogen breakdown and mannose efflux from the liver [16,18]. In healthy adults, oral glucose administration increased both glucose and insulin concentrations yet mannose concentrations were still decreased, arguing for insulin stimulated mannose disposal independent of glucose [17,19].

Table 3 One-Wk fetal insulin infusion

Measurement	d0	d4	d7
Fetal plasma arterial insulin, ng/mL	0.26 ± 0.03	0.92 ± 0.20**	0.92 ± 0.10**
Fetal plasma arterial carbohydrates, μmol/L			
Glucose	1,338 ± 96	998 ± 76	1,201 ± 79
Mannose	33.9 ± 2.1	18.9 ± 2.4**	16.8 ± 2.1**
Inositol	838.2 ± 165.2	703.1 ± 144.3	673.3 ± 85.0
Sorbitol	163.4 ± 62.7	134.2 ± 76.1	82.5 ± 27.1
Fructose	5,751 ± 884	3,214 ± 673**	2,658 ± 538**
Erythritol	402.1 ± 38.2	356.5 ± 32.2*	330.3 ± 32.0**
Arabinol	193.9 ± 24.1	176.5 ± 28.4	178.3 ± 23.8
Ribitol	153.1 ± 21.4	148.9 ± 22.0	133.9 ± 17.1
Mannitol	69.4 ± 5.8	64.5 ± 10.0	60.1 ± 8.8

Values are means ± SEM. *, ** refer to significant differences from d0; $P < 0.05$, 0.01, respectively, by ANOVA. $n = 6$.

From the present study, however, we were able to gain some insight into the independent effects of glucose and insulin on circulating mannose in the fetus. When a direct fetal insulin infusion restored physiological insulin concentrations in 8-wk HG+I fetuses and fetal glucose concentrations were maintained, fetal mannose concentrations were further reduced by 50%. This argues for insulin mediated reductions in fetal plasma mannose to extremely low concentrations (~8 μmol/L) independent of concurrent hypoglycemia. To further determine the independent effects of glucose and insulin on fetal mannose, we infused insulin with a concurrent dextrose infusion to maintain glucose concentrations in a separate group of normal late gestation fetuses. This fetal hyperinsulinemic-euglycemic clamp also resulted in decreased mannose concentrations, confirming a role for insulin in the regulation of fetal mannose independent of glucose concentrations.

We also showed decreased fetal arterial plasma sorbitol and fructose concentrations in 8-wk HG fetuses. Contrary to our hypothesis, fetal plasma inositol concentrations were maintained during restricted fetal glucose supply. Maternal glucose has several potential fates once it enters the placenta. It can be directly transferred to the fetus or oxidized for fuel production [10,33]. Additionally, the placenta can convert glucose to inositol or sorbitol [9,22,32]. Glucose is converted to inositol by glucose-6-phosphate:1-phosphate cyclase and to sorbitol by aldose reductase [34-36]. The balance between these two pathways is regulated by NADPH derived from the placental uptake of glutamate from the fetus [37-40]. Although we did not directly measure placental glutamate uptake in this study, fetal arterial plasma glutamate concentrations are reduced by 50% in HG fetuses, which might lead to a reduction in placental NADPH availability [41]. This would, in turn, limit placental production of sorbitol and preserve or increase the production of

inositol. This also is consistent with our findings of decreased fetal plasma fructose and decreased hepatic sorbitol and fructose, as the major fate for fetal sorbitol in the sheep is conversion to fructose in the liver by sorbitol dehydrogenase.

Interestingly, fetal fructose concentrations also are decreased by fetal insulin infusion independent of glucose. Relationships between fructose and insulin have been previously reported, such that pancreatectomized fetal sheep have increased fructose concentrations [42], and fructose infusion into the sheep fetus can stimulate insulin secretion [43]. The role of fructose in human fetal and neonatal development remains to be determined, though postulated roles include alternative pathways in glucose metabolism, redox balance, and lipid synthesis [6,44].

The results of our study are limited only to changes in maternal and fetal arterial plasma concentrations of carbohydrates and polyols, thus conclusions cannot be made regarding their uptake and utilization rates by the fetus. However, the results are consistent with previous studies in human fetuses showing placental transport of maternal mannose, fetal production of inositol, and placental export to the fetus of sorbitol [9,10]. Future studies are warranted to determine the effects of fetal glucose and insulin on uptake, production, and utilization rates of these carbohydrates, thus providing a more in depth understanding of their metabolism in the pregnant mother and fetus.

Conclusions

In summary, the results of this study show that a chronic and constant maternal insulin infusion suppresses both maternal and fetal mannose concentrations by approximately 50%. Additionally, insulin can suppress fetal mannose and fructose concentrations independent of glucose availability. The functional implications of this degree of fetal hypomannosemia are unclear, but a recent report demonstrating embryonic lethality in mice with a hypomorphic phosphomannomutase 2 gene defect shows the critical role of mannose in normal fetal development [4]. Our results also show a significant reduction in fetal sorbitol concentrations, likely due to decreased placental sorbitol production and transfer to the fetus following increased shuttling of placental glucose into inositol production. Taken together, our findings demonstrate the potential for other carbohydrates, in addition to glucose, to be adversely affected by alterations in maternal and/or fetal insulin concentrations and glucose supply. Given the important role that many of these non glucose carbohydrates have in fetal development, future research on pathological conditions in pregnancy should include further investigation into carbohydrate metabolism beyond glucose, especially when conditions of fetal hyperinsulinemia are considered.

Abbreviations
dGA: Days gestational age; HG: Hypoglycemic; C: Control; HG+I: lypoglycemic plus insulin.

Competing interests
The authors declare that they have no competing interests.

Authors' contributions
LDB and PJR conceived of the study, designed the study, acquired and interpreted data, and wrote the first draft of the manuscript. ST conceived of the study, acquired and interpreted data, and reviewed the manuscript. JL acquired and interpreted data and reviewed the manuscript. AC designed the study, acquired and interpreted data, and reviewed the manuscript. FB conceived of the study, designed the study, interpreted data, and reviewed the manuscript. All authors read and approved the final manuscript.

Acknowledgments
We thank Karen Trembler, David Caprio, and Gates Roe for their technical support. JRL was supported by National Institutes of Health training grant T32 HD007186-32 (W Hay, PI and PD). This work was supported by NIH Grants R01DK088139 and K08HD060688, as well as American Diabetes Association Junior Faculty Award 7-08-JF-51(PJR, PI). A Pilot and Feasibility Award to PJR was provided by the UC Denver DERC (P30DK57516; J. Hutton, PI). LDB was supported as a Scholar by NIH Building Interdisciplinary Careers in Women's Health Scholar Award K12HD057022 (J. Regensteiner, PI) and a Children's Hospital Colorado Research Institute Research Scholar Award (PI). SRT was supported by NIH K01DK090199 (PI) and as a trainee on NIH training grant T32 HD007186-32 (W Hay, PI and PD). The content is solely the responsibility of the authors and does not necessarily represent the official views of the NIDDK or NICHD.

References
1. Groenen PM, Peer PG, Wevers RA, Swinkels DW, Franke B, Mariman EC, Steegers-Theunissen RP: Maternal myo-inositol, glucose, and zinc status is associated with the risk of offspring with spina bifida. Am J Obstet Gynecol 2003, 189:1713–1719.
2. Hallman M, Saugstad OD, Porreco RP, Epstein BL, Gluck L: Role of myoinositol in regulation of surfactant phospholipids in the newborn. Early Hum Dev 1985, 10:245–254.
3. Reece EA, Khandelwal M, Wu YK, Borenstein M: Dietary intake of myo-inositol and neural tube defects in offspring of diabetic rats. Am J Obstet Gynecol 1997, 176:536–539.
4. Schneider A, Thiel C, Rindermann J, DeRossi C, Popovici D, Hoffmann GF, Grone HJ, Korner C: Successful prenatal mannose treatment for congenital disorder of glycosylation-Ia in mice. Nat Med 2012, 18:71–73.
5. Davis JA, Freeze HH: Studies of mannose metabolism and effects of long-term mannose ingestion in the mouse. Biochim Biophys Acta 2001, 1528:116–126.
6. Trindade CE, Barreiros RC, Kurokawa C, Bossolan G: Fructose in fetal cord blood and its relationship with maternal and 48-hour-newborn blood concentrations. Early Hum Dev 2011, 87:193–197.
7. Greene ND, Copp AJ: Mouse models of neural tube defects: investigating preventive mechanisms. Am J Med Genet C: Semin Med Genet 2005, 135C:31–41.
8. Jaeken J, Matthijs G: Congenital disorders of glycosylation: a rapidly expanding disease family. Annu Rev Genomics Hum Genet 2007, 8:261–278.
9. Brusati V, Jozwik M, Jozwik M, Teng C, Paolini C, Marconi AM, Battaglia FC: Fetal and maternal non-glucose carbohydrates and polyols concentrations in normal human pregnancies at term. Pediatr Res 2005, 58:700–704.
10. Staat BC, Galan HL, Harwood JE, Lee G, Marconi AM, Paolini CL, Cheung A, Battaglia FC: Transplacental supply of mannose and inositol in uncomplicated pregnancies using stable isotopes. J Clin Endocrinol Metab 2012, 97:2497–2502.
11. Cavalli C, Teng C, Battaglia FC, Bevilacqua G: Free sugar and sugar alcohol concentrations in human breast milk. J Pediatr Gastroenterol Nutr 2006, 42:215–221.
12. Brown LD, Cheung A, Harwood JE, Battaglia FC: Inositol and mannose utilization rates in term and late-preterm infants exceed nutritional intakes. J Nutr 2009, 139:1648–1652.
13. De Blasio MJ, Gatford KL, McMillen IC, Robinson JS, Owens JA: Placental restriction of fetal growth increases insulin action, growth, and adiposity in the young lamb. Endocrinology 2007, 148:1350–1358.
14. Long NM, Tousley CB, Underwood KR, Paisley SI, Means WJ, Hess BW, Du M, Ford SP: Effects of early- to mid-gestational undernutrition with or without protein supplementation on offspring growth, carcass characteristics, and adipocyte size in beef cattle. J Anim Sci 2012, 90:197–206.
15. Schwartz R, Gruppuso PA, Petzold K, Brambilla D, Hiilesmaa V, Teramo KA: Hyperinsulinemia and macrosomia in the fetus of the diabetic mother. Diabetes Care 1994, 17:640–648.
16. Sharma V, Freeze HH: Mannose efflux from the cells: a potential source of mannose in blood. J Biol Chem 2011, 286:10193–10200.
17. Sone H, Shimano H, Ebinuma H, Takahashi A, Yano Y, Iida KT, Suzuki H, Toyoshima H, Kawakami Y, Okuda Y, Noguchi Y, Ushizawa K, Saito K, Yamada N: Physiological changes in circulating mannose levels in normal, glucose-intolerant, and diabetic subjects. Metabolism 2003, 52:1019–1027.
18. Taguchi T, Yamashita E, Mizutani T, Nakajima H, Yabuuchi M, Asano N, Miwa I: Hepatic glycogen breakdown is implicated in the maintenance of plasma mannose concentration. Am J Physiol Endocrinol Metab 2005, 288:E534–E540.
19. Wood FC Jr, Cahill GF Jr: Mannose utilization in man. J Clin Invest 1963, 42:1300–1312.
20. Nicolini U, Hubinont C, Santolaya J, Fisk NM, Rodeck CH: Effects of fetal intravenous glucose challenge in normal and growth retarded fetuses. Horm Metab Res 1990, 22:426–430.
21. Limesand SW, Rozance PJ, Smith D, Hay WW Jr: Increased insulin sensitivity and maintenance of glucose utilization rates in fetal sheep with placental insufficiency and intrauterine growth restriction. Am J Physiol Endocrinol Metab 2007, 293:E1716–E1725.
22. Regnault TR, Teng C, de Vrijer B, Galan HL, Wilkening RB, Battaglia FC: The tissue and plasma concentration of polyols and sugars in sheep intrauterine growth retardation. Exp Biol Med (Maywood) 2010, 235:999–1006.
23. Rozance PJ, Limesand SW, Hay WW Jr: Decreased nutrient-stimulated insulin secretion in chronically hypoglycemic late-gestation fetal sheep is due to an intrinsic islet defect. Am J Physiol Endocrinol Metab 2006, 291:E404–E411.
24. Maliszewski AM, Gadhia MM, O'Meara MC, Thorn SR, Rozance PJ, Brown LD: Prolonged infusion of amino acids increases leucine oxidation in fetal sheep. Am J Physiol Endocrinol Metab 2012, 302:E1483–E1492.
25. Thorn SR, Sekar SM, Lavezzi JR, O'Meara MC, Brown LD, Hay WW Jr, Rozance PJ: A physiological increase in insulin suppresses gluconeogenic gene

activation in fetal sheep with sustained hypoglycemia. *Am J Physiol Regul Integr Comp Physiol* 2012, **303**:R861–R869.

26. Limesand SW, Rozance PJ, Brown LD, Hay WW Jr: Effects of chronic hypoglycemia and euglycemic correction on lysine metabolism in fetal sheep. *Am J Physiol Endocrinol Metab* 2009, **296**:E879–E887.

27. Rozance PJ, Limesand SW, Zerbe GO, Hay WW Jr: Chronic fetal hypoglycemia inhibits the later steps of stimulus-secretion coupling in pancreatic beta-cells. *Am J Physiol Endocrinol Metab* 2007, **292**:E1256–E1264.

28. Thorn SR, Regnault TRH, Brown LD, Rozance PJ, Keng J, Roper M, Wilkening RB, Hay WW Jr, Friedman JE: Intrauterine growth restriction increases fetal hepatic gluconeogenic capacity and reduces messenger ribonucleic acid translation initiation and nutrient sensing in fetal liver and skeletal muscle. *Endocrinology* 2009, **150**:3021–3030.

29. DiGiacomo JE, Hay WW Jr: Fetal glucose metabolism and oxygen consumption during sustained hypoglycemia. *Metabolism* 1990, **39**:193–202.

30. Carver TD, Hay WW Jr: Uteroplacental carbon substrate metabolism and O2 consumption after long-term hypoglycemia in pregnant sheep. *Am J Physiol* 1995, **269**:E299–E308.

31. Carver TD, Quick AA, Teng CC, Pike AW, Fennessey PV, Hay WW Jr: Leucine metabolism in chronically hypoglycemic hypoinsulinemic growth-restricted fetal sheep. *Am J Physiol Endocrinol Metab* 1997, **272**: E107–E117.

32. Teng CC, Tjoa S, Fennessey PV, Wilkening RB, Battaglia FC: Transplacental carbohydrate and sugar alcohol concentrations and their uptakes in ovine pregnancy. *Exp Biol Med (Maywood)* 2002, **227**:189–195.

33. Aldoretta PW, Hay WW Jr: Effect of glucose supply on ovine uteroplacental glucose metabolism. *Am J Physiol Regul Integr Comp Physiol* 1999, **277**:R947–R958.

34. Brachet EA: Presence of the complete sorbitol pathway in the human normal umbilical cord tissue. *Biol Neonate* 1973, **23**:314–323.

35. Mango D, Scirpa P, Menini E: Effects of dehydroepiandrosterone and 16 alpha-hydroxydehydroepiandrosterone on the reduction of glucose to glucitol by the human placenta. *Horm Metab Res* 1976, **8**:302–307.

36. Quirk JG Jr, Bleasdale JE: Myo-inositol homeostasis in the human fetus. *Obstet Gynecol* 1983, **62**:41–44.

37. Ginsburg J, Jeacock MK: Pathways of glucose metabolism in human placental tissue. *Biochim Biophys Acta* 1964, **90**:166–168.

38. Makarewicz W, Swierczynski J: Phosphate-dependent glutaminase in the human term placental mitochondria. *Biochem Med Metab Biol* 1988, **39**:273–278.

39. Moores RR Jr, Vaughn PR, Battaglia FC, Fennessey PV, Wilkening RB, Meschia G: Glutamate metabolism in fetus and placenta of late-gestation sheep. *Am J Physiol* 1994, **267**:R89–R96.

40. Sakurai T, Takagi H, Hosoya N: Metabolic pathways of glucose in human placenta. Changes with gestation and with added 17-beta-estradiol. *Am J Obstet Gynecol* 1969, **105**:1044–1054.

41. Limesand SW, Hay WW Jr: Adaptation of ovine fetal pancreatic insulin secretion to chronic hypoglycaemia and euglycaemic correction. *J Physiol Lond* 2003, **547**:95–105.

42. Fowden AL, Comline RS: The effects of pancreatectomy on the sheep fetus in utero. *Q J Exp Physiol* 1984, **69**:319–330.

43. Philipps AF, Carson BS, Meschia G, Battaglia FC: Insulin secretion in fetal and newborn sheep. *Am J Physiol Endocrinol Metab* 1978, **235**:E467–E474.

44. Meznarich HK, Hay WW Jr, Sparks JW, Meschia G, Battaglia FC: Fructose disposal and oxidation rates in the ovine fetus. *Q J Exp Physiol* 1987, **72**:617–625.

8

Factors controlling nutrient availability to the developing fetus in ruminants

Kathrin A Dunlap[*], Jacob D Brown, Ashley B Keith and M Carey Satterfield

Abstract

Inadequate delivery of nutrients results in intrauterine growth restriction (IUGR), which is a leading cause of neonatal morbidity and mortality in livestock. In ruminants, inadequate nutrition during pregnancy is often prevalent due to frequent utilization of exensive forage based grazing systems, making them highly susceptible to changes in nutrient quality and availability. Delivery of nutrients to the fetus is dependent on a number of critical factors including placental growth and development, utero-placental blood flow, nutrient availability, and placental metabolism and transport capacity. Previous findings from our laboratory and others, highlight essential roles for amino acids and their metabolites in supporting normal fetal growth and development, as well as the critical role for amino acid transporters in nutrient delivery to the fetus. The focus of this review will be on the role of maternal nutrition on placental form and function as a regulator of fetal development in ruminants.

Keywords: Intrauterine growth restriction, Nutrient transport, Placenta, Ruminant

Introduction

It is widely accepted that maternal nutrient restriction during gestation results in offspring that are smaller at birth than counterparts from adequately fed mothers [1-4]. Sub-optimal fetal growth and development is a significant problem to livestock producers, contributing to lower productivity especially during periods of extreme nutritional hardship such as drought. The ability of the placenta to adapt to such environmental challenges influences nutrient transport to and development of the fetus. Understanding such differences will lead to elucidation of mechanisms for enhancing placental nutrient transport that will be necessary for generation of management strategies to combat fetal and neonatal loss.

In sheep, increasing the efficiency of livestock production can be more readily achieved by increasing the average number of offspring weaned than by improving growth rate or body composition [5]. Nearly one-half of all pre-weaning lamb deaths occur on the day of birth [6] with birth weight being the single greatest contributor to lamb mortality [7]. Even under moderate levels of management, death losses of more than 65% for lambs

weighing less than 4 pounds at birth have been observed [8]. In comparison, lambs born weighing between 9 and 12 pounds exhibited losses of 6.4 to 8.1%. In cattle, analyses of over 83,000 births indicated that birth weight >1.5 standard deviations below the mean, within breed and sex, doubled the likelihood for perinatal mortality in both complicated and uncomplicated pregnancies [9]. The prenatal growth trajectory of all eutherians (placental mammals) is sensitive to the direct and indirect effects of maternal nutrition at all stages between oocyte maturation and birth [10-12].

Fetal nutrient restriction

Pregnancy is a particularly sensitive period to environmental challenges that lead to suboptimal nutritional availability to the fetus due to the increased nutrient requirements for the dam and fetus. In cases of suboptimal nutrition the fetus may never be able to reach its maximal genetic potential [13]. Epidemiological studies have demonstrated links between maternal undernutrition/overnutrition and the susceptibility to chronic metabolic disease in adult offspring [14]. Metabolic syndrome has been defined as a cluster of disorders, including obesity, hyperglycemia [fasting serum glucose (>6.1 mmol/L)], hyperinsulinemia, hyperlipidemia, hypertension, and insulin resistance (impaired response of cells or tissues to

* Correspondence: kdunlap@tamu.edu
Department of Animal Science, Texas A&M University, 2471 TAMU, College Station, Texas 77843, USA

physiological concentrations of insulin) [15]. These epidemiological observations have been recapitulated in controlled experimental studies using a variety of model systems, and have collectively given rise to the concept of fetal programming [16]. Fetal programming proposes that alterations in fetal nutrition and endocrine status impacts development, permanently changes structure and metabolism and, as a result, influences an individual's susceptibility to disease [17]. Many of these changes are potentially mediated at the level of the epigenome, and are observed as stable and sometimes heritable alterations of genes through covalent modifications of DNA and core histones without changes in DNA sequences [18].

Maternal nutrition is arguably the most important and potentially manageable factor that contributes to pregnancy and maternal health [19]. Livestock species are produced in different environmental conditions. Swine are commonly produced in a confinement setting and are fed specifically formulated plant based diets (intensive systems). On the other hand ruminants, such as cattle and sheep, are more commonly managed under grazing systems. However, the quality of the grazing system is directly dependent on availability of nutrients, which is usually limited during the winter months and times of dry conditions. Thus nutrient availability may be limited for adequate growth and development as well as proper gestation and lactation [20]. The sheep, in particular, is a seasonal breeder that enters estrus in fall and early winter. Therefore, most of gestation occurs in the winter months, a time of low forage quality and limited nutrient availability [21]. In fact, Thomas and Knott reported, that without supplementation, the nutrient consumption of grazing ewes in the Western United States is often less than 50% of the National Research Council (NRC) recommendations (Council NR. Nutrient Requirements of Sheep. Washington, DC: Natl. Acad. Press; 1985.) In ruminants, IUGR remains a critical problem due to the traditional utilization of extensive forage based grazing systems which leaves the producer susceptible to large swings in nutrient quality and availability, coupled with challenges in feeding of supplemental nutrients in range based production systems [22]. Researchers and producers alike are working to find efficient and non-invasive means for preventing the continuous occurrence of IUGR and premature delivery. Thus nutritional means are an attractive option for non-invasive prevention. In order to fully utilize nutritional intervention one must understand the mechanisms by which nutrition mediates fetal growth.

Simplistically, it is easy to predict that more offspring lead to more total livestock product, and ultimately greater profit for the producer. Certainly, recent advances in the field of reproductive technologies, such as embryo transfer, have tested this theory. One would be quick to hypothesize that producing twins in cattle would be an easy way to increase profit; as the overhead costs for maintaining a single-calving cow accounts for more than 50% of the total costs of production [23]. Thus, one would assume that by doubling the amount of offspring produced by a single cow this would offer positive financial gains. However, this hypothesis does not recognize the impact of maternal reproductive health and adequate uterine space as well as management of a cow with two calves after birth. Cows that carry twins lose on average 12% body weight during the last trimester of gestation and twinning also reduces fetal growth and consequent birth weight of offspring. This could lead to decrease weight at sale. Thus, careful consideration must be demonstrated in analyzing enhanced production technologies. The sheep industry displays a similar situational paradigm. Under ideal environmental and management conditions multiple offspring pregnancies increase the prolificacy of the species. However, it is important to note, that increased number of fetuses can contribute to placental insufficiencies and thus lower birth weights [24]. As mentioned, prolificacy is an important factor in intensive sheep production, however the economic benefits of high prolificacy ewes are usually not fully exploited because multi-fetal pregnancies are associated with IUGR, increasing perinatal mortality and altering postnatal growth trajectories.

Other natural factors can affect birth weight in sheep, such as maternal age. Lambs born to relatively young ewes and relatively old ewes generally possess lower birth weights [25]. In young animals the fetus is in direct competition for nutrients with the growing dam as opposed to older ewe, which typically have reduced body condition scores, and thus may struggle to balance the nutritional needs for survival as opposed to growth. Female body condition at mating may also influence fetal development, as ewes that were lighter and thinner at mating, and thus exhibited low nutrient intake, had lambs with an associated 13% decreased birth weight [26]. Seasonality can also be an indicator of birth weight. Lambs born in the fall and summer are generally lighter than those born in the spring and winter; which can possibly be explained by alterations in melatonin and seasonal effects on gestation. Additionally, gestation occurring in the warmer months where ewes shuttle blood flow to the periphery of the body to dissipate heat rather than to the core could ultimately decrease placental blood flow and placental growth [27].

Altitude is another factor that affects birth weights. Offspring that are born at high altitude may experience hypobaric hypoxia and be lighter than those born at a lower altitude. Heat stress is a primary concern in cattle production, particularly relevant to the dairy industry.

Heat stress has been shown to reduce placental weight and mass [28-30] while also negatively impacting the hormonal properties of the placenta, having decreased levels of placental lactogen and pregnancy associated glycoprotein [29,31].

All of these situations can ultimately lead to IUGR offspring who possess reduced survival ability [25]. Ultimately, the hypothesis is that exposure to any of these challenges may cause alterations in fetal growth stemming from a hindrance of the placental growth trajectory.

Placental formation

The placenta mediates the transport of nutrients, gases, and waste products of their metabolism between the maternal and fetal circulation [22]. Placental growth precedes fetal growth and its development is crucial for optimal fetal growth [32-34]. The placenta has two evolutionary roles conserved among all species. The first is it generates a large surface area for nutrient exchange by the epithelial barrier and fetal blood vessels, all of which make up the chorioallantoic placenta. The second basic function is that the trophoblast cells interact closely with the uterus, which produces histotroph that provides key growth factors, nutrients, and immune cell regulators necessary to facilitate blood flow and nutrient delivery to the fetus [35]. In addition to the presence of secretory uterine glands, the ruminant uterus is also home to a number of aglandular areas of stroma that are covered by a single layer of luminal epithelial cells, which are termed caruncles. The number of caruncles varies between species, with sheep having approximately 100 and deer less than 10 and cattle ranging between 75–125 [32,36]. During pregnancy the carunclar crypts will interdigitate with the fetal cotyledonary villi to give rise to a feto-maternal structure known as a placentome, which is the structure responsible for high-throughput nutrient transfer between the uterus and the fetus [37]. Failure of placentome development results in loss of the fetus [38], because they provide the primary source of hematotrophic nutrition as maternal and fetal blood vessels are in very close proximity for exchanging oxygen and micronutrients [39]. Vascularization of the placentomes is established early in gestation, which is to be expected as the majority of placental growth occurs during the first half of pregnancy [34]. Increased microvascular density, interdigitation and alterations in villous shape are potential mechanisms by which the placentome adapts to the maternal environment to meet the growing demands of the fetus [40]. As these structures are responsible for an increasing percentage of blood flow throughout gestation it is not surprising that they continue to experience modest changes in capillary area density from mid to late gestation [41,42]. As stated, failure of placentome formation results in loss of pregnancy, however a

surgical reduction in the number of caruncles results in an increase in the average size of placentomes [43]. This may prove to be a compensatory mechanism by which the uterus and placentomes work to provide support required for fetal development. In the bovine uterus a relatively small number (20) of cotyledons are first visible on the fetal placental membrane as early as day 37 of pregnancy, however that number has tripled by day 50 and the beginning of a caruncular- cotyledonary interrelationship is clearly present [44,45]. By day 90 of pregnancy there are greater than 100 placentomes present in the bovine uterus, each possessing a characteristic mushroom-like shape, rooted with a stalk like structure stemming from the original caruncle [46]. Similarly, by gestational day 40 in the sheep and the goat, there is maximal juxtaposition of endometrial and placental microvasculatures [35,47]. As pregnancy progresses and the placentome continues to develop, the principal layers of the fetal capillary, trophoblast epithelium, crypt lining, and maternal capillary remain distinctive [47]. The general morphological characteristics of the placentomes are maintained as pregnancy progresses with the degree of interdigitation of the tissues becoming greater.

A great deal of research has been conducted in the ewe with respect to the interaction with nutrition and placental development and morphology. Recognizing that placental development is a major indicator of birth weight, researchers have found an ever occurring commonality among studies: the morphology of the placenta is altered under nutrient scarce conditions, with the most notable change being increased development of the fetal cotyledon. This change is thought to reflect an adaptive compensatory response to maximize transplacental exchange and thus fetal growth [48-50]. In ovine models of nutritionally induced fetal growth restriction there is an inverse relationship between nutritional plane and fetal vascular development [3,51,52]. These morphological changes are expected to diminish fetal development through altered availability of hematotrophic support.

Placental blood flow

During pregnancy, transport of nutrients from the mother to the fetus is predominantly dependent on uterine blood flow. Indeed, fetal and placental weights and uterine blood flow are highly correlative. In addition, factors that alter fetal weight such as genotype, maternal nutrient intake, environmental heat stress, high altitude, and fecundity, result in parallel alterations in placental weight and uterine blood flow [53]. In the sheep, uterine and placental blood flows progress in a non-linear pattern with an early rise in blood flow between days 40–70 as the placental cotyledon develops and a second increase in flow between days 120–140 [54]. This early rise in blood flow is likely critical in the development of a functional, efficient placenta. Importantly, 85% of blood flow to the gravid uterus is directed to the placentome [55]. A critical regulator for vascular permeability, blood flow, and capillary growth is the angiogenic factor, vascular endothelial growth factor (VEGF). In the developing bovine placentome, VEGF and its receptors FLT1 and KDR are abundantly expressed [56]. Specifically VEGF is present in the trophoblast giant cells, fetal endothelia and maternal endothelia, which also expresses high amounts of FLT and KDR antigen. The sheep exhibits a similar pattern of expression with placental VEGF mRNA present in both the fetal and maternal endothelia as well as the chorionic epithelium [57]. Placental expression of VEGF and the receptors in a model of placental insufficiency-intrauterine growth retardation (PI-IUGR) indicates that placental growth, as indicated by weight, ceases at day 90 of pregnancy in the sheep and declines to term [58]. However, placental transport capacity continues to increase to meet the requirements of the exponentially growing fetus as evidenced by the late rise in uterine blood flow during the latter stages of gestation [59]. Importantly, the ability to respond to the increased demand for nutrients during late gestation can be undermined by poor placental development earlier in gestation. In the sheep, maternal undernutrition results in a 14% decrease in caruncular capillary area density in late pregnancy [53]. Midgestational nutrient restriction in the sheep results in a decreased abundance of mRNA encoding VEGF in placentomes [60]. These morphological and vascular changes reduce hematotrophic support and diminish fetal development. Changes in microvascular density and vessel diameter are also commonly found in compromised pregnancy [51,61,62].

Placental fluids
The ruminant placenta is comprised of two distinct fluid-filled membranes: the amnion and allantois, which act as reservoirs for nutrients that are essential for fetal growth and development [63-65]. Maternal nutrient restriction during mid pregnancy in sheep results in decreased allantoic and amniotic fluid volume [60]. In sheep, fetal growth retardation, induced by maternal nutrient restriction is associated with decreased quantities of serine, α-amino acids, arginine family amino acids, and branched chain amino acids in both amniotic and allantoic fluid [1,2]. Arginine and its precursor, citrulline, are abundant in ovine uterine fluids [66]. Swallowing of amniotic fluid provides a rich source of nutrients for utilization by the fetal intestine and other tissues [67]. Prevention of amniotic fluid entry into the small intestine by esophageal ligation results in fetal IUGR in sheep [63]. Allantoic fluid is primarily derived from placental transport mechanisms [68] and in sheep, the volume

increases in early gestation (days 25–40) before decreasing briefly (days 40–70) prior to increasing steadily until near term (day 140) [32]. It is now clear that allantoic fluid nutrients may be absorbed by the allantoic epithelium into the fetal-placental circulation and utilized by fetal-placental tissues [68]. The allantoic fluid during early to mid-gestation in the sheep is an abundant source of arginine and polyamines [69]. Allantoic fluid is also a reservoir for both glucose and fructose in the sheep, and both concentrations and total levels of fructose are much greater than that of glucose and change significantly with day of gestation [32].

Placental nutrient transport
Placental nutrient transport is required for fetal growth [70,71]. Realizing that most fetal growth occurs during the final third of pregnancy, the development of the placental networks regulating transfer must precede that time point [40]. Although placental blood flow is imperative for optimal nutrient delivery, expression and/or activity of specific transporters is the rate limiting step for delivery of many nutrients, including glucose and amino acids [72]. Numerous environmental factors, such as, under- and over- nutrition, hypoxia, heat stress, and hormone exposure may regulate activity of both glucose and amino acid transporters in the placenta [73-79]. In other species, compromised pregnancies are associated with specific alterations in transporter availability and function [77,80]. In rats maternal dietary protein deprivation results in down regulation of placental amino acid transport systems prior to the emergence of fetal growth restriction [81]. In sheep, maternal infusion of mixed amino acids results in variable changes in umbilical uptake depending on the amino acid [71] highlighting a complex system of amino acid transport likely involving transporter availability, capacity, promiscuity, affinity, and competition. Sheep models of placental insufficiency induced by heat stress, indicate that umbilical uptake of essential amino acids is reduced, however fetal amino acid concentrations are unaltered compared to controls [82]. To date, the literature regarding the effect of maternal nutrient restriction on placental transporter availability and function is limited.

Select nutrients and fetal growth
Fetal growth and metabolism is dependent upon glucose as it is the primary energy substrate for the placenta and fetus and supplied by the maternal bloodstream [83]. As the fetus grows, so does its rate of absolute glucose utilization and thereby the placental rate of glucose transfer must increase accordingly. Changes in maternal glucose levels and/or rate of placental transfer require the fetus to adapt metabolically to maintain consistent levels for its trajectory of development [83]. Fructose,

while not as intensely investigated in the ruminant placenta, also appears to play a role in pregnancy in the sheep, perhaps as a mediator of the mammalian target of rapamycin (MTOR) cell signaling pathway [32,84,85].

It is established that amino acids play a vital role in development of the conceptus (embryo/fetus and associated placental membranes). In addition to serving as building blocks for tissue protein synthesis, amino acids function as antioxidants, regulators of hormone secretion, major fuels for fetal growth, and cell signaling molecules [86,87]. Amino acids are also essential precursors for the synthesis of non-protein substances with biological importance, including nitric oxide, polyamines, neurotransmitters, amino sugars, purine and pyrimidine nucleotides, creatine, carnitine, porphyrins, melatonin, melanin, and sphingolipids [88,89]. Nitric oxide (NO), a product of arginine catabolism, plays a crucial role in regulating placental angiogenesis and fetal-placental blood flow during gestation [90-92]. Polyamines (polycationic molecules synthesized from ornithine) regulate gene expression, signal transduction, ion channel function, and DNA and protein synthesis, as well as cell proliferation, differentiation, and function [87]. It should be noted that a comprehensive series of studies investigating the role of select nutrients and their transporters in the developing ruminant conceptus during the early peri-implantation period have been published [66,93-98], however discussion of these studies is beyond the scope of this review.

Studies from our laboratories have previously shown that maternal nutrient restriction results in the reduction of amino acids and polyamines in fetal umbilical venous plasma as well as amniotic and allantoic fluid [1,2]. Interestingly, inhibition of ornithine decarboxylase (ODC), which converts ornithine to putrescine (a polyamine), during mid-pregnancy in mice results in embryonic growth arrest and impaired development of the yolk sac and placentae [99], while gene ablation results in embryonic lethality prior to gastrulation [100]. In rats, inhibition of ODC activity results in fetal IUGR as well as a reduction in placental weight and placental DNA content [101]. Interestingly, despite a wealth of information highlighting the importance of polyamines in physiology the transport mechanisms of these molecules have not been elucidated in mammals. As a first step in ameliorating IUGR using intervention strategies we administrated sildenafil citrate (Viagra) to pregnant sheep, which resulted in increased fetal growth in both nutrient-restricted and well fed controls [2]. This increase in fetal growth was coordinate with increased concentrations of amino acids and polyamines in the amniotic and allantoic fluids as well as the fetal circulation [2]. As sildenafil citrate is not a viable strategy for production agriculture we performed subsequent studies with nutritional intervention. Intravenous administration

of arginine similarly prevented IUGR in underfed ewes and increased percentage of lambs born alive in ewes carrying multiple fetuses [102-104]. These data indicate that amino acids and polyamines are essential for enhanced fetal growth during maternal undernutrition. It is hypothesized that arginine could be increasing both NO for vasodilation as well as providing a substrate for polyamine production. Further it may play an essential role in activation of the MTOR cell signaling pathways, stimulation of placental growth and mediation of placental blood flow [105].

Localization of glucose transporters within the placentome

A multitude of membrane transporters, primarily those belonging to the solute carrier (SLC) superfamily, may be found on the placenta to facilitate nutrient transport to the fetus. Classification of these transporters is based on their structure and substrate preference [97,106,107]. Transporter localization at both the maternal and fetal surfaces for transport across plasma membranes is essential in the delivery of glucose, amino acids, polyamines, and other nutrients from maternal circulation to umbilical circulation [108]. Work in various species, such as sheep, humans, and rodents, has illustrated the necessity for placental nutrient transporters throughout gestation [66,71,77,81,95-97,106,108,109].

A comprehensive microscopy study of ruminant placentomes (including: sheep, goats, cattle and deer) revealed the specific localization of two primary glucose transporter isoforms, SLC2A1 and SLC2A3 [110]. SLC2A1 is expressed on the inner and outermost membranes of the placenta between the fetal and maternal endothelia while SLC2A3 is expressed only by trophoblast microvilli, suggesting that in order for successful glucose transport between the mother and the fetus both SLC2A1 and SLC2A3 must be utilized sequentially [110]. While localization remains consistent, there is a temporal shift in the abundance of expression of SLC2A1 and SLC2A3 in the sheep placenta, with SLC2A3 levels increasing late in gestation [111]. The change in relative levels of glucose transporter expression suggests that SLC2A3 may play a critical role in mediation of glucose transport late in ruminant gestation. Interestingly mid-gestational maternal nutrient restriction does not influence relative abundance of SLC2A1 nor SLC2A3 mRNA in the sheep placenta [60]. Lactation also places a strain on the maternal metabolic system. In the case of lactating dairy cows, blood glucose levels are lower in comparison to their non-lactating contemporaries [112]. Regardless of lactational status, expression of SLC2A1 and SLC2A3 was most abundant in the placenta as opposed to liver or endometrial tissues in early gestation, with relative levels of SLC2A3 in the placenta progressing with pregnancy [112].

Expression of amino acid transporters within the placentome

Although placental blood flow is imperative for optimal nutrient delivery, expression and/or activity of specific transporters is the rate limiting step for delivery of many nutrients, including glucose and amino acids [70,72]. In both humans and rats, compromised pregnancies are associated with specific alterations in transporter availability and function [77,80]. In rats, maternal dietary protein deprivation results in down regulation of placental amino acid transport systems prior to the emergence of fetal growth restriction [77,81]. The sheep has been a well-utilized model for investigation of amino acid transporter profile in early gestation [66,95,96] as well as manipulated models of pregnancy [113,114]. The observations in the literature support results from our laboratory investigating changes in placentome amino acid transporter expression over the duration of pregnancy as well as in response to maternal nutrient restriction. Using an experimental model developed in our laboratory [102] and methods for quantification of steady-state mRNA levels previously published [102] we investigated expression of the amino acid transporters shown in Table 1. Our results show that stage of pregnancy impacts relative expression of multiple amino acid transporters. Specifically, steady-state mRNA levels of the large neutral amino acid transporter, SLC7A5, (Figure 1) demonstrate that expression is impacted by day of pregnancy rather than maternal nutrient status. Levels of SLC7A5 are higher ($P < 0.05$) at day 50 than at days 100

and 125, with the most abundant level in the nutrient restricted ewes occurring at day 75. Steady-state mRNA levels of cationic amino acid transporters are shown in Figure 2. SLC7A2 mRNA levels increased ($P < 0.05$) in placentomes from days 50 to 100 and were not impacted by maternal nutrient restriction. SLC7A6 mRNA levels were higher ($P < 0.05$) in placentomes from days 75 and 100 of pregnancy as compared to days 50 and 125 (Figure 2). SLC7A7, SLC7A8, and SLC7A1 mRNA levels in placentomes did not differ ($P > 0.10$) over the course of gestation or in response to maternal nutrient restriction. Unlike the previously described amino acid transporters a sodium coupled neutral amino acid transporter (SNAT) SLC38A2 exhibited a day by diet interaction (Figure 3). Specifically, SLC38A2 mRNA levels in placentomes of well- fed ewes increased from days 50 to 75, decreased to day 100 and then increased again to day 100 ($P < 0.05$). In nutrient restricted ewes there was no difference thoughout the course of gestation. In contrast, SLC38A4 mRNA levels in placentomes were not different ($P > 0.10$) between groups.

These data suggest that while there is a marked temporal change in expression the impact of maternal nutrition on transporter expression is more subtle. We have observed in previous studies that the existence of a sub-population of animals that support normal rates of fetal growth despite maternal nutrient restriction. This would suggest the ability of the placenta to adapt, in certain cases, to maternal nutrient restriction. Specifically that

Table 1 Primers utilized for quantitative real-time PCR analysis

Target[1]	Forward/revers primers, (5'→3')[2]	Length of amplicon, bp	GenBank accession No[3]
SLC7A1	CCTAGCGCTCCTGGTCATCA	56	AF212146
	GGGCGTCCTTGCCAAGTA		
SLC7A2	GCAGAGCAGCGCTGTCTTT	62	XM_002698665
	ACTGTCCAGAGTGACGATTTTCC		
SLC7A5	GGTGAACCCTGGTACGAATTTAGT	64	NM_174613
	TCCACGCTCGAGAGGTATCTG		
SLC7A6	CATTTGTGAACTGCGCCTATGT	72	NM_001075937
	CCAGGACCTTGGCATAAGTGA		
SLC7A7	TCAGGCTTGCCCTTCTACTTCT	64	NM_001075151
	GGAGCCAAAGAGGTCGTTTG		
SLC7A8	GGCCATGATCCACGTGAAG	65	NM_001192889
	GGGTGGAGATGCATG⁻GAAGA		
SLC38A2	CAGCTATAGTTCCAACAGCGACTTC	77	NM_001082424
	CATCGGCATAATGGCTTTTCA		
SLC38A4	TGCTTCATGCTTACAGCAAAGTG	63	NM_001205943
	CAGCCAGGCGTACCATGAG		

[1]The amplification target.
[2]The forward and reverse DNA oligos used in the amplification of the target. Forward and reverse primers do not necessarily indicate the in vivo direction of transcription.
[3]The accession number to the ovine or bovine sequence that was used during primer design.

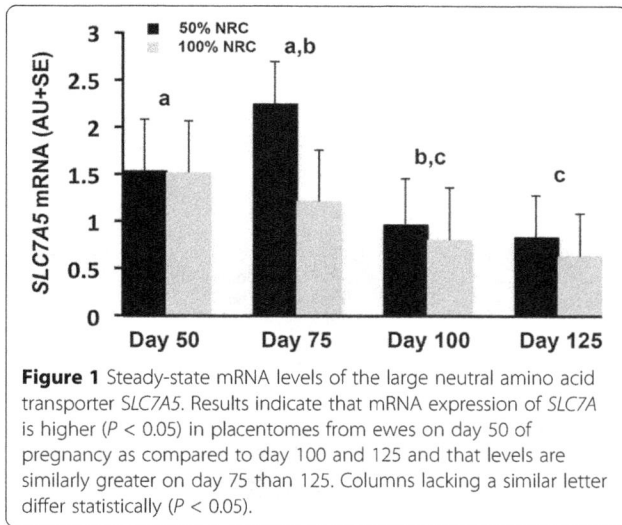

Figure 1 Steady-state mRNA levels of the large neutral amino acid transporter SLC7A5. Results indicate that mRNA expression of SLC7A is higher ($P < 0.05$) in placentomes from ewes on day 50 of pregnancy as compared to day 100 and 125 and that levels are similarly greater on day 75 than 125. Columns lacking a similar letter differ statistically ($P < 0.05$).

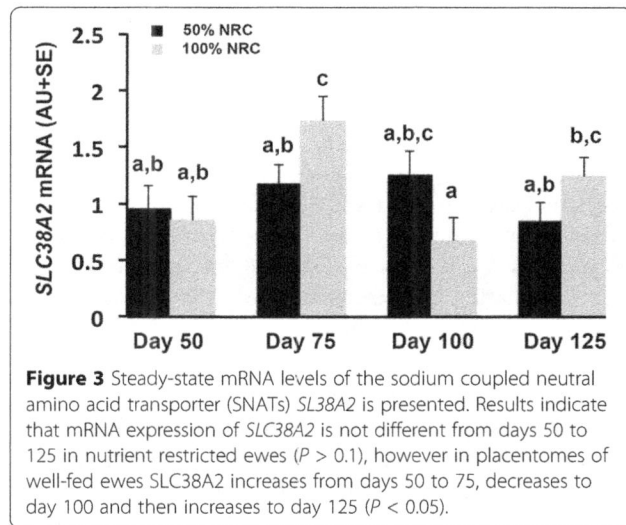

Figure 3 Steady-state mRNA levels of the sodium coupled neutral amino acid transporter (SNATs) SL38A2 is presented. Results indicate that mRNA expression of SLC38A2 is not different from days 50 to 125 in nutrient restricted ewes ($P > 0.1$), however in placentomes of well-fed ewes SLC38A2 increases from days 50 to 75, decreases to day 100 and then increases to day 125 ($P < 0.05$).

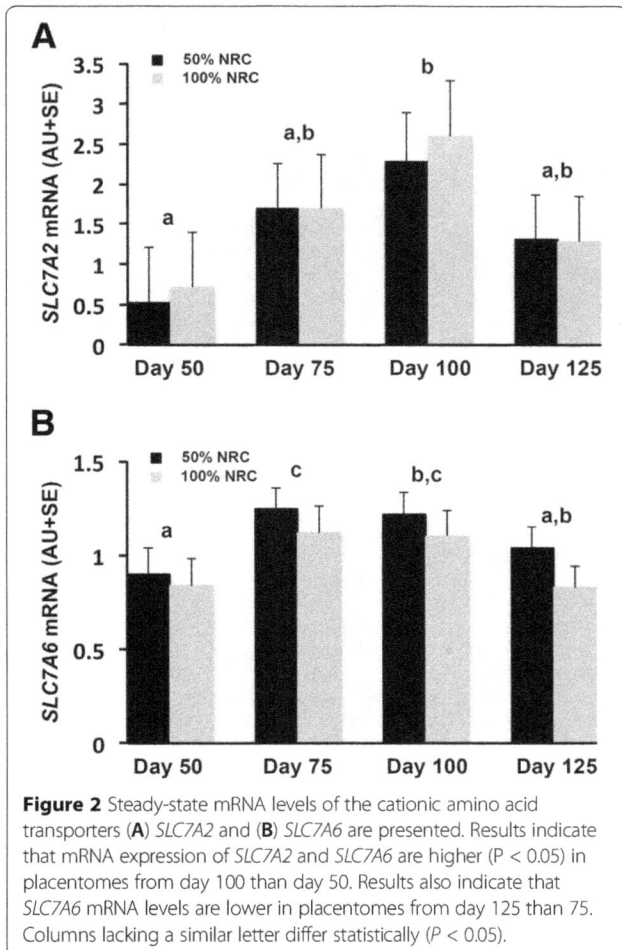

Figure 2 Steady-state mRNA levels of the cationic amino acid transporters (**A**) SLC7A2 and (**B**) SLC7A6 are presented. Results indicate that mRNA expression of SLC7A2 and SLC7A6 are higher ($P < 0.05$) in placentomes from day 100 than day 50. Results also indicate that SLC7A6 mRNA levels are lower in placentomes from day 125 than 75. Columns lacking a similar letter differ statistically ($P < 0.05$).

there is a sub-population of animals that even in cases of maternal undernutrition undergoes an adaptive placental response that supports development of fetal weight at a level similar to well fed controls. Indeed, based on prior observations, our laboratory conducted a series of studies to identify a population of IUGR and non-IUGR offspring from a similar cohort of nutrient-restricted ewes as a first step to assess adaptive mechanisms of placental nutrient transport. Briefly, we have observed that the range of fetal weights is greater from ewes receiving only 50% of their NRC requirements than for ewes fed to meet 100% of their nutrient requirements, when confounding variables such as maternal size, genotype, and fecundity are controlled (Satterfield and Dunlap, unpublished results). Suggesting that there is a mechanism by which the placenta adapts to maternal nutrient restriction to support normal versus restricted fetal growth in an outbred population as opposed to laboratory animals with less heterogeneity. These results are similar to those observed in beef cattle whereby maternal nutrient restriction from early to mid gestation resulted in two distinct groups of IUGR and non-IUGR fetuses at mid gestation [115]. Those IUGR pregnancies were also characterized by smaller cotyledonary weights and reduced placentomal surface area [115] which is similar to results of our studies, and further supporting a large body of literature indicating that placental weight is positively correlated with fetal weight. Further supporting the impact of population variation in maternal response to undernutrition are studies conducted using the heterogenous population of Western white-faced ewes adapted to harsh climates. In these ewes fetal growth to day 78 was not affected by maternal nutrient restriction [49]. Amino acid and polyamine concentrations in the fetal circulation of these sheep are maintained despite maternal nutrient restriction [116].

Assessment of amino acid transporter expression comparing the IUGR to non-IUGR pregnancies identified a number of amino acid transporters that were differentially expressed between the two groups, including *SLC7A6, SLC7A7, SLC7A8, SLC38A2*. Expression of those transporters were greater in IUGR than non-IUGR pregnancies (Satterfield and Dunlap, unpublished observations). These data indicate that while amino acids and polyamines are necessary for fetal growth the mechanisms regulating their transport and utilization may vary greatly amongst populations.

Conclusions

Optimal fetal growth requires the efficient delivery of nutrients to the fetus and corresponding removal of waste products associated with fetal metabolism and growth. The mechanisms by which fetal nutrient delivery is achieved are well accepted and include hematotrophic nutrition, histotrophic nutrition (secretions emanating from the uterine glands), placental metabolism of substrates for use by the fetus, and the activity of nutrient transport systems within the placenta. These mechanisms work in concert to provide sufficient quantities of nutrients to the fetus for growth. Perturbations in any of these mechanisms have significant impacts on the growth and well-being of the fetus. Collectively, results of the presented studies coupled with data from the existing literature suggests that the placenta is a dynamic organ, whose form and function can be regulated by a myriad of factors. Further, results support previous findings from our laboratory and others, and highlight roles for amino acids and their metabolites in supporting normal fetal growth and development, as well as the critical role for amino acid transporters in nutrient delivery to the fetus. Further research is needed to address a series of mechanistic questions in order to increase understanding of appropriate nutrient delivery to the fetus. In the case of the adaptive ruminant placenta, is the difference in fetal growth a response of increased uterine blood flow, an alteration in early placental development, an adaptive recruitment of additional nutrient transporters or an increase in their activity, or some combination of these factors? It is important to utilize such models for future investigation of placental adaptation employed in an effort to increase nutrient delivery to the fetus despite limited maternal nutritional intake.

Abbreviations
IUGR: Intrauterine growth restriction; NRC: National Research Council; VEGF: Vascular endothelial growth factor; FLT: Vascular endothelial growth factor receptor −1; KDR: Vascular endothelial growth factor receptor −2; PI-IUGR: Placental insufficiency intrauterine growth restriction; NO: Nitric oxide; ODC: Ornithine decarboxylase; MTOR: Mammalian target of rapamycin; SLC2A1: Solute carrier family 2 member 1; SLC2A3: Solute carrier family 2 member 3; SLC7A2: Solute carrier family 7 member 2; SLC7A5: solute carrier family 7 member 5; SLC7A6: Solute carrier family 7 member 6;

SLC7A7: Solute carrier family 7 member 7; SLC7A8: Solute carrier family 7 member 8; SNAT: Sodium coupled neutral amino acid transporter; SLC38A1: Solute carrier family 38 member 4; SLC38A4: Solute carrier family 38 member 4.

Competing interests
The authors declare that they have no competing interests.

Authors' contributions
KAD and MCS wrote the review. KAD and MCS designed the study. ABK and JDB performed analysis. MCS, KAD, and ABK read and approved the final manuscript. All authors read and approved the final manuscript.

References
1. Kwon H, Ford SP, Bazer FW, Spencer TE, Nathanielsz PW, Nijland MJ, et al. Maternal nutrient restriction reduces concentrations of amino acids and polyamines in ovine maternal and fetal plasma and fetal fluids. Biol Reprod. 2004;71:901–8.
2. Satterfield MC, Bazer FW, Spencer TE, Wu G. Sildenafil citrate treatment enhances amino acid availability in the conceptus and fetal growth in an ovine model of intrauterine growth restriction. J Nutr. 2010;140:251–8.
3. Vonnahme KA, Hess BW, Hansen TR, McCormick RJ, Rule DC, Moss GE, et al. Maternal undernutrition from early- to mid-gestation leads to growth retardation, cardiac ventricular hypertrophy, and increased liver weight in the fetal sheep. Biol Reprod. 2003;69:133–40.
4. Scheaffer AN, Caton JS, Redmer DA, Reynolds LP. The effect of dietary restriction, pregnancy, and fetal type in different ewe types on fetal weight, maternal body weight, and visceral organ mass in ewes. J Anim Sci. 2004;82:1826–38.
5. Dickerson GE. Animal size and efficiency: basic concepts. Anim Prod. 1978;27:367–79.
6. Huffman EM, Kirk JH, Pappaioanou M. Factors associated with neonatal lamb mortality. Theriogenology. 1985;24:163–71.
7. Dwyer CM. The welfare of the neonatal lamb. Small Rumin Res. 2008;76:31–41.
8. Shelton M. Relation of birth weight to death losses and to certain productive characters of fall-born lambs. J Anim Sci. 1964;23:355–9.
9. Azzam SM, Kinder JE, Nielsen MK, Werth LA, Gregory KE, Cundiff LV, et al. Environmental effects on neonatal mortality of beef calves. J Anim Sci. 1993;71:282–90.
10. Robinson JJ, Sinclair KD, McEvoy TG. Nutritional effects on foetal growth. Anim Sci. 1999;68:315–31.
11. Rehfeldt C, Nissen PM, Kuhn G, Vestergaard M, Ender K, Oksbjerg N. Effects of maternal nutrition and porcine growth hormone (pGH) treatment during gestation on endocrine and metabolic factors in sows, fetuses and pigs, skeletal muscle development, and postnatal growth. Domest Anim Endocrinol. 2004;27:267–85.
12. Ferguson JD. Nutrition and reproduction in dairy herds. Vet Clin Food Anim. 2005;21:325–47.
13. Gluckman PD, Beedle AS, Hanson MA, Low FM. Human growth: evolutionary and life history perspectives. Nestle Nutr Inst Workshop Ser. 2013;71:89–102.
14. Maulik N, Maulik G. Nutrition, epigenetic mechanisms, and human disease. New York: CRC Press; 2011.
15. Jobgen WS, Fried SK, Fu WJ, Meininger CJ, Wu G. Regulatory role for the arginine-nitric oxide pathway in metabolism of energy substrates. J Nutr Biochem. 2006;17:571–88.
16. Seki Y, Williams L, Vuguin PM, Charron MJ. Minireview: epigenetic programming of diabetes and obesity: animal models. Endocrinology. 2012;153:1031–8.
17. Low FM, Gluckman PD, Hanson MA. Developmental plasticity and epigenetic mechanisms underpinning metabolic and cardiovascular diseases. Epigenomics. 2011;3:279–94.
18. Evertts AG, Zee BM, Garcia BA. Modern approaches for investigating epigenetic signaling pathways. J Appl Physiol (1985). 2010;109:927–33.
19. Picciano MF. Pregnancy and lactation: physiological adjustments, nutritional requirements and the role of dietary supplements. J Nutr. 2003;133:1997S–2002.
20. Paterson J, Funston R, Cash D. Forage quality influences beef cow performance. Feedstuffs. 2002;74:11–2.

21. Hoaglund CM, Thomas VM, Petersen MK, Kott RW. Effects of supplemental protein source and metabolizable energy intake on nutritional status in pregnant ewes. J Anim Sci. 1992;70:273–80.

22. Wu G, Bazer FW, Wallace JM, Spencer TE. Board-invited review: intrauterine growth retardation: implications for the animal sciences. J Anim Sci. 2006;84:2316–37.

23. Guerra-Martinez P, Dickerson GE, Anderson GB, Green RD. Embryo-transfer twinning and performance efficiency in beef production. J Anim Sci. 1990;68:4039–50.

24. Gootwine E. Variability in the rate of decline in birth weight as litter size increases in sheep. Anim Sci. 2005;81:393–8.

25. Gootwine E, Spencer TE, Bazer FW. Litter-size-dependent intrauterine growth restriction in sheep. Animal. 2007;1:547–64.

26. McCrabb GJ, Hosking BJ, Egan AR. Changes in the maternal body and feto-placental growth following various lengths of feed restriction during mid-pregnancy in sheep. Aust J Agr Res. 1992;43:1429–40.

27. Kawamura T, Gilbert RD, Power GG. Effect of cooling and heating on the regional distribution of blood flow in fetal sheep. J Dev Physiol. 1986;8:11–21.

28. Collier RJ, Doelger SG, Head HH, Thatcher WW, Wilcox CJ. Effects of heat stress during pregnancy on maternal hormone concentrations, calf birth weight and postpartum milk yield of Holstein cows. J Anim Sci. 1982;54:309–19.

29. Bell AW, McBride BW, Slepetis R, Early RJ, Currie WB. Chronic heat stress and prenatal development in sheep: I. Conceptus growth and maternal plasma hormones and metabolites. J Anim Sci. 1989;67:3289–99.

30. Tao S, Dahl GE. Invited review: heat stress effects during late gestation on dry cows and their calves. J Dairy Sci. 2013;96:4079–93.

31. Thompson IM, Tao S, Branen J, Ealy AD, Dahl GE. Environmental regulation of pregnancy-specific protein B concentrations during late pregnancy in dairy cattle. J Anim Sci. 2013;91:168–73.

32. Bazer FW, Spencer TE, Thatcher WW. Growth and development of the ovine conceptus. J Anim Sci. 2012;90:159–70.

33. Gootwine E. Placental hormones and fetal-placental development. Anim Reprod Sci. 2004;82–83:551–66.

34. Reynolds LP, Borowicz PP, Vonnahme KA, Johnson ML, Grazul-Bilska AT, Redmer DA, et al. Placental angiogenesis in sheep models of compromised pregnancy. J Physiol. 2005;565:43–58.

35. Cross JC, Baczyk D, Dobric N, Hemberger M, Hughes M, Simmons DG, et al. Genes, development and evolution of the placenta. Placenta. 2003;24:123–30.

36. Furukawa S, Kuroda Y, Sugiyama A. A comparison of the histological structure of the placenta in experimental animals. J Toxicol Pathol. 2014;27:11–8.

37. Mott JC. Control of the foetal circulation. J Exp Biol. 1982;100:129–46.

38. Mellor DJ, Mitchell B, Matheson IC. Reductions in lamb weight caused by pre-mating carunclectomy and mid-pregnancy placental ablation. J Comp Pathol. 1977;87:629–33.

39. Reynolds LP, Biondini ME, Borowicz PP, Vonnahme KA, Caton JS, Grazul-Bilska AT, et al. Functional significance of developmental changes in placental microvascular architecture. Endothelium. 2005;12:11–9.

40. Borowicz PP, Arnold DR, Johnson ML, Grazul-Bilska AT, Redmer DA, Reynolds LP. Placental growth throughout the last two thirds of pregnancy in sheep: vascular development and angiogenic factor expression. Biol Reprod. 2007;76:259–67.

41. Vonnahme KA, Zhu MJ, Borowicz PP, Geary TW, Hess BW, Reynolds LP, et al. Effect of early gestational undernutrition on angiogenic factor expression and vascularity in the bovine placentome. J Anim Sci. 2007;85:2464–72.

42. Reynolds LP, Borowicz PP, Caton JS, Vonnahme KA, Luther JS, Buchanan DS, et al. Uteroplacental vascular development and placental function: an update. Int J Dev Biol. 2010;54:355–66.

43. Meyer KM, Koch JM, Ramadoss J, Kling PJ, Magness RR. Ovine surgical model of uterine space restriction: interactive effects of uterine anomalies and multifetal gestations on fetal and placental growth. Biol Reprod. 2010;83:799–806.

44. Mossman HA. Vertebrate fetal membranes. New Brunswick, NJ: Rutgers University Press; 1987.

45. Greenstein JS, Murray RW, Foley RC. Observations on the morphogenesis and histochemistry of the bovine preattachment placenta between 16 and 33 days of gestation. Anat Rec. 1958;132:321–41.

46. Pfarrer C, Ebert B, Miglino MA, Klisch K, Leiser R. The three-dimensional feto-maternal vascular interrelationship during early bovine placental development: a scanning electron microscopical study. J Anat. 2001;198:591–602.

47. Lawn AM, Chiquoine AD, Amoroso EC. The development of the placenta in the sheep and goat: an electron microscope study. J Anat. 1969;105:557–78.

48. Luther JS, Redmer DA, Reynolds LP, Wallace JM. Nutritional paradigms of ovine fetal growth restriction: implications for human pregnancy. Hum Fertil (Camb). 2005;8:179–87.

49. Vonnahme KA, Hess BW, Nijland MJ, Nathanielsz PW, Ford SP. Placentomal differentiation may compensate for maternal nutrient restriction in ewes adapted to harsh range conditions. J Anim Sci. 2006;84:3451–9.

50. Osgerby JC, Wathes DC, Howard D, Gadd TS. The effect of maternal undernutrition on the placental growth trajectory and the uterine insulin-like growth factor axis in the pregnant ewe. J Endocrinol. 2004;182:89–103.

51. Steyn C, Hawkins P, Saito T, Noakes DE, Kingdom JC, Hanson MA. Undernutrition during the first half of gestation increases the predominance of fetal tissue in late-gestation ovine placentomes. Eur J Obstet Gynecol Reprod Biol. 2001;98:165–70.

52. Zhu MJ, Du M, Hess BW, Nathanielsz PW, Ford SP. Periconceptional nutrient restriction in the ewe alters MAPK/ERK1/2 and PI3K/Akt growth signaling pathways and vascularity in the placentome. Placenta. 2007;28:1192–9.

53. Reynolds LP, Caton JS, Redmer DA, Grazul-Bilska AT, Vonnahme KA, Borowicz PP, et al. Evidence for altered placental blood flow and vascularity in compromised pregnancies. J Physiol. 2006;572:51–8.

54. Rosenfeld CR, Morriss Jr FH, Makowski EL, Meschia G, Battaglia FC. Circulatory changes in the reproductive tissues of ewes during pregnancy. Gynecol Invest. 1974;5:252–68.

55. Makowski EL, Meschia G, Droegemueller W, Battaglia FC. Measurement of umbilical arterial blood flow to the sheep placenta and fetus in utero. Distribution to cotyledons and the intercotyledonary chorion. Circ Res. 1968;23:623–31.

56. Pfarrer CD, Ruziwa SD, Winther H, Callesen H, Leiser R, Schams D, et al. Localization of vascular endothelial growth factor (VEGF) and its receptors VEGFR-1 and VEGFR-2 in bovine placentomes from implantation until term. Placenta. 2006;27:889–98.

57. Regnault TR, Orbus RJ, de Vrijer B, Davidsen ML, Galan HL, Wilkening RB, et al. Placental expression of VEGF, PlGF and their receptors in a model of placental insufficiency-intrauterine growth restriction (PI-IUGR). Placenta. 2002;23:132–44.

58. Alexander G. Studies on the placenta of the sheep (Ovis Aries L.). Effect of surgical reduction in the number of caruncles. J Reprod Fertil. 1964;7:307–22.

59. Reynolds LP, Redmer DA. Utero-placental vascular development and placental function. J Anim Sci. 1995;73:1839–51.

60. McMullen S, Osgerby JC, Milne JS, Wallace JM, Wathes DC. The effects of acute nutrient restriction in the mid-gestational ewe on maternal and fetal nutrient status, the expression of placental growth factors and fetal growth. Placenta. 2005;26:25–33.

61. Thomas DM, Clapp JF, Shernce S. A foetal energy balance equation based on maternal exercise and diet. J R Soc Interface. 2008;5:449–55.

62. Mayhew TM, Ohadike C, Baker PN, Crocker IP, Mitchell C, Ong SS. Stereological investigation of placental morphology in pregnancies complicated by pre-eclampsia with and without intrauterine growth restriction. Placenta. 2003;24:219–26.

63. Trahair JF, Harding R. Restitution of swallowing in the fetal sheep restores intestinal growth after midgestation esophageal obstruction. J Pediatr Gastroenterol Nutr. 1995;20:156–61.

64. Bloomfield FH, van Zijl PL, Bauer MK, Harding JE. Effects of intrauterine growth restriction and intraamniotic insulin-like growth factor-I treatment on blood and amniotic fluid concentrations and on fetal gut uptake of amino acids in late-gestation ovine fetuses. J Pediatr Gastroenterol Nutr. 2002;35:287–97.

65. Ross MG, Nijland MJ. Development of ingestive behavior. Am J Physiol. 1998;274:R879–93.

66. Gao H, Wu G, Spencer TE, Johnson GA, Li X, Bazer FW. Select nutrients in the ovine uterine lumen. I. Amino acids, glucose, and ions in uterine lumenal flushings of cyclic and pregnant ewes. Biol Reprod. 2009;80:86–93.

67. Sagawa N, Nishimura T, Ogawa M, Inouye A. Electrogenic absorption of sugars and amino acids in the small intestine of the human fetus. Membr Biochem. 1979;2:393–404.

68. Bazer FW. Allantoic fluid: regulation of volume and composition. In: Brace RA, Ross MG, Robillard JE, editors. Reproductive and perinatal medicine. Vol. 11: fetal and neonatal body fluids. Ithaca: Perinatology Press; 1989. p. 135–55.

69. Kwon H, Spencer TE, Bazer FW, Wu G. Developmental changes of amino acids in ovine fetal fluids. Biol Reprod. 2003;68:1813–20.

70. Regnault TR, Marconi AM, Smith CH, Glazier JD, Novak DA, Sibley CP, et al. Placental amino acid transport systems and fetal growth restriction–a workshop report. Placenta. 2005;26(Suppl A):S76–80.

71. Battaglia FC. In vivo characteristics of placental amino acid transport and metabolism in ovine pregnancy–a review. Placenta. 2002;23(Suppl A):S3–8.

72. Jones HN, Powell TL, Jansson T. Regulation of placental nutrient transport–a review. Placenta. 2007;28:763–74.

73. Fowden AL, Ward JW, Wooding FB, Forhead AJ. Developmental programming of the ovine placenta. Soc Reprod Fertil Suppl. 2010;67:41–57.

74. Carver TD, Hay Jr WW. Uteroplacental carbon substrate metabolism and O2 consumption after long-term hypoglycemia in pregnant sheep. Am J Physiol. 1995;269:E299–308.

75. Ross JC, Fennessey PV, Wilkening RB, Battaglia FC, Meschia G. Placental transport and fetal utilization of leucine in a model of fetal growth retardation. Am J Physiol. 1996;270:E491–503.

76. Wallace JM, Milne JS, Aitken RP. The effect of overnourishing singleton-bearing adult ewes on nutrient partitioning to the gravid uterus. Br J Nutr. 2005;94:533–9.

77. Jansson N, Pettersson J, Haafiz A, Ericsson A, Palmberg I, Tranberg M, et al. Down-regulation of placental transport of amino acids precedes the development of intrauterine growth restriction in rats fed a low protein diet. J Physiol. 2006;576:935–46.

78. Jansson T, Powell TL. Placental nutrient transfer and fetal growth. Nutrition. 2000;16:500–2.

79. Belkacemi L, Nelson DM, Desai M, Ross MG. Maternal undernutrition influences placental-fetal development. Biol Reprod. 2010;83:325–31.

80. Jansson T, Powell TL. IFPA 2005 award in placentology lecture. Human placental transport in altered fetal growth: does the placenta function as a nutrient sensor? – a review. Placenta. 2006;27(Suppl A):S91–7.

81. Malandro MS, Beveridge MJ, Kilberg MS, Novak DA. Effect of low-protein diet-induced intrauterine growth retardation on rat placental amino acid transport. Am J Physiol. 1996;271:C295–303.

82. Regnault TR, de Vrijer B, Battaglia FC, Meschia G, Wilkening RB. Relationship of transplacental oxygen gradient, fetal lactate concentration and amino acid uptake in fetal growth restriction. J Soc Gynecol Investig. 2004;11:A823.

83. Hay Jr WW. Placental-fetal glucose exchange and fetal glucose metabolism. Trans Am Clin Climatol Assoc. 2006;117:321–39. discussion 339–340.

84. Wen HY, Abbasi S, Kellems RE, Xia Y. mTOR: a placental growth signaling sensor. Placenta. 2005;26(Suppl A):S63–9.

85. Kim J, Song G, Wu G, Bazer FW. Functional roles of fructose. Proc Natl Acad Sci U S A. 2012;109:E1619–28.

86. Stipanuk MH, Watford M. Amino acid metabolism. In: Stipanuk MH, editor. Biochemical and physiological aspects of human nutrition. Philadelphia: W. B. Saunders; 2000. p. 233–86.

87. Flynn NE, Meininger CJ, Haynes TE, Wu G. The metabolic basis of arginine nutrition and pharmacotherapy. Biomed Pharmacother. 2002;56:427–38.

88. Wu G, Morris Jr SM. Arginine metabolism: nitric oxide and beyond. Biochem J. 1998;336(Pt 1):1–17.

89. Wu G, Self JT. Amino acids: metabolism and functions. In: Pond WG, Bell AW (eds.), Encyclopedia of Animal Science. New York: Marcel Dekker; 2004.

90. Reynolds LP, Redmer DA. Angiogenesis in the placenta. Biol Reprod. 2001;64:1033–40.

91. Sladek SM, Magness RR, Conrad KP. Nitric oxide and pregnancy. Am J Physiol. 1997;272:R441–63.

92. Bird IM, Zhang L, Magness RR. Possible mechanisms underlying pregnancy-induced changes in uterine artery endothelial function. Am J Physiol Regul Integr Comp Physiol. 2003;284:R245–58.

93. Gao H, Wu G, Spencer TE, Johnson GA, Bazer FW. Select nutrients in the ovine uterine lumen. VI. Expression of FK506-binding protein 12-rapamycin complex-associated protein 1 (FRAP1) and regulators and effectors of mTORC1 and mTORC2 complexes in ovine uteri and conceptuses. Biol Reprod. 2009;81:87–100.

94. Gao H, Wu G, Spencer TE, Johnson GA, Bazer FW. Select nutrients in the ovine uterine lumen. V. Nitric oxide synthase, GTP cyclohydrolase, and ornithine decarboxylase in ovine uteri and peri-implantation conceptuses. Biol Reprod. 2009;81:67–76.

95. Gao H, Wu G, Spencer TE, Johnson GA, Bazer FW. Select nutrients in the ovine uterine lumen. IV. Expression of neutral and acidic amino acid

96. transporters in ovine uteri and peri-implantation conceptuses. Biol Reprod. 2009;80:1196–208.

97. Gao H, Wu G, Spencer TE, Johnson GA, Bazer FW. Select nutrients in the ovine uterine lumen. III. Cationic amino acid transporters in the ovine uterus and peri-implantation conceptuses. Biol Reprod. 2009;80:602–9.

97. Gao H, Wu G, Spencer TE, Johnson GA, Bazer FW. Select nutrients in the ovine uterine lumen. ii. glucose transporters in the uterus and peri-implantation conceptuses. Biol Reprod. 2009;80:94–104.

98. Wang X, Frank JW, Little DR, Dunlap KA, Satterfield MC, Burghardt RC, et al. Functional role of arginine during the peri-implantation period of pregnancy. I. Consequences of loss of function of arginine transporter SLC7A1 mRNA in ovine conceptus trophectoderm. FASEB J. 2014;28:2852–63.

99. Lopez-Garcia C, Lopez-Contreras AJ, Cremades A, Castells MT, Marin F, Schreiber F, et al. Molecular and morphological changes in placenta and embryo development associated with the inhibition of polyamine synthesis during midpregnancy in mice. Endocrinology. 2008;149:5012–23.

100. Pendeville H, Carpino N, Marine JC, Takahashi Y, Muller M, Martial JA, et al. The ornithine decarboxylase gene is essential for cell survival during early murine development. Mol Cell Biol. 2001;21:6549–58.

101. Ishida M, Hiramatsu Y, Masuyama H, Mizutani Y, Kudo T. Inhibition of placental ornithine decarboxylase by DL-alpha-difluoro- methyl ornithine causes fetal growth restriction in rat. Life Sci. 2002;70:1395–405.

102. Satterfield MC, Dunlap KA, Keisler DH, Bazer FW, Wu G. Arginine nutrition and fetal brown adipose tissue development in nutrient-restricted sheep. Amino Acids. 2013;45:489–99.

103. Lassala A, Bazer FW, Cudd TA, Datta S, Keisler DH, Satterfield MC, et al. Parenteral administration of L-arginine prevents fetal growth restriction in undernourished ewes. J Nutr. 2010;140:1242–8.

104. Lassala A, Bazer FW, Cudd TA, Datta S, Keisler DH, Satterfield MC, et al. Parenteral administration of L-arginine enhances fetal survival and growth in sheep carrying multiple fetuses. J Nutr. 2011;141:849–55.

105. Wu G, Bazer FW, Satterfield MC, Li X, Wang X, Johnson GA, et al. Impacts of arginine nutrition on embryonic and fetal development in mammals. Amino Acids. 2013;45:241–56.

106. Grillo MA, Lanza A, Colombatto S. Transport of amino acids through the placenta and their role. Amino Acids. 2008;34:517–23.

107. Bazer FW, Gao H, Johnson GA, Wu G, Bailey DW, Burghardt RC. Select nutrients and glucose transporters in pig uteri and conceptuses. Soc Reprod Fertil Suppl. 2009;66:335–6.

108. Battaglia FC, Regnault TR. Placental transport and metabolism of amino acids. Placenta. 2001;22:145–61.

109. Coan PM, Vaughan OR, Sekita Y, Finn SL, Burton GJ, Constancia M, et al. Adaptations in placental phenotype support fetal growth during undernutrition of pregnant mice. J Physiol. 2010;588:527–38.

110. Wooding FB, Fowden AL, Bell AW, Ehrhardt RA, Limesand SW, Hay WW. Localisation of glucose transport in the ruminant placenta: implications for sequential use of transporter isoforms. Placenta. 2005;26:626–40.

111. Ehrhardt RA, Bell AW. Developmental increases in glucose transporter concentration in the sheep placenta. Am J Physiol. 1997;273:R1132–41.

112. Lucy MC, Green JC, Meyer JP, Williams AM, Newsom EM, Keisler DH. Short communication: glucose and fructose concentrations and expression of glucose transporters in 4- to 6-week pregnancies collected from Holstein cows that were either lactating or not lactating. J Dairy Sci. 2012;95:5095–101.

113. Regnault TR, de Vrijer B, Galan HL, Wilkening RB, Battaglia FC, Meschia G. Umbilical uptakes and transplacental concentration ratios of amino acids in severe fetal growth restriction. Pediatr Res. 2013;73:602–11.

114. Wali JA, de Boo HA, Derraik JG, Phua HH, Oliver MH, Bloomfield FH, et al. Weekly intra-amniotic IGF-1 treatment increases growth of growth-restricted ovine fetuses and up-regulates placental amino acid transporters. PLoS One. 2012;7:e37899.

115. Long NM, Vonnahme KA, Hess BW, Nathanielsz PW, Ford SP. Effects of early gestational undernutrition on fetal growth, organ development, and placentomal composition in the bovine. J Anim Sci. 2009;87:1950–9.

116. Jobgen WS, Ford SP, Jobgen SC, Feng CP, Hess BW, Nathanielsz PW, et al. Baggs ewes adapt to maternal undernutrition and maintain conceptus growth by maintaining fetal plasma concentrations of amino acids. J Anim Sci. 2008;86:820–6.

9

Cytokines from the pig conceptus: roles in conceptus development in pigs

Rodney D Geisert[1*], Matthew C Lucy[1], Jeffrey J Whyte[1], Jason W Ross[2] and Daniel J Mathew[1]

Abstract

Establishment of pregnancy in pigs involves maintaining progesterone secretion from the corpora lutea in addition to regulating a sensitive interplay between the maternal immune system and attachment of the rapidly expanding trophoblast for nutrient absorption. The peri-implantation period of rapid trophoblastic elongation followed by attachment to the maternal uterine endometrium is critical for establishing a sufficient placental-uterine interface for subsequent nutrient transport for fetal survival to term, but is also marked by the required conceptus release of factors involved with stimulating uterine secretion of histotroph and modulation of the maternal immune system. Many endometrial genes activated by the conceptus secretory factors stimulate a tightly controlled proinflammatory response within the uterus. A number of the cytokines released by the elongating conceptuses stimulate inducible transcription factors such as nuclear factor kappa B (NFKB) potentially regulating the maternal uterine proinflammatory and immune response. This review will establish the current knowledge for the role of conceptus cytokine production and release in early development and establishment of pregnancy in the pig.

Keywords: Cytokines, Embryo development, Porcine conceptus, Pregnancy, Prostaglandins uterus

Introduction

Establishment of pregnancy by the pre-implantation porcine conceptuses (embryo and extraembryonic membranes) requires extending the lifespan and progesterone secretion from the corpora lutea (CL) and appropriately contributing to the intricate interplay between the maternal immune system and attachment of the rapidly expanding trophoblast. Rapid (less than 1 h) elongation of the pig conceptuses across the uterine epithelial surface provides the physiological mechanism for the release of conceptus estrogens (maternal recognition of pregnancy signal) to rapidly redirect endometrial release of luteolytic prostaglandin $F_{2\alpha}$ away from endocrine movement (towards the uterine vasculature) to an exocrine secretion (into the uterine lumen) to enable CL maintenance. Porcine conceptuses are proteolytic and highly invasive outside the luminal environment of the uterus [1] but *in utero* the conceptuses are non-invasive (invasiveness controlled by the release of numerous endometrial protease inhibitors) resulting in the superficial epitheliolchorial type of placentation. The

peri-implantation period of rapid trophoblastic elongation (Days 11 to 12) and attachment to the maternal uterine surface (Day 13 to 18) is essential for establishing sufficient placental uterine area for subsequent nutrient transport for piglet survival to term. Additionally, conceptus release of factors during this critical phase of pregnancy establishment also involves the stimulation of uterine secretion of histotroph and modulation of the maternal immune system. The semiallogeneic conceptuses ability to modify the maternal uterine environment into an environment favorable for growth and survival occurs through the activation of inducible transcription factors within the conceptus and uterine endometrium. Many genes activated by the conceptuses stimulate a tightly controlled proinflammatory response within the uterus [2-4]. A number of the cytokines released by the elongating conceptuses stimulate inducible transcription factors, such as nuclear factor kappa B (NFKB), which are thought to contribute to the maternal uterine proinflammatory and immune response [5]. Activation of NFKB is not limited to the immune system but can regulate cell differentiation, proliferation and survival. A number of recent reviews have described the complex nature for the role of growth factors and cytokines during implantation [5-9]. The following review will establish our

* Correspondence: geisertr@missouri.edu
[1]Animal Sciences Research Center, University of Missouri, 920 East Campus Drive, Columbia, MO 65211, USA
Full list of author information is available at the end of the article

current knowledge of the role of conceptus cytokine production and release in early development and establishment of pregnancy in the pig.

Window of implantation

To fully appreciate the intricate interplay between the conceptus and uterus during the peri-implantation period requires a thorough understanding of the cellular localization and shifts in endometrial steroid receptors regulating the release of growth factors involved with conceptus development [4,8]. Opening of the "window of receptivity" for trophoblastic elongation and attachment to the uterine luminal epithelium is regulated through ovarian estrogen and progesterone release and cell specific expression of steroid receptors within the uterine luminal (LE) and glandular (GE) epithelia and stroma. Although ovarian estrogen from the developing ovulatory follicles during proestrous and estrus is critical for priming the endometrium, progesterone and localization of its receptor play an essential role with cellular communication between the uterine epithelium and stroma in establishing a proper uterine environment for conceptus attachment and early development [10-12]. Progesterone's role in opening the window for implantation during early pregnancy is associated with cell-specific changes in expression of endometrial progesterone receptor (PGR). Epithelial PGR (specifically PGRA) has been demonstrated to be a key regulator of uterine epithelial-stromal crosstalk essential for uterine development and function [13]. While uterine stromal and myometrial cells express PGR throughout pregnancy, a clear spatiotemporal association exists between the down-regulation of PGR in the endometrial LE and GE, and receptivity for conceptus implantation [11-16]. Down-regulation of PGR in endometrial epithelia is a conserved event among most mammals [14-20] and is associated with the down-regulation of high molecular weight mucin O-linked glycoproteins such as mucin 1 which serve as steric transmembrane inhibitors of trophoblast attachment [21-24]. A uterine environment permissive for peri-implantation conceptus development and activation of implantation is established through the loss of PGR from LE and GE cells. Maintenance of PGR in the stromal cell layer stimulates expression and secretion of progestamedins such as fibroblast growth factor 7 (FGF7) and hepatocyte growth factor [4,10,25] which in turn activate multiple uterine genes involved with growth, morphogenesis, synthesis of enzymes and enzyme inhibitors, extracellular matrix and cell adhesion prior to trophoblast attachment to the uterine surface [8,12,26,27]. With cell specific loss of PGR from the LE and GE, estrogen receptor (specifically ESR1) is up-regulated in the uterine epithelium [28-30]. Establishment of a receptive endometrium for conceptus attachment is thus regulated through progesterone induction of epithelial PGR loss allowing finely

synchronized alterations in the LE extracellular matrix exposing attachment factors such as transmembrane integrin heterodimer receptors and release of the matricellular protein, secreted phosphoprotein 1 (SPP1; also referred to as osteopontin) [3,31] and balanced secretion of numerous growth factors, cytokines, prostaglandins, enzymes and their inhibitors which are enhanced by conceptus estrogen synthesis and release during the peri-implantation period [11,27,32]. Conceptus attachment and secretions also increase endometrial folding and LE proliferation (Figure 1) during early implantation in the pig [33]. The increase in endometrial folding and immune cell trafficking to the uterine surface may be induced by conceptus secretion of cytokines like interleukin 1β, interferons, estrogens or a combination of the conceptus release factors.

Conceptus development

Opening the window of receptivity for conceptus attachment to the uterine endometrium (Day 10 to 14) following down-regulation of the uterine epithelial PGR marks a period of conceptus growth, development and change in morphology stimulated by the release of multiple uterine growth factors and cytokines [2-4]. During the early peri-implantation period, the endometrium increases the release of epidermal growth factor (EGF) [34-37], insulin-like growth factor-1 (IGF-1) [38-42], FGF7 [43,44], vascular endothelial growth factor (VEGF) [45-47], interleukin 6 (IL6) [48-50], transforming growth factor beta (TGFB) [51-53], and leukemia inhibitory factor (LIF) [48-50] for which the developing conceptus trophectoderm expresses EGF-receptor (EGFR) [36], IGF1R [54], FGFR2 [55], VEGFR1 and 2 [45,47], IL6R [50], TGFBR1 and 2 [52], and LIFR [50]. The increased endometrial release of EGF, FGF7, LIF, and IGF-1 are enhanced in the epithelium during the period of conceptus elongation and estrogen release [42,44,50,51,55]. Receptor activation by many of the uterine secreted factors has been shown to occur through multiple signaling pathways such as phosphatidylinositol 3-kinase (P13K)/ AKT1 and mitogen-activated protein kinase ERK1/ 2MAPK [36,47,54] which are cell signaling pathways linked to stimulating trophectoderm proliferation, migration and survival. In addition to stimulating proliferation of trophoblast cells, TGFB, LIF and IL6 increase cell viability and attachment *in vitro* [50-52].

Growth of the early developing porcine conceptuses stimulated through the release of uterine growth factors is essential for achieving a critical developmental threshold that triggers rapid trophoblast expansion within the uterine lumen. Timing for the increased release of growth factors is dependent upon the length of progesterone stimulation which facilitates down-regulation of epithelial PGR in the endometrium [2,3]. Several studies have elegantly demonstrated the impact of the duration

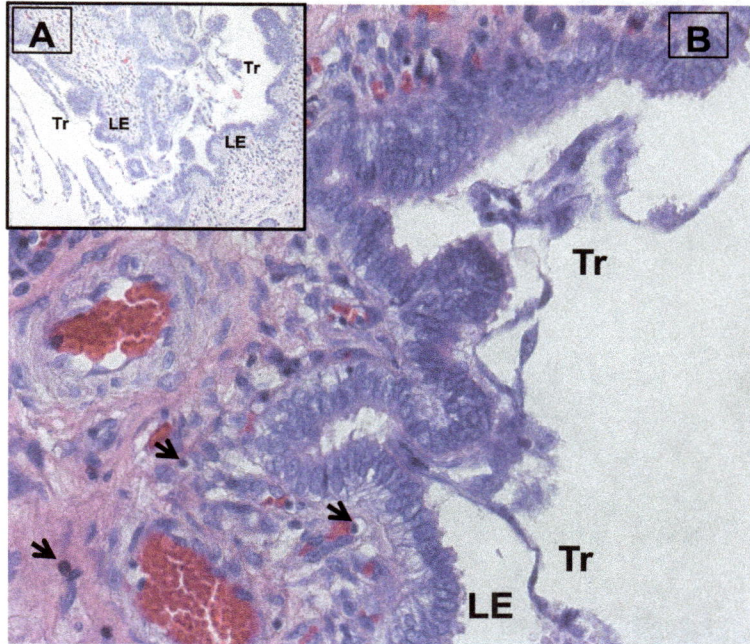

Figure 1 Endometrial folding during pig conceptus attachment. Following rapid trophoblast elongation on Day 12 of pregnancy, conceptus attachment to the endometrial surface epithelium induces a localized increase in endometrial surface folding on Day 14 of pregnancy **(A)**. Local conceptus release of IL1BE, IFN, estrogens or combination of the factors released by the conceptus to alter the uterine surface architecture (attachment and folding) to increase the surface area needed to support the epithiochorial type of placentation in the pig and alter immune cell trafficking to the uterine surface **(B)**. (Tr = trophectoderm, LE = luminal epithelium, arrows = lymphocytes in the underlying stratum compactum).

of progesterone priming in that exogenous progesterone immediately following ovulation accelerates early conceptus growth in both sheep [56,57] and cattle [58-60]. Administration of progesterone shortly after ovulation advances down-regulation of epithelial PGR by two days during the normal estrous cycle and pregnancy [56-60]. The advancement of epithelial PGR down regulation accelerates the release of uterine growth factors for the developing sheep conceptus [61].

Release of the uterine growth factors is clearly involved with growth and differentiation of the porcine conceptuses following hatching from the zona pellucida on Days 6–7 of gestation. Following hatching, peri-implantation development in the pig is unique in that conceptuses develop from a 1–2 mm sphere to a 9–10 mm long ovoid shape between Days 10 to 12 of pregnancy and then rapidly transition to tubular and filamentous forms by elongating at 30–40 mm/h to >100 mm in length (Figure 2) in 1 to 2 h [12,33,62]. Rapid conceptus elongation provides the mechanism for delivery of estrogen across the uterine surface to maintain CL function, stimulate secretions from the uterine LE and GE which are closely linked to initiation of trophoblast attachment to the uterine LE and establish individual placental surface area for nutrient absorption from the underlying endometrium for individual conceptuses [3,23,63].

The specific factor(s) involved with triggering the rapid morphological transformation of the ovoid conceptus to its filamentous shape is currently unknown. Although endometrial release of growth factors is involved with conceptus growth and development, variation in stages of development prior to and during the time of trophoblast elongation (spherical, ovoid, tubular and filamentous conceptuses present within the same litter) indicate that elongation is not necessarily triggered by a uterine-stimulated event but rather a specific stage of conceptus differentiation and development [33,62,64-68]. Rapid conceptus elongation does not occur through cellular hyperplasia but rather cellular remodeling [62]. The morphological alteration in shape of the trophectoderm and transformation of the underlying endoderm forming filapodia provides a mechanism to physically move cells into the elongation zone [62]. The focal point for the cellular restructuring occurs from the ends of the epiblast forming an extended band of cells (elongation zone) to the elongating tips of the conceptus trophectoderm [3,62]. The force necessary for the cellular restructuring of the trophoblast during elongation occurs through modifications in microfilaments and junctional complexes [3,62,69,70]. Elongation of the conceptuses may involve interaction of integrins on the endometrial LE apical surface [71].

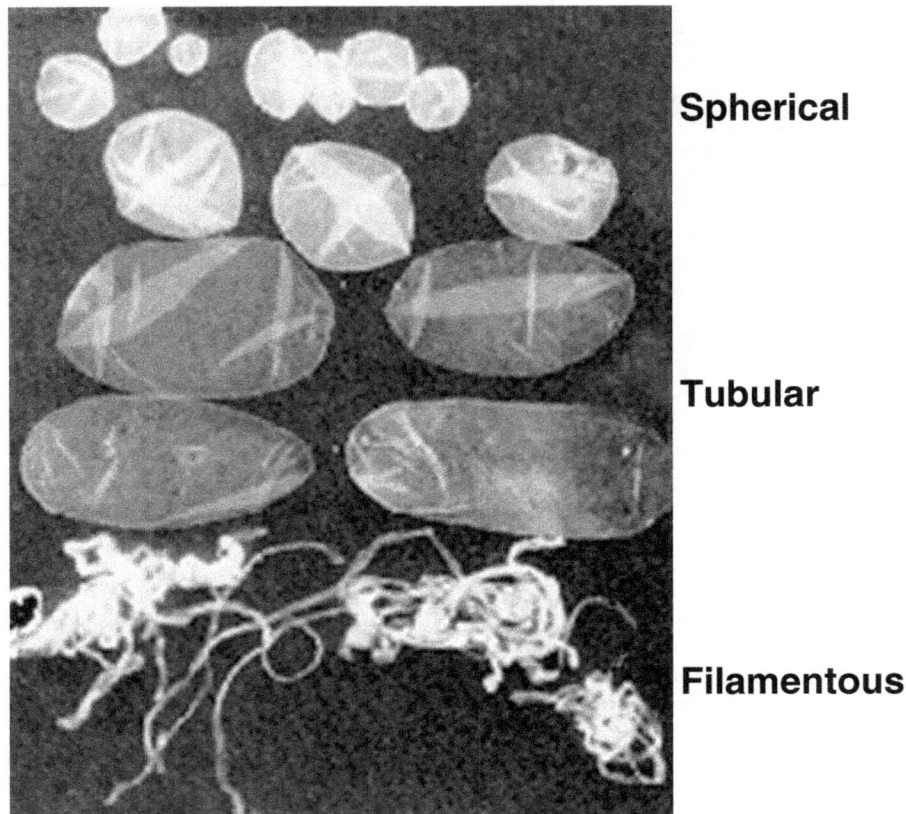

Figure 2 Morphological stages of early conceptus development between Days 10 to 12 of pregnancy. Upon reaching a spherical diameter of appropriately 10 mm, conceptuses rapidly transition to ovoid, tubular and filamentous morphologies within 2 to 3 h.

As previously stated, timing of rapid conceptus elongation is established by the conceptus achieving a specific stage of development which is temporally associated with gastrulation and formation of the extraembryonic mesoderm [65,72-74]. Yelich et al. [72] first indicated that 6 mm spherical conceptuses expressed gene transcripts for brachyury (marker for mesoderm formation) which precedes the initial detection of mesodermal outgrowth in 10 mm ovoid conceptuses. The increase in brachyury expression is associated with an alteration in steroidogenesis in the developing conceptuses [75]. Valdez Magana et al. [68] recently reported that epiblast development and differentiation provides the paracrine signaling between the epiblast and trophectoderm for trophoblast proliferation and mesoderm differentiation. Transcripts for FGF4 are highly detectable in the porcine epiblast but absent/low in the trophectoderm [68,76]. However in ovoid conceptuses, FGFR2 is expressed in trophectoderm cells where there is abundance of FGF4 ligand which activates MAPK phosphorylation [68]. In addition, bone morphogenetic protein 4 (BMP4) expression in the developing extraembryonic mesoderm outgrowth from the epiblast that occurs between trophectoderm and endoderm

stimulates BMPR2 in trophectoderm (absent in epiblast and hypoblast). Valdez Magaña et al. [68] suggested that increased epiblast production of FGF4 and expression of FGFR2 in the adjacent trophectoderm cells trigger the signaling cascade for trophoblast elongation. The novel suggestion that FGF4 is involved in the initial response of the conceptus is supported by information which indicates that FGF4 is not normally released into the extracellular fluid but moves in a gradient only over a short distance of a few cells [77,78]. Induction of FGF4 in the epiblast stimulating MAPK in the trophectoderm through FGFR2 could coordinate with the extraembryonic mesoderm production of BMP4 to initiate the cascade of events involved with modifying microfilaments and junctional complexes necessary for the elongation process.

Although formation of the extraembryonic mesoderm in the conceptus is clearly a marker for the time of rapid trophoblast elongation and the cellular alternations involved, the conceptus factor triggering elongation of the porcine conceptus is unknown. Although conceptus elongation has not been achieved *in vitro*, it is clear that the conceptus activates elongation at a specific stage of development. Presence of spherical with filamentous conceptuses within the same litter [12] and the failure to advance elongation *in vivo* through estrogen administration prior to a stage of development for elongation [62,67] demonstrate that initiation of trophoblast elongation is regulated by conceptus development. However, alterations in uterine secretion do have a direct impact on the rate of conceptus development to reach the stage for elongation.

A number of studies have evaluated the transcriptome of developing spherical, ovoid, tubular and filamentous pig conceptuses prior to and during elongation [64-66,72,79,80]. These studies described a multitude of transcripts involved with steroidogenesis, lipid metabolism, cell morphogenesis, calcium binding, protein binding and nucleotide binding. Specific transcripts involved in steroidogenesis, such as steroidogeneic acute regulatory protein, cytochrome P450 side chain cleavage protein, 17α-hydrolase and aromatase all increase in abundance as the pig conceptuses approach and initiate the elongation process [64,65,72]. However, although administration of estrogen can advance uterine gene expression and secretions associated with the increase in conceptus estrogen production at elongation; it does not induce premature elongation of the conceptuses [81]. A number of transcripts involved with embryonic development, attachment and immune cell regulation such as s-adenosylhomocysteine hydrolase [79], retinoic acid receptors and retinol binding protein [72], TGFB [64,72], LIFR [72], interferon-γ (IFNγ), B-cell linker, and chemokine ligand 14 [66] are altered during early conceptus development. The most striking change in the conceptus transcriptome during the transition from ovoid to filamentous morphology is the increase in expression of interleukin 1β (*IL1B*) [79,80]. The increase in *IL1B* during transition to the filamentous form of porcine conceptus development was first described by Tuo et al. [82]. Interleukin 1β is a proinflammatory cytokine which is dependent on the expression of members of the IL-1 system belonging to the IL1B/Toll-like receptor (TLR) superfamily. The IL-1 system consists of two agonists (IL1A and IL1B), two receptors (IL1R1 (functional) and IL1R2 (pseudo-receptor)), converting enzymes, a receptor accessory protein (IL1RAP), and multiple isoforms of receptor antagonists (IL1Rant) [5,83] which are all present in the porcine endometrium and conceptuses [79,84,85].

Conceptus IL-1β

Conceptus *IL1B2* mRNA abundance rapidly increases during trophoblast elongation, but decreases over 2000-fold immediately following completion of the elongation process [86]. Based on the timing and pattern of conceptus IL1B release and the presence of the IL-1 system in the conceptuses and endometrium, Ross et al. [86] proposed that conceptus IL1B secretion was the signal to

initiate the cascade of events involved with the rapid elongation process.

Recently, analyses of pig genome sequences and expressed sequence tags (EST) indicate that gene duplication resulted in two *IL1B* genes on *Sus scrofa* chromosome 3. The classical *IL1B1* is expressed in macrophages and endometrial tissue while the embryonic form (*IL1B2*) is only detected in the early porcine conceptus prior to attachment to the uterine LE [2,87]. *IL1B2* is considered novel because the sequence is not expressed in other mammals [88]. The two predicted protein sequences are 85% identical and are least homologous near the N-terminus as caspase-1 cleaves this portion of the peptide resulting in a functional protein (D.J. Mathew, M.C. Lucy and R.D. Geisert unpublished results). Interestingly, in the embryonic form there is a proline inserted 2 amino acids from the predicted caspase-1 cleavage site. While the two genes are very similar from exon 2 to exon 7, exon 1 and the active promoter regions are different between the two genes. The promoter differences may partially explain variation in mRNA expression between the two forms. Activity and cell specificity of the two forms may also differ as recombinant IL1B2 can activate NFKB in alveolar macrophages and uterine surface epithelium but has reduced activity compared to recombinant IL1B1 (D.J. Mathew, R.D. Geisert and M.C. Lucy unpublished results).

Porcine IL1B2 is secreted only within a brief window associated with the morphological and functional changes that take place in conceptus development and elongation on Days 10 to 12 of pregnancy [86]. It has been postulated that one function of IL1B2 is to act as an inflammatory mediator in the endometrium [89]. Following synthesis and secretion by the conceptus, IL1B2 may trigger a cascade of signaling events that activate the transcription factor, NFKB in the LE of the endometrium. NFKB activation is an important component in opening the implantation window in pigs and other mammals [90]. Genes transcriptionally regulated by NFKB are involved in inflammation, immune function, cell adhesion, and release of cytokines, growth factors, anti-apoptotic factors and immunoreceptors [91]. The activation of inflammatory pathways in the endometrium likely enhances progesterone-induced uterine receptivity for conceptus implantation. It is important, however, that the inflammation cascade triggered by IL1B2 be tightly regulated in order to prevent rejection of the semi-allogeneic conceptus [9]. Conceptus estrogen release during elongation may play a key role in counterbalancing the increased inflammatory response by activating estrogen receptor (ESR1) which can affect the transcriptional activity of NFKB [90]. Thus, conceptus expression of IL1B2 would be consistent with the continued activation of NFKB, whereas the synchronous estrogen secretion by pig conceptuses may pose a suppressive effect to prevent an inflammatory reaction that would be detrimental to conceptus survival [2]. Interleukin-1β increases aromatase expression within human cytotrophoblast [92] and the increased synthesis of IL1B2 by pig conceptuses is temporally associated with elevated conceptus aromatase expression and the acute release of estrogen into the uterine lumen [72,86]. Thus the increase in expression of both IL1B2 and estrogen by individual conceptuses that are expanding through the uterine lumen would counter-balance stimulation of the pro-inflammatory and immune response within the uterus.

IL1B2 may have other roles in rapid conceptus elongation and the regulation of maternal recognition. IL1B is an inducer of phospholipase A2 [93] and thus up-regulates cell membrane arachidonic acid release, thereby increasing membrane fluidity that is necessary for remodeling of the trophectoderm during elongation [2,94]. The arachidonic acid could also be converted to prostaglandins which are needed for placental attachment during the establishment of pregnancy. Recent results from studies with ewes suggest that IL1B could play a role in regulating prostaglandin-endoperoxide synthase 2 (PTGS2) and the subsequent synthesis of prostaglandins that control conceptus elongation [95]. Pig conceptus IL1B2 secretion, therefore, may be of pivotal importance in the rapid morphological transformation of the pig conceptuses on Day 12 of pregnancy.

IL1B2 activation of NFKB stimulates prostaglandin synthesis through induction of PTGS2. IL1B1 increases endometrial IL1R1 and in conjunction with estrogen, IL1RAP, suggesting that IL1B2 and estrogen regulate endometrial transcriptional activity of NFKB during elongation [85,86,96]. IL1B has a stimulatory effect on endometrial prostaglandin E_2 (PGE_2) secretion and PTGS1 and PTGS2 mRNA expression from Days 10 to 13 of pregnancy [85,97-99]. The presence of PGE_2 receptors in the CL and endometrium [98] suggests that conceptus PGE_2 secretion could also affect maintenance of the CL and directly stimulate adhesion and attachment of the trophoblast to the uterine epithelium [100]. Conceptus secretion of IL1B2 into the uterine lumen may also enhance endometrial expression of LIF and IL6 [50] possibly through activation of NFKB within the uterine LE and GE. IL1B1 induces human endometrial expression of LIF [101-103] and IL6 in placental villous core mesenchymal cells in vitro [104]. Suppression of NFKB activity in the endometrium alters the timing of implantation in the mouse which can be partially rescued by LIF supplementation [105]. LIF and IL1B stimulate expression of fucosyltransferase enzymes which are involved with embryo attachment to the uterine surface epithelium in the mouse [106]. During and following rapid conceptus elongation in the pig, there is increased endometrial secretion of LIF and IL6 [48-50]. Both LIFR and IL6R mRNA are detected in porcine conceptus [49,50] suggesting that endometrial

secretion of LIF and IL6 may play an important role in conceptus development and attachment to the uterine surface. Blitek et al. [50] indicated that LIF and IL6 stimulated proliferation and attachment of porcine trophoblast cells in vitro. Conceptus estrogen and IL1B2 secretion serve as major components in the embryo-uterine crosstalk to stimulate endometrial LIF and IL6 to contribute to the pathway for conceptus attachment to the uterine luminal surface.

Several papers have investigated endometrial differential gene expression between cyclic and pregnant pigs which provide numerous endometrial genes and pathways that the conceptus stimulates during the period of conceptus elongation and attachment [107-110] which will not be covered in this review. One interesting gene differentially expressed during pregnancy is IL11RA [110]. IL11 and its receptor (IL11RA) is proposed to prevent the invasion of trophoblast cells in the mouse [111] and human [112]. Although gene expression IL-11RA is lower in endometria of pregnant pigs, there was a pregnancy-specific increase in IL11RA on the surface epithelium [110]. As previous indicated porcine conceptuses are proteolytic and highly invasive outside the luminal environment of the uterus [1]. Therefore in addition to endometrial release of protease inhibitors during trophoblast attachment, porcine endometrial expression of IL11RA may serve to help inhibit the proteolytic trophoblast invasion through the surface epithelium during attachment [110].

Switch to endometrial IL-18

Porcine conceptus IL1B2 gene expression and secretion is clearly temporally associated with the rapid conceptus elongation as a dramatic reduction in mRNA abundance is soon followed by a depletion of IL1B2 protein in the uterine lumen following conceptus elongation on Day 12 [86]. The loss of conceptus IL1B2 secretion following elongation suggests that another closely related cytokine may function at the conceptus-uterine interface to continue regulation of the immunological interactions necessary for establishment of pregnancy in the pig. Interleukin 18 (IL18), also referred to as interferon-γ inducing factor [113], is a member of the IL-1 family of pro-inflammatory cytokines believed to play a significant role in implantation. Following the loss of conceptus IL1B2 stimulation, there is a switch to endometrial IL18 production and release during placental attachment in the pig [114]. Porcine endometrial IL18 mRNA expression increases from Days 10 to 15 of the estrous cycle with mRNA expression increasing 10-fold on Day 18 of pregnancy. However, there is a pregnancy-specific increase in uterine luminal content of IL18 between Days 15 and 18 due to an increase in caspase-1 expression induced by the developing conceptuses [114]. Caspase-1 cleaves and activates the proforms of both IL1B and IL18 [115]. Pro-IL18, which has structural similarities to pro-IL1B, is involved with modulation of the immune system through induction of interferon-γ [116]. Conceptus secretion of IFNG increases immediately following trophoblast elongation in the pig [117], suggesting that the conceptuses may induce endometrial IL18 release to assist in development and placental attachment during early pregnancy. Interestingly, unlike IL1B which is stimulated by inflammatory responses in cells, IL18 is stored in healthy cells and its biological activity is dependent upon its release through caspase-1 processing [118]. Although similar to IL1B, IL18 binds to a unique IL18 receptor which consists of two receptor chains, ligand-binding chain IL18RA and a co-receptor IL18RB chain (similar to IL1B accessory protein), which are required for cellular signaling [119]. The conceptus factor that stimulates the increase in caspase-1 in the uterine epithelium is unknown, although IL1B2 could stimulate release IL18 from the uterine epithelial cells through increasing caspase-1 activity [120]. Biological activity of IL18 is regulated through release of an IL18 binding protein (IL18BP) which functions as a negative feedback loop to suppress IFNG production and limit Th1 cell responses.

The increased endometrial expression of caspase-1, and release of IL18 into the uterine lumen may stimulate expression and secretion of IFNG by conceptuses [117] to modulate the maternal immune system through signal transducer and activator of transcription 1 (STAT1) at the interface between trophectoderm and uterine LE [121]. The loss of conceptus IL1B2 stimulation and switch to endometrial IL18 production during placental attachment in the pig would decrease the potential pro-inflammatory stimulation of the conceptuses following trophoblast elongation which maybe important to control cytokine and immune functions following implantation [122]. Increased secretion of IL18 at the uterine/trophoblast interface is associated with increased pregnancy rates in one line of abortion-prone mice [123].

Conceptus interferons (IFN)

During the peri-implantation period of conceptus attachment to the uterine LE following trophoblastic elongation, pig conceptuses secrete of IFNG (Type II IFN) and IFND (Type I IFN) between Days 12 to 20 of gestation [117,121,124]. Trophoblastic production and secretion of two IFNs, of which IFNG is the predominate form [125,126], is unique compared with other mammalian species. Trophoblast secretion of IFNG and IFND would enable activation of a distinct gene set through two different receptors that may provide a uniquely regulated stimulation within the endometrium [127]. With the abrupt decline in conceptus expression of IL1B2 following rapid elongation, there is a tremendous increase in the filamentous conceptus trophoblastic expression of specifically

IFNG during initiation of attachment to the uterine LE on Day 13 [66,117,121]. Unlike IFNT produced by the conceptus of ruminant species, pig trophoblastic IFNs do not directly function as a maternal recognition signal for CL maintenance [3]. However, pig IFNG and IFND can increase endometrial PGE$_2$ secretion [128] and induce cell-specific endometrial IFN-stimulated genes [127,129].

Joyce et al. [121] suggested that conceptus estrogens and IFNs regulate endometrial IFN-stimulated genes through a cell-type-specific manner. Conceptus secretion of estrogen increases STAT1 in LE to initiate the signal for pregnancy recognition and CL maintenance as well as inducing changes to the apical surface glycocalyx of LE to allow conceptus attachment. Conceptus IFNG and IFND induced increases of STAT1 are limited to the underlying endometrial stromal cells that express interferon regulatory factor 1, IFNG/STAT1-responsive gene, that is absent in the LE [121]. Pig conceptuses secrete estrogen during the peri-implantation period of pregnancy which increases uterine LE expression of interferon regulatory factor 2 (IRF2), a transcriptional repressor of classical IFN-stimulated genes, which would also restrict IFNG and IFND stimulation to the underlying stroma. Thus expression of classical IFN responsive genes such as MX1, interferon stimulated gene 15 (ISG15), IRF1, STAT1 and STAT2 are localized in the stroma and GE in pigs [121]. The cell specific activation by the pig trophoblastic IFNs may play an essential role in regulating the immunological barrier for attachment of the semi-allogeneic conceptuses [3,130]. MHC class I molecules such as SLA and β2-microgobulin which are involved with recognition of foreign cells and pathogens are not expressed on the trophoblast and are absent in early pregnancy of the pig [127]. The increase in uterine angiogenesis which occurs during the peri-implantation period between Days 13 to18 of pregnancy [131] could also be stimulated through the trophoblast secretion of IFNs in addition to other conceptus and uterine angiogenic factors such as VEGF.

Conclusion

Proper timing for conceptus growth and development is proposed to be regulated through the initial down-regulation of PR in the uterine LE which stimulates growth factors to promote mesodermal differentiation and expression of FGF4 and BMP4 that initiate conceptus IL1B2 expression and release to stimulate rapid elongation of the conceptuses throughout the uterine lumen (Figure 3). Expansion of the conceptuses throughout the uterine horns provides the mechanism for estrogen to cover the uterine surface for maternal recognition of pregnancy, initiate trophoblast attachment to the LE and regulate the maternal lymphocyte response to conceptus IFNs which stimulate vascular changes and increases angiogenesis for the proper microenvironment for placentation.

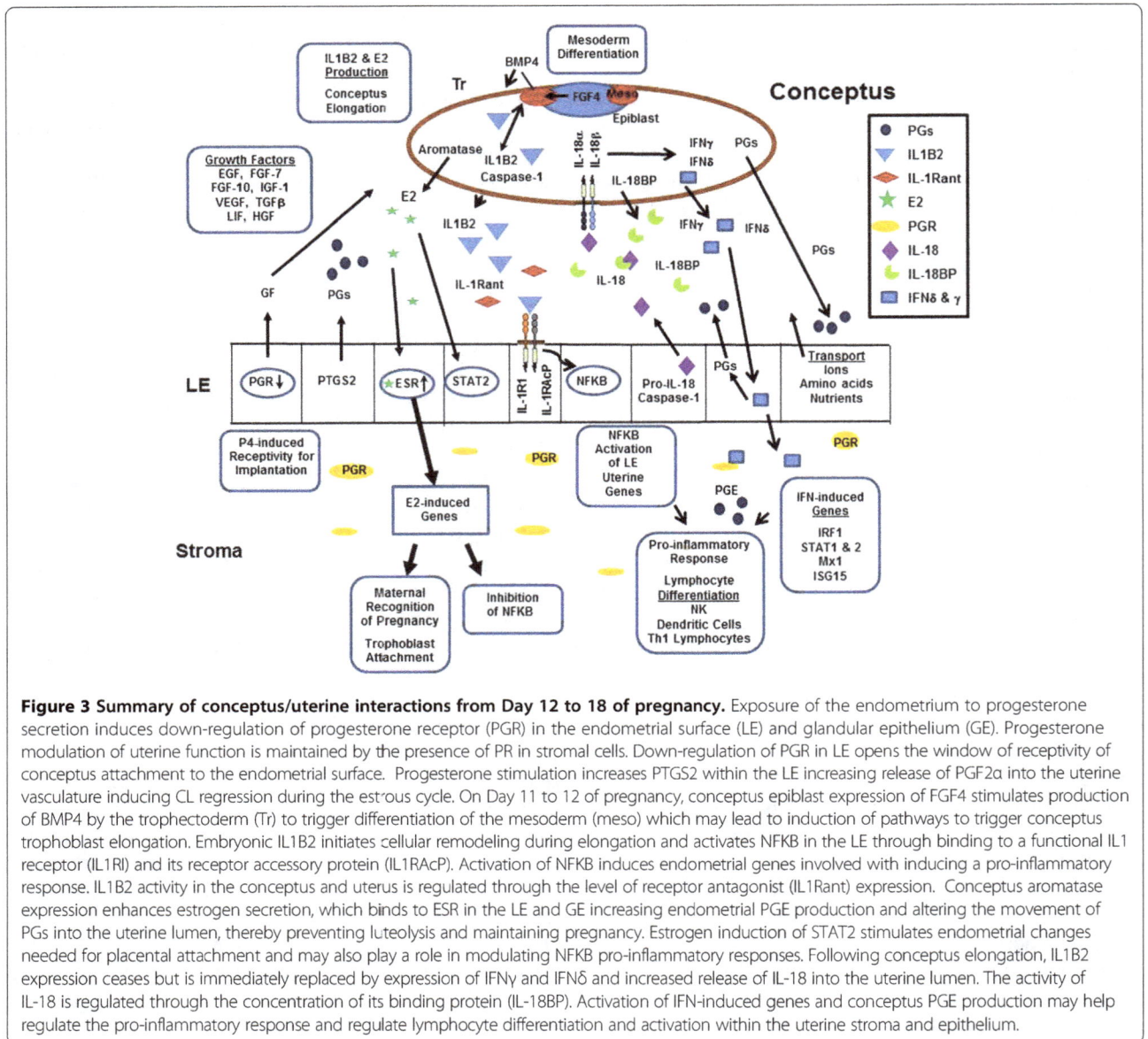

Figure 3 Summary of conceptus/uterine interactions from Day 12 to 18 of pregnancy. Exposure of the endometrium to progesterone secretion induces down-regulation of progesterone receptor (PGR) in the endometrial surface (LE) and glandular epithelium (GE). Progesterone modulation of uterine function is maintained by the presence of PR in stromal cells. Down-regulation of PGR in LE opens the window of receptivity of conceptus attachment to the endometrial surface. Progesterone stimulation increases PTGS2 within the LE increasing release of PGF2α into the uterine vasculature inducing CL regression during the estrous cycle. On Day 11 to 12 of pregnancy, conceptus epiblast expression of FGF4 stimulates production of BMP4 by the trophectoderm (Tr) to trigger differentiation of the mesoderm (meso) which may lead to induction of pathways to trigger conceptus trophoblast elongation. Embryonic IL1B2 initiates cellular remodeling during elongation and activates NFKB in the LE through binding to a functional IL1 receptor (IL1RI) and its receptor accessory protein (IL1RAcP). Activation of NFKB induces endometrial genes involved with inducing a pro-inflammatory response. IL1B2 activity in the conceptus and uterus is regulated through the level of receptor antagonist (IL1Rant) expression. Conceptus aromatase expression enhances estrogen secretion, which binds to ESR in the LE and GE increasing endometrial PGE production and altering the movement of PGs into the uterine lumen, thereby preventing luteolysis and maintaining pregnancy. Estrogen induction of STAT2 stimulates endometrial changes needed for placental attachment and may also play a role in modulating NFKB pro-inflammatory responses. Following conceptus elongation, IL1B2 expression ceases but is immediately replaced by expression of IFNγ and IFNδ and increased release of IL-18 into the uterine lumen. The activity of IL-18 is regulated through the concentration of its binding protein (IL-18BP). Activation of IFN-induced genes and conceptus PGE production may help regulate the pro-inflammatory response and regulate lymphocyte differentiation and activation within the uterine stroma and epithelium.

The role of IL18 and IFNG in regulating Th1 lymphocytes and natural killer (NK) cell responses in tissues suggests that pig conceptus secretion of estrogens, IL1B2, prostaglandins, IFNs and endometrial release of IL18 serve to not only induce cell surface adhesion factors for trophoblast attachment, but also play a critical role in balancing the immune cell migration and recognition of receptors to support or reject the developing embryos and their extraembryonic membranes. The IL-1 family of cytokines plays a critical role in the regulation of immune cell differentiation and activity during pregnancy as well as many inflammatory diseases [132]. During pregnancy in the pig, the conceptus recruits uterine natural killer lymphocytes, dendritic cells and other immune cells at the sites of trophoblast attachment which induce major changes in the endometrial vasculature and angiogenesis to support the developing conceptus [133,134]. Although not demonstrated in the uterus of the pig, the increase in PGE$_2$ from the conceptuses and endometrium may play a role in minimizing pro-inflammatory tissue damage through switching from leukotriene B$_2$ synthesis to lipoxin A$_4$ and release of the anti-inflammatory resolvins and protectins [135]. Clearly pig conceptuses release a number of paracrine factors at the maternal/placental interface to regulate the vascular, angiogenic and immune changes needed to establish pregnancy (Figure 3). The conceptus IL-1 family of cytokines is but one component of a larger group of signaling pathways involved with successful survival of developing embryos. However, pregnancy is not only dependent upon the presence of the various cytokines during implantation but also in the appropriate timing of their release.

It is well established that exposure of pregnant gilts to exogenous estrogen 48 h prior to normal conceptus release at elongation on Day 12 results in fragmentation of the conceptus between Days 15 to 18 of pregnancy [136,137]. Premature exposure of the endometrium to estrogen advances expression of multiple genes during the period of trophoblast elongation and attachment [138]. Most of the aberrantly expressed endometrial genes are those involved with immune cell regulation and cell adhesion. Early estrogen exposure (Days 9 and 10) of pregnant gilts does not affect endometrial *IL18* mRNA expression but disrupts the normal LE release of IL18 into the uterine lumen [114]. Although caspase-1 increases between Days 12 to 18 in estrogen-treated gilts, there is no increase in the luminal content of IL18 as occurs in untreated pregnant gilts. Lack of IL18 release from the LE may directly affect conceptus expression of IFNG. Although STAT1 expression is present in the LE, stromal expression of STAT1 is absent in estrogen-treated gilts [121]. These data indicate a temporally regulated presence of intricate interactions between conceptus estrogen, IL1B2, IFNG and uterine IL18 release in programing downstream transcription factors needed to establish pregnancy in the pig.

Abbreviations
BMP4: Bone morphogenic protein 4; BMPR2: Bone morphogenic protein receptor 2; CL: Corpora lutea; EGE: Epidermal growth factor; EGFR: Epidermal growth factor receptor; ESR: Estrogen receptor; EST: Expressed sequence tags; FGF: Fibroblast growth factor; FGFR2: Fibroblast growth factor 2 receptor; GE: Glandular epithelium; IGF: Insulin-like growth factor; IGF-1R: Insulin-like growth factor 1 receptor; IFN: Interferon; IRF: Interferon regulatory factor; IL: Interleukin; IL-6R: Interleukin 6 receptor; IL-1B2: Interleukin 1β conceptus form; IL-1RAP: Interleukin 1 receptor accessory protein; IL-1Rant: Interleukin 1receptor antagonist; IL-1RT1: Interleukin 1 receptor type 1; LIF: Leukemia inhibitory factor; LIFR: Leukemia inhibitory factor receptor; LE: Luminal epithelium; Mx1: Interferon-induced GTP-binding protein; NFKB: Nuclear factor κB; PR: Progesterone receptor; PG: Prostaglandin; PTGS2: Prostaglandin endoperoxide synthase 2; STAT: Signal transducer and activator of transcript; TGFβ: Transforming growth factor beta; TGFBR: Transforming growth factor beta receptor 1; VEGF: Vascular endothelial growth factor; VEGFR: Vascular endothelial growth factor receptor.

Competing interests
The authors declare that they have no competing interests.

Authors' contributions
RDG, JW, MCL and DJM jointly wrote the review. JWR contributed to research published in RDG laboratory and critical review of the manuscript. All authors read and approved the final manuscript.

Authors' information
RDG is a Reproductive Physiologist in Division of Animal Science at the University of Missouri, Columbia where his research program over the past 30 years has investigated the interaction between the early developing porcine conceptuses and uterus. MCL is a Reproductive Physiologist in Division of Animal Science at the University of Missouri, Columbia where his research program is focused on dairy cattle reproduction and estrous synchronization. DJM is currently a doctoral graduate student completing his research program on porcine conceptus elongation and establishment of

pregnancy in the pig. JWR is a Reproductive Physiologist in the Department of Animal Science at Iowa State University, Ames where he has established a research program on small RNA regulation of reproductive function and the effects of heats stress on pig development. JW is a Reproductive Physiologist in Division of Animal Science at the University of Missouri, Columbia where his research program is involved with the development of transgenic pigs for research in development and disease models.

Acknowledgements
Research from the author's laboratory provide in the review has been supported in part by Agriculture and Food Research Initiative Competitive Grant no. 2007-35203-17836 and Grant no. 2013-67015-21023 from the USDA National Institute of Food and Agriculture.

Author details
[1]Animal Sciences Research Center, University of Missouri, 920 East Campus Drive, Columbia, MO 65211, USA. [2]Department of Animal Science, Iowa State University, 2356 Kildee Hall, Ames, IA 50011, USA.

References
1. Samuel CA, Perry JS: The ultrastructure of pig trophoblast transplanted to an ectopic site in the uterine wall. *J Anat* 1972, 113:139–149.
2. Geisert R, Fazleabas A, Lucy M, Mathew D: Interaction of the conceptus and endometrium to establish pregnancy in mammals: role of interleukin-1β. *Cell Tissue Res* 2012, 349:825–838.
3. Bazer FW, Johnson GA: Pig blastocyst-uterine interactions. *Differentiation* 2014, 87:52–65.
4. Bazer FW, Song G, Kim J, Dunlap KA, Satterfield MC, Johnson GA, Burghardt RC, Wu G: Uterine biology in pigs and sheep. *J Anim Sci Biotech* 2012, 3:23.
5. Hayden MS, Ghosh S: NF-κB, the first quarter-century: remarkable progress and outstanding questions. *Genes Dev* 2014, 26:203–234.
6. Chaouat G, Dubanchet S, Ledée N: Cytokines: Important for implantation? *J Assist Reprod Genet* 2007, 24:491–505.
7. Guzeloglu-Kayisli O, Kayisli UA, Taylor HS: The role of growth factors and cytokines during implantation: endocrine and paracrine interactions. *Semin Reprod Med* 2009, 27:62–79.
8. Bazer FW, Spencer TE, Johnson GA, Burghardt RC, Wu G: Comparative aspects of implantation. *Reprod* 2009, 138:195–209.
9. Warning JC, McCracken SA, Morris JM: A balancing act: mechanisms by which the fetus avoids rejection by the maternal immune system. *Reprod* 2011, 141:715–724.
10. Spencer TE, Bazer FW: Biology of progesterone action during pregnancy recognition and maintenance of pregnancy. *Front Biosci* 2002, 7:1879–1898.
11. Spencer TE, Bazer FW: Uterine and placental factors regulating conceptus growth in domestic animals. *J Anim Sci* 2004, 82(E-Suppl):E4–E13.
12. Geisert RD, Ross JW, Ashworth MD, White FJ, Johnson GA, DeSilva: Maternal recognition of pregnancy signal or endocrine disruptor: The two faces of oestrogen during establishment of pregnancy in the pig. *Soc Reprod Fertil* 2006, 62(Suppl):131–145.
13. Franco HL, Rubel CA, Large MJ, Wetendorf M, Fernandez-Valdivia R, Jeong JW, Spencer TE, Behringer RR, Lydon JP, Demayo FJ: Epithelial progesterone receptor exhibits pleiotropic roles in uterine development and function. *FASEB J* 2012, 26:1218–1227.
14. Tan J, Paria BC, Dey SK, Das SK: Differential uterine expression of estrogen and progesterone receptors correlates with uterine preparation for implantation and decidualization in the mouse. *Endocrinology* 1999, 140:5310–5321.
15. Geisert RD, Pratt T, Bazer FW, Mayes JS, Watson GH: Immunocytochemical localization and changes in endometrial progestin receptor protein during the porcine oestrous cycle and early pregnancy. *Reprod Fert Dev* 1994, 6:749–760.
16. Mathew DJ, Sellner EM, Green JC, Okamura CS, Anderson LL, Lucy MC, Geisert RD: Uterine progesterone receptor expression, conceptus development, and ovarian function in pigs treated with RU 486 during early pregnancy. *Biol Reprod* 2011, 84:130–139.
17. Spencer TE, Bazer FW: Temporal and spatial alterations in uterine estrogen receptor and progesterone receptor gene expression during

the estrous cycle and early pregnancy in the ewe. *Biol Reprod* 1995, 53:1527–1545.

18. Hart LS, Carling SJ, Joyce MM, Johnson GA, Vanderwall DK, Ott TL: Temporal and spatial associations of oestrogen receptor alpha and progesterone receptor in the endometrium of cyclic and early pregnant mares. *Reproduction* 2005, 130:241–250.

19. Fazleabas AT, Kim JJ, Srinivasan S, Donnelly KM, Brudney A, Jaffe RC: Implantation in the baboon: endometrial responses. *Semin Reprod Endocrinol* 1999, 17:257–265.

20. Lessey BA, Yeh I, Castelbaum AJ, Fritz MA, Ilesanmi AO, Korzeniowski P, Sun J, Chwalisz K: Endometrial progesterone receptors and markers of uterine receptivity in the window of implantation. *Fertil Steril* 1996, 65:477–483.

21. Bowen JA, Bazer FW, Burghardt RC: Spatial and temporal analysis of integrin and Muc-1 expression in porcine uterine epithelium and trophectoderm in vivo. *Biol Reprod* 1996, 55:1098–1106.

22. Burghardt RC, Bowen JA, Newton GR, Bazer FW: Extracellular matrix and the implantation cascade in pigs. *J Reprod Fertil* 997, 52(Suppl):151–164.

23. Burghardt RC, Johnson GA, Jaeger LA, Ka H, Garlow JE, Spencer TE, Bazer FW: Integrins and extracellular matrix proteins at the maternal-fetal interface in domestic animals. *Cells Tissues Organs* 2002, 172:202–217.

24. Singh H, Aplin JD: Adhesion molecules in endometrial epithelium: tissue integrity and embryo implantation. *J Anat* 2009, 215:3–13.

25. Cunha GR, Cooke PS, Kurita T: Role of stromal-epithelial interactions in hormonal responses. *Arch Histol Cytol* 2004, 67:417–434.

26. van Mourik MS, Macklon NS, Heijnen CJ: Embryonic implantation: cytokines, adhesion molecules, and immune cells in establishing an implantation environment. *J Leukoc Biol* 2009, 85:4–19.

27. Bazer FW, Wu G, Spencer TE, Johnson GA, Burghardt RC, Bayless K: Novel pathways for implantation and establishment and maintenance of pregnancy in mammals. *Mol Hum Reprod* 2010, 16:135–152.

28. Geisert RD, Brenner RM, Moffatt JR, Harney JP, Yellin T, Bazer FW: Changes in estrogen receptor protein, mRNA expression and localization in the endometrium of cyclic and pregnant gilts. *Reprod Fertil Dev* 1993, 5:247–260.

29. Sukjumlong S, Kaeoket K, Dalin A-M, Persson E: Immunohistochemical studies on oestrogen receptor alpha (ERα) and the proliferative marker Ki-67 in the sow uterus at different stages of the oestrous cycle. *Reprod Domest Anim* 2003, 3(8):5–12.

30. Knapczyk-Stwora K, Durlej M, Bilinska B, Slomczynska M: Immunohistochemical studies on the proliferative marker Ki-67 and estrogen receptor alpha (ERα) in the uterus of neonatal and immature pigs following exposure to flutamide. *Acta Histochem* 2011, 113:534–541.

31. Burghardt RC, Burghardt JR, Taylor JD 2nd, Reeder AT, Nguen BT, Spencer TE, Bayless KJ, Johnson GA: Enhanced focal adhesion assembly reflects increased mechanosensation and mechanotransduction at maternal-conceptus interface and uterine wall during ovine pregnancy. *Reproduction* 2009, 137:567–582.

32. Wang H, Dey SK: Roadmap to embryo implantation: clues from mouse models. *Nat Rev Genet* 2006, 7:185–199.

33. Stroband HW, Van der Lende T: Embryonic and uterine development during early pregnancy in pigs. *J Reprod Fertil* 1990, 40(Suppl):261–277.

34. Vaughan TJ, James PS, Pascall JC, Brown KD: Expression of the genes for TGFα, EGF and EGF receptor during early pig development. *Development* 1992, 116:663–669.

35. Kim GY, Besner GE, Steffen CL, McCarthy DW, Downing MT, Luquette MH, Abad MS, Brigstock DR: Purification of heparin-binding epidermal growth factor-like growth factor from pig uterine luminal flushings, and its production by endometrial tissues. *Biol Reprod* 1995, 52:561–571.

36. Jeong W, Kim J, Bazer FW, Song G: Epidermal growth factor stimulates proliferation and migration of porcine trophectoderm cells through protooncogenic protein kinase 1 and extracellular-signal-regulated kinases 1/2 mitogen-activated protein kinase signal transduction cascades during early pregnancy. *Mol Cell Endocrinol* 2013, 381:302–311.

37. Kennedy TG, Brown KD, Vaughan TJ: Expression of the genes for the epidermal growth factor receptor and its ligands in porcine oviduct and endometrium. *Biol Reprod* 1994, 50:751–756.

38. Tavakkol A, Simmen FA, Simmen RC: Porcine insulin-like growth factor-I (pIGF-I): complementary deoxyribonucleic acid cloning and uterine expression of messenger ribonucleic acid encoding evolutionarily conserved IGF-I peptides. *Mol Endocrinol* 1988, 2:674–681.

39. Simmen FA, Simmen RC, Geisert RD, Martinat-Botte F, Bazer FW, Terqui M: Differential expression, during the estrous cycle and pre- and postimplantation conceptus development, of messenger ribonucleic acids encoding components of the pig uterine insulin-like growth factor system. *Endocrinology* 1992, 130:1547–1556.

40. Green ML, Simmen RCM, Simmen FA: Developmental regulation of steroidogenic enzyme gene expression in the preimplantation porcine conceptus: a paracrine role for insulin-like growth factor-I. *Endocrinology* 1995, 136:3961–3970.

41. Lee CY, Green ML, Simmen RCM, Simmen FA: Proteolysis of insulin-like growth factor-binding proteins (IGFBPs) within the pig uterine lumen associated with peri-implantation conceptus development. *J Reprod Fertil* 1998, 112:369–377.

42. Ashworth MD, Ross JW, Allen DT, Stein DR, Spicer LJ, Geisert RD: Endocrine disruption of uterine insulin-like growth factor (IGF) expression in the pregnant gilt. *Reprod* 2005, 130:545–551.

43. Ka H, Spencer TE, Johnson GA, Bazer FW: Keratinocyte growth factor: Expression by endometrial epithelia of the porcine uterus. *Biol Reprod* 2000, 62:1772–1778.

44. Ka H, Al-Ramadan S, Erikson DW, Johnson GA, Burghardt RC, Spencer TE, Jaeger LA, Bazer FW: Regulation of expression of fibroblast growth factor 7 in the pig uterus by progesterone and estradiol. *Biol Reprod* 2007, 77:172–180.

45. Kaczmarek MM, Blitek A, Kaminska K, Bodek G, Zygmunt M, Schams D, Ziecik AJ: Assessment of VEGF-receptor system expression in the porcine endometrial stromal cells in response to insulin-like growth factor-I, relaxin, oxytocin and prostaglandin E2. *Mol Cell Endocrinol* 2008, 291:33–41.

46. Kaczmarek MM, Kiewisz J, Schams D, Ziecik AJ: Expression of VEGF-receptor system in conceptus during peri-implantation period and endometrial and luteal expression of soluble VEGFR-1 in the pig. *Theriogenology* 2009, 71:1298–1306.

47. Jeong W, Kim J, Bazer FW, Song G: Stimulatory effect of vascular endothelial growth factor on proliferation and migration of porcine trophectoderm cells and their regulation by the phosphatidylinositol-3-Kinase-AKT and mitogen-activated protein kinase cell signaling pathways. *Biol Reprod* 2014, 90:1–10.

48. Anegon I, Cuturi MC, Godard A, Moreau M, Terqui M, Martinat-Botte F, Soulillou JP: Presence of leukaemia inhibitory factor and interleukin 6 in porcine uterine secretions prior to conceptus attachment. *Cytokine* 1994, 6:493–499.

49. Modric T, Kowalski AA, Green ML, Simmen RCM, Simmen FA: Pregnancy-dependent expression of leukaemia inhibitory factor (LIF), LIF receptor-β and interleukin 6 (IL-6) messenger ribonucleic acids in the porcine female reproductive tract. *Placenta* 2000, 21:345–353.

50. Blitek A, Morawska E, Ziecik AJ: Regulation of expression and role of leukemia inhibitory factor and interleukin-6 in the uterus of early pregnant pigs. *Theriogenology* 2012, 78:951–964.

51. Blitek A, Morawska-Pucinska E, Szymanska M, Kiewisz J, Waclawik A: Effect of conceptus on transforming growth factor (TGF) β1 mRNA expression and protein concentration in the porcine endometrium–in vivo and in vitro studies. *J Reprod Dev* 2013, 59:512–519.

52. Jaeger LA, Spiegel AK, Ing NH, Johnson GA, Bazer FW, Burghardt RC: Functional effects of transforming growth factor beta on adhesive properties of porcine trophectoderm. *Endocrinology* 2005, 146:3933–3942.

53. Massuto DA, Kneese EC, Johnson GA, Burghardt RC, Hooper RN, Ing NH, Jaeger LA: Transforming growth factor beta (TGFB) signaling is activated during porcine implantation: proposed role for latency-associated peptide interactions with integrins at the conceptus-maternal interface. *Reproduction* 2010, 139:465–478.

54. Jeong W, Song G, Bazer FW, Kim J: Insulin-like growth factor I induces proliferation and migration of porcine trophectoderm cells through multiple cell signaling pathways, including protooncogenic protein kinase 1 and mitogen-activated protein kinase. *Mol Cell Endocrinol* 2014, 384:175–184.

55. Ka H, Jaeger LA, Johnson GA, Spencer TE, Bazer FW: Keratinocyte growth factor is up-regulated by estrogen in the porcine uterine endometrium and functions in trophectoderm cell proliferation and differentiation. *Endocrinology* 2001, 142:2303–2310.

56. Lawson RA, Cahill LP: Modification of the embryo-maternal relationship in ewes by progesterone treatment early in the oestrous cycle. *J Reprod Fertil* 1983, 67:473–475.

57. Satterfield MC, Bazer FW, Spencer TE: Progesterone regulation of preimplantation conceptus growth and galectin 15 (LGALS15) in the ovine uterus. *Biol Reprod* 2006, 75:289–296.

58. Garrett JE, Geisert RD, Zavy MT, Morgan GL: **Evidence for maternal regulation of early conceptus growth and development in beef cattle.** *J Reprod Fertil* 1988, **84**:437–446.

59. Mann GE, Fray MD, Lamming GE: **Effects of time of progesterone supplementation on embryo development and interferon-tau production in the cow.** *Vet J* 2006, **171**:500–503.

60. Carter F, Forde N, Duffy P, Wade M, Fair T, Crowe MA, Evans AC, Kenny DA, Roche JF, Lonergan P: **Effect of increasing progesterone concentration from Day 3 of pregnancy on subsequent embryo survival and development in beef heifers.** *Reprod Fertil Dev* 2008, **20**:368–375.

61. Satterfield MC, Hayashi K, Song G, Black SG, Bazer FW, Spencer TE: **Progesterone regulates FGF10, MET, IGFBP1, and IGFBP3 in the endometrium of the ovine uterus.** *Biol Reprod* 2008, **79**:1226–1236.

62. Geisert RD, Brookbank JW, Roberts RM, Bazer FW: **Establishment of pregnancy in the pig: II. Cellular remodeling of the porcine blastocysts during elongation on day 12 of pregnancy.** *Biol Reprod* 1982, **27**:941–955.

63. White FJ, Ross JW, Joyce MM, Geisert RD, Burghardt RC, Johnson GA: **Steroid regulation of cell specific secreted phosphoprotein 1 (osteopontin) expression the in the pregnant porcine uterus.** *Biol Reprod* 2005, **73**:1294–1313.

64. Blomberg LA, Schreier L, Li RW: **Characteristics of peri-implantation porcine concepti population and maternal milieu influence the transcriptome profile.** *Mol Reprod Dev* 2010, **77**:978–989.

65. Blomberg LA, Garrett WM, Guillomot M, Miles JR, Sonstegard TS, Van Tassell CP, Zuelke KA: **Transcription profiling of the tubular porcine conceptus identifies the differential regulation of growth and developmentally associated genes.** *Mol Reprod Dev* 2006, **73**:1491–1502.

66. Ross JW, Ashworth MD, Stein DR, Couture OP, Tuggle CK, Geisert RD: **Identification of differential gene expression during porcine conceptus rapid trophoblastic elongation and attachment to uterine luminal epithelium.** *Physiol Genomics* 2009, **36**:140–148.

67. Anderson LL: **Growth, protein content and distribution of early pig embryos.** *Anat Rec* 1978, **190**:143–153.

68. Valdez Magaña G, Rodríguez A, Zhang H, Webb R, Alberio R: **Paracrine effects of embryo-derived FGF4 and BMP4 during pig trophoblast elongation.** *Dev Biol* 2014, **387**:15–27.

69. Albertini DF, Overstrom EW, Ebert KM: **Changes in the organization of the actin cytoskeleton during preimplantation development of the pig embryo.** *Biol Reprod* 1987, **37**:441–451.

70. Mattson BA, Overstrom EW, Albertini DF: **Transitions in trophectoderm cellular shape and cytoskeletal organization in the elongating pig blastocyst.** *Biol Reprod* 1990, **42**:195–205.

71. Erikson DW, Burghardt RC, Bayless KJ, Johnson GA: **Secreted phosphoprotein 1 (SPP1, osteopontin) binds to integrin alpha v beta 6 on porcine trophectoderm cells and integrin alpha v beta 3 on uterine luminal epithelial cells, and promotes trophectoderm cell adhesion and migration.** *Biol Reprod* 2009, **81**:814–25.

72. Yelich JV, Pomp D, Geisert RD: **Ontogeny of elongation and gene expression in the early developing porcine conceptus.** *Biol Reprod* 1997, **57**:1256–1265.

73. Vejlsted M, Offenberg H, Thorup F, Maddox-Hyttel P: **Confinement and clearance of OCT4 in the porcine embryo at stereomicroscopically defined stages around gastrulation.** *Mol Reprod Dev* 2006, **73**:709–718.

74. Vejlsted M, Du Y, Vajta G, Maddox-Hyttel P: **Post-hatching development of the porcine and bovine embryo-defining criteria for expected development in vivo and in vitro.** *Theriogenology* 2006, **65**:153–165.

75. Conley AJ, Christenson LK, Ford SP, Christenson RK: **Immunocytochemical localization of cytochromes P450 17 alpha-hydroxylase and aromatase in embryonic cell layers of elongating porcine blastocysts.** *Endocrinology* 1994, **135**:2248–2254.

76. Fujii T, Sakurai N, Osaki T, Iwagami G, Hirayama H, Minamihashi A, Hashizume T, Sawai K: **Changes in the expression patterns of the genes involved in the segregation and function of inner cell mass and trophectoderm lineages during porcine preimplantation development.** *J Reprod Dev* 2013, **59**:151–158.

77. Shimokawa K, Kimura-Yoshida C, Nagai N, Mukai K, Matsubara K, Watanabe H, Matsuda Y, Mochida K, Matsuo I: **Cell surface heparan sulfate chains regulate local reception of FGF signaling in the mouse embryo.** *Dev Cell* 2011, **21**:257–272.

78. Ornitz DM: **FGFs, heparan sulfate and FGFRs: complex interactions essential for development.** *Bioessays* 2000, **22**:108–112.

79. Ross JW, Ashworth MD, Hurst AG, Malayer JR, Geisert RD: **Analysis and characterization of differential gene expression during rapid trophoblastic elongation in the pig using suppression subtractive hybridization.** *Reprod Biol Endocrinol* 2003, **1**:23.

80. Blomberg LA, Long EL, Sonstegard TS, Van Tassell CP, Dobrinsky JR, Zuelke KA: **Serial analysis of gene expression during elongation of the peri-implantation porcine trophectoderm (conceptus).** *Physiol Genomics* 2005, **20**:188–194.

81. Morgan GL, Geisert RD, Zavy MT, Fazleabas AT: **Development and survival of pig blastocysts after oestrogen administration on day 9 or days 9 and 10 of pregnancy.** *J Reprod Fertil* 1987, **80**:133–141.

82. Tou W, Harney JP, Bazer FW: **Developmentally regulated expression of interleukin-1β by peri-implantation conceptuses in swine.** *J Reprod Immunol* 1996, **31**:185–198.

83. Mantovani A, Muzio M, Ghessi P, Colotta C, Introna M: **Regulation of inhibitory pathways of the interleukin-1 system.** *Ann N Y Acad Sci* 1998, **840**:338–351.

84. Seo H, Kim M, Choi Y, Ka H: **Salivary lipocalin is uniquely expressed in the uterine endometrial glands at the time of conceptus implantation and induced by interleukin 1beta in pigs.** *Biol Reprod* 2011, **84**:279–287.

85. Seo H, Choi Y, Shim J, Choi Y, Ka H: **Regulatory mechanism for expression of IL1B receptors in the uterine endometrium and effects of IL1B on prostaglandin synthetic enzymes during the implantation period in pigs.** *Biol Reprod* 2012, **87**:1–11.

86. Ross JW, Malayer JR, Ritchey JW, Geisert RD: **Characterization of the interleukin-1beta system during porcine trophoblastic elongation and early placental attachment.** *Biol Reprod* 2003, **69**:1251–1259.

87. Mathew DJ, Newsom EM, Geisert RD, Green JA, Tuggle CK, Lucy MC: **Characterization of nucleotide and predicted amino acid sequence of a porcine Interleukin-1 variant expressed in elongated porcine embryos.** *J Anim Sci* 2011, **89**(E-suppl):2.

88. Groenen MA, Archibald AL, Uenishi H, Tuggle CK, Takeuchi Y, Rothschild MF, Rogel-Gaillard C, Park C, Milan D, Megens HJ, Li S, Larkin DM, Kim H, Frantz LA, Caccamo M, Ahn H, Aken BL, Anselmo A, Anthon C, Auvil L, Badaoui B, Beattie CW, Bendixen C, Berman D, Blecha F, Blomberg J, Bolund L, Bosse M, Botti S, Bujie Z, *et al*: **Analyses of pig genomes provide insight into porcine demography and evolution.** *Nature* 2012, **491**:393–398.

89. Waclawik A, Kaczynski P, Jabbour HN: **Autocrine and paracrine mechanisms of prostaglandin E₂ action on trophoblast/conceptus cells through the prostaglandin E₂ receptor (PTGER2) during implantation.** *Endocrinology* 2013, **154**:3864–3876.

90. Ross JW, Ashworth MD, Mathew D, Reagan P, Ritchey JW, Hayashi K, Spencer TE, Lucy M, Geisert RD: **Activation of the transcription factor, nuclear factor kappa-B, during the estrous cycle and early pregnancy in the pig.** *Reprod Biol Endocrinol* 2010, **8**:39.

91. Ali S, Mann DA: **Signal transduction via the NF-κB pathway: targeted treatment modality for infection, inflammation and repair.** *Cell Biochem Funct* 2004, **22**:67–79.

92. Nestler JE: **Interleukin-1 stimulates the aromatase activity of human placental cytotrophoblasts.** *Endocrinology* 1993, **132**:566–570.

93. Kol S, Kehat I, Adashi EY: **Ovarian interleukin-1-induced gene expression: privileged genes threshold theory.** *Med Hypotheses* 2002, **58**:6–8.

94. Geisert RD, Rasby RJ, Minton JE, Wetteman RP: **Role of prostaglandins in development of porcine blastocysts.** *Prostaglandins* 1986, **31**:191–204.

95. Dorniak P, Bazer FW, Spencer TE: **Prostaglandins regulate conceptus elongation and mediate effects of interferon tau on the ovine uterine endometrium.** *Biol Reprod* 2011, **84**:1119–1127.

96. White FJ, Kimball EM, Wyman G, Stein DR, Ross JW, Ashworth MD, Geisert RD: **Estrogen and interleukin-1beta regulation of trophinin, osteopontin, cyclooxygenase-1, cyclooxygenase-2, and interleukin-1beta system in the porcine uterus.** *Soc Reprod Fertil Suppl* 2009, **66**:203–204.

97. Wilson ME, Fahrenkrug SC, Smith TPL, Rohrer GA, Ford SP: **Differential expression of cyclooxygenase-2 around the time of elongation in the pig conceptus.** *Anim Reprod Sci* 2002, **71**:229–237.

98. Franczak A, Zmijewska A, Kurowicka B, Wojciechowicz B, Kotwica G: **Interleukin 1β-induced synthesis and secretion of prostaglandin E₂ in the porcine uterus during various periods of pregnancy and the estrous cycle.** *J Physiol Pharmacol* 2010, **61**:733–742.

99. Franczak A, Zmijewska A, Kurowicka B, Wojciechowicz B, Petroff BK, Kotwica G: **The effect of tumor necrosis factor α (TNFα), interleukin 1β (IL1β) and interleukin 6 (IL6) on endometrial PGF2α synthesis, metabolism and release in early-pregnant pigs.** *Theriogenology* 2012, **77**:155–165.

100. Waclawik A, Blitek A, Kaczmarek MM, Kiewisz J, Ziecik AJ: **Antiluteolytic mechanisms and the establishment of pregnancy in the pig.** *Soc Reprod Fertil Suppl* 2009, **66**:307–320.

101. Sawai K, Matsuzaki N, Okada T, Shimoya K, Koyama M, Azuma C, Saji F, Murata Y: **Human decidual cell biosynthesis of leukemia inhibitory factor: regulation by decidual cytokines and steroid hormones.** *Biol Reprod* 1997, **56**:1274–1280.

102. Perrier D'Hauterive S, Charlet-Renard C, Berndt S, Dubois M, Munaut C, Goffin F, Hagelstein MT, Noël A, Hazout A, Foidart JM, Geenen V: **Human chorionic gonadotropin and growth factors at the embryonic-endometrial interface control leukemia inhibitory factor (LIF) and interleukin 6 (IL-6) secretion by human endometrial epithelium.** *Hum Reprod* 2004, **19**:2633–2643.

103. Gonzalez RR, Leary K, Petrozza JC, Leavis PC: **Leptin regulation of the interleukin-1 system in human endometrial cells.** *Mol Hum Reprod* 2003, **9**:151–158.

104. Kauma SW, Turner TT, Harty JR: **Interleukin-1 beta stimulates interleukin-6 production in placental villous core mesenchymal cells.** *Endocrinology* 1994, **134**:457–460.

105. Nakamura H, Kimura T, Ogita K, Koyama S, Tsujie T, Tsutsui T, Shimoya K, Koyama M, Kaneda Y, Murata Y: **Alteration of the timing of implantation by in vivo gene transfer: delay of implantation by suppression of nuclear factor kappaB activity and partial rescue by leukemia inhibitory factor.** *Biochem Biophys Res Commun* 2004, **321**:886–892.

106. Jasper MJ, Care AS, Sullivan B, Ingman WV, Aplin JD, Robertson SA: **Macrophage-derived LIF and IL1B regulate alpha(1,2)fucosyltransferase 2 (Fut2) expression in mouse uterine epithelial cells during early pregnancy.** *Biol Reprod* 2011, **84**:179–188.

107. Samborski A, Graf A, Krebs S, Kessler B, Reichenbach M, Reichenbach HD, Ulbrich SE, Bauersachs S: **Transcriptome changes in the porcine endometrium during the preattachment phase.** *Biol Reprod* 2013, **89**:134.

108. Franczak A, Wojciechowicz B, Kotwica G: **Transcriptomic analysis of the porcine endometrium during early pregnancy and the estrous cycle.** *Reprod Biol* 2013, **13**:229–237.

109. Kiewisz J, Krawczynski K, Lisowski P, Blitek A, Zwierzchowski L, Ziecik AJ, Kaczmarek MM: **Global gene expression profiling of porcine endometria on Days 12 and 16 of the estrous cycle and pregnancy.** *Theriogenology* 2014, **82**:897–909.

110. Østrup E, Bauersachs S, Blum H, Wolf E, Hyttel P: **Differential endometrial gene expression in pregnant and nonpregnant sows.** *Biol Reprod* 2010, **83**:277–285.

111. Robb L, Li R, Hartley L, Nandurkar HH, Koentgen F, Begley CG: **Infertility in female mice lacking the receptor for interleukin 11 is due to a defective uterine response to implantation.** *Nat Med* 1998, **4**:303–308.

112. Paiva P, Salamonsen LA, Manuelpillai U, Dimitriadis E: **Interleukin 11 inhibits human trophoblast invasion indicating a likely role in the decidual restraint of trophoblast invasion during placentation.** *Biol Reprod* 2009, **80**:302–310.

113. Okamura H, Tsutsi H, Komatsu T, Yutsudo M, Hakura A, Tanimoto T, Torigoe K, Okura T, Nukada Y, Hattori K: **Cloning of a new cytokine that induces IFN-gamma production by T-cells.** *Nature* 1995, **15**:379–384.

114. Ashworth MD, Ross JW, Stein DR, White FJ, Desilva UW, Geisert RD: **Endometrial caspase 1 and interleukin-18 expression during the estrous cycle and peri-implantation period of porcine pregnancy and response to early exogenous estrogen administration.** *Reprod Biol Endocrinol* 2010, **8**:33.

115. Hentze H, Lin XY, Choi MS, Porter AG: **Critical role for cathepsin B in mediating caspase-1 dependant IL18 maturation and caspase-1 independent necrosis triggered by the microbial toxin nigericin.** *Cell Death Differ* 2003, **10**:956–968.

116. Fantuzzi G, Dinarello CA: **Interleukin-18 and interleukin-1β: two cytokine substrates for ICE (caspase-1).** *J Clin Immunol* 1999, **19**:1–11.

117. Cencic A, La Bonnardière C: **Trophoblastic interferon-gamma: current knowledge and possible role(s) in early pig pregnancy.** *Vet Res* 2002, **33**:139–157.

118. Arend WP, Palmer G, Gabay C: **IL-1, IL-18, and IL-33 families of cytokines.** *Immunol Rev* 2008, **223**:20–38.

119. Lee J-K, Kim S-H, Lewis EC, Azam T, Reznikov LL, Dinarello CA: **Differences in signaling pathways by IL1B and IL18.** *Proc Nat Acad Sci* 2004, **101**:8815–8820.

120. Pomerantz BJ, Reznikov LL, Harken AH, Dinarello CA: **Inhibition of caspase 1 reduces human myocardial ischemic dysfunction via inhibition of IL-18 and IL-1beta.** *Proc Natl Acad Sci U S A* 2001, **98**:2871–2876.

121. Joyce MM, Burghardt RC, Geisert RD, Burghardt JR, Hooper RN, Ross JW, Ashworth MD, Johnson GA: **Pig conceptuses secrete estrogen and IFN-γ to differentially regulate uterine STAT 1 in a temporal and cell type-specific manner.** *Endocrinology* 2007, **198**:4420–4431.

122. Hadfield KA, McCracken SA, Ashton AW, Nguyen TG, Morris JM: **Regulated suppression of NF-κB throughout pregnancy maintains a favourable cytokine environment necessary for pregnancy success.** *J Reprod Immunol* 2011, **89**:1–9.

123. Ostojic S, Dubanchet S, Chaouat G, Abdelkarim M, Truyens C, Capron F: **Demonstration of the presence of IL-16, IL-17 and IL-18 at the murine fetomaternal interface during murine pregnancy.** *Am J Reprod Immunol* 2003, **49**:101–112.

124. Joyce MM, Burghardt JR, Burghardt RC, Hooper RN, Jaeger LA, Spencer TE, Bazer FW, Johnson GA: **Pig conceptuses increase uterine interferon-regulatory factor 1 (IRF1), but restrict expression to stroma through estrogen-induced IRF2 in luminal epithelium.** *Biol Reprod* 2007, **77**:292–302.

125. La Bonnardière C, Martinat-Botté F, Terqui M, Lefèvre F, Zouari K, Martal J, Bazer FW: **Production of two species of interferon by Large White and Meishan pig conceptuses during the peri-attachment period.** *J Reprod Fertil* 1991, **91**:469–478.

126. Lefèvre F, Martinat-Botté F, Guillomot M, Zouari K, Charley B, La Bonnardière C: **Interferon-gamma gene and protein are spontaneously expressed by the porcine trophectoderm early in gestation.** *Eur J Immunol* 1990, **20**:2485–2490.

127. Joyce MM, Burghardt JR, Burghardt RC, Hooper RN, Bazer FW, Johnson GA: **Uterine MHC class I molecules and beta 2-microglobulin are regulated by progesterone and conceptus interferons during pig pregnancy.** *J Immunol* 2008, **181**:2494–2505.

128. Harney JP, Bazer FW: **Effect of porcine conceptus secretory proteins on interestrous interval and uterine secretion of prostaglandins.** *Biol Reprod* 1989, **41**:277–284.

129. Hicks BA, Etter SJ, Carnahan KG, Joyce MM, Assiri AA, Carling SJ, Kodali K, Johnson GA, Hansen TR, Mirando MA, Woods GL, Vanderwall DK, Ott TL: **Expression of the uterine Mx protein in cyclic and pregnant cows, gilts, and mares.** *J Anim Sci* 2003, **81**:1552–1561.

130. Cencic A, Guillomot M, Koren S, La Bonnardière C: **Trophoblastic interferons: do they modulate uterine cellular markers at the time of conceptus attachment in the pig?** *Placenta* 2003, **24**:862–869.

131. Keys JL, King GJ, Kennedy TG: **Increased uterine vascular permeability at the time of embryonic attachment in the pig.** *Biol Reprod* 1986, **34**:405–411.

132. Garlanda C, Dinarello CA, Mantovani A: **The interleukin-1 family: back to the future.** *Immunity* 2013, **39**:1003–1018.

133. Edwards AK, Wessels JM, Kerr A, Tayade C: **An overview of molecular and cellular mechanisms associated with porcine pregnancy success or failure.** *Reprod Domest Anim* 2012, **47**(Suppl 4):394–401.

134. Wessels JM, Linton NF, Croy BA, Tayade C: **A review of molecular contrasts between arresting and viable porcine attachment sites.** *Am J Reprod Immunol* 2007, **58**:470–480.

135. Serhan CN: **Pro-resolving lipid mediators are leads for resolution physiology.** *Nature* 2014, **510**:92–101.

136. Morgan GL, Geisert RD, Zavy MT, Shawley RV, Fazleabas AT: **Development of porcine blastocysts in an uterine environment advanced by exogenous oestrogen.** *J Reprod Fert* 1987, **80**:125–131.

137. Gries LK, Geisert RD, Zavy MT, Garrett JE, Morgan GL: **Uterine secretory alterations coincident with embryonic mortality in the gilt after exogenous estrogen administration.** *J Anim Sci* 1989, **67**:276–284.

138. Ross JW, Ashworth MD, White FJ, Johnson GA, Ayoubi PJ, DeSilva U, Whitworth KM, Prather RS, Geisert RD: **Premature estrogen exposure alters endometrial gene expression to disrupt pregnancy in the pig.** *Endocrinology* 2007, **148**:4761–4773.

10

Effects of magnesium on the performance of sows and their piglets

Jianjun Zang[1†], Jingshu Chen[1†], Ji Tian[1], Aina Wang[2], Hong Liu[1], Shengdi Hu[1], Xiangrong Che[3], Yongxi Ma[1], Junjun Wang[1], Chunlin Wang[1], Guanghua Du[3] and Xi Ma[1*]

Abstract

The objective of this study was to evaluate the effects of supplemental magnesium (Mg) on the performance of gilts and parity 3 sows and their piglets. Fifty-six gilts (Trial 1) and 56 sows (Trial 2) were assigned to one of 4 treatments according to their mating weight, respectively. The treatments comprised corn-soybean meal based gestation and lactation diets (0.21% magnesium) supplemented with 0, 0.015, 0.03, or 0.045% Mg from mating until weaning. The results showed that magnesium supplementation significantly ($P < 0.05$) reduced the weaning to estrus interval in both gilts and sows. There were significant effects ($P < 0.05$) of supplemental magnesium on the total number of piglets born, born alive and weaned in sows. In late gestation and lactation, the digestibility of crude fiber (quadratic effects, $P < 0.05$), and crude protein ($P < 0.05$), were significantly influenced by magnesium in gilts and sows, respectively. There were differences among the 4 groups in terms of the apparent digestibility of dry matter and crude fiber in sows ($P < 0.05$) during both early and late gestation. The apparent digestibility of gross energy was increased for sows in late gestation ($P < 0.05$), and lactation (quadratic effects, $P < 0.05$). At farrowing and weaning, serum prolactin levels and alkaline phosphate activities linearly increased in sows as the Mg supplementation increased ($P < 0.05$). Serum Mg of sows at farrowing and serum urea nitrogen of sows at weaning was significantly influenced by Mg supplementation ($P < 0.05$). The Mg concentration in sow colostrum and the serum of their piglets were increased by supplemental magnesium ($P < 0.05$). In addition, growth hormone levels were linearly elevated ($P < 0.05$) in the serum of piglets suckling sows. Our data demonstrated that supplemental magnesium has the potential to improve the reproduction performance of sows, and the suitable supplemental dose ranged from 0.015% to 0.03%.

Keywords: Gilts, Magnesium, Piglets, Reproduction, Sows

Background

The reproductive performance of high producing sows has increased dramatically over the past decades [1] which may contribute to the changes in nutritional requirement of high producing sows [2]. In addition, confinement feeding of swine has resulted in the removal of nutrients and minerals from the soil to animals [3].

It is an important cofactor of several enzymes involved in protein and energy metabolism which also is a constituent of bone (NRC, 1998) [4]. It is considered one of the essential macro-minerals for swine [5]. However, researchers

generally thought that it was unnecessary to supplement magnesium in diets of sows [6], since common corn-soybean meal diets can supply magnesium at levels 0.14 to 0.18% [7], which at least are 3 times more than the NRC (1998) recommendation of magnesium for sows (0.04%) [4].

The precise determination of Mg requirements of farm animals is necessary, depending on the stage of growth, performance and reproduction of the animals [8]. Magnesium supplementation improved the conception rate of sows by 11-15% and reduced the wean to service interval by 9 days. The improved conception rate may also have been influenced by Mg supplementation of boars used to service the sows [9]. However, studies on the relationship between magnesium and sows performance are scarce. The accurate effects of supplemental magnesium

* Correspondence: maxi@cau.edu.cn
†Equal contributors
[1]State Key Laboratory of Animal Nutrition, Ministry of Agriculture Feed Industry Centre, China Agricultural University, Beijing 100193, China
Full list of author information is available at the end of the article

in sow diets on the performance are still unknown, especially the dose-effect relationship. Thus, this study was conducted to reveal the effects of magnesium supplementation in gestation and lactation diets on the performance of sows and gilts as well as their piglets.

Materials and methods

Ethics statement
The experimental protocol was approved by the Animal Care and Use Ethics Committee of China Agricultural University (Beijing, China).

Experimental design and animals
Fifty-six gilts (Trial 1) and 56 sows (Trial 2) were either assigned into one of 4 treatments according to mating weight. The treatments contained corn-soybean meal based gestation and lactation diets (0.21% magnesium). The basal diets were supplemented with 0, 0.015%, 0.03%, or 0.045% magnesium from mating until weaning. Magnesium, in the form of $MgSO_4 \cdot 7H_2O$ (17.4% Mg) was added at the expense of corn (Table 1). Fourteen gilts (mean BW = 141 ± 5.7 kg) or sows (mean BW = 237 ± 8.8 kg) were fed each diet. All diets were fortified to meet or exceed the nutrient requirements recommended by the NRC (1998) [4]. The guidelines of the China Agricultural University Animal Care and Use Ethics Committee is referred to the Regulations of Laboratory Animal of China published in 1988 [10].

Experimental procedures
Sows were individually housed in 2.20 m × 0.65 m gestation crates during the first 107 ± 1 d of gestation and were then moved to farrowing crates (2.2 m × 1.8 m) 7 d prior to farrowing. Gilts were offered 1.8 kg of gestation feed in 2 feedings per day from mating to d 84 of gestation and 2.5 kg of gestation feed in 2 feedings per day from d 85 of gestation until movement into lactation, while sows were offered 2 kg in 2 feedings per day from mating to d 84 of gestation and 3 kg feed per day in 2 feedings from d 85 of gestation to movement into lactation. After farrowing, gilts and sows were fed the lactation diets 3 times per day *ad libitum* and litters were standardized to 11 piglets by cross fostering within treatments during the first 48 h postpartum. Piglets were weaned at 28 d of age and had no access to creep feed. Sows and piglets were given free access to water throughout the experimental period. Ambient temperature in the gestation and farrowing rooms was maintained between 18–22°C.

The numbers and weights of individual pigs were recorded at birth and weaning. Only those pigs alive at first weighing were regarded as born alive. Defacation was observed to find out if magnesium had any effect on lowering the incidence of constipation. Blood samples were collected from ear vein of 7 gilts or 7 sows per treatment

Table 1 Ingredient composition and chemical analysis of the gestation and lactation diets fed to gilts and sows (as-fed basis)

Item	Gestation	Lactation
Ingredients, %		
Corn	62.94	60.38
Soybean meal	16.78	20.10
Extruded soybean	-	14.00
Glucose	-	1.50
Wheat bran	16.27	-
Dicalcium phosphate	2.05	1.96
Limestone	1.24	1.21
Sodium chloride	0.32	0.45
Vitamin mix[1]	0.04	0.04
Mineral mix[2]	0.20	0.20
Choline chloride (50%)	0.16	0.16
Analyzed nutrient levels, %		
Dry matter	88.00	88.00
Crude protein	16.60	17.10
Crude fiber	4.90	5.39
Calcium	1.14	1.23
Total phosphorus	0.59	0.60
Magnesium	0.21	0.21
Calculated nutrient levels		
Metabolizable energy, kcal/kg	3,263	3,270
Lysine, %	0.56	0.98
Methionine + cystine, %	0.37	0.45
Threonine, %	0.46	0.60
Tryptophan, %	0.11	0.18

[1]Supplied per kilogram of complete diet: vitamin A, 4,000 IU; vitamin D_3,1,000 IU; vitamin E, 75 IU; vitamin K_3, 0.86 mg; riboflavin, 6.0 mg; pantothenic acid, 15.0 mg; niacin, 30 mg; cobalamin, 0.025 mg; biotin, 0.2 mg; folic acid, 0.6 mg; thiamine, 5.0 mg; pyridoxine, 6 mg.
[2]Supplied per kilogram of complete diet: iron, 45.0 mg; zinc, 35.0 mg; manganese, 10.0 mg; iodine, 0.2 mg; copper, 6.0 mg; selenium, 0.17 mg.

and from precaval vein of their piglets (4 males and 4 females per litter) 12 h after farrowing, also at 28-d weaning gilts and sows without fasting, and therefore the values reflect the dietary supply of nitrogen and amino acids, or the dietary imbalance of nitrogen and amino acids. Then the blood was put into heparinized tubes. Samples were centrifuged at 3,000 × g for 15 min at 4°C, and serum was stored at −20°C for assays of biochemical parameters. Gilts and sows serum urea nitrogen, reproductive hormone activities, as well as gilts, sows, and their piglets' serum calcium, magnesium concentrations and growth hormone activities were determined. Analysis of gilts and sows serum urea nitrogen concentration and hormone activities in the collected materials was performed using enzyme-linked immunosorbent assay (ELISA) according to the instructions of the kits (R & D Systems Company, Minneapolis, MN).

Colostrum or milk samples were taken manually from the same mammary glands (first, third and fifth teat on both sides) of 7 examined sows in each group within or after 6 h postpartum, respectively, in order to analyze the level of calcium and magnesium. Calcium was determined following the methods of AOAC (2005) [11] while magnesium levels were tested using the method suggested by Miller et al. [12].

Chromic oxide (0.20%) was supplemented to each of diets as an inert marker for a period of 7 d before fecal collection to determine apparent nutrient digestibility. Fresh feces samples were collected by rectal massage on d 45 to 48 and d 95 to 98 of gestation as well as d 18–22 of lactation. Feces were immediately stored at –20°C until analysis. Samples were thawed, dried and mixed uniformly within each sow and sub-samples were finely ground through a 1 mm sieve for chemical analysis. Apparent total tract digestibility (ATTD) was calculated and the indigestible marker method using the following formula:

$$ATTD\% = [1-(N_F/N_D) \times (Cr_D/Cr_F)] \times 100$$

Where N_F and N_D represent the nutrient concentration (%) in feces and diet dry matter respectively, and Cr_D and Cr_F represent the chromium concentrations (%) in diet and fecal dry matter respectively.

Samples of diets and feces were analyzed for gross energy using bomb calorimetry (Parr Instruments, Moline, IL; AOAC, 2005) [11]. All feces and experimental diets were analyzed for dry matter (AOAC method 930.15, 2000) [13], crude protein (AOAC method 990.03, 2000) [13], ether extract (Thiex et al., 2003) [14], crude fiber (AOAC method 978.10, 2000) [13], ash (AOAC method 942.05, 2000) [13], calcium (AOAC method 927.02, 2000) [13] and total phosphorous (AOAC method 965.17, 2000) [13].

Statistical analysis

All experimental data (n = 16) were statistically evaluated by one-way ANOVA using the GLM procedure of SAS (SAS 8.01 Institute, Cary, NC). In addition, polynomial contrasts were made to determine the linear and quadratic effects of magnesium supplementation on the various parameters measured. $P < 0.05$ was considered as significantly.

Results

Performance

As shown in Table 2, magnesium supplementation significantly reduced the weaning to estrus interval in gilts (linear and quadratic effect $P < 0.05$). However, the total number of piglets born and born alive, the number of piglets weaned, birth weight, and weaning weight were unaffected by magnesium supplementation for gilts ($P > 0.05$). These results agreed with a previous report which revealed that supplemental magnesium did not influence the number or weight of pigs at birth or weaning in gilts [6].

In contrast, there were significant differences among the treatments in terms of sow performance (Table 3). Supplemental magnesium significantly increased the total number of piglets born, born alive, and weaned ($P < 0.05$). The increase was particularly evident for sows fed 0.015 and 0.03% magnesium (quadratic effect, $P < 0.05$). In addition, the weaning to estrus interval was shortened (quadratic effect, $P < 0.05$) for these treatments. Katalin et al. also indicated that magnesium supplementation improved reproduction performance (conception rate and litter size) and shortened the weaning to service interval of sows fed magnesium supplemented diets from farrowing to subsequent mating [15].

Apparent total tract digestibility

The effects of magnesium on the apparent total tract digestibility (ATTD) of various chemical constituents in

Table 2 Effects of supplemental magnesium on performance of gilts and their progeny (Exp. 1)

Item	Supplemental magnesium, %				SEM[1]	P value	
	0	0.015	0.03	0.045		Linear	Quadratic
Number born per litter[2]	11.2	11.7	12.0	11.6	0.48	0.46	0.51
Number of pigs born alive per litter	10.3	10.8	10.9	10.4	0.38	0.76	0.44
Average pig weight at birth, kg	1.40	1.42	1.43	1.32	0.06	0.42	0.36
Pigs weaned per litter	9.4	9.6	9.7	9.3	0.26	0.88	0.44
Average pig weight at weaning, kg	7.02	6.91	7.01	6.89	0.34	0.24	0.96
Wean to estrus interval, d	7.4	6.6	6.9	6.3	0.25	0.01	0.04
Constipation rate[3], %	57.1	50.0	32.1	21.4	0.36	0.03	0.35
Average daily lactation feed intake of gilts, kg	4.82	4.86	4.81	4.79	0.70	0.72	0.51

[1] SEM = Standard error of mean.
[2] Number of litters represented per dietary treatment.
[3] Constipation rate (%) was defined as numbers of pigs on a treatment with constipation/(total number of pigs × 28d) × 100%.

Table 3 Effects of magnesium on performance in parity 3 sows and their piglets (Exp. 2)

Item	Supplemental magnesium, %				SEM[1]	P value	
	0	0.015	0.03	0.045		Linear	Quadratic
Number born per litter[2]	12.1	13.7	12.8	12.4	0.37	0.93	0.04
Number of pigs born alive per litter	10.4	12.5	11.4	10.8	0.35	0.91	0.01
Average pig weight at birth, kg	1.27	1.30	1.46	1.42	0.06	0.04	0.09
Pigs weaned per litter	9.21	10.64	10.0	9.43	0.21	1.00	0.01
Average pig weight at weaning, kg	7.1	7.7	7.3	6.9	0.18	0.46	0.03
Wean to estrus interval, d	7.1	6.3	6.1	6.8	0.19	0.30	0.01
Constipation rate[3], %	64.2	42.8	35.7	21.4	0.06	0.02	0.36
Average daily lactation feed intake of sows, kg	5.12	5.13	5.17	5.10	0.62	0.56	0.82

[1]SEM = Standard error of mean.
[2]Number of litters represented per dietary treatment.
[3]Constipation rate (%) was defined as numbers of pigs on a treatment with constipation/(total number of pigs × 28d) × 100%.

gilts and sows were shown in Tables 4 and 5. For gilts, the ATTD of crude protein and crude fiber were quadratically ($P < 0.05$) affected by magnesium level in late gestation, and meanwhile the ATTD of crude fiber was quadratically ($P < 0.05$) affected by magnesium level in lactation (Table 4). In each case, the ATTD of gilts fed with 0.015 or 0.03% magnesium were higher than those for gilts fed with 0.0 and 0.045% magnesium. The ATTD of gross energy was linearly ($P < 0.05$) increased by supplemental magnesium in late gestation. There was no

effect of magnesium level on the ATTD of any nutrient during early gestation ($P < 0.05$)

For sows during early gestation, the ATTD of dry matter and crude fiber, was linearly and quadratically affected ($P < 0.05$) by magnesium supplementation (Table 5). The ATTD of sows fed the 0.015, 0.03, and 0.045% magnesium treatments were higher than those for sows fed the 0.0 magnesium treatments.

In late gestation, the ATTD of dry matter, gross energy, crude protein, and crude fiber were linearly and

Table 4 Apparent nutrient digestibility (%)[2] of gestation or lactation diets for gilts containing various dietary levels of magnesium (Exp. 1)

Item	Supplemental magnesium, %				SEM[1]	P value	
	0	0.015	0.03	0.045		Linear	Quadratic
Early gestation (d 45–48)							
Dry matter	85.5	86.6	86.5	86.2	0.23	0.49	0.67
Gross energy	83.1	82.6	81.8	81.4	0.20	0.70	0.75
Crude protein	86.4	86.8	88.3	84.9	0.14	0.12	0.54
Ether extract	86.3	80.1	87.7	85.0	0.21	0.81	0.77
Crude fiber	57.3	58.8	58.2	55.5	0.13	0.41	0.47
Late gestation (d 95–98)							
Dry matter	79.3	82.9	81.2	78.3	1.49	0.49	0.08
Gross energy	76.8	77.4	83.4	81.5	1.67	0.03	0.07
Crude protein	77.7	84.5	80.5	79.3	1.47	0.90	0.04
Ether extract	82.9	84.7	82.6	82.5	1.35	0.57	0.68
Crude fiber	48.9	58.3	56.4	52.0	1.08	0.44	0.01
Lactation (d 18–21)							
Dry matter	85.3	84.3	84.7	84.2	0.88	0.44	0.71
Gross energy	82.2	84.3	83.2	83.7	1.03	0.48	0.56
Crude protein	84.0	84.5	85.5	82.9	0.85	0.59	0.21
Ether extract	85.9	85.7	86.1	86.1	1.16	0.86	0.97
Crude fiber	48.2	53.6	58.7	50.7	0.92	0.18	0.01

[1]SEM = Standard error of mean.
[2]Digestibility was determined using chromium oxide as marker.

Table 5 Apparent nutrient digestibility (%)[2] of gestation or lactation diets for parity 3 sows containing various dietary levels of magnesium (Exp. 2)

Item	Supplemental magnesium, %				SEM[1]	P value	
	0	0.015	0.03	0.045		Linear	Quadratic
Early gestation (d 45–48)							
Dry matter	80.0	86.9	85.7	85.6	0.99	0.02	0.02
Gross energy	83.5	86.3	86.2	89.4	1.53	0.83	0.08
Crude protein	83.6	86.6	87.1	86.6	1.06	0.81	0.61
Ether extract	86.4	86.5	85.8	86.0	1.20	0.71	0.31
Crude fiber	51.7	58.6	57.4	60.1	0.98	0.01	<0.01
Late gestation (d 95–98)							
Dry matter	78.1	83.8	83.5	83.9	1.05	0.01	0.01
Gross energy	81.5	84.1	84.0	81.1	1.24	0.01	0.04
Crude protein	83.8	84.0	85.3	83.7	0.88	0.05	0.04
Ether extract	85.9	89.1	86.2	86.4	0.92	0.71	0.93
Crude fiber	50.3	59.9	61.9	58.5	0.81	0.01	0.03
Lactation (d 18–21)							
Dry matter	84.6	86.1	88.2	85.2	0.60	0.41	0.08
Gross energy	77.8	82.1	82.2	75.9	0.72	0.45	0.01
Crude protein	80.9	84.9	84.8	85.5	1.03	0.02	0.03
Ether extract	74.58	79.47	84.14	82.36	0.56	0.14	0.11
Crude fiber	78.9	79.6	84.2	78.6	1.64	0.69	0.34

[1]SEM = Standard error of mean.
[2]Digestibility was determined using chromium oxide as marker.

Table 6 Effects of magnesium on serum urea nitrogen, calcium, and magnesium concentrations as well as reproductive hormone activities in gilts (Exp. 1)

Item	Supplemental magnesium, %				SEM[1]	P value	
	0	0.015	0.03	0.045		Linear	Quadratic
Farrowing							
Urea nitrogen, pmol/L	8,693.8	8,257.9	8,001.4	8,087.1	255.7	0.07	0.12
Estrogen, pmol/L	68.3	69.0	70.0	73.9	2.94	0.18	0.35
Progesterone, pmol/L	4,598.4	4,340.6	4,566.0	4,273.9	140.6	0.24	0.50
Prolactin, ng/L	133.8	161.4	149.9	138.7	8.82	0.94	0.10
Alkaline phosphate, U/L	8.72	9.97	9.36	9.45	0.37	0.11	0.29
Magnesium, mg/100 mL	1.83	2.00	1.85	1.89	0.05	0.89	0.10
Calcium, mg/100 mL	8.22	8.78	8.61	8.59	0.21	0.33	0.25
Weaning (d 28)							
Urea nitrogen, pmol/L	8,728.6	8,524.1	8,308.1	8,591.6	214.6	0.79	0.84
Estrogen, pmol/L	12.94	13.72	13.13	13.99	0.75	0.44	0.74
Progesterone, pmol/L	243.1	221.4	239.1	255.1	4.14	0.34	0.31
Prolactin, ng/L	122.9	135.5	130.3	127.8	7.64	0.78	0.59
Alkaline phosphate, U/L	13.62	13.80	14.26	14.79	0.63	0.16	0.36
Magnesium, mg/100 mL	1.36	1.44	1.34	1.42	0.09	0.94	0.99
Calcium, mg/100 mL	5.66	5.00	5.15	5.71	0.28	0.82	0.11

[1]SEM = Standard error of means.

quadratically affected ($P < 0.05$) by magnesium supplementation. The ATTD of crude protein (linear and quadratic effects, $P < 0.05$), and gross energy (quadratic effects, $P < 0.05$) were also affected by supplemental magnesium in lactation. However, there was no significant ($P > 0.05$) difference in terms of crude fiber in lactation. Further research is needed to clarify the role of magnesium in influencing digestibility in gilts and sows.

Serum parameters

Magnesium level had no effect on the serum levels of magnesium in gilts at either farrowing or weaning (Tables 6). However, the serum levels of magnesium in farrowing sows, was significantly affected ($P < 0.05$) by magnesium supplementation (Tables 7). Magnesium level had no effect on the serum levels of calcium in gilts or sows at farrowing or weaning (Tables 6 and 7). Serum prolactin concentrations and alkaline phosphate activities were linearly increased ($P < 0.05$) by increasing the dietary magnesium level in sows but not gilts at both farrowing and weaning (Table 7).

Calcium, magnesium, and growth hormone levels in colostrums, milk and piglet serum

Levels of calcium, magnesium, and growth hormone in gilt and sow milk and piglet serum were presented in Tables 8 and 9, respectively. Magnesium supplementation had no significant effect on magnesium and calcium concentration in colostrum and milk in gilt and progeny serum. Growth hormone was unaffected as well.

With increasing magnesium levels in sow lactation diets, the concentration of magnesium in colostrum was increased (linear and quadratic effect, $P < 0.05$). Serum magnesium concentration in the serum of piglets suckling sows were also increased by increased levels of dietary magnesium (linear and quadratic effect, $P < 0.05$). In addition, growth hormone levels were linearly increased in nursing progeny from sows or gilts supplemented with Mg.

Discussion

The reproductive performance of high producing sows has increased dramatically in the past decade [1] and it is possible that the nutritional requirement of high producing sows has been altered [2]. Our results indicated that the sows appeared to respond to supplemental magnesium. It is well known that body stores of minerals become increasingly depleted in high producing sows with advancing parity [16]. Therefore, it is possible that magnesium body storage declines as the sow ages, which may increase the sow's reliance on magnesium provided in diets. Another possible explanation for the improved reproductive performance of sows supplemented with magnesium may be related to a reduced incidence of constipation. Treatments of magnesium had lower incidence of constipation than the control group. Constipation has been shown to negatively affect the reproductive performance

Table 7 Effects of magnesium on serum urea nitrogen, calcium and magnesium concentrations as well as reproductive hormone activities in sows (Exp. 2)

Item	Supplemental magnesium, %				SEM[1]	P value	
	0	0.015	0.03	0.045		Linear	Quadratic
Farrowing							
Urea nitrogen, pmol/L	8,970.2	8,695.2	8,001.4	7,890.1	102.2	0.91	0.43
Estrogen, pmol/L	83.4	90.5	85.5	78.8	5.42	0.44	0.33
Progesterone, pmol/L	9,235.8	8,000.2	8,247.7	8,365.6	107.5	0.17	0.08
Prolactin, ng/L	138.5	152.5	185.5	195.7	7.54	<0.01	<0.01
Alkaline phosphate, U/L	8.69	9.06	8.92	10.58	0.44	0.01	0.01
Magnesium, mg/100 mL	1.76	1.93	1.99	1.95	0.06	0.02	0.02
Calcium, mg/100 mL	8.63	9.07	9.99	9.02	0.38	0.52	0.70
Weaning (d 28)							
Urea nitrogen, pmol/L	9,423.0	8,834.4	7,557.9	8,227.9	38.4	0.01	0.01
Estrogen, pmol/L	27.2	26.3	27.3	25.2	1.38	0.39	0.63
Progesterone, pmol/L	206.3	227.2	199.9	228.0	14.4	0.56	0.82
Prolactin, ng/L	139.3	141.2	184.6	196.8	6.1	<0.01	<0.01
Alkaline phosphate, U/L	10.7	14.2	14.8	14.9	0.65	<0.01	<0.01
Magnesium, mg/100 mL	1.73	1.94	1.97	1.86	0.09	0.27	0.10
Calcium, mg/100 mL	8.48	8.62	7.51	8.58	0.42	0.67	0.52

[1]SEM = Standard error of mean.

Table 8 The effects of magnesium on calcium, magnesium and growth hormone levels in colostrum and milk of gilts and piglet serum (Exp. 1)

Item	Supplemental magnesium, %				SEM[1]	P value	
	0	0.015	0.03	0.045		Linear	Quadratic
Colostrum							
Magnesium, mg/100 mL	9.47	8.43	7.56	7.73	0.68	0.31	0.54
Calcium, mg/100 mL	85.67	82.12	69.16	83.23	8.11	0.72	0.76
Milk							
Magnesium, mg/100 mL	11.20	12.61	11.90	11.47	0.27	0.83	0.29
Calcium, mg/100 mL	368.50	383.69	359.10	406.37	23.56	0.65	0.41
Piglet serum							
Magnesium, mg/100 mL	2.23	2.37	2.22	2.68	0.29	0.91	0.76
Calcium, mg/100 mL	8.37	9.53	9.87	11.85	3.38	0.78	0.87
Growth hormone, μg/L	6.81	14.47	9.71	12.71	1.28	0.53	0.33

[1]SEM = Standard error of mean.

of sows [17] and magnesium sulfate has been successfully used as a laxative to prevent constipation in gestating and lactating sows [18]. Although it is well known that the lowest dietary level of Mg was 3.5 fold higher than NRC (2012) and 5 fold higher than NRC (1998), our data demonstrated that supplemental magnesium has the potential to improve the reproduction performance of sows, and the suitable supplemental dose ranged from 0.015% to 0.03%. Further researches will reveal the actual requirements.

Magnesium was an important cofactor of several enzymes involved in protein and energy metabolism and was involved in many biochemical processes including activation of phosphates and participation in carbohydrate metabolism [15]. Where positive effects of magnesium supplementation on ATTD were observed, it is likely that the improvements were mediated by some effect of magnesium on the activity of some of these

enzymes. Further investigation will be conducted to analyze what kinds of enzymes are mediated by adding magnesium. Apart from this, effect of Mg supplementation on rate of passage and water/electrolyte balance should also be considered in the following research.

Previous reports on the effects of dietary magnesium levels on serum magnesium levels in swine are inconsistent. Harmon et al. [6] reported that an increase in dietary magnesium increased serum magnesium levels in weaned pigs [6]. However, Svajgr et al. reported that supplemental magnesium had a negative influence on serum magnesium levels in growing and finishing pigs [7]. Nuoranne et al. concluded that serum magnesium is not a reliable index of body magnesium status [19]. Excess magnesium antagonizes calcium leading to a greater excretion and lower absorption of calcium [20]. Therefore, it was expected that serum calcium levels would be reduced by magnesium supplementation. However, in

Table 9 The effect of magnesium on calcium, magnesium and growth hormone levels in colostrum and milk of sows and piglet serum (Exp. 2)

Item	Supplemental magnesium, %				SEM[1]	P value	
	0	0.015	0.03	0.045		Linear	Quadratic
Colostrum							
Magnesium, mg/100 mL	7.45	7.93	8.24	9.68	0.25	0.01	0.03
Calcium, mg/100 mL	81.73	91.58	77.57	101.73	7.89	0.17	0.39
Milk							
Magnesium, mg/100 mL	12.61	11.07	10.32	12.13	0.51	0.60	0.21
Calcium, mg/100 mL	271.27	274.43	230.46	384.53	39.41	0.50	0.46
Piglet serum							
Magnesium, mg/100 mL	1.82	2.90	2.95	2.90	0.18	0.02	0.04
Calcium, mg/100 mL	8.63	8.53	8.30	10.53	0.41	0.32	0.39
Growth hormone, μg/L	9.55	10.68	8.95	10.29	1.69	0.04	0.11

[1]SEM = Standard error of mean.

agreement with the results of the current study, Harmon et al. reported that serum calcium levels were not influenced by dietary magnesium level [6]. It is well known that the main function of prolactin is to induce the mammary gland to produce milk [21]. The more milk, the faster the growth of piglets [22], which might be the reason that litter weight gain was increased by magnesium supplementation in diets fed to sows. Additionally, with increasing magnesium levels in sow lactation diets, the concentration of magnesium in colostrum was increased, as well as serum magnesium concentration in the serum of piglets suckling sows. Growth hormone levels were linearly increased in nursing progeny from sows or gilts supplemented with Mg. On the other hand, our data indicated that 0.015% Mg supplementation had a higher litter gain than the control and the other treatments (0.03 or 0.045%), although there was a statistical increase in mean piglet weight at birth between treatments (0.03 or 0.045%) and the control. Therefore, this increase in serum growth hormone levels may only partially explain the increase in litter weight gain for piglets suckling sows fed supplemental magnesium, and the dose-effect relationship between the Mg supplementation and litter gain were not linear. Further researches are needed to perform to reveal the potential mechanism.

Conclusion

In conclusion, our data indicated that increased oral administration with magnesium could reduce the weaning to estrus interval in both gilts and sows. In addition, magnesium supplementation improved other reproductive parameters of sows, but not gilts. Supplemental Mg ranging from 0.015% to 0.03% was suitable for high producing sows. The effect appeared to be age related which may be due to depleted body stores of minerals in high producing sows as they age [16]. Therefore, it is possible that as the sow ages, magnesium stores in the body decline, increasing the sow's reliance on the diet to provide magnesium.

Competing interests
All the authors declare that they have no competing interests in the present work.

Authors' contributions
Conceived and designed the experiments: JJZ XRC XM; Performed the research JJZ JSC ANW JT SDH YXM CLW GHD; Analyzed the data: JJZ JSC HL JT SDH; Wrote and edit the manuscript: JJZ JJW XM. All authors read and approved the final manuscript.

Acknowledgements
We thank Dr. Jamie Lynn Kazenstein, University of Georgia Institute of Technology, for excellent assistance in editing the manuscript. The financial support from the Chinese Universities Scientific Fund (No. 15059102, 2014JD017, 2012QJ102, 2012QJ105), National "Twelfth Five-Year" Science & Technology Pillar Program (No. 2011BAD26B02) and National Department Public Benefit Research Foundation (201403047) are gratefully acknowledged.

Author details
[1]State Key Laboratory of Animal Nutrition, Ministry of Agriculture Feed Industry Centre, China Agricultural University, Beijing 100193, China. [2]Weifang Business Vocational College, Zhucheng, Shandong 262234, China. [3]College of Animal Science and Veterinary Medicine, Shanxi Agricultural University, Taigu, Shanxi 030801, China.

References
1. Ronald OB, Samuel RS, Moehn S: Nutrient Requirements of Prolific Sows. Adv Pork Prod 2008, 19:223–236.
2. Ebert AR, Berman AS, Harrell RJ, Kessler AM, Cornelius SG, Odle J: Vegetable proteins enhance the growth of milk-fed piglets, despite lower apparent ileal digestibility. J Nutr 2005, 135:2137–2143.
3. Mahan DC, Kim YY: The role of vitamins and minerals in the production of high quality pork. A review Asian-Aust J Anim Sci 1999, 12:287–294.
4. National Research Council (NRC): Nutrient Requirements of Swine. Washington, DC: 10th ed. National Academy Press; 1998.
5. Miller ER, Kornegay ET: Mineral and vitamin nutrition of swine. J Anim Sci 1983, 57:315–329.
6. Harmon BG, Liu CT, Jensen AH, Baker DH: Dietary magnesium levels for sows during gestation and lactation. J Anim Sci 1976, 42:860–865.
7. Svajgr AJ, Peo ER, Vipperman PE: Effects of dietary levels of manganese and magnesium on performance of growing-finishing swine raised in confinement and on pasture. J Anim Sci 1969, 29:439–443.
8. Gaál KK, Sáfár O, Gulyás L, Stadler P: Magnesium in animal nutrition. J Am Coll Nutr 2004, 23:754S–757S.
9. Kova'csne' Gaa'l K, Szerdajelyi A: Effect of magnesium supplementation on the reproduction performance of young sows. A' llattenye'szte's e's takarma'nyoza's 1987, 36:123.
10. Song: Regulations of Laboratory Animal of the People's Republic of China. Natl Sci Tech Comm 1988, 2.
11. Association of Official Analytical Chemists (AOAC: Official Methods of Analysis. Gaithersburg, MD: 18th ed. Association of Official Analytical Chemists; 2005.
12. Miller ER, Ullrey DE, Zutaut CL, Hoefer JA, Luecke RW: Mineral balance studies with the baby pig: Effects of dietary vitamin D2 level upon calcium, phosphorus and magnesium balance. J Nutr 1965, 86:209–212.
13. Association of Official Analytical Chemists (AOAC): Official Methods of Analysis. 18th edition. Gaithersburg, MD: Association of Official Analytical Chemists; 2000.
14. Thiex NJ, Anderson S, Gildemeister B: Crude fat, diethyl ester extraction, in feed, cereal grain, and forage (Randall/Soxtec/submersion method): Collaborative study. J AOAC Int 2003, 86:888–898.
15. Katalin KG, Sa'fa'r O, Gulya's L, Stadler P: Magnesium in animal nutrition. J Amer Coll Nutr 2004, 23:754S–757S.
16. Mahan D, Taylor-Pickard J: Meeting the mineral needs of highly prolific sows. Pig Progr 2008, 24:21–23.
17. Oliviero C, Kokkonen T, Heinonen M, Sankari S, Peltoniemi O: Feeding sows with high fibre diet around farrowing and early lactation: impact on intestinal activity, energy balance related parameters and litter performance. Res Vet Sci 2009, 86:314–319.
18. Young LG, King GJ, McGirr L, Sutton JC: Moldy corn in diets of gestating and lactating swine. J Anim Sci 1982, 54:976–982.
19. Nuoranne PJ, Raunio RP, Saukko P, Karppanen H: Metabolic effects of a low-magnesium diet in pigs. Br J Nutr 1980, 44:53–60.
20. Tillman AD: Recent developments in beef cattle feeding. Proc Pfizer Res Conf 1966, 8:14–15.
21. Delouis C: Physiology of colostrum production. Ann Rech Vet 1978, 9:193–203.
22. Quesnel H, Meunier-Salaün MC, Hamard A, Guillemet R, Etienne M, Farmer C, Dourmad JY, Père MC: Dietary fiber for pregnant sows: influence on sow physiology and performance during lactation. J Anim Sci 2009, 87:532–543.

11

Effects of different dietary energy and protein levels and sex on growth performance, carcass characteristics and meat quality of F1 Angus × Chinese Xiangxi yellow cattle

Lingyan Li[1], Yuankui Zhu[2], Xianyou Wang[1], Yang He[1] and Binghai Cao[1*]

Abstract

Background: The experiment evaluated the effect of nutrition levels and sex on the growth performance, carcass characteristics and meat quality of F1 Angus × Chinese Xiangxi yellow cattle.

Methods: During the background period of 184 d,23 steers and 24 heifers were fed the same ration,then put into a 2 × 2 × 2 factorial arrangement under two levels of - dietary energy (TDN: 70/80% DM), protein (CP: 11.9/14.3% DM) and sex (S: male/female) during the finishing phase of 146 d. The treatments were - (1) high energy/low protein (HELP), (2) high energy/high protein (HEHP), (3) low energy/low protein (LELP) and (4) low energy/high protein (LEHP). Each treatment used 6 steers and 6 heifers, except for HELP- 5 steers and 6 heifers.

Results: Growth rate and final carcass weight were unaffected by dietary energy and protein levels or by sex. Compared with the LE diet group, the HE group had significantly lower dry matter intake (DMI, 6.76 vs. 7.48 kg DM/d), greater chest girth increments (46.1 vs. 36.8 cm), higher carcass fat (19.9 vs.16.3%) and intramuscular fat content (29.9 vs. 22.8% DM). The HE group also had improved yields of top and medium top grade commercial meat cuts (39.9 vs.36.5%). The dressing percentage was higher for the HP group than the LP group (53.4 vs. 54.9%). Steers had a greater length increment (9.0 vs. 8.3 cm), but lower carcass fat content (16.8 vs. 19.4%) than heifers. The meat quality traits (shear force value, drip loss, cooking loss and water holding capacity) were not affected by treatments or sex, averaging 3.14 kg, 2.5, 31.5 and 52.9%, respectively. The nutritive profiles (both fatty and amino acid composition) were not influenced by the energy or protein levels or by sex.

Conclusions: The dietary energy and protein levels and sex significantly influenced the carcass characteristics and chemical composition of meat but not thegrowth performance, meat quality traits and nutritive profiles.

Keywords: Carcass characteristics, Energy, F1 Angus × Chinese Xiangxi yellow cattle, Growth performance, Meat quality, Protein, Sex

Background

Angus is one of the most popular breeds of cattle used in beef production because it has a considerable growth rate, high carcass yield and well-marbled meat. Xiangxi yellow cattle area breed of Chinese indigenous yellow cattle that is bred in the northwest of Hunan Province. It was included in *the National Protection List of Livestock and Poultry Genetic Resources* of China in 2006 [1]. The breed is well-adapted to low-quality roughage and high temperature environments, its mature weight does not exceed 400 kg, its growth rate is under 0.5 kg/d and the dressing percentage and longissimus muscle (LM) area are 49.48% and 46.75 cm^2, respectively [2]. In the past, the low level of agricultural mechanization has meant that the yellow cattle in China were only used as draft animals. However as a result of rapid economic development, the standard of living in many communities has increased, which has produced a higher demand for beef in terms of

* Correspondence: caobhchina@163.com
[1]National Beef Cattle Industry and Technology System, College of Animal Science and Technology, China Agricultural University, Beijing 100193, China
Full list of author information is available at the end of the article

both quantity and quality. Therefore, to meet the current market demands, methods to increase beef quality and quantity have been introduced, such as crossing superior foreign breeds with native breeds and manipulating nutrition. Dietary nutrition has important roles in growth performance, carcass quality and meat quality traits [3-5]. Previous studies have assessed the effect of nutrition on the growth performance, carcass characteristics and meat quality of Angus [6] or Angus crossbred cattle, such as Angus × Holstein–Friesian, Angus × Gelbvieh and Angus × Limousin cattle [7-9]. However, little research has been performed on the crossbred progeny of Angus and Chinese yellow cattle. Therefore, the aim of this study was to assess the effects of different dietary energy and protein levels and sex on the growth performance, carcass characteristics and meat quality of F1 Angus × Chinese Xiangxi yellow cattle.

Materials and methods
Animals and management
Animal care and procedures were approved and conducted under established standards of the College of Animal Science & Technology, China Agricultural University.

Twenty-three (23) male and twenty-four (24) female weaning calves of F1 Angus × Chinese Xiangxi yellow cattle were selected and transferred from the breeding centre to the fattening farm of Hunan Tin Wah Industrial Co., Ltd. The calves were the F1 progeny of purebred Angus bulls bred to dams of purebred Chinese Xiangxi yellow cows. After arriving at the fattening farm, all male calves were castrated and dewormed. The cattle at an average age of 6.5 mon were weighed and fed the same ration as during the background period (184 d). The animals were then placed in a $2 \times 2 \times 2$ factorial arrangement to study the effects of two levels of dietary energy (TDN (Total Digestible Nutrients): 70%, 80% DM (Dry Matter)) and protein (CP (Crude Protein): 11.9%, 14.3% DM) and sex (S: male, female) on the growth performance, carcass characteristics and meat quality during the finishing phase. The cattle were divided into four treatment groups based on age, body weight (BW) and growth rate during the background period and body size. The treatment groups were:

(1) high energy and low protein(HELP; TDN: 80% DM, CP: 11.9% DM),
(2) high energy and high protein(HEHP; TDN: 80% DM, CP: 14.3% DM),
(3) low energy and low protein(LELP; TDN: 70% DM, CP: 11.9% DM),
(4) low energy and high protein(LEHP; TDN: 70% DM, CP: 14.3% DM).

Six steers and 6 heifers were placed in each treatment group, except for the HELP group, which contained 5 steers and 6 heifers. At the start of the finishing phase, a 14-day adaptation period was used for transition between rations which consisted of mixing 1/3 of the finishing ration with the previous ration for 7 d, then adding 2/3 of the finishing ration for the following 7 d and finally switching to the entirely new ration. The cattle were not implanted with any steroid hormones and were fed for 146 d until slaughter. All of the cattle were held in eight sheltered pens at a stocking density of 5 m^2 per animal during the background period and were tied up to feed during the finishing period. The animals were fed twice a day at 0700 h and 1700 h and allowed to drink water freely. The nutrition levels of different phases and treatment groups are shown in Table 1.

Growth performance
The cattle were weighed at the beginning and end of the background and finishing phases. The dry matter intake was recorded every 2 wk to calculate the average DMI during the background and finishing phases. Samples of the diet and refusals (uneaten feed) were collected every month for analysis using the standard methods of AOAC(2000) for DM, CP, Ca and P [10]. NDF (Neutral Detergent Fiber) and ADF (Acid Detergent Fiber) were determined following a

Table 1 Feed ingredients and nutrition levels for background period and different treatment groups during finishing

Item	Background	Finishing			
		HE		LE	
		LP	HP	LP	HP
Ingredient, % DM					
Hybrid penisetum	20.00	—	—	—	—
Rice straw	20.00	20.00	20.00	30.00	30.00
Ground corn	29.90	66.54	61.03	41.23	36.83
Soybean meal	19.18	10.46	15.97	3.70	7.86
Cottonseed meal	7.45	—	—	4.32	7.56
Wheat bran	—	—	—	18.00	15.00
Limestone	0.67	1.00	1.00	1.00	1.00
Sodium bicarbonate	1.00	—	—	—	—
Salt	1.20	1.20	1.20	1.05	1.05
Premix[1]	0.60	0.80	0.80	0.70	0.70
Nutrient level, % DM					
DM	74.19	86.31	86.45	86.69	86.85
CP	16.52	11.96	14.34	11.90	14.30
TDN	68.22	79.78	79.87	69.95	70.03
NDF	34.27	21.60	21.81	33.09	33.24
ADF	20.53	11.27	11.59	17.75	18.40
Ca	0.42	0.39	0.40	0.41	0.42
P	0.31	0.23	0.24	0.35	0.38

[1]Vitamin and mineral premix contained per kilogram DM: Vitamin A, 154,000 IU; Vitamin D, 38,500 IU; Vitamin E, 3,500 IU; Fe, 9.0 g; Zn, 7.0 g; Mn, 14.0 g; Cu, 1.0 g; I, 138.0 mg; Se 30.0 mg; Co, 60 mg; Monensin, 30 g/1,000 kg.

modification of the procedure of Van Soest et al. [11]. Body measurements, including withers height, body length, chest girth and shin circumference were taken at the beginning and end of the finishing phase. The first two measurements were recorded using calipers and the latter two were recorded using a metal measuring tape.

Carcass characteristics

At the end of the trial, all of the cattle were slaughtered. The hot carcass weight (HCW) and cold carcass weight (CCW, hot carcass × 0.98) were recorded to calculate the dressing percentage and the carcass composition. At the same time, a sample meat cut(2 cm × 5 cm × 3 cm), free of external fat and connective tissue, was also taken between the 6th and 7th ribs of the LM (*Longissimus dorsi* muscle) from the left side of each carcass. The sample was weighed, hung by a nylon cordin a plastic bag at 4°C for 48 h, then dried on absorbent paper before reweighing to ascertain the drip loss percentage, which was calculated by (initial weight-final weight)/initial weight. The carcasses were then put into a chiller at 0–4°C and aged for 7 d. The ultimate pH of the LM (12–13th rib) was measured on the left body side at 48 h post-mortem using a pH electrode probe (Testo 205, Testo AG, Lenzkirch, Germany), and the following carcass linear measurements were recorded: length of carcass, depth of chest, length of leg [12], maximum girth of leg, lean thickness (muscle and subcutaneous fat) of leg, rib (between the 5th and 6th rib) and loin (between the 3rd and 4th lumbar vertebrae). The carcass composition (bone, fat and meat) was assessed by dissection of the 8th rib, cut on the 8th day post-mortem [13]. The LM area and fat thickness were measured between the 12th and 13th ribs of the LM using a plastic grid and Vernier caliper. Commercial meat cuts were dissected and named following the standard method by Chen [14] and were weighed after trimming. Based on the most popular and economic meat cuts in Chinese markets, the highrib, ribeye, striploin and tenderloin were considered as the top grade cuts, and the chunk tender, topside, outside flat, eye round, rump and knuckle were considered as the medium top grade cuts. The top and medium top grade cut yields were then calculated.

Meat quality

The meat quality was determined from the sample (6.0 cm thick) removed from the longissimus muscle between the 12th and 13th ribs on the left sideof the body after carcass dissection, with no external fat or connective tissue. The meat samples were then frozen (−24°C) and transported to the China Agriculture University until analysis could be conducted. The meat samples were cut into three steaks using a saw before thawing. The first sample (2.54 cm thick) was used for calculating cooking loss by measuring the difference in weight before and after a period of heating

to an internal sample temperature of 70°C in a 75°C water bath six 1.27-cm cores parallel to the muscle fiber orientation were then removed from the cooked sample for the instrumental measurement of tenderness by a texture analyzer (TA.XT plus, SMS, Godalming, Surrey, UK). The second sample (1.5 g) was allowed to thaw for 12 h at 1–2°C and then its water holding capacity (WHC) was measured by holding the sample under pressure (35 kg) for 5 min by a texture analyzer (TA.XT plus) fitted with a compression platen (diameter 7.5 cm) and reweighing. The following equation was used to calculate WHC.

$$X = \frac{M_1 A - (M_1 - M_2)}{M_1 A}$$

Where X =% WHC, M_1 = weight of sample before compression (g), M_2 = weight of sample after compression (g) and A = total water content in the sample (%). The third sample was freeze-dried and then DM, crude protein, intramuscular fat, fatty acids (FA) and amino acids (AA) were measured.

The one-step extractive methylation procedure for fatty acids gain [15] was performed in a gas chromatograph (GC-2014, Shimadzu, Kyoto, Japan) with a capillary column (HP-88 100 m long, 0.25 mm diameter, 0.20 μm film. Agilent Santa Clara, California, America) using margaric acid (C17:0) as an internal standard. The oven temperature was programmed to provide three consecutive ramps, the first had an initial temperature of 120°C maintained for 1 min then increasedby 10°C/min until it reached 175°C, where it was maintained for 10 min; the second increased by 5°C/min until it reached 210°C, where it was maintained for 5 min and the third ramp increased by 5°C/min to 230°C, where it was maintained for 5 min. The carrier gas was helium at a flow rate of 2 mL/min. An automatic split/splitless injector with a 1/50 split and a temperature of 250°C was used. The injection volume was 1 μL. A flame ionization detector (FID) was used with an air flow of 450 mL/min, hydrogen flow of 40 mL/min and a detector temperature of 280°C. Fatty acids were expressed in gravimetric concentrations (mg/g of freeze dried sample).

Amino acids were determined in the dried, fat-free meat samples using a Shimadzu (10A VP DAD) high-performance liquid chromatograph (HPLC) following the procedure described by Wu [16].

Statistical analysis

There were three major factors in the experimental design: dietary energy, protein and sex with each factor having two levels: energy (TDN: 70%, 80% DM), protein (CP: 11.9%, 14.3% DM) and sex (S: male, female). Therefore the effects of dietary energy, protein and sex and their interactions with growth performance, carcass characteristics and meat quality traits were analyzed using a $2 \times 2 \times 2$ factorial arrangement (energy × protein × sex) using the GLM procedure of SAS (version 9.0, SAS Inst. Inc., Cary, NC, USA). When a significant effect of treatment was detected ($P < 0.05$), differences between the means were tested using Tukey's multiple comparison test.

Results

Background performance
The growth performance results from the cattle during the background phase are shown in Table 2. The animals were fed the same rations during the background phase; the initial weight was no different between the steers and heifers but the ADG and final BW for the steers was higher than those for the heifers, but not significantly so ($P > 0.05$).

Finishing performance
The growth performance results from the cattle during the finishing phase are shown in Table 3. The initial BW values did not differ according to treatments or sex. The ADG and final BW were not affected by the energy or protein levels or by sex. The cattle fed an HE diet had a significantly lower dry matter intake (6.76 vs. 7.48 kgDM/d, $P < 0.01$) and FCR (Feed conversion ratio) (9.38 vs. 11.13, $P < 0.01$) than those fed an LE diet.

The body measurements were not significantly different between treatments or sex at the start of the finishing phase (Table 4). Compared with cattle fed an LE diet, a greater chest girth increment (46.1 vs. 36.8 cm, $P < 0.01$) and larger chest girth (190.2 vs. 182.6 cm, $P < 0.05$) were found in cattle fed an HE diet with steers having a greater body length increment than heifers (9.0 vs. 8.3 cm, $P < 0.01$).

Carcass characteristics
The carcass quality traits are shown in Table 5. The hot carcass weights were not affected by the energy or protein levels or by sex. The dressing percentage was higher with higher protein levels ($P < 0.05$) and the cattle in the LE treatment contained 3.4% units more lean meat and 3.6% units less fat than HE treatment ($P < 0.05$), which indicated that the meat:fat ratio was 23.6% higher. There

Table 3 Effects of dietary treatments and sex on finishing growth performance of cattle

Item		HE		LE		SEM	E	P	S
		LP	HP	LP	HP				
Initial BW, kg	M	319.75	286.50	299.60	294.00	16.86	ns	ns	ns
	F	293.50	283.66	295.20	282.83				
Final BW, kg	M	435.00	391.50	391.80	391.20	23.86	ns	ns	ns
	F	396.75	418.50	412.00	401.66				
DMI, kg/d	M	6.73[a]	6.74[a]	7.46[b]	7.47[b]	0.020	**	ns	ns
	F	6.74[a]	6.77[a]	7.50[b]	7.45[b]				
ADG, kg	M	0.77	0.70	0.62	0.65	0.079	ns	ns	ns
	F	0.69	0.90	0.78	0.79				
FCR	M	8.79	10.34	12.72	11.91	1.23	**	ns	ns
	F	10.13	8.23	9.72	10.15				

Significance: **($P < 0.01$), ns not significant ($P > 0.05$).
M = male, steer, F = female, heifer, E = energy, P = protein, S = sex.
Interactions between energy, protein levels and sex were not significant so not shown in the table.
[a,b]Means within the same row with the same superscript letter are not significantly different ($P > 0.05$).

was an energy × sex interaction for the 12th rib fat thickness ($P < 0.05$). The bone content, meat:bone ratio and LM area were not affected by any of the factors. No differences were observed for the weight of most of the top and medium top grade cuts except for the highrib, striploin and chunk tender cuts which were affected by the energy or energy × protein interaction (Table 6). The yields of the top and medium top grade cuts were higher in the HE than LE treatments ($P < 0.05$).

The carcass measurements are shown in Table 7. The steers had a greater chest depth and maximum leg girth ($P < 0.05$) and a thinner rib lean thickness ($P < 0.01$) than the heifers. The cattle fed an HE diet had a greater leg and rib lean thickness than those fed an LE diet ($P < 0.01$).

Meat quality
The LM chemical composition and quality traits are shown in Tables 8 and 9. The dry matter was higher for heifers than for steers (29.2 vs. 27.6%, $P < 0.05$). The ultimate pH, protein and intramuscular fat content were significantly affected by the energy level, with a lower ultimate pH (5.71 vs. 5.79, $P < 0.01$), protein content (68.6 vs. 74.9% DM, $P < 0.05$) and higher intramuscular fat content (29.9 vs. 22.8% DM, $P < 0.01$) detected in the HE treatment compared with the LE treatment. The shear force value, drip loss, cooking loss and water holding capacity were not affected by the energy or protein levels or by sex and averaged 3.14 kg, 2.5, 31.5 and 52.9%, respectively. The fatty acid composition (Table 10) and amino acid composition (Tables 11 and 12) were not influenced by the energy or protein levels or by sex. ($P > 0.05$), but the ratio of unsaturated fatty acids to saturated fatty acids was higher in the HE

Table 2 Effects of sex on background growth performance of cattle

Item	Steer	Heifer	SEM	P
Initial BW, kg	149.19	149.10	5.24	ns
Final BW, kg	299.61	288.05	7.99	ns
DMI, kg/d	5.04	5.10	0.02	ns
ADG, kg	0.82	0.74	0.03	ns
FCR	6.33	7.39	0.49	ns

Significance: ns not significant ($P > 0.05$).

Table 4 Effects of dietary treatments and sex on body measurementsof cattle

Item		HE		LE		SEM	E	P	S
		LP	HP	LE	HP				
Start of finishing									
Chest girth, cm	M	144.9	145.7	147.1	147.0	2.00	ns	ns	ns
	F	142.3	143.4	146.5	142.8				
Withers height, cm	M	106,2	108.7	108.6	112.6	2.65	ns	ns	ns
	F	106.0	109.8	111.9	109.8				
Shin circumference, cm	M	18.4	16.6	17.8	16.1	0.88	ns	ns	ns
	F	15.3	17.8	16.3	15.4				
Body length, cm	M	123.9	116.3	122.1	122.2	3.81	ns	ns	ns
	F	119.3	121.1	122.5	118.5				
End of finishing									
Chest girth, cm	M	197.4	187.1	182.8	180.0	4.27	*	ns	ns
	F	184.0	192.2	183.7	183.9				
Withers height, cm	M	116.4	119.2	118.3	123.3	3.03	ns	ns	ns
	F	118.5	120.8	120.4	119.3				
Shin circumference, cm	M	19.9	18.5	19.7	18.2	0.92	ns	ns	ns
	F	17.2	19.3	17.8	17.1				
Body length, cm	M	132.8	125.0	130.8	131.6	3.71	ns	ns	ns
	F	127.2	129.5	130.7	127.0				
Increment									
Chest girth, cm	M	52.5	41.4	35.8	33.0	4.56	**	ns	ns
	F	41.7	48.7	37.2	41.1				
Withers height, cm	M	10.2	10.4	9.7	10.7	1.48	ns	ns	ns
	F	12.5	11.0	8.5	9.4				
Shin circumference, cm	M	1.5	1.9	1.9	2.1	0.21	ns	ns	ns
	F	1.8	1.6	1.6	1.8				
Body length, cm	M	8.9	8.7	8.8	9.4	0.34	ns	ns	**
	F	7.9	8.4	8.2	8.5				

Significance: *($P < 0.05$), **($P < 0.01$), ns not significant ($P > 0.05$).
M = male, steer, F = female, heifer, E = energy, P = protein, S = sex.
Effects of interactions between energy,protein levels and sexwere not significant ($P > 0.05$).

treatment compared with the LE treatment ($P < 0.05$). The percentage of amino acids producing an umami, sour or sweet taste was more than 60% with the percentage producing a bitter taste at about 29%. The essential and non-essential amino acids were approximately 28.4 and 71.7%, respectively.

Discussion
Growth performance
The initial BW (293.4 kg) did not differ between treatments or sex at the beginning of the finishing phase, and the ADG was not affected by the energy or protein levels or sex averaging 0.74 kg/d. The cattle fed an HE diet had a lower dry matter intake (6.76 vs. 7.48 kgDM/d, $P < 0.01$) and FCR (9.38 vs. 11.13, $P < 0.01$) compared with those fed an LE diet. The DMI was lower in the HE

treatment group, which can be explained by the theory of satiety limit intake where in the metabolic needs are completely met [17]. Compared with the current study, Angus crossbred cattle such as Angus × Gelbvieh gained 1.76 kg/d and their FCR was 5.36 when they were fed for 180 d with a similar dietary nutrition level at nearly the same initial BW of approximately 293.6 kg [8]. The contrast could be considered the result of the genetic influence of the Xiangxi yellow cattle because the rate of gain is usually positively related to the mature size [18]. The mature weight and withers height of the Xiangxi yellow bulls were 334.3 kg and 117.1 cm, respectively, and for cows were 240.2 kg and 106.1 cm, respectively [2]. However, the mature weight and withers height of the Gelbvieh bulls were 1,100–1,300 kg and 148–156 cm, respectively, and of the cows, 650–850 kg and 140 cm,

Table 5 Effects of dietary treatments and sex oncarcass quality traits of cattle

Item		HE		LE		SEM	E	P	S	E × P	E × S
		LP	HP	LP	HP						
HCW, kg	M	235.50	214.75	205.00	210.00	12.284	ns	ns	ns	ns	ns
	F	207.25	233.75	223.60	220.83						
Dressing percentage, %	M	54.00	54.75	52.00	53.80	0.009	ns	*	ns	ns	ns
	F	52.75	55.83	54.60	55.00						
CCW, kg	M	231.00	210.50	201.00	205.80	11.80	ns	ns	ns	ns	ns
	F	203.00	229.16	219.20	216.33						
Carcass composition, %											
Meat	M	64.29	67.65	71.08	70.47	0.018	*	ns	ns	ns	ns
	F	62.78	67.21	69.05	67.48						
Fat	M	19.51	19.56	14.05	13.99	0.017	*	ns	*	ns	ns
	F	22.23	18.11	17.34	19.84						
Bone	M	15.29	11.89	14.15	14.68	0.143	ns	ns	ns	ns	ns
	F	13.53	14.38	12.73	12.22						
Meat:fat ratio	M	3.30	3.45	5.04	5.04	0.466	*	ns	*	ns	*
	F	2.83	3.71	3.99	3.41						
Meat:bone ratio	M	4.20	5.69	5.01	4.80	0.421	ns	ns	ns	ns	ns
	F	4.65	4.67	5.44	5.53						
Fat thickness, cm	M	1.10	0.98	0.72	0.71	0.118	ns	ns	ns	ns	*
	F	0.76	1.04	1.01	0.86						
LM area, cm²	M	65.71	66.59	54.24	60.47	3.660	ns	ns	ns	ns	ns
	F	59.63	59.04	59.59	58.31						

Significance: *($P < 0.05$), ns not significant ($P > 0.05$).
M = male, steer, F = female, heifer, E = energy, P = protein, S = sex.
Effects of P × S and E × P × Sinteractions were not significant ($P > 0.05$).

respectively [19]. Therefore, the extremely large difference between the mature size ofthe Xiangxi yellow cattle and Gelbvieh cattle resulted in a lower ADG for the Angus × Chinese Xiangxi yellow cattle compared with the Angus × Gelbvieh. However, growth stimulants were not used for the cattle in the present study, whereas Synovex-S was implanted in the Angus × Gelbvieh cross, which might have improved the ADG in the research of Ludden et al. [8].

The average withers height at 12 and 18mon for the cattle in the present study was 109.2 and 119.5 cm, respectively, which was shorter than the Angus × Hereford steers that had a yearling height of 112.0 and 122.4 cm at the age of 16 mon [20]. Angus bulls can reach a height of 120.2 cm at 12 mon [21]. Withers height, shin circumference and body length are mainly determined by the composition of the bones, which are an early maturing part of the body; however the chest girth is a relatively late maturing part of the body and is mainly determined by meat and fat. Therefore the chest increment revealed that the higher energy in the diet may have resulted in additional protein deposition and fat cover.

Carcass characteristics

Carcass quality traits are shown in Table 5. The hot carcass weights were not affected byenergy or protein levels or by sex and had a mean value of 219.0 kg. This result cannot be compared with data obtained from Angus or other Angus crossbred cattle, because they have a greater growth rate resulting in a heavier slaughter and carcass weight at the age of 17–19 mon [6,21] or even at 14 mon [22]. The authors reported carcass weights of 292.3 and 335.7 kg for the Angus bulls and 293.8 kg for 19 various Angus crossbred steers. The dressing percentage was higher with increasing protein levels (53.4 vs. 54.9%, $P < 0.05$), which might have been caused by theincreased water concentrations in tissues as a result of the hydrophilic characteristics of systemic ammonium ions leading to higher dressing percentage [23]. The mean dressing percentage for all of the cattle was 54.2%, which was lower than the values of 55.0, 56.2 and 58.2% found by Cuvelier et al. [6], Albertí et al. [21] and Laborde et al. [22], respectively. In China, a dressing percentage of 52% is set at a threshold value for the gain or loss of 0.3 Yuan RMB per kg for one percent higher or lower [24].

Table 6 Effects of diets and sex on yield of Top and Medium top grade commercial cuts

Item	HE		LE		SEM	E	P	E×P
	LP	HP	LP	HP				
Total meat, kg	138.3	148.9	146.9	145.6	11.6	ns	ns	ns
Top grade cuts								
Highrib, kg	8.1	8.3	6.9	7.2	0.38	*	ns	ns
Ribeye, kg	8.3	8.3	7.8	7.5	0.50	ns	ns	ns
Striploin, kg	5.4	6.4	6.2	5.7	0.30	ns	ns	*
Tenderloin, kg	2.7	2.4	2.4	2.3	0.14	ns	ns	ns
Medium top grade cuts								
Chunk tender, kg	2.1	1.8	1.7	2.0	0.10	ns	ns	*
Topside, kg	9.6	9.3	8.2	8.4	0.59	ns	ns	ns
Outside flat, kg	5.0	4.8	4.6	4.7	0.28	ns	ns	ns
Eye round, kg	3.0	2.9	2.8	2.8	0.18	ns	ns	ns
Rump, kg	5.5	6.1	5.4	4.7	0.36	ns	ns	ns
Knuckle, kg	6.9	6.3	6.9	7.0	0.37	ns	ns	ns
Top grade cuts yield, %	18.0	17.3	15.9	15.6	0.006	*	ns	ns
Medium top grade cuts yield, %	23.5	21.0	20.1	20.3	0.009	*	ns	ns
Total, %	41.4	38.3	36.3	36.7	0.01	*	ns	ns

Significance: *($P < 0.05$), ns not significant ($P > 0.05$).
M = male, steer, F = female, heifer, E = energy, P = protein.
Sex had no effect on yield of Top and Medium top grade commercial ($P > 0.05$).

Regarding the carcass composition, the cattle in the LE diet treatment contained more lean meat (69.6 vs. 65.5%, $P < 0.05$) and a lower fat content (16.3 vs. 19.9%, $P < 0.05$) than in the HE treatment, which might have resulted from the higher glucose content in the HE diet which increased the fat deposition. In the present study, a higher meat content (67.6 vs. 62.2%, 61.6%) and lower fat content (18.1 vs. 23.6%, 21.7%) was observed compared with that found by Cuvelier et al. [6] and Albertí et al. [21] at a slaughter age of 17-19 mon because the

Table 7 Effects of dietary treatments and sex on carcass measurements of cattle

Item		HE		LE		SEM	E	P	S
		LP	HP	LP	HP				
Carcass length, cm	M	137.0	127.5	131.8	135.0	3.18	ns	ns	ns
	F	128.0	136.3	133.7	132.5				
Chest depth, cm	M	70.3	67.9	66.9	68.0	1.31	ns	ns	*
	F	64.2	68.0	64.9	66.8				
Maximum leg, cm	M	73.4	72.2	73.0	73.6	1.64	ns	ns	*
	F	66.8	71.0	71.6	70.9				
Leg length, cm	M	62.9	62.8	63.9	64.3	1.57	ns	ns	ns
	F	61.8	63.3	63.6	62.5				
Leg lean thickness, cm	M	10.2	11.2	10.4	9.7	0.40	**	ns	ns
	F	10.4	10.9	9.3	9.4				
Loin lean thickness, cm	M	6.0	6.1	6.8	6.0	0.34	ns	ns	ns
	F	6.0	7.0	5.9	6.4				
Rib lean thickness, cm	M	4.4[ab]	5.0[ab]	4.1[b]	4.0[ab]	0.29	**	ns	**
	F	5.0[ab]	5.8[a]	5.0[ab]	4.8[b]				

Significance: *($P < 0.05$), **($P < 0.01$), ns not significant ($P > 0.05$).
M = male, steer, F = female, heifer, E = energy, P = protein, S = sex.
Effects of interactions between energy, protein level and sex were not significant ($P > 0.05$).
[ab]Means within same row with the same superscript letter are not significantly different ($P > 0.05$).

Table 8 Effects of dietary treatments and sex onchemical composition of LM

Item		HE		LE		SEM	E	P	S
		LP	HP	LP	HP				
Dry matter, %	M	28.6	28.0	27.7	25.2	0.010	ns	ns	*
	F	30.2	29.3	29.0	23.3				
Crude protein, % DM	M	70.2	70.8	76.8	75.8	0.035	*	ns	ns
	F	65.2	68.0	72.7	74.4				
Intramuscular fat, % DM	M	28.4	27.7	18.5	22.4	0.038	**	ns	ns
	F	32.3	31.2	25.7	24.6				

Significance: *(P < 0.05), **(P < 0.01), ns not significant (P > 0.05).
M = male, steer, F = female, heifer, E = energy, P = protein, S = sex.
Effects of interactions between energy, protein level and sex were not significant (P > 0.05).

age of puberty for Angus cattle is 295 d [25], which is less than that of Chinese Xiangxi yellow cattle at 497 d [2]. Therefore, pure Angus cattle deposit fat at a younger age and the high growth rates of 1.66 kg/d and 1.9 kg/d, reported by Cuvelier et al. [6] and Albertí et al. [21], accelerates fat deposition. Thus, the fat contents in these previous experiments were higher than that of the Angus × Chinese Xiangxi yellow cattle in the present study.

Heifers had a higher fat content than steers (16.8 vs. 19.4%, respectively) and under the conditions of an LE diet, heifers had a greater 12th rib fat thickness than steers. This suggests that heifers deposit fat more easily, which is possibly related to hormonal effects [26]. The LM area was not different between treatments or sex and averaged 60.4 cm^2 which was within the range of

Table 9 Effects of dietary treatments and sex onquality traits of LM

Item		HE		LE		SEM	E	P	S
		LP	HP	LP	HP				
pH	M	5.67^b	5.77^{ab}	5.74^{ab}	5.80^{ab}	0.042	**	ns	ns
	F	5.75^{ab}	5.65^b	5.76^{ab}	5.86^a				
Shear force, kg	M	3.38	3.31	2.81	3.11	0.370	ns	ns	ns
	F	3.34	2.98	3.40	2.82				
Cooking loss, %	M	30.40	31.33	32.50	33.20	0.016	ns	ns	ns
	F	30.33	32.50	30.33	31.14				
Drip loss, %	M	2.60	2.00	2.50	2.13	0.013	ns	ns	ns
	F	2.00	2.66	2.66	3.24				
WHC, %	M	53.00	51.16	54.00	53.80	0.019	ns	ns	ns
	F	50.66	52.50	53.83	54.43				

Significance: **(P < 0.01), ns not significant (P > 0.05).
M = male, steer, F = female, heifer, E = energy, P = protein, S = sex.
Effects of interactions between energy, protein level and sex were not significant (P > 0.05).
^{ab}Means within same row with the same superscript letter are not significantly different (P > 0.05).

Table 10 Effects of dietary treatments and sex onfatty acid composition of LM (mg/g DM)

Item	HE		LE		SEM	E	P	E × P
	LP	HP	LP	HP				
C14:0	6.70	5.45	4.56	5.88	1.31	ns	ns	ns
C14:1	1.97	2.12	1.71	2.26	0.36	ns	ns	ns
C16:0	61.68	51.55	47.46	55.03	9.28	ns	ns	ns
C16:1	13.86	10.40	9.18	10.72	1.96	ns	ns	ns
C18:0	22.92	19.72	17.06	21.42	3.66	ns	ns	ns
C18:1$^{trans-9}$	2.14	1.26	1.35	1.24	0.37	ns	ns	ns
C18:1 $^{cis-9}$	91.91	75.95	63.24	76.70	13.79	ns	ns	ns
C18:2 $^{cis-9,12}$	4.27	4.61	3.32	3.22	0.80	ns	ns	ns
C18:3n-3	0.30	0.53	0.23	0.46	0.14	ns	ns	ns
C20:3n-6	2.75	2.22	2.00	2.95	0.24	ns	ns	**
SFA	91.30	76.72	69.08	82.33	14.11	ns	ns	ns
MUFA	109.9	89.72	75.48	90.93	16.19	ns	ns	ns
PUFA	7.16	7.08	5.39	6.38	0.85	ns	ns	ns
n-6:n-3	25.94	19.74	14.95	17.36	7.99	ns	ns	ns
P:S	0.11	0.14	0.09	0.11	0.03	ns	ns	ns
UFA:SFA	1.31	1.27	1.17	1.20	0.05	*	ns	ns

Significance: *(P < 0.05), **(P < 0.01), ns not significant (P > 0.05).
M = male, steer, F = female, heifer, E = energy, P = protein, S = sex.
Sex had no effect on fatty acid composition of LM (P > 0.05).

58.7–70.3 cm^2 for pure Angus or Angus crossbred cattle with a slaughter weight of approximately 400 kg [27-29].

The yields of top and medium top grade cuts were higher in the HE treatment than in the LE treatment, which suggests that the high plane of nutrition, especially for the dietary energy level, contributed to the higher yields of the top and medium top grade cuts.

Steers had a greater chest depth and maximum leg girth, but thinner rib lean thickness (P < 0.01) than heifers. Although the male cattle had been castrated before the experiment, they still had more development in the fore body and legs than the heifers. The increasing energy level contributed to a greater leg and rib lean thickness.

Meat quality

The DM of the LM was higher for heifers than steers, which can be explained by the heifers having a greater intramuscular fat content, because fat tissues contain little water, so the DM of the LM was higher [30]. A higher intramuscular fat content (29.9 vs. 22.8%, P < 0.01) and lower protein content (68.6 vs. 74.9%, P < 0.05) were observed in the HE treatment compared with the LE diet treatment, this could have resulted from the intramuscular fat being derived from a glucose substrate that is absorbed in the small intestine and stimulates a greater activity of ATP citrate lyase, which synthesizes fat from glucose [31]. A maize-based diet could enhance the glucose absorbed in the small intestine. In the present study, a greater

Table 11 Effects of dietary treatments and sex on amino acid composition of LM (mg/100 mg DM basis)

Item	HE		LE		SEM	E	P	E × P
	LP	HP	LP	HP				
Essential								
Lysine	4.61	3.72	4.06	4.25	0.46	ns	ns	ns
Valine	3.53	3.05	2.94	2.94	0.35	ns	ns	ns
Histidine	1.26	0.64	0.74	0.64	0.27	ns	ns	ns
Leucine	4.82	4.65	4.61	4.76	0.19	ns	ns	ns
Isoleucine	3.19	3.37	3.05	3.14	0.18	ns	ns	ns
Methionine	4.28	4.11	3.90	4.03	0.17	ns	ns	ns
Phenylalanine	1.59	1.58	1.61	1.56	0.12	ns	ns	ns
Threonine	4.52	4.09	4.16	4.18	0.27	ns	ns	ns
Total E	27.80	25.21	25.09	25.50	0.97	ns	ns	ns
Non essential								
Aspartic acid	13.86	13.76	13.35	12.15	1.44	ns	ns	ns
Glutamic acid	21.93	19.54	18.67	20.41	1.19	ns	ns	*
Cysteine	4.79	4.59	4.50	4.45	0.33	ns	ns	ns
Alanine	5.99	5.53	5.36	5.70	0.32	ns	ns	ns
Glycine	4.92	4.32	4.29	4.38	0.32	ns	ns	ns
Serine	9.21	6.23	5.86	5.76	1.37	ns	ns	ns
Proline	2.14	2.13	2.02	2.08	0.21	ns	ns	ns
Arginine	6.26	5.20	5.23	5.28	0.48	ns	ns	ns
Tyrosine	3.82	2.94	3.41	3.54	0.42	ns	ns	ns
Total NE	72.93	64.25	62.69	63.75	2.76	ns	ns	ns
Total AA	100.73	89.46	87.79	89.25	3.51	ns	ns	ns
E/NE, %	38.12	40.00	40.02	40.58	0.015	ns	ns	ns
E/TAA, %	27.59	28.38	28.57	28.77	0.007	ns	ns	ns

Significance: *(P < 0.05), ns not significant (P > 0.05).
M = male, steer, F = female, heifer, E = energy, P = protein.
Sex had no effect on amino acid composition of LM (P > 0.05).

Table 12 Effects of dietary treatments and sex on flavor amino acid composition of LM (mg/100 mg DM basis)

Item	HE		LE		SEM	E	P	E × P
	LP	HP	LP	HP				
Lysine	4.61	3.72	4.06	4.25	0.46	ns	ns	ns
Cysteine	4.79	4.59	4.50	4.45	0.33	ns	ns	ns
Umami taste								
Aspartic acid	13.86	13.76	13.35	12.15	1.44	ns	ns	ns
Glutamic acid	21.93	19.54	18.67	20.41	1.19	ns	ns	*
Total AA(U)	35.79	33.30	32.02	32.56	1.71	ns	ns	ns
Sweet taste								
Threonine	4.52	4.09	4.16	4.18	0.27	ns	ns	ns
Alanine	5.99	5.53	5.36	5.70	0.32	ns	ns	ns
Glycine	4.92	4.32	4.29	4.38	0.32	ns	ns	ns
Serine	9.21	6.23	5.86	5.76	1.37	ns	ns	ns
Proline	2.14	2.13	2.02	2.08	0.21	ns	ns	ns
Total AA(S)	26.79	22.30	21.69	22.10	1.41	ns	ns	ns
Bitter taste								
Arginine	6.26	5.20	5.23	5.28	0.48	ns	ns	ns
Histidine	1.26	0.64	0.74	0.64	0.27	ns	ns	ns
Leucine	4.82	4.65	4.61	4.76	0.19	ns	ns	ns
Isoleucine	3.19	3.37	3.05	3.14	0.18	ns	ns	ns
Methionine	4.28	4.11	3.90	4.03	0.17	ns	ns	ns
Phenylalanine	1.59	1.58	1.61	1.56	0.12	ns	ns	ns
Tyrosine	3.82	2.94	3.41	3.54	0.42	ns	ns	ns
Valine	3.53	3.05	2.94	2.94	0.35	ns	ns	ns
Total AA(B)	28.74	25.54	25.52	25.89	1.21	ns	ns	ns
Total AA	100.73	89.46	87.79	89.25	3.51	ns	ns	ns
AA(U)/TAA, %	36.33	36.82	36.48	36.14	0.01	ns	ns	ns
AA(S) /TAA, %	26.09	25.12	24.71	24.86	0.007	ns	ns	ns
AA(B) /TAA, %	28.33	28.79	29.06	29.30	0.009	ns	ns	ns

Significance: *(P < 0.05), ns not significant (P > 0.05).
M = male, steer, F = female, heifer, E = energy, P = protein.
Sex had no effect on amino acid composition of LM (P > 0.05).

amount of ground corn was included in the ration in the HE compared with the LE treatment (63.8 vs.39.0%), which resulted in a higher intramuscular fat content from the HE diet treatment. The intramuscular fat content was 26.4% in the present study, which was higher than the value of 21% for Angus steers [32] and 9.3% for Angus × Limousin steers found in previous studies [7]. However, the fat thickness was less than that found for Angus and Angus × Limousin (0.90 vs. 1.15 cm, 0.98 cm). This result suggests that F1 Angus × Chinese Xiangxi yellow cattle develop intramuscular fat more strongly at lower levels of subcutaneous fat.

Post-slaughter, glycogen is converted to lactic acid and there is an associated reduction in muscle pH from the neutral value of 7.2 [33]. The ultimate pH in this experiment was lower (5.71 vs. 5.79, P < 0.01) for the cattle fed an HE diet compared with an LE diet, because there is an increasing effect of energy level with increased glycogen availability. The ultimate pH can also affect meat tenderness, with a pH of 5.4–5.8 found in normal, tender meat, a pH value of 5.8–6.2 in inconsistently tender meat (moderate DFD) and pH > 6.2 found in tender meat with microbial spoilage (DFD meat) [34]. Therefore, the meat observed in the present study can be considered as normal, tender meat with values of meat pH values similar to those reported by Cuvelier et al. [6] and Faucitano et al. [7].

Meat tenderness is the most important quality trait for the consumer and consumers prefer and will pay more for tender beef meat [35]. A threshold shear force of 4.6 kg has been used to distinguish tough and tender steaks [36]. A shear force value from 2.27 to 3.58 kg is considered tender; 4.08–5.40 kg intermediate; and

5.90–7.21 kg tough [37]. The shear force value was not affected by the energy or protein levels or sex with a mean value of 3.14 kg, so can be classified as "very tender" meat. One study has found that the shear force value of meat was 3.62 kg when Angus steers were slaughtered at a younger age of 14 mon with a 14 d post-mortem [38], this value was higher than that found in the present study. A younger slaughter age and longer post-mortem ageing time could produce more tender meat [39,40]. Therefore, if F1 Angus × Chinese Xiangxi yellow cattle were slaughtered at the age of 14 mon with 14 d post-mortem ageing, the meat would be more tender.

Drip loss can be categorized as follows: low drip loss ≤2.60%, medium drip loss: 2.60–4.00%, and high drip loss ≥4.00% [41]. The average drip loss in the present study was 2.5%, which is thus classified as low. The cooking loss was 31.5%, a value similar to 33.8% for Angus bulls [6] and 29.5% for Angus × Limousin cattle [7]. The WHC increased slightly in the LE diet treatment, which could be related to the "sponge effect" hypothesis, in which the higher ultimate pH of the LE treatment accelerates the breakdown of meat structure and results in a reduction of water loss from the channels [42].

Nutritive profile (intramuscular fatty acid and amino acid composition)

The intramuscular fatty acid content (FA) was not significantly different between treatments or sex and was dominated by MUFA at 51.4%, followed by SFA at approximately 44.9% and PUFA at approximately 3.7%. The mean value of the n-6:n-3 ratio was 19.5 which was much higher than the <4.0 value from nutritional advice [43]. When the cattle were grain-fed, the concentrate diet could have improved the proportion of PUFA, which was dominated by n-6, especially C18:2n-6. Forages such as fresh grass or grass silage are rich in C18:3n-3 [44]. Therefore, grass-fed cattle have a higher amount of C18:3n-3 and a lower amount of C18:2n-6 in their muscles compared with concentrate-fed cattle [45]. The cattle in the present study were fed a high concentrate diet and the roughage was yellow rice straw instead of fresh grass or silage, thus a higher ratio of n-6:n-3 was observed. The mean value of the P:S ratio was 0.11, which is normal for beef [46]. These results were consistent with the results of Warren et al. [9] and Ludden et al. [8]. The UFA/SFA ratio was significantly higher in the HE treatment, which was verified in this study and is related to a loss of efficiency of rumen biohydrogenation because less fibrous diets pass through the rumen at a faster rate. From the perspective of meat flavor, 'sweet', 'oily', 'chemical-like' and 'perfume-like' are induced by a high C18:2n-6 content in the meat [47] and 'fishy' and 'grassy' flavors by higher n-3 content [48]. Therefore, the meat in the present study would taste 'sweet', 'oily',

'chemical-like' and 'perfume-like' and not include 'fishy' and 'grassy' flavors.

The amino acid composition in this study was not affected by the energy or protein levels or by sex. The percentage of amino acids producing the tastes of umami, sour and sweet was more than 60% with the percentage producing a bitter taste at approximately 29%. The essential and non-essential amino acid requirements of an adult man are 0.18 g/kg per day (EAA) and 0.48 g/kg per day (NEAA), respectively, which equals EAA/NEAA = 37.5% and EAA/TAA = 27.3% [49]. In the present study, the mean ratios of EAA/NEAA and EAA/TAA of the meat samples were 39.7 and 28.4%, which were a little higher than those recommended by FAO/WHO/UNU [48] but can meet an adult man's needs appropriately, therefore the meat appears to be an excellent source of high biological value protein.

Conclusions

The cattle carcass characteristics and chemical composition of the meat were significantly influenced by dietary energy and protein levels and by sex. The growth performance, meat quality traits and nutritive profiles were not affected by energy or protein levels or by sex.

The meat quality of the F1 Angus × Chinese Xiangxi yellow cattle was high based on its tenderness, flavor and nutritional value. However, there is considerable potential to obtain higher daily gains and meat production for this kind of crossbred cattle. However, feeding and breeding techniques must be developed to determine the best methods for improving beef products in both quantitative and qualitative terms.

Competing interests
The authors declare that they have no competing interests.

Authors' contributions
LYL carried out the experiments, finished the data analysis and drafted the manuscript. YKZ, XYW and YH participated in the experiments and helped with data collection and analysis. BHC conceived the experiment and finished the manuscript. All authors approved the final version of the manuscript for publication.

Acknowledgments
The authors would like to acknowledge the National Beef Cattle Industry and Technology System for their financial support. Special thanks are given to Tin Wah Industrial Co.Ltd of Hunan province for providing the cattle and meat samples.

Author details
[1]National Beef Cattle Industry and Technology System, College of Animal Science and Technology, China Agricultural University, Beijing 100193, China. [2]Tin Wah Industrial Co.Ltd, Shimen Industrial Area of Lian Yuan 417100, Hunan Province, China.

References
1.   MOA: **National Protection List of Livestock and Poultry Genetic Resources.** *Ministry of Agriculture of the People's Republic of China* 2006, **662**:245.

2. Ouyang SJ: *Livestock and poultry breeds of Hunan Province.* Changsha: Hunan Science and Technology Press; 1984:54–60.

3. Dunshea FR, D'Souza DN, Pethick DW, Harper GS, Warner RD: **Effects of dietary factors and other metabolic modifiers on quality and nutritional value of meat.** *Meat Science* 2005, **71**:8–38.

4. Arthaud VH, Mandigo RW, Koch RM, Kotula AW: **Carcass composition, quality and palatability attributes of bulls and steers fed different energy levels and killed at four ages.** *J Anim Sci* 1970, **44**:53–64.

5. Kannan G, Gadiyaram KM, Galipalli S, Carmichael A, Kouakou B, Pringle TD, McMillin KW, Gelaye S: **Meat quality in goats as influenced by dietary protein and energy levels, and postmortem aging.** *Small Ruminant Research* 2006, **61**:45–52.

6. Cuvelier C, Cabaraux JF, Dufrasne I, Clinquart A, Hocquette JF, Istasse L, Hornick JL: **Performance, slaughter characteristics and meat quality of young bulls from Belgian Blue, Limousin and Aberdeen Angus breeds fattened with a sugar-beet pulp or a cereal-based diet.** *Animal Science* 2007, **82**:125–132.

7. Faucitano L, Berthiaume R, D'Amours M, Pellerin D, Ouellet DR: **Effects of corn grain particle size and treated soybean meal on carcass and meat quality characteristics of beef steers finished on a corn silage diet.** *Meat Sci* 2011, **88**:750–754.

8. Ludden PA, Kucuk O, Rule DC, Hess BW: **Growth and carcass fatty acid composition of beef steers fed soybean oil for increasing duration before slaughter.** *Meat Sci* 2009, **82**:185–192.

9. Warren HE, Scollan ND, Enser M, Hughes SI, Richardson RI, Wood JD: **Effects of breed and a concentrate or grass silage diet on beef quality in cattle of 3 ages. I: Animal performance, carcass quality and muscle fatty acid composition.** *Meat Sci* 2008, **78**:256–269.

10. AOAC: **Official methods of analysis.** In *Association of Official Analytical Chemists.* 17th edition. Arlington, VA; 2000.

11. Van Soest PJ, Robertson JB, Lewis BA: **Methods for Dietary Fiber, Neutral Detergent Fiber, and Nonstarch Polysaccharides in Relation to Animal Nutrition.** *J Dairy Sci* 1991, **74**:3583–3597.

12. Boer HD, Dumont B, Pomeroy R, Weniger J: **Manual on E.A.A.P. Reference Methods for the assessment of carcass characteristics in cattle.** *Livestock Production Science* 1974, **1**:151–164.

13. Verbeke R, Van de Voorde G: **Détermination de la composition de demi-carcasses de bovins par la dissection d'une seule côte.** *Revue de l'Agriculture* 1978, **31**:875–880.

14. Chen YC: **Discussions on the names of beef carcass high-grade cuts.** *J Yellow Cattle Sci* 2003, **29**:1–3.

15. Lepage G, Roy C: **Direct transesterification of all classes of lipids in a one-step reaction.** *J Lipid Res* 1986, **27**:114–120.

16. Wu G, Davis PK, Flynn NE, Knabe DA, Davidson JT: **Endogenous synthesis of arginine plays an important role in maintaining arginine homeostasis in postweaning growing pigs.** *J Nutr* 1997, **127**:2342–2349.

17. Grovum WL: **Mechanisms explaining the effects of short chain fatty acids on feed intake in ruminants-osmotic pressure, insulin and glucagon.** In *Ruminant physiology: digestion, metabolism, growth and reproduction. Proceedings 8th International Symposium on Ruminant Physiology;* 1995:173–197.

18. Klosterman EW: **Beef cattle size for maximum efficiency.** *J Anim Sci* 1972, **34**:875–880.

19. Felius M: **Cattle breeds - an encyclopedia.** In *Misset.,Doetinchem.* Netherlands: 1995; 1995.

20. Paschal J, Sanders J, Kerr J, Lunt D, Herring A: **Postweaning and feedlot growth and carcass characteristics of Angus-, gray Brahman-, Gir-, Indu-Brazil-, Nellore-, and red Brahman-sired F1 calves.** *J Anim Sci* 1995, **73**:373–380.

21. Alberti P, Panea B, Sañudo C, Olleta JL, Ripoll G, Ertbjerg P, Christensen M, Gigli S, Failla S, Concetti S, Hocquette JF, Jailler R, Rudel S, Renand G, Nute GR, Richardson RI, Williams JL: **Live weight, body size and carcass characteristics of young bulls of fifteen European breeds.** *Livestock Science* 2008, **114**:19–30.

22. Laborde FL, Mandell IB, Tosh JJ, Wilton JW, Buchanan-Smith JG: **Breed effects on growth performance, carcass characteristics, fatty acid composition, and palatability attributes in finishing steers.** *J Anim Sci* 2001, **79**:355–365.

23. Gleghorn JF, Elam NA, Galyean ML, Duff GC, Cole NA, Rivera JD: **Effects of crude protein concentration and degradability on performance, carcass characteristics, and serum urea nitrogen concentrations in finishing beef steers.** *J Anim Sci* 2004, **82**:2705–2717.

24. Zhou GH, Liu L, Xiu XL, Jian HM, Wang LZ, Sun BZ, Tong BS: **Productivity and carcass characteristics of pure and crossbred Chinese Yellow Cattle.** *Meat Sci* 2001, **58**:359–362.

25. Lunstra DD, Echternkamp SE: **Puberty in beef bulls: acrosome morphology and semen quality in bulls of different breeds.** *J Anim Sci* 1982, **55**:638.

26. Morgan JB, Wheeler TL, Koohmaraie M, Crouse JD, Savell JW: **Effect of castration on myofibrillar protein turnover, endogenous proteinase activities, and muscle growth in bovine skeletal muscle.** *J Anim Sci* 1993, **71**:408–414.

27. Arthur P, Archer J, Johnston D, Herd R, Richardson E, Parnell P: **Genetic and phenotypic variance and covariance components for feed intake, feed efficiency, and other postweaning traits in Angus cattle.** *J Anim Sci* 2001, **79**:2805–2811.

28. Urick JJ, MacNEIL MD, Reynolds WL: **Biological type effects on postweaning growth, feed efficiency and carcass characteristics of steers.** *J Anim Sci* 1991, **69**:490–497.

29. Henricks DM, Jenkins TC, Ward JR, Krishnan CS, Grimes L: **Endocrine responses and body composition changes during feed restriction and realimentation in young bulls.** *J Anim Sci* 1994, **72**:2289–2297.

30. Dinh TTN, Blanton JR, Riley DG, Chase CC, Coleman SW, Phillips WA, Brooks JC, Miller MF, Thompson LD: **Intramuscular fat and fatty acid composition of longissimus muscle from divergent pure breeds of cattle.** *J Anim Sci* 2009, **88**:756–766.

31. Pethick DW, McIntyre BL, Tudor G, Rowe JB: **The partitioning of fat in ruminants: can nutrition be used as a tool to regulate marbling?** *Recent Advances in Animal Nutrition in Australia* 1997, **11**:151–158.

32. Baker S, Szasz J, Klein T, Kuber P, Hunt C, Glaze J, Falk D, Richard R, Miller J, Battaglia R: **Residual feed intake of purebred Angus steers: Effects on meat quality and palatability.** *J Anim Sci* 2006, **84**:938–945.

33. Muir PD, Deaker JM, Bown MD: **Effects of forage- and grain-based feeding systems on beef quality: A review.** *New Zealand Journal of Agricultural Research* 1998, **41**:623–635.

34. Silva J, Patarata L, Martins C: **Influence of ultimate pH on bovine meat tenderness during ageing.** *Meat Sci* 1999, **52**:453–459.

35. Killinger KM, Calkins CR, Umberger WJ, Feuz DM, Eskridge KM: **Consumer sensory acceptance and value for beef steaks of similar tenderness, but differing in marbling level.** *J Anim Sci* 2004, **82**:3294–3301.

36. Shackelford SD, Morgan JB, Cross HR, Savell JW: **Identification of threshold levels for warner-bratzler shear force in beef top loin steaks.** *J Muscle Foods* 1991, **2**:289–296.

37. Boleman S, Boleman S, Miller R, Taylor J, Cross H, Wheeler T, Koohmaraie M, Shackelford S, Miller M, West R: **Consumer evaluation of beef of known categories of tenderness.** *J Anim Sci* 1997, **75**:1521–1524.

38. Loken BA, Maddock RJ, Stamm MM, Schauer CS, Rush I, Quinn S, Lardy GP: **Growing rate of gain on subsequent feedlot performance, meat, and carcass quality of beef steers.** *J Anim Sci* 2009, **87**:3791–3797.

39. Purchas RW, Burnham DL, Morris ST: **Effects of growth potential and growth path on tenderness of beef longissimus muscle from bulls and steers.** *J Anim Sci* 2002, **80**:3211–3221.

40. Marino R, Albenzio M, Malva A d, Santillo A, Loizzo P, Sevi A: **Proteolytic pattern of myofibrillar protein and meat tenderness as affected by breed and aging time.** *Meat Sci* 2013, **95**:281–287.

41. Traore S, Aubry L, Gatellier P, Przybylski W, Jaworska D, Kajak-Siemaszko K, Santé-Lhoutellier V: **Higher drip loss is associated with protein oxidation.** *Meat Sci* 2012, **90**:917–924.

42. Farouk MM, Mustafa NM, Wu G, Krsinic G: **The "sponge effect" hypothesis: An alternative explanation of the improvement in the waterholding capacity of meat with ageing.** *Meat Sci* 2012, **90**:670–677.

43. Scollan N, Hocquette J-F, Nuernberg K, Dannenberger D, Richardson I, Moloney A: **Innovations in beef production systems that enhance the nutritional and health value of beef lipids and their relationship with meat quality.** *Meat Sci* 2006, **74**:17–33.

44. Dewhurst RJ, Shingfield KJ, Lee MRF, Scollan ND: **Increasing the concentrations of beneficial polyunsaturated fatty acids in milk produced by dairy cows in high-forage systems.** *Animal Feed Science and Technology* 2006, **131**:168–206.

45. Varela A, Oliete B, Moreno T, Portela C, Monserrrat L, Carballo JA, Sánchez L: **Effect of pasture finishing on the meat characteristics and intramuscular fatty acid profile of steers of the Rubia Gallega breed.** *Meat Sci* 2004, **67**:515–522.

46. Wood JD, Richardson RI, Nute GR, Fisher AV, Campo MM, Kasapidou E, Sheard PR, Enser M: **Effects of fatty acids on meat quality: a review.** *Meat Sci* 2004, **66:**21–32.

47. Park RJ ALFORD, Ratcliffe D: **Effect on meat flavor of period of feeding a protected lipid supplement to lambs.** *J Food Science* 1975, **40:**1217–1221.

48. Campo MM, Nute GR, Wood JD, Elmore SJ, Mottram DS, Enser M: **Modelling the effect of fatty acids in odour development of cooked meat in vitro: part I—sensory perception.** *Meat Sci* 2003, **63:**367–375.

49. FAO/WHO/UNU: **Protein and amino acid requirements in human nutrition.** *Report of a joint FAO/WHO/UNU expert consultation (WHO Technical Report Series)* 2007, **935:**149–150.

12

Determination of reference intervals for metabolic profile of Hanwoo cows at early, middle and late gestation periods

Da Chuan Piao[1], Tao Wang[2,3], Jae Sung Lee[4], Renato SA Vega[5], Sang Ki Kang[1], Yun Jaie Choi[1*] and Hong Gu Lee[4*]

Abstract

Background: Metabolic profile was initially designed as a presymptomatic diagnostic aid based on statistical analyses of blood metabolites to provide an early warning of certain types of metabolic disorder. However, there is little metabolic profile data available about Korean Hanwoo cows. Therefore, this study aimed to determine the reference intervals of metabolic profile for Korean Hanwoo cows.

Methods: Healthy animals (2,205) were selected and divided into early (day 1 to 95), middle (day 96 to 190) and late (day 191 to 285) period according to their gestating period. Metabolic profile including total protein (TP), albumin (Alb), urea (UREA), glucose (Glu), total cholesterol (T-Cho), long-chain fatty acid (LCFA), aspartate aminotransferase (AST), gamma-glutamyl transpeptidase (GGT), creatinine (Crea), calcium (Ca), inorganic phosphorous (iP) and magnesium (Mg) were analyzed using a TBA-40FR automatic biochemical analyzer. The data of Korean Hanwoo cows were then compared to those of the Japanese Wagyu cows.

Results: Most of the data of the Korean Hanwoo cows were relatively higher than those of Japanese Wagyu cows, with the exception of Glu and GGT. This may indicate that the nutritional level of feed for the Korean Hanwoo cows was higher than that of the Japanese Wagyu cows because of the different feeding system. In particular, relatively higher levels of UREA and LCFA were observed in the Korean Hanwoo cows, and this may also contribute to the low reproduction efficiency.

Conclusions: These findings may provide some theoretical basis for understanding the reproductive and feeding situation of Korean Hanwoo cows.

Keywords: Hanwoo cows, Metabolic profile, Reference intervals, Wagyu cows

Introduction

Blood metabolites reflect the nutritional status as well as the physiological condition of an animal. The physiology of a cow changes in the peripartum period, and it is important to monitor nutritional and physiological status rapidly and precisely, because these cows are prone to peripartum metabolic disorders and reproductive diseases.

The metabolic profile was initially designed as a presymptomatic diagnostic aid based on statistical analyses of blood metabolites to provide an early warning for certain types of metabolic disorders [1]. Subsequently, metabolic profile has been applied to assess nutritional status [2,3], improve feeding management and diagnose metabolic disorders in dairy herds [4,5]. Reference intervals are useful when interpreting a set of metabolic profile results. Payne et al. [6] estimated normal intervals (95% confident interval) using 2,400 blood samples from 13 dairy herds. Kida [4] established a 10-day criteria for metabolic profile by using data from 29,043 cows in 1,130 commercial dairy herds covering dry and lactation periods [5]. The practicability of the criteria was evaluated in herds with

* Correspondence: cyjcow@snu.ac.kr; hglee66@konkuk.ac.kr
[1]Laboratory of Animal Cell Biotechnology, Department of Agricultural Biotechnology, Seoul National University, Shinlim-dong, Kwanak-gu, Seoul 151-742, South Korea
[4]Department of Animal Science and Technology, College of Animal Bioscience & Technology, Konkuk University, 120 Neungdong-ro, Gwangjin-gu, Seoul 143-701, South Korea
Full list of author information is available at the end of the article

peripartum diseases, and the metabolic abnormalities were successfully detected not only in the herd but also in individual cows. However, little metabolic profile data has been found about Korean Hanwoo cows. Therefore, in the current study, twelve blood metabolites were analyzed in order to determine reference intervals of metabolic profile at the early, middle and late reproduction period in Korean Hanwoo cows. Moreover, the data of Korean Hanwoo cows were also compared to those of Japanese Wagyu cows obtained from NOSAI [7].

Materials and methods

Experiment design, animals and sampling

A total of 2,205 healthy Korean Hanwoo cows were selected from various farms in Jangsu-gun Jeollabuk-do, South Korea from 2006 to 2011. All experimental procedures were in accordance with the "Guidelines for the Care and Use of Experimental Animals of Seoul National University". No abnormalities in these animals were observed or monitored. Generally, all cows were housed indoors and some were on pasture in summer. Feeding systems included continuous feeding of a total mixed ration or separate feeding of forage and concentrates. The metabolic profile was conducted through four seasons. The animals were sorted into four groups according to the plasma pregnancy-associated glycoproteins (PAGs) (Table 1). Blood samples were taken via the external jugular vein after the morning meal. Serum was recovered from the blood samples through centrifugation at 3,500 rpm at 4°C for 15 min, and stored at −80°C until required. All samples were analyzed within one week of being collected.

Analysis of metabolic profile

The metabolic profile was performed with a Toshiba Accute Biochemical Analyzer-TBA- 40FR (Toshiba Medical Instruments, Otawara-shi, Tochigi-ken, Japan) according to a previously described method [5]. Indicators for protein metabolism consist of total protein (TP), albumin (Alb) and urea (UREA); for energy metabolism, glucose (Glu), total cholesterol (T-Cho) and long-chain fatty acids (LCFA); for liver and kidney function, aspartate aminotransferase (AST) and γ-glutamyl transpeptidase (GGT) and

Creatinine (Crea); for mineral metabolism, calcium (Ca), inorganic phosphorus (iP) and magnesium (Mg) (Table 2). All the reagents required for this procedure were purchased from Wako Pure Chemical Industries, Ltd. (Chuo-ku, Osaka, Japan). The reference intervals were determined by following the recommendations of the Clinical and Laboratory Standards Institute (CLSI).

Statistical analysis

The metabolic profile data were presented as Mean ± SD and analyzed using a one-way analysis of variance (one-way ANOVA) (SPSS Inc., Chicago, IL, USA). In all cases, differences were considered significant if $P < 0.05$.

Results and discussion

The metabolic profile of Korean Hanwoo cows in different physiological stages were determined according to data collected from 2,205 animals (Table 3). Protein metabolism: It has been well known that TP, Alb and UREA levels are indicators of sufficient protein intake from diets. Alb is not a long-term indicator of protein intake because of its relatively short half-life in the blood. On the other hand, UREA level is a good indicator of long-term intake of dietary protein [8]. TP, Alb and UREA were significantly ($P < 0.05$) increased from the EP to LP periods, suggesting that there was an increased intake of

Table 1 Classification of cows into four stages according to pregnant status

Reproduction period	Duration
Non-pregnant	After calving or before pregnancy
Early-pregnant[1]	Day 1 to 95
Middle-pregnant	Day 96 to 190
Late-pregnant	Day 191 to 285

[1]The pregnancy tests were defined based to the plasma pregnancy-associated glycoproteins (PAGs) levels.

Table 2 Analytical method used for metabolic profile

Components	Method of analysis
Protein metabolism	
Total protein (TP)	Biuret Test
Albumin (Alb)	Bromcresol green (BCG) Method
Urea (UREA)	Urease-Glutamate dehydrogenase (GLDH) Method
Energy metabolism	
Glucose (Glu)	Hexokinase (HK)-
Glucose-6-phosphate dehydrogenase (G6PD) Method	
Total cholesterol (T-Cho)	Cholesterol oxidase
Long-chain fatty acid (LCFA)	Enzymatic colorimetric
Liver function	
Aspartate aminotransferase (AST)	Malate dehydrogenase (MDH) UV
Gamma glutamyl transpeptidase (GGT)	Glu-3-CA-4-NA substrate
Kidney function	
Creatinine (Crea)	Jaffe' Method
Mineral metabolism	
Calcium (Ca)	O-Cresolphthalein Complexone (OCPC) Method
Inorganic phosphorous (iP)	Enzymatic UV
Magnesium (Mg)	Enzymatic UV

Table 3 Serum metabolic profile of Korean Hanwoo cows at different stages of pregnancy

Indicator	NP[1]		EP[1]		MP[1]		LP[1]	
	Mean	SD	Mean	SD	Mean	SD	Mean	SD
TP, g/dL	7.56[a]	0.82	7.61[ab]	0.83	7.73[c]	0.92	7.69[b]	0.79
Alb, g/dL	3.60[a]	0.31	3.62[ab]	0.29	3.65[b]	0.31	3.71[c]	0.30
UREA, mg/dL	10.40[a]	3.50	11.10[b]	3.50	11.40[b]	3.20	11.50[b]	4.10
Glu, mg/dL	39.00[a]	19.00	39.00[a]	18.00	41.00[a]	18.00	45.00[b]	18.00
T-Cho, mg/dL	128.00	36.00	130.00	35.00	126.00	33.00	131.00	36.00
LCFA, µEq/L	175.00[b]	145.00	137.00[a]	106.00	163.00[b]	108.00	225.00[b]	193.00
AST, IU/L	75.00[b]	16.00	74.00[ab]	17.00	72.00[a]	19.00	74.00[ab]	18.00
GGT, IU/L	17.00[a]	6.00	18.00[b]	7.00	17.00[a]	6.00	17.00[a]	6.00
Crea, mg/dL	1.43[a]	0.37	1.46[a]	0.36	1.55[b]	0.35	1.56[c]	0.44
Ca, mg/dL	10.24[bc]	0.87	10.14[b]	0.87	10.14[a]	1.09	10.30[c]	1.14
IP, mg/dL	6.75	1.88	6.57	2.15	6.79	1.63	6.68	2.96
Mg, mg/dL	2.57[a]	0.35	2.60[a]	0.37	2.59[a]	0.38	2.67[b]	0.45

[1]NP: non-pregnant, n = 592; EP: early pregnant, n = 686; MP: middle pregnant, n = 517; LP: late pregnant, n = 410.
[a-c]Values followed by different letters within each component are significantly different (*P* < 0.05).

diet. The mean level of TP was slightly higher than in previous reported studies [9]. Energy metabolism: In the past three decades, the question of whether Glu can be used as indicator for energy metabolism has been discussed [8]. Blood Glu has a moderate diagnostic value in the assessment of nutritional status of cattle, as it varies moderately in blood [10]. It has also been reported that glucose can be used in combination with other indicators to assess energy metabolism [9]. The serum Glu concentration detected in this study was similar with one previously reported [9], but lower than some other reported studies in other species [11,12]. These differences may be attributed to the species differences or the feeding system. Significantly higher levels of LCFA were observed at the EP and LP stage (*P* < 0.05). Liver function and kidney function: AST and GGT are enzymes that indicate liver cell damage and biliary obstruction, respectively. In this study, the mean level of AST was higher than in previously reported studies [9]. Therefore, it seemed that the high level of AST reveals early signs of liver cell damage in Korean Hanwoo cows. Mineral metabolism: Serum concentrations of Ca and IP reflect dietary calcium and phosphate intake [8], and it is well known that serum Ca is under homeostatic control of the endocrine system. Ca, IP and Mg have a high diagnostic value in determining the nutritional status of animals due to their low variability in blood [10]. In this study, low variability of the three indicators was also

Table 4 Reference intervals of serum components of Korean Hanwoo cows at different stages of pregnancy

Indicator	NP[1]		EP[1]		MP[1]		LP[1]	
	D[2]	Intervals	D	Intervals	D	Intervals	D	Intervals
TP, g/dL	G[2]	5.9 ~ 9.2	N	6.4 ~ 9.6	N	6.5 ~ 10.4	N	6.4 ~ 9.8
Alb, g/dL	G	2.97 ~ 4.22	G	3.04 ~ 4.20	G	3.03 ~ 4.27	G	3.11 ~ 4.31
UREA, mg/dL	G	3.4 ~ 17.4	G	4.0 ~ 18.1	G	4.9 ~ 17.8	N	4.7 ~ 17.9
Glu, mg/dL	N[2]	9 ~ 57	N	7 ~ 57	N	7 ~ 57	N	4 ~ 69
T-Cho, mg/dL	G	56 ~ 200	G	60 ~ 200	G	60 ~ 192	G	59 ~ 203
LCFA, µEq/L	N	37 ~ 525	N	30 ~ 430	N	35 ~ 417	N	45 ~ 699
AST, IU/L	G	43 ~ 107	N	46 ~ 113	N	49 ~ 116	N	48 ~ 126
GGT, IU/L	N	7 ~ 32	N	6 ~ 38	G	5 ~ 29	N	4 ~ 31
Crea, mg/dL	G	0.69 ~ 2.17	N	0.74 ~ 2.18	G	0.85 ~ 2.25	G	0.68 ~ 2.0
Ca, mg/dL	G	8.5 ~ 11.9	N	8.5 ~ 11.5	N	9.0 ~ 11.8	G	8.9 ~ 12.5
IP, mg/dL	G	3.1 ~ 10.5	N	3.2 ~ 10.8	G	3.5 ~ 10.5	N	3.6 ~ 10.2
Mg, mg/dL	G	1.87 ~ 3.27	G	2.23 ~ 3.34	G	1.80 ~ 3.35	N	2.04 ~ 3.27

[1]NP: non-pregnancy, n = 592; EP: early pregnancy, n = 686; MP: middle pregnancy, n = 517; LP: late pregnancy, n = 410.
[2]D: Distribution of data; G: Gaussian distribution; N: Non-Gaussian distribution.

observed, but the mean values of these three indicators were higher than in previously reported studies. This may be due to the high level of mineral supplementation in the diet of Korean Hanwoo cows.

When interpreting laboratory data for metabolic profile from a herd or individual animal, reference intervals need to be determined for clinical application. In this experiment, we also determined reference intervals for serum components at 4 physiological stages in Korean Hanwoo cows, as are shown in Table 4. It was found that reference intervals were changed according to physiological status. These reference intervals may provide a basis for interpreting data analysis and metabolic disorder. A comparison between Korean Hanwoo and Japanese Wagyu cows was performed. The amounts of serum components are affected by several factors including nutrition, physiological status, breed, season and age.

The mean values of serum components in Korean Hanwoo and Japanese Wagyu cows were compared during four physiological stages (non-pregnant, early pregnancy, middle pregnancy and late pregnancy), because the physiology of these two breeds was known to be very similar. The results of this comparison, which are shown in Figure 1, revealed that the patterns of serum components were very similar in the two breeds. All the serum components were higher in Korean Hanwoo than in Japanese Wagyu cows during all test periods, with the exception of Glu and GGT. TP, Alb and UREA indicate the protein intake of an animal from diet [4]. In this experiment, mean values of these three indicators were higher in Korean Hanwoo than in Japanese Wagyu cows, in all physiological stages. This may indicate that the protein level in the diets of Korean Hanwoo cows was higher than in the diets of Japanese Wagyu cows. Blood Glu, T-cho and LCFA are the most commonly

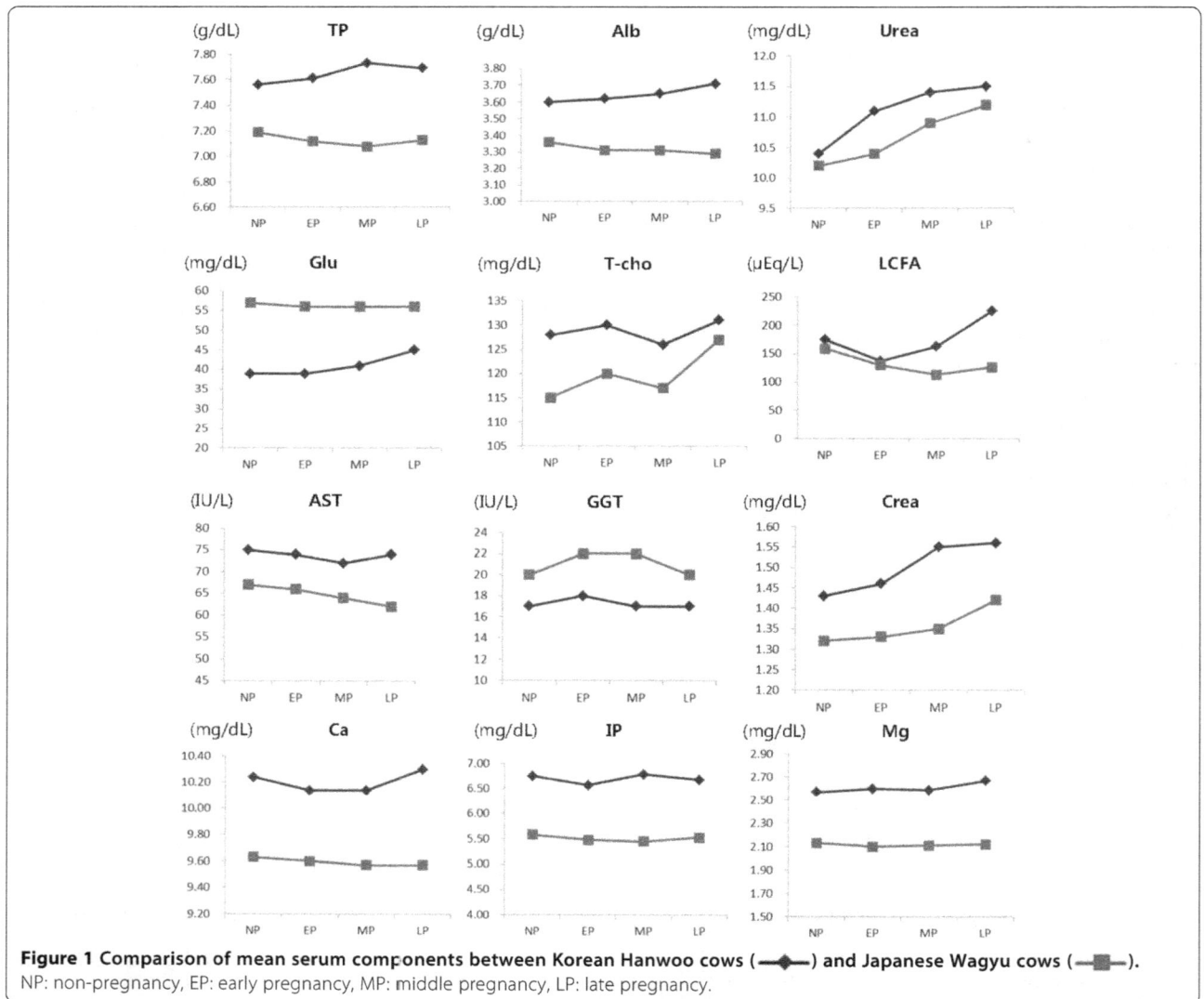

Figure 1 Comparison of mean serum components between Korean Hanwoo cows (—◆—) and Japanese Wagyu cows (—■—).
NP: non-pregnancy, EP: early pregnancy, MP: middle pregnancy, LP: late pregnancy.

used blood metabolites to assess the energy metabolism. The physiological status of an animal affects the serum concentration of these related metabolites in energy metabolism. The elevated level of LCFA at MP and LP periods may indicate that there was a negative energy balance during pregnancy periods in Korean Hanwoo cows. On the other hand, the mean value of serum Glu was lower in Korean Hanwoo than in Japanese Wagyu cows. For this reason, we suggest that the elevated level of LCFA and the lowered level of Glu may be caused by an insufficient energy intake. Serum activity of AST and GGT indicate liver function, and Crea indicates kidney function. AST is an enzyme that expresses in many tissues, particularly in liver and cardiac muscle [13]. In this study, AST and Crea were higher in Korean Hanwoo than in Japanese Wagyu cows; however, GGT was lower in Korean Hanwoo than in Japanese Wagyu cows. Serum Ca, Mg and IP were higher in the Korean Hanwoo than in the Japanese Wagyu cows, indicating that the mineral level was higher in Korean Hanwoo cows' diets than in Japanese Wagyu cows. In this study, the patterns of serum components were highly similar in the two breeds.

Conclusions

In this study, the metabolic profile of the Korean Hanwoo cows was determined. The differences in the metabolic profile between Korean Hanwoo cows and Japanese Wagyu cows were also verified. Our findings may provide some basis for understanding the reproductive and feeding situations of Korean Hanwoo cows.

Competing interests
The authors declare that they have no competing interests.

Authors' contributions
YJ and HG conceived and designed the Experiments. DC executed the experiment and analyzed the samples. TW and RSA revised the manuscript. JS and SK provided help during the animal experiment. All authors interpreted the data, critically revised the manuscript for content and approved the final version.

Acknowledgements
This research was supported by Bioindustry Technology Development Program (313020041SB010) for Ministry of Agriculture, Food and Rural Affairs, Republic of Korea and Concentrated Research Professor Program for Konkuk University, Seoul, Republic of Korea.

Author details
[1]Laboratory of Animal Cell Biotechnology, Department of Agricultural Biotechnology, Seoul National University, Shinlim-dong, Kwanak-gu, Seoul 151-742, South Korea. [2]College of Animal Science and Technology, Jilin Agricultural University, 2888 Xincheng Street, Nan-guan District, Changchun 130118, People's Republic of China. [3]Key Laboratory of Animal Nutrition and Feed Science, Jilin Province, Jilin Agricultural University, 2888 Xincheng Street, Nan-guan District, Changchun 130118, People' Republic of China. [4]Department of Animal Science and Technology, College of Animal Bioscience & Technology, Konkuk University, 120 Neungdong-ro, Gwangjin-gu, Seoul 143-701, South Korea. [5]Animal & Dairy Sciences Cluster, College of Agriculture, University of the Philippines Los Bañose, Los Baños 4031, Laguna, Philippines.

References
1. Payne JM. The Compton metabolic profile test. Proc R Soc Med. 1972;65:181–3.
2. Blowey RW, Wood DW, Davis JR. A nutritional monitoring system for dairy herds based on blood glucose, urea and albumin levels. Vet Rec. 1970;92:691–6.
3. Adams RS, Stout WL, Kradel DC, Guss Jr SB, Moser BL, Jung GA. Use and limitations of profiles in assessing health or nutritional status of dairy herds. J Dairy Sci. 1978;61:1671–9.
4. Kida K. The metabolic profile test: its practicability in assessing feeding management and periparturient diseases in high yielding commercial dairy herds. J Vet Med Sci. 2002;64:557–63. X.
5. Deluyker HA, Gay JM, Weaver JD, Azari AS. Change of milk yield with clinical diseases for a high producing dairy herd. J Dairy Sci. 1991;74:436–45.
6. Payne JM, Dew SM, Manston R, Faulks M. The use of a metabolic profile test in dairy herds. Vet Rec. 1970;87:150–8.
7. NOSAI, Japan National Agricultural Insurance Association, http://www.nosai.or.jp/.
8. Kida K. Relationships of metabolic profiles to milk production and feeding in dairy cows. J Vet Med Sci. 2003;65:671–7.
9. Wee SH, Park SJ. Studies on the pure-bred Korean-native cattle of Chonnam area. Korean J Vet Serv. 1990;13:75–9.
10. Ndlovu T, Chimonyo M, Okoh AI, Muchenje V, Dzama K, Raats JG. Assessing the nutritional status of beef cattle: current practices and future prospects. Afri J Biotech. 2007;6:2727–34.
11. Lee AJ, Twardock AR, Bubar RH, Hall JE, Davis CL. Blood metabolic profiles: their use and relation to nutritional status of dairy cows. J Dairy Sci. 1978;61:1652–70.
12. Jones GM, Wildman EE, Troutt Jr HF, Lesch TN, Wagner PE, Boman RL. Metabolic profiles in Virginia dairy herds of different milk yields. J Dairy Sci. 1982;65:683–8.
13. Otto F, Baggasse P, Bogin E, Harun M, Vilela F. Biochemical profile of Angoni cattle in Mozambique. Israel Vet Med Assoc. 2000;55:1–9.

13

Current strategies for reproductive management of gilts and sows in North America

Robert R Kraeling[1*] and Stephen K Webel[2]

Abstract

Many advances in genetic selection, nutrition, housing and disease control have been incorporated into modern pork production since the 1950s resulting in highly prolific females and practices and technologies, which significantly increased efficiency of reproduction in the breeding herd. The objective of this manuscript is to review the literature and current industry practices employed for reproductive management. In particular the authors focus on assisted reproduction technologies and their application for enhanced productivity. Modern maternal line genotypes have lower appetites and exceptional lean growth potential compared to females of 20 yr ago. Thus, nutrient requirements and management techniques and technologies, which affect gilt development and sow longevity, require continuous updating. Failure to detect estrus accurately has the greatest impact on farrowing rate and litter size. Yet, even accurate estrus detection will not compensate for the variability in the interval between onset of estrus and actual time of ovulation. However, administration of GnRH analogs in weaned sows and in gilts after withdrawal of altrenogest do overcome this variability and thereby synchronize ovulation, which makes fixed-time AI practical. Seasonal infertility, mediated by temperature and photoperiod, is a persistent problem. Training workers in the art of stockmanship is of increasing importance as consumers become more interested in humane animal care. Altrenogest, is used to synchronize the estrous cycle of gilts, to prolong gestation for 2–3 d to synchronize farrowing and to postpone post-weaning estrus. P.G. 600® is used for induction of estrus in pre-pubertal gilts and as a treatment to overcome seasonal anestrous. Sperm cell numbers/dose of semen is significantly less for post cervical AI than for cervical AI. Real-time ultrasonography is used to determine pregnancy during wk 3–5. $PGF_{2\alpha}$ effectively induces farrowing when administered within two d of normal gestation length. Ovulation synchronization, single fixed-time AI and induced parturition may lead to farrowing synchronization, which facilitates supervision and reduces stillbirths and piglet mortality. Attendance and assistance at farrowing is important especially to ensure adequate colostrum consumption by piglets immediately after birth. New performance terminologies are presented.

Keywords: Gilts, Management, Nutrition, Reproductive technology, Sows

Introduction

Basic and applied research in physiology, nutrition, genetics, animal behavior, environment and housing over the last 40 yr provided the foundation for development of highly prolific females and various management practices and technologies, which have significantly increased efficiency of reproduction in the breeding herd. An ovulation rate of 20 is not uncommon in contemporary highly prolific females [1]. Therefore, if one assumes a gestation length of 115 d, a lactation length of 21 d, a weaning-to-estrus interval (WEI) of 5 d, 100% conception rate and zero embryonic and preweaning

mortality, sows have the potential to farrow 2.6 times/yr and to produce 52 pigs weaned/mated sow/yr. However, due to numerous factors, such as season, nutrition, disease, embryo mortality before d 30 of pregnancy and piglet preweaning mortality, the potential of 52 weaned pigs/mated sow/yr has not been reached. In 2012, the number of liveborn pigs/litter was 11.8–12.3, pigs weaned/mated sow was 10.3–10.5 and the average number of litters/mated female/yr was 2.3 [2,3]. Thus, the average number of pigs/mated female/yr was approximately 24 in 2012. In 2012, according to PigChamp [2], the average Canadian farrowing rate and total born were 86.6% and 14.0, respectively, and in the U.S.A. they were 83.6% and 13.4, respectively. Comparable data from 2001 for Canada were 74.9%

---

* Correspondence: rrkraeling@bellsouth.net
[1]L&R Research Associates, Watkinsville, GA, USA
Full list of author information is available at the end of the article

and 11.5 and for the U.S.A. were 69.0% and 11.3. In 2014, Ketchem and Rix [4] reported the following data for the highest 10%, the average and the lowest 30% of producers, respectively: 30.1; 25.3; 21.9 weaned pigs/mated female/yr and 36.3; 32.7; 29.6 total pigs born per litter/mated female/yr.

Figure 1 presents the authors' vision for incorporating the discussed technologies into future pig production. The orally active progestin, altrenogest, is used to synchronize the estrous cycle of gilts. GnRH analogs synchronize ovulation thereby making fixed-time AI practical. A single fixed-time AI of every female in a group on one d enables producers to reduce the cost of semen, eliminate weekend inseminations and focus resources on other tasks on the remaining d of the week. Benefits of induced farrowing with $PGF_{2\alpha}$ are: 1) a high proportion of farrowings occur during normal working h, 2) no farrowing on weekends, 3) reduced age and weight range within batches of growing pigs and 4) efficient use of facilities and batching of routine tasks. Ovulation synchronization, single fixed-time AI and induced parturition with $PGF_{2\alpha}$ leads to farrowing synchronization, which facilitates supervision of sows and piglets. Attendance and assistance at farrowing is especially important to ensure adequate colostrum consumption by piglets immediately after birth. These technologies save a significant amount of time, which allows redistribution of labor (i.e. focusing more on facility maintenance, gilt development, evaluating sows' body condition, adjusting gestation feeders, assisting in farrowing and training workers in the art of stockmanship). In addition, they maximize the leverage of high index boars, which will improve overall pork production efficiency.

The purpose of this paper is to summarize results from basic and applied research that may be applied to management of gilts and sows in the breeding herd and to present the authors' perspectives on the current strategies actually used by pork producers. Knox and co-workers [5] recently published an analysis of survey data, which documented current reproductive management practices of North American swine farms.

Gilt development and management

The average sow replacement rate was 45% in 2012 [2]. This high rate is due to failure of postpartum sows to return to estrus and conceive, poor reproductive performance, poor feet and leg soundness, and introduction of new genetic lines [6-9]. Excellent reviews of the literature regarding gilt management were published by Foxcroft [10,11], Gill [7], Williams et al. [12], Bortolozzo et al. [13], Wiedmann [14] and Whitney and Masker [15]. The authors suggest referring to one or more of these excellent manuscripts for a comprehensive review and discussion on gilt management, development and nutrition since these topics are not discussed in detail in this manuscript.

Breeding and selection of maternal line gilts is generally conducted by breeding stock suppliers based on growth rate, body composition, disease status, sexual development and dam's reproductive history. The ability to express estrus and continue to cycle should be the key reproductive trait for selection of replacement gilts. Sterning et al. [16] reported that heritability of the ability to display estrus at puberty and ovulate within 10 d after weaning a litter is 0.31. Gilts not displaying estrus at puberty also had a higher incidence of ovulation without estrus within 10 d after weaning their first litter.

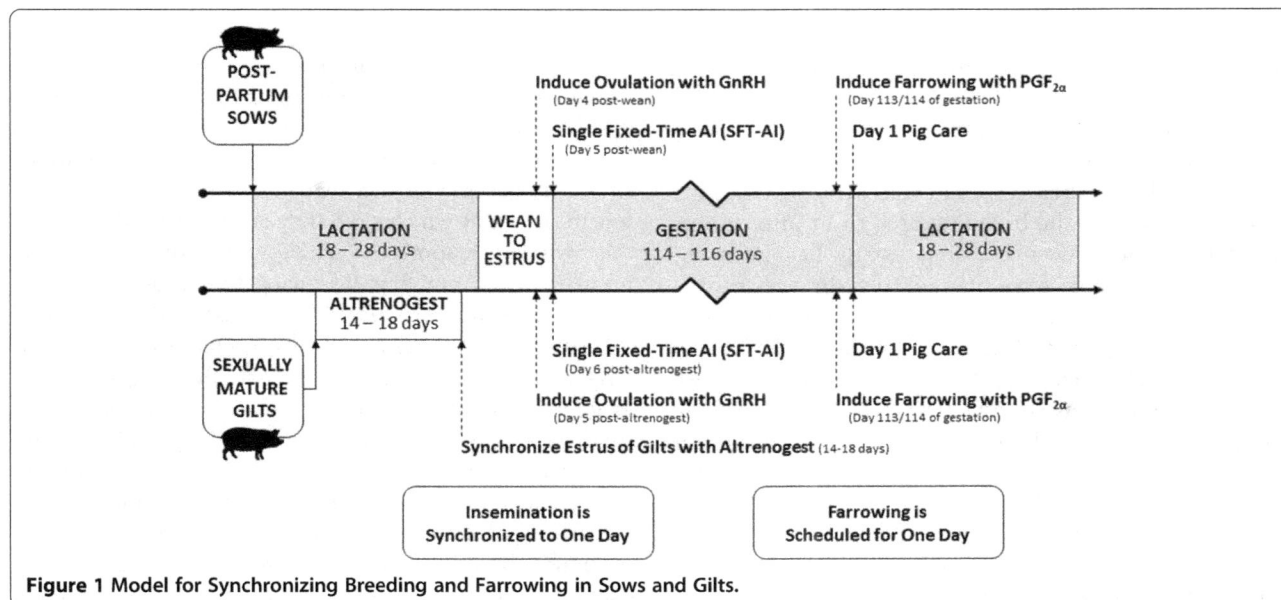

Figure 1 Model for Synchronizing Breeding and Farrowing in Sows and Gilts.

Because of the large number of piglets per litter farrowed by modern females, often there are not enough functional nipples for all the piglets. Therefore, the number and functionality of the mammary glands and teats are critical for survival of piglets. Feet and legs are important because sows are expected to farrow more than 2 litters per yr, nurse a large litter for approximately 3 weeks, breed back in 5–7 d after weaning and live on solid concrete or slatted floors [17]. Lameness due to incorrect structure of feet and legs hinder sows from getting up and down in the farrowing crate, which results in reduced feed intake [17,18].

In general, selected gilts are moved from a growing and finishing facility to a development facility at 150–180 d of age when daily boar exposure begins. Generally age at puberty is positively associated with age at onset of boar exposure [19]. Exposure of peripubertal gilts to boars for 20 min/d stimulates expression of estrus. Boars must be mature (>10 mo of age) and express the full complement of male mating characteristics. For best results, gilts are brought to boars where initially, gilts experience the sight, sound and odor of the boar with fence line contact. Direct physical contact is best. However, constant exposure to boar sounds and scent causes habituation which hinders heat detection, but not necessarily onset of sexual maturity. Gilts expressing estrus are then removed and the boar is allowed full access to all non-cyclic gilts for 10–15 min per d. Moving, mixing, transport and boar exposure typically induce first estrus in a high percentage of gilts within 10–20 d. Gilts that respond to boar exposure at an early age tend to remain in production longer than gilts that respond at a later age [11,20,21]. Stimulating gilts to cycle and breed on the second or third estrus is a well-established practice. The terminology, HNS (Heat-No-Serve), is frequently used to describe this important management practice for introducing gilts into the breeding herd.

Cyclic gilts are then moved into the breeding barn for acclimation to facilities and management routines before breeding. Since estrous cycles are known, gilts may be staged into the breeding area to fit into groups of weaned sows. After the first estrus has been recorded, gilts should be acclimated to stalls or breeding and gestation housing at least 16 d prior to breeding.

Most producers breed gilts on the second estrus if they have cycled before 200 d of age, whereas those which express estrus for the first time after 200 d of age are often bred at the first estrus. It is important at first mating that adequate fat stores are available for good lactation and a short WEI represented by a back fat measurement of 12–18 mm. They should be maintained in small groups of approximately 10 with a minimum of 1.4 square meters per gilt [14,22].

Gilt nutrition

Modern maternal line genotypes are more sensitive to nutritional management because their appetite is lower and they have exceptional lean growth potential compared to females of 20 yr ago [10,23,13]. Replacement gilts are typically fed ad lib a diet lower in energy than diets fed to slaughter pigs in order to avoid excessive body fat [15]. This also allows for slightly slower growth, which limits mature body size, thereby preventing feet and leg problems and excessive fat gain. An estimate of their genetic potential for growth can be made at this time. Subsequently, diets for replacement gilts should contain higher concentrations of vitamin A and E, calcium, phosphorus, selenium, chromium and zinc than the typical finishing diet because highly prolific gilts reach puberty with limited reserves of protein and body fat and they continue to grow during their first gestation. Concentrations of Ca and P must be high enough for maximum bone mineralization, which is mobilized for fetal growth and lactation [15,23]. Also, protein and amino acid deficiencies lead to delayed puberty. Older literature indicates that selected replacement gilts should be limit fed energy from 100 to 104 kg BW or until 2 weeks prior to mating so they will not become too fat. However, Foxcroft and coworkers [10], Williams et al. [12] and Gill [7] presented evidence that fatness is not an issue with modern lean maternal line genotype females, which deposit and mobilize lean tissue with little impact on fat tissue depots [10]. Therefore, lean tissue mass is a key consideration for correct management of the gilt [10]. Gill [7] proposed that a nutrition program should result in a body condition score of 3 at first service.

Sow management

Sow longevity is important because litter size and piglet weights increase until the fourth or fifth parities, and the number of pigs weaned per sow per yr increases until the sixth and seventh parities. Mature, structurally sound replacement gilts will most likely reach their fourth parity, at which time they are most productive for the swine operation [6,7,24,25]. Sow longevity is the number of d from first farrowing to removal from the herd or total number of pigs produced in the lifetime of the sow [21]. Numerous observational studies demonstrated that multiple factors impact sow longevity, such as genetics, nutrition, housing, disease, lameness, age at first mating, assistance at farrowing, length of lactation and growth rate, body condition and performance of parity one sows [20,21,26-28]. Specific cultural environments and consumer attitudes in the U.S.A. and Europe influence breeding herd management [14,29]. Some examples are use of prostaglandin $F_{2\alpha}$ for inducing farrowing, limiting the use of antibiotics and moving away from individual gestation stalls.

Sow nutrition

Modern highly prolific females have more defined nutrient requirements than females 20 yr ago [30], thus continuous updating of nutrient requirements and management techniques and technologies are required [10,23,30-36]. The effect of environmental factors, such as temperature, humidity and building design on feed intake during lactation should be considered.

Gestation: It is well documented that "flushing" by increasing feed by 50–100% or feeding sources of high energy, such as dextrose for 10–14 d before first service, increases ovulation rate and litter size. However, there are conflicting reports in the literature regarding feeding programs during the remainder of gestation [37]. Vignola [38] reviewed research indicating that feed intake should then be decreased after mating to an appropriate gestation diet because sows that are overfed throughout gestation, especially during the first two weeks after breeding, frequently have high embryonic mortality and produce small litters. Sows, which are "too fat", have farrowing problems, crush piglets, eat poorly during subsequent lactation and are less prolific at the next parity [38]. Sows with back fat depths of 23 mm or more at farrowing have depressed appetite during lactation [38]. Peltoniemi et al. [35] reviewed research indicating that feed restriction after mating may only apply to the first 4 d in gilts and not at all for sows. Love and coworkers [39] and Virolainen and coworkers [40] reported that abundant feeding during early pregnancy increased embryo survival and failed to influence maintenance of pregnancy. Martineau and Badouard [29] proposed that two major characteristics of the hyperprolific sow are lack of early embryonic death with overfeeding after ovulation and a positive influence of overfeeding during the last weeks of pregnancy on piglet birth weights.

As indicated from the above discussion, nutrient needs of sows change significantly as pregnancy progresses. A phase feeding program is used by many producers to accommodate these changes [31,41-43]. Three stages (i.e. phases of gestation) which justify different feeding strategies are: 1) early gestation (d 0–30), in which embryo survival and implantation are impacted, 2) mid-gestation (d 30–75), in which body growth in young sows and recovery of body reserves lost during lactation in older sows are impacted and 3) late gestation (approximately the last 45 d), in which fetal and mammary growth are impacted. Conceptus protein content and weight gain increases rapidly after d 68 of gestation and has greater priority for nutrient supply than maternal weight gain. Fetal weight, fetal protein content and mammary protein content increase 5, 18, and 27 times, respectively, in the last 45 d of gestation. Therefore, amino acid and energy requirements are greater in late gestation than in early gestation. Amino acid requirements increase to a higher degree than energy requirements in late gestation. If consumption of the same diet increases to meet amino acid requirements, the sows will consume excessive energy, which result in sows being too fat at farrowing. Moehn and Ball [34] recommended a strategy of formulating just two diets; one to meet the highest and the other to meet the lowest amino acid requirements. The two diets would be mixed in appropriate ratios to meet the entire range of amino acid requirements from late gestating gilts to early gestating sows. In practice, feed intake of pregnant sows is actually restricted to control body weight and prevent excess weight gain. Therefore, energy is the limiting factor for gestating sows, and thus, the feed allowance necessary to provide energy requirements must be considered first when formulating a sow feeding regimen. Segregated phase feeding should be considered because maternal growth rate decreases with age so that mature sows have lower nutrient requirements than gilts and young sows, which are still growing.

To regulate feed intake, sows should be fed based on an objective measure of individual body weight, body condition and, ideally, measurement of back fat depth [44]. Feed intake during the last 2 to 3 weeks should be adjusted to at least avoid a negative energy balance prior to farrowing and to promote higher feed intake in early lactation, easier farrowing and adequate birth weights of newborn pigs [38]. However, restriction of feeding prior to parturition significantly reduces the risk of postpartum dysgalactia syndrome (PDS). Peltoniemi and coworkers [35] reported that not only feed restriction before term, but also keeping the feed low in energy and high in fiber through parturition and into the first few d of lactation appears to improve intestinal function and initiation of lactation.

Lactation: Highly prolific sows of today produce large litters of lean and fast growing pigs [33,45]. Litter size increased by three pigs over approximately the last 40 yr [46]. Because appetite is often deficient after farrowing, increased nutrients needed for milk production generally come from mobilization of body reserves [38]. Thus, lactation puts a great nutritional demand on sows. Using body reserves could lead to excessive weight loss, which then results in reduced litter weight gain due to lowered milk production and subsequent reproduction problems for sows. Adequate feed intake, especially during the first 7 to 10 d of lactation is important to replenish body reserves, and re-establish secretion of hormones which control subsequent reproductive performance [47-51]. In addition, numerous studies demonstrated that high ambient temperatures experienced in summer are detrimental to feed intake and milk production [52-57]. Feed restriction at any time results in prolonged weaning

WEI and reduced pregnancy rates and litter size. Proper feeder design is critical and has received much attention in recent yr. Most lactation feeders today include a reservoir that will hold a minimum of 9 kg of feed. Ensuring sows have access to full feed 24 h/d results in optimum return to estrus and piglet weight at weaning. Getting sows up 2–3 times per d stimulates sows to urinate and defecate, resulting in drinking and eating, thus optimizing feed intake, lactation performance and return to estrus.

The practice of feeding fat is controversial. This additional source of energy is used principally by the mammary gland to produce very rich milk and it will not be an efficient source of energy for sows [58]. High fat addition could improve piglet weaning weight, but could also impair subsequent reproductive performance by reducing LH secretion in early lactation [51]. Fat as a high density energy source is often incorporated into lactation diets to compensate for depressed appetite during heat stress.

It is essential to have good quality water [59,60]. Water quality should be checked annually. Lack of water limits milk production. Water available at time of feeding is important with a flow rate of 1.0 liter/minute. High performing sows have a water intake of up to 40 liters/d when milk production is at its highest three weeks after farrowing.

Post-weaning: It is well documented that WEI is inversely related to lactation length [61,62]. Edgerton [61] reviewed research results, which showed that the minimum WEI of 4 d is reached at approximately 3 weeks of lactation. Therefore, each additional d of lactation beyond 3 weeks adds a non-productive d (NPD), resulting in decreased pigs/sow/yr. Mabry and coworkers [63] analyzed records from 178,519 litters in 13 commercial herds around the U.S.A. from 1985–1995. The WEI was minimized at a lactation length of 22–27 d; WEI increased significantly at a lactation length of either less than 22 d or greater than 27 d. Knox et al. [5] reported that, in North America, the most frequently reported average WEI is 5 d and that 88% of farms surveyed indicated that greater than 80% of weaned sows were mated within 7 d. Soede et al. [62] concluded from reviewing the literature that lactation length of less than 3 weeks leads to suboptimal reproductive performance. Currently, sows are weaned at approximately 20.5 to 21 d in the U.S.A. and approximately 22 d in Canada, which is at the height of milk production [3,64].

Sows experience the stress of piglet removal and change of location, as well as the transition of the mammary tissue into the dry period and follicular development and subsequent ovulation all within 4–5 d. These events require a high level of energy and nutrients. Maintaining ad lib feed and water consumption optimizes these events as measured by subsequent fertility. Boar exposure of sows should start the d of weaning by allowing a boar in front of sows for at least 10 min/d. Weaned sows are typically exposed to boars within 2 d; most commonly once per d [5].

Housing and environment

Research on housing and environment was summarized by Einarsson and coworkers [65], Flowers [27], Jansen [66], Kim et al. [33], Rhodes [67], McGlone [68-72], Vignola [38] and Weidmann [14]. Purchased gilts should be quarantined for at least 4 weeks, during which time they should be observed for and serologically tested for infectious diseases. Introduction of young cull sows, market hogs and manure exchange during the latter part of isolation acclimate replacement gilts to the herd's resident pathogens. Both purchased gilts following quarantine and internally selected gilts should be acclimated in small groups in the breeding barn to allow them to build immunity to organisms present in the breeding herd. Gilts should be vaccinated for diseases, such as E. coli, erysipelas, leptospirosis, mycoplasma and parvovirus, before breeding.

There are conflicting reports in the literature regarding the effects of individual pens versus group housing for gilts and sows. Despite this, consumer attitudes in North America and Europe are persuading regulators to move producers toward group housing of pregnant sows [67,70-73]. Gilts reared in individual pens or groups of 3 reach puberty significantly later than those reared in groups of 10 or more. In addition, gilts reared in individual pens have more silent heats and irregular estrous cycles than gilts reared in group pens. However, gilts reared in groups of 50 or more also had a lower conception rate than those in smaller groups. In North America, 90% of farms house breeding and gestation sows in standard gestation stalls [5]. In Europe, sows are kept in groups from week 4 of gestation until one week before farrowing [65,74]. A primary disadvantage of group housing for gestating sows is the inability to uniformly control sow body condition and sow weight gain because dominant sows consume more than timid ones. In addition, aggression of dominant sows causes physical damage to themselves and others. However, feeding stalls with self-locking or manual-locking doors or electronic sow feeders enable sows to have contact with other sows, but have privacy when eating. A stable social hierarchy and well-designed farrowing crates, floors and ergonomic feeders and water nipples, gut fill and lying comfort are important for sow health, performance and longevity. Any stress in the first three weeks of gestation may result in loss of pregnancy or reduced litter size. Moving sows early in gestation should be done gently in small groups.

Estrus detection and breeding management

Estrus normally lasts 24 to 48 h in gilts and up to 72 h in sows. Approximately 90% of sows express estrus 3–6 d after weaning [11,75] and sows which are mated at estrus 4–6 d after weaning have greater farrowing rates and litter sizes than sows mated later than 6 d after weaning [75]. Ovulation occurs approximately 38–48 h after onset of estrus and females ovulate 2/3 of the way through estrus [76-78]. Obviously, because of the variation in time of ovulation relative to onset of and duration of estrus [76,78-80], the more frequently estrus is checked, the more accurately the time of ovulation can be predicted. Farrowing rates and litter sizes will be lower if insemination occurs more than 24 h before ovulation because sperm live approximately 24 h after insemination and eggs can be fertilized for only 12 h after ovulation [77,78]. Therefore, the best way to predict ovulation is to detect estrus frequently. Conception rates [78] and litter size [81] are unsatisfactory if mating occurs early or late relative to ovulation. The general practice in the U.S.A. is to inseminate on the d of detected estrus and the morning of the following d.

Failure to detect estrus accurately has the greatest impact on farrowing rate and litter size. Efficiency of estrus detection is significantly lower when gilts are in stalls, or the boar is moved to gilt pens for estrus detection than when gilts are moved to boar pens. Less time is required to elicit standing response and a greater percentage of females are detected when gilts are moved to the boar area than with other methods. Boars must be sexually mature (at least 10 mo of age) and emit odor and sound. Some general recommendations are: check estrus after feeding, remove all distractions from the area, detect estrus in the same place and same way each time, keep animals calm, allow sufficient time for interaction and do not interfere with female and boar interaction.

Farrowing management, supervision and induction

Hyperprolific maternal line females of today commonly have 14–16 piglets born alive and piglet pre-weaning mortality ranges from 11 to 24% predominantly in the first five d of age, therefore, there is renewed interest in attendance and assistance at farrowing [45,82]. Moreover, stillbirth rates increase as litter size increases. In addition, a larger litter generally means smaller and weaker pigs [83,84]. The rate of stillborn piglets increases as duration of farrowing and interval between births increase. Baxter and Edwards [82] and Vanderhaeghe et al. [85] reviewed literature demonstrating that stillborn pigs are a multifactorial problem, which includes litter size, parity, sow body condition, and farrowing supervision/birth assistance.

Generally, farrowing supervision/birth assistance includes the following practices: 1) preventing savaging of piglets by the sow, 2) manually delivering piglets when the birth interval becomes longer than 30 min, 3) removing placental envelopes around piglets and clearing airways of piglets to prevent suffocation and crushing of piglets, 4) ligating the umbilical cord, 5) towel drying and positioning piglets under a heat lamp immediately after birth to prevent chilling, 6) placing low weight, low-viability piglets in a heated crib or box away from the sow, 7) feeding low-viability pigs colostrum or milk replacer orally if necessary, 8) "split suckling" or cross-fostering litters to ensure piglets from large litters consume adequate colostrum, 9) administering fluids to dehydrated piglets, either orally or subcutaneously, and 10) taping legs of splay-legged piglets together. Many producers practice the McREBEL™ management program, which minimizes cross-fostering and maximizes supportive care [86-88]. The most important factor for ensuring piglet survival is to ensure adequate colostrum consumption immediately after birth, particularly since colostrum production by the sow occurs for only 24 h after farrowing [89].

Several controlled studies have investigated the benefits of farrowing supervision/birth assistance. Holyoake et al. [90] assigned sows and gilts to 4 treatments in a 2×2 factorial arrangement: 1) induced/supervised, 2) non-induced/supervised, 3) induced/unsupervised and 4) non-induced/unsupervised. Farrowing was induced with 250 µg of $PGF_{2\alpha}$. Each supervised group was supervised from 3 h before the first expected farrowing time until the youngest litter of pigs in the group was 3 d old. Therefore, litters born first within a group were supervised for longer than 3 d. Number of stillbirths/litter and number of pre-weaning deaths/litter were significantly greater ($P < 0.05$) for unsupervised sows (0.68 ± 0.08 and 1.29 ± 0.13, respectively) than for supervised sows (0.26 ± 0.08 and 0.86 ± 0.13, respectively), whereas total weaned/litter was less ($P < 0.05$) for unsupervised sows (9.44 ± 0.19) than for supervised sows (10.17 ± 0.2). Of the 274 piglets which died in the pre-weaning period, 47% died during the first 3 d after birth and 62% died during the first 4 d after birth. White et al. treated 30 sows each as follows: Group 1 (Control)–farrowings were not attended and Group 2–farrowings were attended and piglets assisted on the d of and the d after farrowing. The percentage of stillborn piglets was 6.8 for Control sows and 1.6 for Group 2 sows ($P < 0.05$). Overall, preweaning mortality was 18.2 for Control sows and 10.1 for Group 2 sows ($P < 0.05$). Nguyen et al. [91] assigned multiparous sows to 2 treatments: Group 1 - farrowing was induced with $PGF_{2\alpha}$ and farrowing was supervised and piglets were assisted as needed on the d of farrowing only and Group 2 - farrowing was not induced and not supervised. The percentage of stillbirths was lower ($P < 0.001$) for

Group 1 (0.4 ± 0.09/litter) than for Group 2 (1.0 ± 0.17/litter). Twenty-seven percent of Group 1 sows had at least one stillborn piglet whereas 49% of Group 2 sows had at least one stillborn piglet. However, there was no effect of treatment on overall pre-weaning mortality. Nguyen and coworkers [91] concluded that sows and piglets require more than one d of supervision to reduce overall pre-weaning mortality.

Seasonal infertility

The effect of season on fertility is mediated by temperature and photoperiod [92-97]. Puberty is delayed in summer mo and the WEI and duration of estrus are longer and ovulation rate, conception rate and litter size are lower in summer than in late autumn and winter. Parity one sows are more susceptible to reduced reproductive performance than older sows.

As noted by Claus and Weiler [93], photoperiod is the only environmental factor which is highly repeatable from yr to yr. Artificially altering photoperiod failed to influence WEI, conception rate, farrowing rate or litter size when light/dark ratios were abruptly changed and held constant [97,98]. Kraeling and coworkers [99] reported that exposure of lactating sows and ovariectomized gilts to 8 h light/16 h dark or 16 h light/8 h dark failed to affect prolactin, growth hormone or luteinizing hormone secretion. Nevertheless, when photoperiod was extended to 16 h light/8 h dark, milk yield increased by 20–24%, thus piglet survival rate and body weight improved, which was explained by differences in suckling frequency of the piglets [100]. Pigs may not be able to respond to sudden changes in photoperiod. However, Auvigne and coworkers [101] analyzed ultrasound diagnosis results from farms located in four regions of France for 5 yr (2003–2007). Seasonal infertility was significantly higher during 2003 than in the other four yr, which did not differ among each other. In all regions, the highest number of hot d was in 2003 with the least number of hot d in 2007. They concluded that photoperiod has a prominent role in seasonal infertility with an additional influence of heat stress during the hottest yr.

High environmental temperature decreases lactation feed intake, delays puberty, disrupts behavioral estrus, lowers ovulation rate, increases embryonic mortality, decreases milk production and prolongs the WEI in sows. Heat stress is most detrimental to reproductive performance during the first 30 d due to increased embryonic death and last 30 d of gestation due to increased stillborn piglets. Management and nutrition determine the degree of impact of season on reproduction. Strategies to reduce heat stress are: 1) feed high energy diets with lower fiber and crude protein content, 2) feed at night, 3) feed multiple times per d, 4) use air cooling or water dripping equipment, and 5) decrease group size to 15 or

fewer in gestation and use individual gestation stalls to reduce social stress. Most farms in North America experience seasonal infertility caused by estrus failure in gilts and weaned sows and pregnancy failure [5].

Photoperiod may modulate the impact of other management factors unless it is extremely skewed to either all light or all dark, but by itself photoperiod likely has a minimal impact and is not a major factor in seasonal infertility. Decreasing photoperiod and high temperatures generally occur at the same seasonal time frame. The wild boar is not selected for continued reproduction, yet remains a seasonal breeder [93,96]. To optimize sow production producers should manage sow herds to minimize heat stress and adapt light and dark cycles to avoid either excessive light or dark periods.

Stockmanship

Hemsworth and coworkers [102,103] reported that pigs, which displayed a high level of fear of humans, had sustained elevation in plasma concentrations of corticosteroids associated with poor conception rate and litter size. In a study of 19 commercial farms in Australia, there were highly significant negative correlations between sows' level of fear of humans and reproductive performance of the farm, and the stockperson's behavior was significantly correlated with both the sows' level of fear of humans and productivity of the farm. Kirkden and coworkers [104] examined the effect of the stockperson's skill and attitude on reproductive performance. As expected, most studies showed that positively handled pigs are less fearful of humans than pigs exposed to electrical prod and that occasional negative experiences have a significant impact on the way pigs perceive the stockperson. Sows which are fearful of humans during gestation are more likely to savage their piglets and repeated aversive handling of sows during late gestation results in increased piglet morbidity. Training farm workers in the art of stockmanship continues to be a challenge for swine farm managers and is of increasing importance as consumers become more interested in humane farm animal care.

Strategies for use of new and current management technologies

Knox [75] and Estill [105] reviewed the impact of reproductive technologies on pig production. These technologies dramatically changed the way pigs are raised and made the pig the most efficient livestock species for food production in the world. Many of the technologies developed during the past 2–4 decades have been incorporated into modern pork production systems. In many cases, producers have adapted and are utilizing the technologies for applications or objectives that differ from the original or approved use or claim. The authors' intent is to describe and discuss how the technologies are actually being applied and not to endorse or advocate product use that may differ from the regulatory approval in various countries.

Altrenogest

Gilts: The search for an effective and acceptable method to synchronize estrus and ovulation in post-pubertal gilts began over 50 yr ago [106]. The corpus luteum (CL) of the pig has an inherent life span of 14–16 d [107,108] and is resistant to the luteolytic action of uterine $PGF_{2\alpha}$ secretion before d 12 of the estrous cycle [109,110], thereby making $PGF_{2\alpha}$ ineffective for estrous synchronization. Therefore, the predominant approach to estrous synchronization in the gilt was to administer a treatment, which suppresses pituitary gonadotropin secretion for 14 to 20 d to allow time for the CL to regress, and at the same time, prohibit growth of new follicles and ovulation. Upon withdrawal of such a treatment, it was expected that gonadotropin secretion would resume synchronously among the treated animals. Orally active, synthetic progesterone-like compounds were the most commonly investigated [111]. Unfortunately, the post-treatment estrus was often accompanied by development of ovarian cysts, decreased fertility and/or poor synchronization of estrus and ovulation. However, based on exhaustive studies of the progestin, altrenogest (17α-allyl-estratiene-4-9-11,17-β-ol-3-one), in the 1970s and 1980s, as reviewed by Webel and Day [111,112] and Estill [105], the Food and Drug Administration approved its use for estrous synchronization in sexually mature gilts in 2003 (Federal Register, October 31, 2003). Altrenogest is marketed by Merck Animal Health under the trade name, MATRIX® in the U.S.A. and by several other companies under other trade names outside of the U.S.A. In controlled studies [111,113-115], and in applied studies at large commercial farms in the U.S.A. [116,117], approximately 85% of gilts fed 15 mg of altrenogest/gilt/d for 14 d displayed estrus within 4 to 9 d after withdrawal of altrenogest. For maximum effectiveness, post-pubertal gilts must have displayed at least one estrus before feeding altrenogest.

Sows: Because farrowing time within a group of sows is spread over as much as a 10 d period, "batch farrowing" and "all-in-all-out" production practices are facilitated by delaying parturition in the sows which were mated earliest in the group and inducing farrowing in those mated later in the group. Allowing pigs to stay in utero an extra 2–3 d improves birth weight and colostrum antibodies increase in the sow as gestation length increases. Farrowing on weekends is avoided when parturition is precisely controlled. Guthrie [118] noted that the effectiveness of an orally active progestogen to delay parturition in the sow was reported over 50 yr ago. Numerous researchers demonstrated that intramuscular injections of progesterone or feeding altrenogest for 2–3 d, starting several d before the time of normal parturition, prolongs

gestation without affecting the incidence of stillbirths, piglet mortality or dystocia in the sow [118-122]. However, to prevent increased stillbirth rates, the length of gestation should not be prolonged by more than two d beyond the normal herd average. Guthrie et al. [121] demonstrated that farrowing is even more precisely synchronized after administration of $PGF_{2\alpha}$ the d after the last progestogen treatment.

Numerous studies demonstrated that postponing postweaning estrus by administering altrenogest improves reproductive performance [73,123-125]. Although not approved for this use in the U.S.A., feeding altrenogest for 5–7 d delays onset of postpartum estrus, which gives extra time for sows to recover body condition lost during lactation, to establish batch farrowing groups and to establish sow groups following piglet death from diseases such as porcine epidemic diarrhea virus (PEDV). In addition, subsequent conception rate and litter size increase. The most prevalent use of altrenogest in commercial herds is for delaying early farrowing before d 115 of gestation and for the transition from continuous to batch farrowing. Typically, producers employ altrenogest for delaying estrus in weaned sows for one to two weeks to assemble groups or batches of sows.

Gonadotropins

Prepubertal and peripubertal gilts: The only commercially available hormone preparation for inducing estrus and ovulation in prepubertal gilts and postpartum sows in the U.S.A. is P.G. 600®, each dose of which contains 400 IU of pregnant mare's serum gonadotropin (eCG) and 200 IU of human chorionic gonadotropin (hCG). Numerous studies demonstrated that i.m. injection of P.G. 600® in prepubertal and peripubertal gilts induces estrus in 50–90% within 5 d [73,77,126-130]. Up to 30% of those exhibiting estrus have an irregular return to a subsequent estrus.

Postpartum sows: Many studies demonstrated that i.m. injection of PMSG or P.G. 600® in sows at weaning induces estrus in sows within 5 d [73,77,126,128,130-132]. Weaning to estrus interval is shorter, but the estrus synchronization rate, subsequent farrowing rate and litter size are similar compared to untreated sows. Gonadotropins are often misused on production farms by attempting to induce estrus in sows or gilts in the presence of CL or perhaps cystic follicles. For example, treating sows that have not expressed estrus by 10–12 d following weaning is a common practice, but is often ineffective because many sows experienced a silent estrus and the gonadotropin treatment is ineffective. The most common and effective use of gonadotropin treatment is for parity one sows and during seasonal anestrous. For inducing estrus, P.G. 600® is typically given on the d of weaning to induce a more synchronous return to post-weaning estrus.

Fixed-time artificial insemination (AI)

As noted above, feeding 15 mg of altrenogest/gilt/d for 14 d results in approximately 85% of the gilts displaying estrus within 4–9 d after withdrawal of altrenogest. Weaning a group of sows, when all have reached three weeks or greater of lactation, results in a high percentage of these sows being bred within a 3 d period beginning about 4 d after weaning. Therefore, a treatment, which more precisely synchronizes ovulation, is needed to facilitate a single fixed-time AI of all gilts and sows in a group on the same d without regard to onset of estrus, thereby eliminating the need for estrus detection.

Brüssow et al. [128] and Driancourt [133] summarized literature which demonstrated the effectiveness of hCG, LH and GnRH analogs to synchronize ovulation in weaned sows and in gilts after withdrawal of altrenogest. Zak et al [134,135] reported that i.m. administration of 5 mg of pLH to weaned sows at onset of behavioral estrus followed by a double fixed-time AI resulted in a farrowing rate comparable to controls inseminated multiple times while in estrus. JBS United Animal Health launched OvuGel®, a FDA licensed proprietary gel formulation containing a GnRH-analogue in 2013. OvuGel® is the first product approved for synchronizing ovulation followed by a single fixed-time AI in weaned sows. OvuGel® is administered intravaginally to sows 96 h after weaning. Because ovulation starts in some sows 32–36 h after OvuGel® administration and a high percentage of sows complete ovulation 40–48 h after OvuGel® administration, all sows are inseminated with a single dose of semen without regard to estrus 22–24 h after OvuGel® administration to optimize fertility [136]. Thus, there is no need for heat detection, which effectively decreases labor costs and increases throughput and utilization of inventory. Fertility after OvuGel® treatment followed by a single fixed-time AI are equivalent to those of untreated sows inseminated on each d in standing estrus [137].

Driancourt and coworkers [133,138] administered buserelin acetate 115–120 h after last feeding of altrenogest in gilts and 83–89 h after weaning in sows via intramuscular or subcutaneous injection. A single AI was performed 30–33 h after buserelin treatment in only females which had displayed estrus. Fertility was equivalent to those of untreated animals.

Time of farrowing among a group of sows, all of which receive a single fixed-time AI on the same d is less variable than those receiving a AI on each d they are in estrus. This breeding precision facilitates a synchronized farrowing for the majority of a breeding group, which provides an opportunity for careful attention to d one pig care and reduction in stillborn rates. An even more precise synchronized farrowing is achieved by induction of parturition with prostaglandin $F_{2\alpha}$.

Cervical artificial insemination (CAI) and post-cervical artificial insemination (PCAI)

Cervical artificial insemination (CAI) is the predominant breeding method on farms of all sizes [5]. Benefits of CAI and PCAI are introduction of improved genetics, reduced risk of disease transmission and improved performance of reproductive tasks, which improves time management compared to natural service [139]. CAI technician fatigue should be avoided. Farrowing rates decrease from 85 to 78% when technicians perform more than 10 CAIs before taking a break. Farrowing rate decreases to 71% when more than 15 CAIs are performed without a break.

Recent interest in PCAI is primarily because sperm cell numbers/dose of semen is significantly less than that of CAI [140-142]. Therefore, more sows are inseminated with superior genetics and fewer sires are required to produce such semen. PCAI bypasses the cervix and deposits the majority of semen directly into the uterine body. PCAI, also known as intrauterine AI and transcervical AI, is not new [143-145]. Some reported disadvantages of PCAI are that it is not easily applied to gilts, the catheter is expensive and it could be harmful to the female, if not performed correctly. The benefits are reduced labor, decreased time performing AI, more sows bred with semen from superior sires and fewer sires needed to produce that semen [140-142]. Additional time and labor are saved because no boar should be present during PCAI. Traditional CAI takes 7–10 min, whereas PCAI takes one to two min. With PCAI, $1-2 \times 10^9$ sperm cells/dose [140,142] or even 0.5×10^9 sperm cells/dose [146] are used compared with 3×10^9 sperm cells/dose used for CAI.

Several workers reported that reproductive performance was compromised after PCAI in primaparous sows [147]. However, Sbardella and coworkers [148] compared the reproductive performance of primaparous sows, which were mated by PCAI with 1.5×10^9 sperm cells in 45 mL, and those which were mated by cervical AI with 3×10^9 sperm cells in 90 mL. There was no difference between treatments in farrowing rate and litter size. Passage of the intrauterine catheter was possible in 86.8% of the PCAI sows. PCAI is particularly beneficial when using frozen-thawed semen and sex sorted sperm, which reduce the number of sperm cells available and/or the lifespan of sperm cells.

Knox et al. [5] documented the following practices for farms in North America. Fifty-five percent of farms used the interval from weaning to estrus to time the first AI, after detection of estrus, 62% of farms timed the first AI to occur within min or a few h of estrus, whereas 30% delayed AI until the next AM or PM period. Seventy-six percent of farms planned for two doses of semen for each sow, whereas only 14% planned for three doses of semen per sow. Prominent procedures during AI were back pressure (93%), boar exposure (89%), flank rubbing (80%) and gravity semen flow (81%). PCAI is practiced on only 6% of farms; 61% having no experience with PCAI and 25% having tried it, but not used it since. However, during the past two yr, interest in and implementation of PCAI has experienced a dramatic increase with a high success rate and may soon become the most prevalent AI technique.

Ultrasound for pregnancy detection

According to Flowers and coworkers [149] and Knox and coworkers [73], the most common strategy for identifying non-pregnant females is detection of estrus by daily boar exposure from 17 to 23 d after breeding followed by examination by either amplitude-depth (A-mode) or Doppler ultrasonography between d 28 and 45 of gestation. In addition, Flowers and coworkers presented data demonstrating that real-time (B-mode) ultrasonography is accurate when used after the first 3 weeks of gestation. In North America, most medium and large farms use real-time ultrasonography to determine pregnancy during week 3 to 5 [5].

Induced farrowing

Guthrie [118], Kirkden and coworkers [104,150], Kirkwood [129] and Nguyen et al. [91] published literature reviews on the impact of induction of synchronized farrowing on piglet mortality. Farrowing is usually induced by administration of $PGF_{2\alpha}$ or a $PGF_{2\alpha}$ analog. Numerous studies indicate approximately 92% of sows farrow within the working d following $PGF_{2\alpha}$ injection compared to 38% for untreated controls. Induction of farrowing substantially decreases piglet mortality because 1) a high proportion of farrowings occur during normal working h, 2) farrowing can be closely supervised, which provides opportunity to save and cross-foster piglets, 3) farrowing is avoided on weekends, 4) batch farrowing reduces variation in piglet age at weaning, and 5) batching of routine tasks result in efficient use of facilities. Weaning a group of piglets at a more narrow age range, results in less variation in subsequent market weights. The negative experiences some producers have had with induction of farrowing were most likely due to incorrect use of PGF products. $PGF_{2\alpha}$ must be administered no earlier than 2 d before expected farrowing based on the average expected farrowing date of the herd, since mean gestation length between herds varies from 113 to 117 d. Induction too early will result in low piglet birth weight, increased duration of farrowing and increased stillborn and live born mortality rates. Induction of farrowing has not been widely practiced in the USA, except for inducing sows with the longest gestation to ensure they farrow with a particular group.

Generally, natural farrowing time within a group of sows is spread over approximately a 10 d period due to variation between sows in weaning to estrus interval and length of gestation. However, due to increased interest in biosecurity and piglet health issues, U.S.A. producers have expressed renewed interest in "batch farrowing" and "all-in-all-out" production practices. The introduction of ovulation synchronization and single fixed-time AI inspired a vision for closer farrowing synchronization to facilitate supervision and reduce stillbirths. If a "batch" or group of sows are induced to ovulate and inseminated once at the same h on the same d, then farrowing induction with reduced time and labor required for providing supervision for the batch becomes practical. Recent trials support this theory in that the time and variability of farrowing was reduced for sows inseminated following treatment with OvuGel® and a single fixed-time AI compared to contemporary control sows inseminated without ovulation synchronization [149]. Furthermore, addition of $PGF_{2\alpha}$ to induce farrowing on d 113 of gestation resulted in highly synchronized farrowing, reduced variation in and increased age of piglets at weaning. Greater than 80% of the sows inseminated with a single fixed-time AI, then treated with $PGF_{2\alpha}$ on the same d (113 of gestation), farrowed on the same d. Ninety percent of the piglets were the same d of age at weaning and were 1.3 d older than those from non-induced sows. In these studies, 92% of treated sows farrowed within 2 d compared to 38% for controls [151,152]. The precision observed following single fixed-time AI and treatment for induction of farrowing on the same d now permits farrowing managers to attend farrowing and provide intensive piglet care. As the practice of batch farrowing, fixed-time AI and intensive birthing care become more prevalent, the pre-weaning survival of piglets is expected to increase. Figure 1 schematically depicts the authors' vision for utilization of currently available technologies. These combined technologies will result in production benefits discussed above.

New performance terminology

Figure 1 schematically depicts the authors' vision for utilization of currently available technologies. These combined technologies will result in production benefits discussed above. Because all weaned sows are inseminated once (Figure 1) without regard to estrus in a fixed-timed AI program, the conventional farrowing rate terminology of number farrowed ÷ number bred is not appropriate. Webel and coworkers [151] suggested that weaned sow farrowing rate (number farrowed ÷ number weaned) is a more appropriate measure of sow utilization when comparing farrowing performance. Also, the commonly used metric of pigs per mated female may be a misleading indicator of sow farm productivity because it does not account for sows that do not return to estrus and are not mated promptly following weaning. For a group of weaned sows, total live pigs produced per 100 weaned sows (piglet index) provides a more valuable economic measure of sow farm efficiency [151,152].

Conclusion

Many advances in genetic selection, nutrition, housing and disease control since the 1950s have been incorporated into modern pork production. Genetics, nutrition, housing, disease, lameness, age at first mating, assistance at farrowing, length of lactation and growth rate, body condition and performance of parity one impact sow longevity. Seasonal infertility, mediated by temperature and photoperiod, is a persistent problem. The following technologies have been adopted by many swine producers over the past 2–4 decades. The orally active progestin, altrenogest is used to: 1) synchronize the estrous cycle of gilts, 2) prolong gestation to synchronize farrowing and 3) postpone post-weaning estrus to give extra time for sows to recover body condition lost during lactation. The only commercially available preparation for inducing estrus and ovulation in the U.S.A. is P.G. 600®. GnRH analogs synchronize ovulation thereby making fixed-time AI practical. A single fixed-time AI of every female in a group on one d enables producers to plan precisely how much semen to have available on a particular d and to focus resources on other tasks the other 6 d of the week. It is also possible to eliminate weekend inseminations, and if breeding is performed on a weekend, the AI technician knows exactly what needs to be done, thereby reducing errors. Therefore, less semen is wasted and less old semen is used from previous orders. Post-cervical AI uses significantly fewer sperm cells/dose of semen than cervical AI. Real-time ultrasonography is used to determine pregnancy during weeks 3–5. Benefits of induced farrowing with $PGF_{2\alpha}$ are: 1) a high proportion of farrowing occur during normal working h, 2) close supervision at farrowing, 3) no farrowing on weekends, 4) reduce age range within batches of growing pigs and 5) efficient use of facilities and batching of routine tasks. Ovulation synchronization, single fixed-time AI and induced parturition with $PGF_{2\alpha}$ leads to farrowing synchronization, which facilitates supervision of sows and piglets. Attendance and assistance at farrowing is especially important to ensure adequate colostrum consumption by piglets immediately after birth. These technologies save a significant amount of time, which allows redistribution of labor (i.e. focusing more on facility maintenance, gilt development, evaluating sows' body condition, adjusting gestation feeders, assisting in farrowing and training workers in the art of stockmanship, which is important for humane farm animal care). In addition, they maximize the leverage of high index boars, which will improve overall pork production efficiency. New performance terminologies were proposed. Because all weaned sows are inseminated once without regard to estrus in a fixed-timed AI program, the conventional farrowing rate terminology (number farrowed ÷ number bred) is inappropriate. Weaned sow farrowing rate (number farrowed ÷ number weaned) is a more appropriate measure of sow utilization when comparing farrowing performance. Also, the commonly used metric of pigs per mated female is a misleading indicator of sow farm productivity because it does not account for sows that do not return to estrus and are not mated promptly following weaning. Therefore, total live pigs produced per 100 weaned sows (piglet index) provides a more valuable economic measure of sow farm efficiency.

Competing interests

RRK received compensation from JBS United Animal Health who markets nutrition products and OvuGel®. SKW is employed by JBS Animal Health who markets nutrition products and OvuGel®.

Authors' contributions

The authors shared equally in preparing this review paper. Both authors read and approved the final manuscript.

Acknowledgements

The authors thank Michael Johnston and Christopher Anderson, JBS United Animal Health for providing review and guidance during preparation of this manuscript. The authors also thank Robert Knox, University of Illinois, Wayne Singleton, Professor Emeritus, Purdue University, Donald Levis, Levis Worldwide Swine Consultancy their review for and suggestions for improving this manuscript.

Author details

[1]L&R Research Associates, Watkinsville, GA, USA. [2]JBS United Animal Health, Sheridan, IN, USA.

References

1. Foxcroft GR, Dixon WT, Novak S, Putnam CT, Town SC, Vinsky MDA. The biological basis for prenatal programming of postnatal performance in pigs. J Anim Sci. 2006;84(E. Suppl):E105–12.
2. PigChamp. USA 2012 - year end summary. 2013. http://benchmark.farms.com/2013_Summary_of_the_2012_data.html.
3. Stalder KJ. Pork industry productivity analysis, research grant report. Des Moines, Iowa: National Pork Board; 2014. p. 1–13.
4. Ketchem R, Rix M. SMS Database Revision Reveals Intriguing Production Trends. National Hog Farmer. 2014. (Online Exclusive http://nationalhogfarmer.com/business/sms-database-revision-reveals-intriguing-production-trends?page=1). 2014. January 6, 2014.
5. Knox RV, Rodriguez Zas SL, Sloter NL, McNamara KA, Gall TJ, Levis DG, et al. An analysis of survey data by size of the breeding herd for the reproductive management practices of North American sow farms. J Anim Sci. 2013;91:433–45.
6. Engblom L, Lundeheim N, Strandberg E, Schneider MP, Dalin AM, Andersson K. Factors affecting length of productive life in Swedish commercial sows. J Anim Sci. 2008;86:432–41.
7. Gill P. Nutritional management of the gilt for lifetime productivity - feeding for fitness or fatness? In: London Swine Conference Proceedings. London, Ontario: Today's Challenges. . . Tomorrow's Opportunities. 2007. p. 83–99.
8. Lucia Jr T, Dial GD, Marsh WE. Lifetime reproductive performance in female pigs having distinct reasons for removal. Livest Prod Sci. 2000;63:213–22.
9. Tomes GK, Nielson HE. Factors affecting reproductive efficiency of the breeding herd. In: Cole DJA, Foxcroft G, editors. Control of pig reproduction. London: Butterworth Scientific; 1982. p. 527–9.
10. Foxcroft G, Beltranena E, Patterson J, Williams N, Pizzarro G. Physiological limits to maximizing sow productivity. In: London Swine Conference

Proceedings. London, Ontario: Production at the leading edge; 2005. p. 29–46.

11. Foxcroft G, Patterson J, Dyck M. Improving production efficiency in a competitive industry. In: 24th Manitoba Swine Seminar. Sharing Ideas and Information for Efficient Pork Production. Winnipeg, Manitoba: Manitoba Pork Council; 2010. p. 81–98.

12. Williams N, Patterson J, Foxcroft GR. Non-negotiables of gilt development. Advances in Pork Production. 2005;16:1–9.

13. Bortolozzo FP, Bernardi ML, Kummer R, Wentz I. Growth, body state and breeding performance in gilts and primiparous sows. Soc Reprod Fertil. 2009;66(Suppl):281–91.

14. Wiedmann R. Advances in sow and gilt management. In: London Swine Conference Proceedings. London, Ontario: Focus on the Future; 2010. p. 53–9.

15. Whitney MH, Masker C. Replacement gilt and boar nutrient recommendations and feeding management. U.S. Pork Center of Excellence: Des Moines, Iowa; 2010.

16. Sterning M, Rydhmer L, Eliasson-Selling L. Relationships between age at puberty and interval from weaning to estrus and between estrus signs at puberty and after the first weaning in pigs. J Anim Sci. 1998;76:353–9.

17. Stalder KJ, Johnson C, Miller DP, Baas TJ, Berry N, Christian AE, et al. Replacement gilt evaluation guide: for the evaluation of structural, feet, leg and reproductive soundness in repalcement gilts. Des Moines, Iowa: National Pork Board; 2009.

18. Wilson ME, Ward TL. Impact of lameness on productive potential of the sow. In: London Swine Conference. A Time for Change: 28 March 2012. 2012. p. 27–33.

19. Filha WSA, Bernardi ML, Wentz I, Bortolozzo FP. Growth rate and age at boar exposure as factors influencing gilt puberty. Livestock Sci. 2009;120:51–7.

20. Rozeboom DW, Pettigrew JE, Moser RL, Cornelius SG, El Kandelgy SM. Influence of gilt age and body composition at first breeding on sow reproductive performance and longevity. J Anim Sci. 1996;74:138–50.

21. Hoge MD, Bates RO. Developmental factors that influence sow longevity. J Anim Sci. 2011;89:1238–45.

22. Gonyou H, Rioja-Lang F, Seddon Y. Group housing systems: floor space allowance and group size. Des Moines, Iowa: National Pork Board; 2013.

23. Southern LL, Olayiwola A, DeLange CFM, Hill GM, Kerr BJ, Lindemann MD, et al. Nutrient requirements of swine. Washington, DC: National Academic Press; 2012.

24. Rodriguez-Zas SL, Davis CB, Ellinger PN, Schnitkey GD, Romine NM, Connor JF, et al. Impact of biological and economic variables on optimal parity for replacement in swine breed-to-wean herds. J Anim Sci. 2006;84:2555–65.

25. Stalder KJ, Lacy C, Cross TL, Conatser GE. Financial impact of average parity of culled females in a breed-to-wean swine operation using replacement gilt net present value analysis. J Swine Health Prod. 2003;11:69–74.

26. Anil SS, Anil L, Deen J. Analysis of periparturient risk factors affecting sow longevity in breeding herds. Can J Anim Sci. 2008;88:381–9.

27. Flowers W. Gilt and sow management considerations in sow longevity. In: Banff Pork Seminar. Advances in Pork Production: 21–23 January 2014. 2014. p. 26–55.

28. Serenius T, Stalder KJ, Baas TJ, Mabry JW, Goodwin RN, Johnson RK, et al. National pork producers council maternal line national genetic evaluation program: a comparison of sow longevity and trait associations with sow longevity. J Anim Sci. 2006;84:2590–5.

29. Martineau G-P, Badouard B. Managing highly prolific sows. In: London Swine Conference. Tools of the Trade. 2014. p. 3–19. 4-1-2009.

30. Ball RO, Samuel RS, Moehn S. Nutrient requirements of prolific sows. Advances in Pork Production. 2008;19:223–36.

31. Boyd RD, Castro GC, Cabrera RA. Nutrition and management of the sow to maximize lifetime productivity. Advances in Pork Production. 2002;13:47–59.

32. Campos PH, Silva BA, Donzele JL, Oliveira RF, Knol EF. Effects of sow nutrition during gestation on within-litter birth weight variation: a review. Animal. 2012;6:797–806.

33. Kim SW, Weaver AC, Shen YB, Zhao Y. Improving efficiency of sow productivity: nutrition and health. J Anim Sci Biotechnol. 2013;4:26.

34. Moehn S, Ball RO. London swine conference. Managing for production. 2013. p. 55–63. 3-27-2013.

35. Peltoniemi OA, Oliviero C, Halli O, Heinonen M. Feeding affects reproductive performance and reproductive endocrinology in the gilt and sow. Acta Vet Scand. 2007;49 Suppl 1:S6.

36. Theil PK, Lauridsen C, Quesnel H. Neonatal piglet survival: impact of sow nutrition around parturition on fetal glycogen deposition and production and composition of colostrum and transient milk. Animal. 2014;8:1–10.

37. Langendijk P, Peltoniemi O. How does nutrition influence luteal function and early embryo survival. Soc Reprod Fertil. 2013;68:145–58.

38. Vignola M. Sow feeding management during lactation. In: London Swine Conference. Tools of the Trade. 2009. p. 107–17. 4-1-2009.

39. Love RJ, Klupiec C, Thornton EJ, Evans G. An interaction between feeding rate and season affects fertility of sows. Anim Reprod Sci. 1995;39:275–84.

40. Virolainen JV, Tast A, Sorsa A, Love RJ, Peltoniemi OA. Changes in feeding level during early pregnancy affect fertility in gilts. Anim Reprod Sci. 2004;80:341–52.

41. Johnston L. Gestating swine nutrient recommendations and feeding management. Des Moines, Iowa, U.S: Pork Center of Excellence. National Swine Nutrition Guide; 2010.

42. Moehn S, Levesque CL, Samuel RS, Ball RO. Applying new research to reduce sow feed costs. Advances in Pork Production. 2009;20:84–94.

43. Moehn S, Franco D, Levesque C, Samuel R, Ball RO. Phase feeding for pregnant sows. Edmonton, Alberta, Canada: Swine Research and Technology Center 4-10 Agriculture/Forestry Centre University of Alberta; 2012.

44. Young MG, Tokach MD, Aherne FX, Main RG, Dritz SS, Goodband RD, et al. Comparison of three methods of feeding sows in gestation and the subsequent effects on lactation performance. J Anim Sci. 2004;82:3058–70.

45. Oliviero C. Management to improve neonate piglet survival. Soc Reprod Fertil. 2013;68:203–10.

46. National Agricultural Statistic Service NASS. Agricultural statistics book by year. Washington DC, USA: Government Printing Office; 2011.

47. Kauffold J, Gottschalk J, Schneider F, Beynon N, Wahner M. Effects of feeding level during lactation on FSH and LH secretion patterns, and follicular development in primiparous sows. Reprod Domest Anim. 2008;43:234–8.

48. Jones DB, Stahly TS. Impact of amino acid nutrition during lactation on luteinizing hormone secretion and return to estrus in primiparous sows. J Anim Sci. 1999;77:1523–31.

49. Jones DB, Stahly TS. Impact of amino acid nutrition during lactation on body nutrient mobilization and milk nutrient output in primiparous sows. J Anim Sci. 1999;77:1513–22.

50. Quesnel H, Pasquier A, Mounier AM, Prunier A. Influence of feed restriction during lactation on gonadotropic hormones and ovarian development in primiparous sows. J Anim Sci. 1998;76:856–63.

51. Kemp B, Soede NM, Helmond FA, Bosch MW. Effects of energy source in the diet on reproductive hormones and insulin during lactation and subsequent estrus in multiparous sows. J Anim Sci. 1995;73:3022–9.

52. Prunier A, Quesnel H, de Braganca MM, Kermabon MY. Environmental and seasonal influences on the return-to-oestrus after weaning in primiparous sows: a review. Livest Prod Sci. 1996;45:103–10.

53. Prunier A, DeBragana MM, LeDividich J. Influence of high ambient temperature on performance of reproductive sows. Livest Prod Sci. 1997;52:123–33.

54. Quiniou N, Noblet J. Influence of high ambient temperatures on performance of multiparous lactating sows. J Anim Sci. 1999;77:2124–34.

55. Renaudeau D, Noblet J. Effects of exposure to high ambient temperature and dietary protein level on sow milk production and performance of piglets. J Anim Sci. 2001;79:1540–8.

56. Silva BA, Noblet J, Donzele JL, Oliveira RF, Primot Y, Gourdine JL, et al. Effects of dietary protein level and amino acid supplementation on performance of mixed-parity lactating sows in a tropical humid climate. J Anim Sci. 2009;87:4003–12.

57. Silva BA, Noblet J, Oliveira RFM, Donzele JL, Fernandez HC, Lima AL, et al. Effect of floor cooling and dietary amino acids content on performance and behaviour of lactating primiparous sows during summer. Livest Sci. 2014;120:25–34.

58. van den Brand H, Kemp B. Dietary fat and reproduction in the post partum sow. Soc Reprod Fertil. 2006;62(Suppl):177–89.

59. Leibbrandt VD, Johnston LJ, Shurson GC, Crenshaw JD, Libal GW, Arthur RD. Effect of nipple drinker water flow rate and season on performance of lactating swine. J Anim Sci. 2001;79:2770–5.

60. Mroz Z, Jongbloed AW, Lenis NP, Vreman K. Water in pig nutrition: physiology, allowances and environmental implications. Nutr Res Rev. 1995;8:137–64.

61. Edgerton LA. Effect of lactation upon the postpartum interval. J Anim Sci. 1980;51 Suppl 2:40–52.

62. Soede NM, Hazeleger W, Gerritsen R, Langendijk P, Kemp B. Ovarian responses to lactation management strategies. Soc Reprod Fertil. 2009;66(Suppl):177–86.

63. Mabry JW, Culbertson MS, Reeves D. Effects of lactation length on weaning-to-first service interval, first-service farrowing rate, and subsequent litter size. Swine Health Prod. 1996;4:185–8.

64. Niell C, Williams N. Milk production and nutritional requirements of modern sows. In: London Swine ConferenceProceedings. London, Ontario: Focus on the Future; 2010. p. 232–31.

65. Einarsson S, Sjunnesson Y, Hulten F, Eliasson-Selling L, Dalin AM, Lundeheim N, et al. A 25 years experience of group-housed sows-reproduction in animal welfare-friendly systems. Acta Vet Scand. 2014;56:37.

66. Jansen J, Kirkwood RN, Zanella AJ, Tempelman RJ. Influence of gestation housing on sow behavior and fertility. J Swine Health Prod. 2007;15:132–6.

67. Rhodes RT, Appleby MC, Chinn K, Douglas L, Firkins LD, Houpt KA, et al. A comprehensive review of housing for pregnant sows. J Am Vet Med Assoc. 2005;227:1580–90.

68. McGlone JJ, Stansbury WF, Tribble LF. Management of lactating sows during heat stress: effects of water drip, snout coolers, floor type and a high energy-density diet. J Anim Sci. 1988;66:885–91.

69. McGlone JJ, Stansbury WF, Tribble LF, Morrow JL. Photoperiod and heat stress influence on lactating sow performance and photoperiod effects on nursery pig performance. J Anim Sci. 1988;66:1915–9.

70. McGlone JJ, von Borell EH, Deen J, Johnson AK, Levis DG, Meunier-Salaun M, et al. Review: compilation of the scientific literature comparing housing systems for gestating sows and gilts using measures of physiology, behavior, performance, and health. Prof Anim Sci. 2004;20:105–17.

71. McGlone JJ. Review: updated scientific evidence on the welfare of gestating sows kept in different housing systems. Prof Anim Sci. 2013;29:189–98.

72. McGlone JJ, Salak-Johnson J. Changing from Sow gestation crates to pens: problem or opportunity? 2008 Manitoba Swine Seminar. In: Proceedings of the Manitoba Swine Seminar. Winnipeg, Manitoba: Manitoba Pork Council; 2008. p. 47–53.

73. Knox RV. Impact of swine reproductive technologies on pig and global food production. Adv Exp Med Biol. 2014;752:131–60.

74. Council Directive 2008/120/EC. Laying down minimum standards for the protection of pigs. 2008.

75. Soede NM, Langendijk P, Kemp B. Reproductive cycles in pigs. Anim Reprod Sci. 2011;124:251–8.

76. Soede NM, Kemp B. Expression of oestrus and timing of ovulation in pigs. J Reprod Fertil. 1997;52(Suppl):91–103.

77. Knox RV, Rodriguez-Zas SL, Miller GM, Willenburg KL, Robb JA. Administration of P.G. 600 to sows at weaning and the time of ovulation as determined by transrectal ultrasound. J Anim Sci. 2001;79:796–802.

78. Soede NM, Wetzels CC, Zondag W, de Koning MA, Kemp B. Effects of time of insemination relative to ovulation, as determined by ultrasonography, on fertilization rate and accessory sperm count in sows. J Reprod Fertil. 1995;104:99–106.

79. Almeida FRCL, Novak S, Foxcroft GR. The time of ovulation in relation to estrus duration in gilts. Theriogenology. 2000;53:1389–96.

80. Kemp B, Soede NM. Consequences of variation in interval from insemination to ovulation on fertilization in pigs. J Reprod Fertil. 1997;52(Suppl):79–89.

81. Rozeboom KJ, Troedsson MH, Shurson GC, Hawton JD, Crabo BG. Late estrus or metestrus insemination after estrual inseminations decreases farrowing rate and litter size in swine. J Anim Sci. 1997;75:2323–7.

82. Baxter EM, Edwards SA. Determining piglet survival. Soc Reprod Fertil. 2013;68:129–43.

83. Foxcroft GR. Reproduction in farm animals in an era of rapid genetic change: will genetic change outpace our knowledge of physiology? Reprod Domest Anim. 2012;47 Suppl 4:313–9.

84. Phillips CE, Farmer C, Anderson JE, Johnston LJ, Shurson GC, Deen J, et al. Pre-weaning mortality in group-housed lactating sows: hormonal differences between high risk and low risk sows. J Anim Sci. 2014;92:2603–11.

85. Vanderhaeghe C, Dewulf J, de Kruif A, Maes D. Non-infectious factors associated with stillbirth in pigs: a review. Anim Reprod Sci. 2013;139:76–88.

86. McCaw MB. McREBEL™ PRRS: management procedures for PRRS control in large herd nurseries. In: Allen D, editor. Leman Swine Conference, vol. 22. 1995. p. 161–2.

87. McCaw MB. Impact of McREBEL™ (minmum cross-fostering) management upon nursery pig sale weight and survival under production conditions in PRRS asymtomatic herds. International Pig Veterinary Society Congress. In: Proceedings of the International Pig Veterinary Society Congress. 2000. p. 333.

88. McCaw MB. Effect of reducing cross-fostering at birth on piglet mortality and performance during an acute outbreak of porcine reproductive respiratory syndrome. Swine Health Prod. 2000;8:15–21.

89. Quesnel H, Gondret F, Merlot E, Loisel F, Farmer C. Sow influence on neonatal survival: a special focus on colostrum. Soc Reprod Fertil. 2013;68:117–28.

90. Holyoake PK, Dial GD, Trigg T, King VL. Reducing pig mortality through supervision during the perinatal period. J Anim Sci. 1995;73:3543–51.

91. Nguyen K, Casser G, Friendship RM, Dewey C, Farzan A, Kirkwood RN. Stillbirth and preweaning mortality in litters of sows induced to farrow with supervision compared to litters of naturally farrowing sows with minimal supervision. J Swine Health Prod. 2011;19:214–7.

92. Britt JH, Szarek VE, Levis DG. Characterization of summer infertility of sows in large confinement units. Theriogenology. 1983;20:133–40.

93. Claus R, Weiler U. Influence of light and photoperiodicity on pig prolificacy. J Reprod Fertil. 1985;33(Suppl):185–97.

94. Love RJ. Seasonal infertility in pigs. Vet Rec. 1981;109:407–9.

95. Love RJ, Evans G, Klupiec C. Seasonal effects on fertility in gilts and sows. J Reprod Fertil. 1993;48(Suppl):191–206.

96. Peltoniemi OA, Virolainen JV. Seasonality of reproduction in gilts and sows. Soc Reprod Fertil. 2005;62(Suppl):205–18.

97. Wettemann RR, Bazer FW. Influence of environmental temperature on prolificacy in pigs. J Reprod Fertil. 1985;33(Suppl):199–208.

98. Mabry JW, Cunningham FJ, Kraeling RR, Rampacek GB. The effect of artificially extended photoperiod during lactation on maternal performance of the sow. J Anim Sci. 1982;54:918–21.

99. Kraeling RR, Rampacek GB, Mabry JW, Cunningham FL, Pinkert CA. Serum concentrations of pituitary and adrenal hormones in female pigs exposed to two photoperiods. J Anim Sci. 1983;57:1243–50.

100. Mabry JW, Coffey MT, Seerley RW. A comparison of an 8- versus 16-h photoperiod during lactation on suckling frequency of the baby pig and maternal performance of the sow. J Anim Sci. 1983;57:292–5.

101. Auvigne V, Leneveu P, Jehannin C, Peltoniemi O, Salle E. Seasonal infertility in sows: a five year field study to analyze the relative roles of heat stress and photoperiod. Theriogenology. 2010;74:60–6.

102. Hemsworth PH, Barnett JL. Behavioural responses affecting gilt and sow reproduction. J Reprod Fertil. 1990;40(Suppl):343–54.

103. Hemsworth PH. Human-animal interactions in livestock production. Appl Anim Behav Sci. 2003;81:185–98.

104. Kirkden RD, Broom DM, Andersen IL. Invited review: piglet mortality: management solutions. J Anim Sci. 2013;91:3361–89.

105. Estill CT. Current concepts in estrus synchronization in swine. Proc Am Soc Anim Sci. 1999;77:1–9.

106. Ulberg LC, Grummer RH, Casida LE. The effects of progesterone upon ovarian function in gilts. J Anim Sci. 1951;10:665–71.

107. Anderson LL, Dyck GW, Mori H, Henricks DM, Melampy RM. Ovarian function in pigs following hypophysial stalk transection or hypophysectomy. Am J Physiol. 1967;212:1188–94.

108. Du Mesnil Du Buisson F, Leglise PC. Effet de l'hypophysectomie sur les corps jaunes de la truie. C R Hebd Seances Acad Sci. 1963;257:261–3.

109. Diehl JR, Day BN. Effect of prostaglandin F2 alpha on luteal function in swine. J Anim Sci. 1974;39:392–6.

110. Hallford DM, Wettemann RP, Turman EJ, Omtvedt IT. Luteal function in gilts after prostaglandin F2alpha. J Anim Sci. 1975;41:1706–10.

111. Webel SK, Day BN. The control of ovulation. In: Cole DJA, Foxcroft G, editors. Control of pig reproduction. London: Butterworth Scientific; 1982. p. 197–210.

112. Webel SK. Ovulation control in the pig. In: Crighton DB, Foxcroft G, Haynes NB, Lamming GE, editors. Control of Ovulation. London: Butterworth Scientific; 1978. p. 421–34.

113. Estienne MJ, Harper AF, Horsley BR, Estienne CE, Knight JW. Effects of P.G. 600 on the onset of estrus and ovulation rate in gilts treated with Regu-mate. J Anim Sci. 2001;79:2757–61.

114. Kraeling RR, Dziuk PJ, Pursel VG, Rampacek GB, Webel SK. Synchronization of estrus in swine with allyl trenbolone (RU-2267). J Anim Sci. 1981;52:831–5.

115. Martinat-Botte F, Bariteau F, Badouard B, Terqui M. Control of pig reproduction in a breeding programme. J Reprod Fertil. 1985;33(Suppl):211–28.

116. Sporke J, Patterson J, Beltranena E, Foxcroft GR. Gilt development unit management and using matrix and pg600 in a commercial swine operation. In: Allen D, editor. Leman Swine Conference. 2005. p. 72–100.

117. Wood CM, Kornegay ET, Shipley CF. Efficacy of altrenogest in synchronizing estrus in two swine breeding programs and effects on subsequent reproductive performance of sows. J Anim Sci. 1992;70:1357–64.

118. Guthrie HD. Control of time of parturition in pigs. J Reprod Fertil. 1985;33(Suppl):229–44.

119. Foisnet A, Farmer C, David C, Quesnel H. Altrenogest treatment during late pregnancy did not reduce colostrum yield in primiparous sows. J Anim Sci. 2010;88:1684–93.

120. Gooneratne A, Hartmann PE, McCauley I, Martin CE. Control of parturition in the sow using progesterone and prostaglandin. Aust J Biol Sci. 1979;32:587–95.

121. Guthrie HD, Meckley PE, Young EP, Hartsock TG. Effect of altrenogest and Lutalyse on parturition control, plasma progesterone, unconjugated estrogen and 13,14-dihydro-15-keto-prostaglandin F2 alpha in sows. J Anim Sci. 1987;65:203–11.

122. Jackson JR, Hurley WL, Easter RA, Jensen AH, Odle J. Effects of induced or delayed parturition and supplemental dietary fat on colostrum and milk composition in sows. J Anim Sci. 1995;73:1906–13.

123. Martinat-Botte F, Bariteau F, Forgerit Y, Macar C, Poirier P, Terqui M. Control of reproduction with a progestagen-altrenogest (Regumate) in gilts and at weaning in primiparous sows: effect on fertility and litter size [Abstract]. Reprod Domest Anim. 1994;29:362–5.

124. van Leeuwen JJ, Martens MR, Jourquin J, Driancourt MA, Kemp B, Soede NM. Effects of altrenogest treatments before and after weaning on follicular development, farrowing rate, and litter size in sows. J Anim Sci. 2011;89:2397–406.

125. Fernandez L, Diez C, Ordonez J, Carbajo M. Reproductive performance in primiparous sows after postweaning treatment with a progestagen. J Swine Health Prod. 2005;13:28–30.

126. Knox RV, Tudor KW, Rodriguez-Zas SL, Robb JA. Effect of subcutaneous vs intramuscular administration of P.G. 600 on estrual and ovulatory responses of prepubertal gilts. J Anim Sci. 2000;78:1732–7.

127. Britt JH, Day BN, Webel SK, Brauer MA. Induction of fertile estrus in prepuberal gilts by treatment with a combination of pregnant mare's serum gonadotropin and human chorionic gonadotropin. J Anim Sci. 1989;67:1148–53.

128. Brussow KP, Schneider F, Kanitz W, Ratky J, Kauffold J, Wahner M. Studies on fixed-time ovulation induction in the pig. Soc Reprod Fertil. 2009;66(Suppl):187–95.

129. Kirkwood RN. Pharmacological intervention in swine reproduction. Swine Health Prod. 1999;7:29–55.

130. Kirkwood RN, Aherne FX, Foxcroft GR. Effect of gonadotropin at weaning on reproductive performance of primiparous sows. Swine Health Prod. 1998;6:51–5.

131. Armstrong TA, Flowers WL, Britt JH. Control of the weaning-to-estrus interval in sows using gonadotropins and prostaglandins during lactation. J Anim Sci. 1999;77:2533–9.

132. Patterson JL, Cameron AC, Smith TA, Kummer AB, Schott R, Greiner LL, et al. The effect of gonadotrophin treatment at weaning on primiparous sow performance. J Swine Health Prod. 2010;18:196–9.

133. Driancourt MA. Fixed time artificial insemination in gilts and sows. Tools, schedules and efficacy. Soc Reprod Fertil. 2013;68(Suppl):89–99.

134. Zak LJ, Patterson J, Hancock J, Hockley J, Rogan D, Foxcroft GR. Benefits of synchronizing ovulation with porcine luteinizing hormone in a fixed-time insemination protocol in weaned multiparous sows. J Swine Health Prod. 2010;18:125–31.

135. Zak LJ, Patterson J, Hancock J, Rogan D, Foxcroft GR. Benefits of synchronizing ovulation with porcine luteinizing hormone (pLH) in a fixed time insemination protocol in weaned multiparous sows. Soc Reprod Fertil. 2009;66:305–6.

136. Francisco C, Johnston M, Kraeling R, Webel S. Field evaluations using OvuGel® for single fixed-time artificial insemination. In: AASV 45th Annual Meeting Proceedings. 2014. p. 251–5. 3-1-2014.

137. Knox RV, Taibl JN, Breen SM, Swanson ME, Webel SK. Effects of altering the dose and timing of triptorelin when given as an intravaginal gel for advancing and synchronizing ovulation in weaned sows. Theriogenology. 2014;82:379–86.

138. Driancourt MA, Cox P, Rubion S, Harnois-Milon G, Kemp B, Soede NM. Induction of an LH surge and ovulation by buserelin (as Receptal) allows breeding of weaned sows with a single fixed-time insemination. Theriogenology. 2013;80:391–9.

139. Flowers WL, Alhusen HD. Reproductive performance and estimates of labor requirements associated with combinations of artificial insemination and natural service in swine. J Anim Sci. 1992;70:615–21.

140. Mickevicius E. Post cervical artificial insemination in sows - what, why, how? In: London Swine Conference. A Time for Change. 2012. p. 37–43. 3-28-2010.

141. Willenburg K, Dyck M, Foxcroft G. Tools, techniques and strategies to improve reproductive performance and genetic progress. In: London Swine Conference Proceedings. A Time for Change. London, Ontario. 2012. p. 9–14. 3-28-2012.

142. Wilson ME. Differences in mating between a boar, traditional artificial insemination, and post cervical insemination. In: London Swine Conference. A Time for Change. 2012. 3-28-2012.

143. Hancock JL. Pig insemination technique. Vet Rec. 1959;71:527.

144. Martinez EA, Vazquez JM, Roca J, Lucas X, Gil MA, Parrilla I, et al. Successful non-surgical deep intrauterine insemination with small numbers of spermatozoa in sows. Reproduction. 2001;122:289–96.

145. Martinez EA, Vazquez JM, Roca J, Lucas X, Gil MA, Parrilla I, et al. Minimum number of spermatozoa required for normal fertility after deep intrauterine insemination in non-sedated sows. Reproduction. 2002;123:163–70.

146. Mezalira A, Dallanora D, Bernardi M, Wentz I, Bortolozzo FP. Influence of sperm cell dose and post-insemination backflow on reproductive performance of intrauterine inseminated sows. Reprod Domest Anim. 2005;40:1–5.

147. Serret CG, Alvarenga MVF, Coria ALP, Corcini CD, Correa MN, Deschamps JC, et al. Intrauterine artificial insemination of swine with different sperm concentrations, parities, and methods for prediction of ovulation. Anim Reprod. 2005;2:250–6.

148. Sbardella PE, Ulguim RR, Fontana DL, Ferrari CV, Bernardi ML, Wentz I, et al. The post-cervical insemination does not impair the reproductive performance of primiparous sows. Reprod Domest Anim. 2014;49:59–64.

149. Flowers WL, Armstrong JD, White SL, Woodward TO, Almond GW. Real-time ultrasonography and pregnancy diagnosis in swine. J Anim Sci. 2000;77:1–7.

150. Kirkden RD, Broom DM, Andersen IL. Piglet mortality: the impact of induction of farrowing using prostaglandins and oxytocin. Anim Reprod Sci. 2013;138:14–24.

151. Webel S, Johnston MF, Francisco CK, Kraeling R. Sow reproductive performance using Triptorelin Gel and fixed-time AI in commercial swine farms. In: 23rd International Pig Veterinary Society Congress. 2014. p. 136. 6-8-2014.

152. Johnston ME. Personal communication. 2014.

14

Osteopontin: a leading candidate adhesion molecule for implantation in pigs and sheep

Greg A Johnson[1*], Robert C Burghardt[1] and Fuller W Bazer[2]

Abstract

Osteopontin (OPN; also known as Secreted Phosphoprotein 1, SPP1) is a secreted extra-cellular matrix (ECM) protein that binds to a variety of cell surface integrins to stimulate cell-cell and cell-ECM adhesion and communication. It is generally accepted that OPN interacts with apically expressed integrin receptors on the uterine luminal epithelium (LE) and conceptus trophectoderm to attach the conceptus to the uterus for implantation. Research conducted with pigs and sheep has significantly advanced understanding of the role(s) of OPN during implantation through exploitation of the prolonged peri-implantation period of pregnancy when elongating conceptuses are free within the uterine lumen requiring extensive paracrine signaling between conceptus and endometrium. This is followed by a protracted and incremental attachment cascade of trophectoderm to uterine LE during implantation, and development of a true epitheliochorial or synepitheliochorial placenta exhibited by pigs and sheep, respectively. In pigs, implanting conceptuses secrete estrogens which induce the synthesis and secretion of OPN in adjacent uterine LE. OPN then binds to $\alpha v \beta 6$ integrin receptors on trophectoderm, and the $\alpha v \beta 3$ integrin receptors on uterine LE to bridge conceptus attachment to uterine LE for implantation. In sheep, implanting conceptuses secrete interferon tau that prolongs the lifespan of CL. Progesterone released by CL then induces OPN synthesis and secretion from the endometrial GE into the uterine lumen where OPN binds integrins expressed on trophectoderm ($\alpha v \beta 3$) and uterine LE (identity of specific integrins unknown) to adhere the conceptus to the uterus for implantation. OPN binding to the $\alpha v \beta 3$ integrin receptor on ovine trophectoderm cells induces *in vitro* focal adhesion assembly, a prerequisite for adhesion and migration of trophectoderm, through activation of: 1) P70S6K via crosstalk between FRAP1/MTOR and MAPK pathways; 2) MTOR, PI3K, MAPK3/MAPK1 (Erk1/2) and MAPK14 (p38) signaling to stimulate trohectoderm cell migration; and 3) focal adhesion assembly and myosin II motor activity to induce migration of trophectoderm cells. Further large *in vivo* focal adhesions assemble at the uterine-placental interface of both pigs and sheep and identify the involvement of sizable mechanical forces at this interface during discrete periods of trophoblast migration, attachment and placentation in both species.

Keywords: Implantation, Integrins, Psteoponti, Pigs, Sheep

Introduction

Domestic animal models for research are generally under-appreciated [1]; however, pigs and sheep offer unique characteristics of pregnancy, as compared to rodent or primate models, and studies of pigs and sheep have provided significant insights into the physiology of implantation including: 1) elongation of the blastocyst into a filamentous conceptus; 2) the protracted peri-implantation period of pregnancy when the conceptus is free within the uterine lumen requiring extensive paracrine signaling between conceptus and endometrium, as well as nutritional support provided by uterine secretions; 3) a protracted and incremental attachment cascade of trophectoderm to endometrial epithelium during implantation; and (4) development of a true epitheliochorial or synepitheliochorial placenta, respectively, that utilizes extensive uterine and placental vasculatures for hematotrophic nutrition, and placental areolae for histotrophic support of the developing fetuses. Our understanding of the complex mechanistic events that underlie successful implantation and placentation across species has been and will likely continue to be advanced by studies of pigs and sheep as biomedical research models

* Correspondence: gjohnson@cvm.tamu.edu
[1]Department of Veterinary Integrative Biosciences, Texas A&M University, College Station, TX 77843-4458, USA
Full list of author information is available at the end of the article

and to increase reproductive success in animal agriculture enterprises providing high quality protein for humans.

Overview of the biology of osteopontin (OPN)

OPN is a secreted extracellular matrix (ECM) protein that binds to a variety of cell surface integrins and several CD44 variants [2-6]. Integrins are transmembrane glycoprotein receptors composed of non-covalently bound α and β subunits that promote cell-cell and cell-ECM adhesion, cause cytoskeletal reorganization to stabilize adhesion, and transduce signals through numerous signaling intermediates [7,8]. Integrin-mediated adhesion is focused within a primary mechanotransduction unit of dynamic structure and composition known as a focal adhesion whose size, composition, cell signaling activity and adhesion strength are force-dependent [2,9]. The intrinsic properties of the ECM in different niches and tissue-level compartments affect the composition and size of focal adhesions that, in turn, modulate cell behavior including gene expression, protein synthesis, secretion, adhesion, migration, proliferation, viability and/or apoptosis [10]. Integrins are dominant glycoproteins in many cell adhesion cascades, including well defined roles in leukocyte adhesion to the apical surface of polarized endothelium for extravasation of leukocytes from the vasculature into tissues [11]. A similar adhesion cascade involving interactions between the ECM and apically expressed integrin receptors on the uterine luminal epithelium (LE) and conceptus (embryo and placental membranes) trophectoderm is proposed as a mechanism for attachment of the conceptus to the uterus for implantation; the initial step for the extensive tissue remodeling that occurs during placentation [12]. OPN is a leading candidate adhesion molecule for implantation in pigs and sheep [13].

OPN is an acidic member of the small integrin-binding ligand N-linked glycoprotein (SIBLING) family of proteins [9]. The breadth of literature pertaining to the diverse functions of OPN is extensive, and OPN has been independently identified as a protein associated with metastatic cancers (2ar), as an ECM protein of bones and teeth (OPN, BSP1, BNSP,SPP1), as a cytokine produced by activated lymphocytes and macrophages (early T-cell activation factor 1, Eta-1), and as a major constituent of the uterus and placenta during pregnancy [13-17]. In general, OPN is a monomer ranging in length from 264 to 301 amino acids. OPN contains a hydrophobic leader sequence characteristic of a secreted protein, a calcium phosphate apatite binding region of consecutive asparagine residues, a GRGDS sequence that interacts with integrins, a thrombin cleavage site, and two glutamine residues that are recognized substrates for transglutaminase-supported multimer formation [3,5,6]. Genes encoding OPN from different species present only moderate sequence conservation, except in the NH_2-terminal region, around the

Arg-Gly-Asp (RGD) integrin-binding sequence, and in the COOH-terminus [3,5,6,18]. OPN undergoes extensive posttranslational modifications that can alter its function in different physiological microenvironments. These modifications include proteolytic cleavage, phosphorylation, glycosylation, sulfation and cross-linking with self and other macromolecules [19-23]. OPN is present on epithelial cells and in secretions of the gastrointestinal tract (including the liver), respiratory tract, kidneys, thyroid, breast, testes, uterus and placenta [24-32]. Other cell types that express OPN include leukocytes, smooth muscle cells, and highly metastatic cancer cells [33-35]. OPN is a multifunctional ECM protein reported to 1) stimulate cell-cell adhesion, 2) increase cell-ECM communication, 3) promote cell migration, 4) decrease cell death, 5) stimulate immunoglobulin production, 6) induce changes in phosphorylation of focal adhesion kinase and paxillin, 7) stimulate phosphotidylinositol 3'-kinase activity, 8) alter intracellular calcium levels, and 9) promote calcium phosphate deposition [36-42].

Timeline of key advancements in understanding the role of OPN as an attachment factor for implantation

OPN was first observed in endometrial tissue when, in 1988, Nomura et al., [43] performed *in situ* hybridization to localize OPN in mouse embryos, the endometrium from the gravid and non-gravid uterine horns of pregnant mice, and the endometrium from mice exposed to intrauterine injection of oil to induce a deciduoma. High levels of OPN mRNA were detected in the LE, but not GE, of the gravid uterine horns. Interestingly, epithelial expression of OPN appeared to be specific to pregnancy because little to no OPN mRNA was observed in the uterine LE of non-gravid or pseudopregnant mice [43]. In addition to the LE, high levels of OPN mRNA were localized to the granulated metrial gland (GMC) cells of decidual and deciduoma tissues, with lower numbers of OPN positive cells in the deciduoma of uteri [43]. It is noteworthy that these investigators were the first to argue that OPN plays a wider role than had previously been assumed, and that its functions are not confined to bone development. The decidual cells that express OPN have since been confirmed to be uterine natural killer (uNK) cells [44,45]. Similar to expression in mice, immunocytochemical studies performed by Young and colleagues in 1990 [25] localized OPN protein to the decidua of women; however, in contrast to mice, OPN was also expressed by the secretory phase endometrial GE. It was suggested that the absence of OPN in GE during the proliferative phase of the menstrual cycle indicated that changes in expression in GE of normal cycling endometrium were the result of hormonal regulation and that the function(s) of OPN in the endometrium might be associated with its ability to enhance cell attachment [25].

A significant conceptual advance regarding the function (s) of epithelia-derived OPN was made by Brown and co-workers [26] in 1992, when OPN mRNA and protein were localized to epithelial cells of a variety of organs, including the hypersecretory endometrial GE associated with pregnancy in women. In the secretory epithelia of all organs examined, OPN protein was associated with the apical domain of the cells, and when the luminal contents were preserved in tissue sections, proteins secreted into the lumen were positive for OPN staining. It was hypothesized that OPN secreted by epithelia, including uterine epithelia, binds integrins on luminal surfaces to effect communication between the surface epithelium and the external environment [26]. Between 1992 and 1996, Lessey and co-workers established that transient uterine expression of αvβ3 and α4β1 integrins defines the window of implantation in women [46-48] and that altered expression of these integrins correlates with human infertility [49,50]. Noting that the αvβ3 and α4β1 integrin heterodimers present during the implantation window bind OPN, these investigators suggested involvement of OPN and integrins in trophoblast-endometrial interactions during the initial attachment phase of implantation [46].

Comprehensive examination of the temporal and spatial expression and hormonal regulation of uterine OPN mRNA and protein and integrin subunit proteins in the uteri and placentae of sheep (discussed in detail later in this review), performed from 1999 through 2002, provided the first strong evidence that OPN is a progesterone-induced secretory product of endometrial glands (histotroph) that binds integrins on apical surfaces of endometrial LE and conceptus trophectoderm to mediate attachment of uterus to trophectoderm for implantation [18,29,51,52]. Indeed, pregnant Day 14 ewes, which lack uterine glands (uterine gland knockout, UGKO phenotype), exhibit an absence of OPN in uterine flushings compared with normal ewes, and do not maintain pregnancy through the peri-implantation period [53]. Similarly, functional intrauterine blockade of αv and β3 integrin subunits, that combine to form a major receptor for OPN, reduces the number of implantation sites in mice and rabbits [54,55]. Further evidence for regulation of uterine OPN by sex steroids was provided by results from studies using human and rabbit models. Progesterone treatment increased OPN expression by human endometrial adenocarcinoma Ishakawa cells (in vitro findings, 2001) as well as endometrium of rabbits (in vivo findings, 2003) [56,57]. In contrast, i.m. injection of estrogen induced expression of OPN in the uterine LE of cyclic pigs (in vivo, 2005) [58]. Results from pigs were the first to suggest that conceptuses can directly regulate the regional expression of OPN in the endometrium at specific sites of implantation through secretion of estrogens [58,59]. Microarray studies from 2002 and 2005 strongly support a role for OPN during implantation [60-62]. Two reports confirmed that OPN is the most highly up-regulated ECM-adhesion molecule in the human uterus as it becomes receptive to implantation [60-62].

Research regarding OPN has begun to focus on its interactions with integrin receptors in the female reproductive tract. In 2009, Burghardt and colleagues [63] reported the in vivo assembly of large focal adhesions containing aggregates of αv, α4, α5, β1, β5, alpha actinin, and focal adhesion kinase (FAK) at the uterine-placental interface of sheep, that expand as pregnancy progresses. It is noteworthy that OPN was present along the surfaces of both uterine LE and trophectoderm, although it was not determined whether it co-localized to the focal adhesions [63]. Similar focal adhesions form during implantation in pigs [64,65]. Affinity chromatography and immunoprecipitation experiments revealed direct in vitro binding of porcine trophectoderm αvβ6 and uterine epithelial cell αvβ3, and ovine trophectoderm αvβ6 integrins to OPN [64,66]. These were the first functional demonstrations that OPN directly binds specific integrins to promote trophectoderm cell migration and attachment to uterine LE that may be critical to conceptus elongation and implantation. Recently (2014), Aplin and co-workers [67] employed three in vitro models of early implantation with Ishakawa cells to demonstrate that OPN potentially interacts with the αvβ3 integrin receptor during implantation in humans.

Key events during the peri-implantation period of pigs and sheep

Communication and reciprocal responses between the conceptus and uterus are essential for conceptus survival during the peri-implantation period of pregnancy. These interactions also lay the critical physiological and anatomical groundwork for subsequent development of functional uterine LE, GE, stroma and placentae required to maintain growth and development of the conceptus throughout pregnancy. In a progesterone dominated uterine environment, establishment and maintenance of pregnancy in pigs and sheep requires; (i) secretion of estrogens or interferon tau, respectively, from the conceptus to signal pregnancy recognition [68-71], (ii) secretions from uterine LE and GE, i.e., histotroph, to support attachment, development and growth of the conceptus [72-74], and (iii) cellular remodeling at the uterine LE-conceptus trophectoderm interface to allow for attachment during implantation [8,75,76]. These events are orchestrated through endocrine, paracrine, autocrine and juxtracrine communication between the conceptus and uterus, and the complexity of these events likely underlies the high rates of conceptus mortality during the peri-implantation period of pregnancy [77,78].

Implantation and placentation are critical events in pregnancy. Implantation failure during the first three weeks of pregnancy is a major cause of infertility in all mammals [77-80]. The process of implantation is highly synchronized, requiring reciprocal secretory and physical interactions between a developmentally competent conceptus and the uterine endometrium during a restricted period of the uterine cycle termed the "window of receptivity". These initial interactions between apical surfaces of uterine LE and conceptus trophectoderm begin with sequential phases i.e., non-adhesive or pre-contact, apposition, and adhesion, and conclude with formation of a placenta that supports fetal-placental development throughout pregnancy [81-83]. Conceptus attachment first requires loss of anti-adhesive molecules in the glycocalyx of uterine LE, comprised largely of mucins that sterically inhibit attachment [52,84,85]. This results in "unmasking" of molecules, including selectins and galectins, which contribute to initial attachment of conceptus trophectoderm to uterine LE [86-88]. These low affinity contacts are then replaced by a repertoire of adhesive interactions between integrins and maternal ECM which appear to be the dominant contributors to stable adhesion at implantation [1,8,52,89-91]. OPN is expressed abundantly within the conceptus-maternal environment in numerous species, including pigs and sheep [17,29,57,59,62,92,93].

Osteopontin is structurally and functionally suited to support implantation of pig and sheep conceptuses

Depending on cell context and species, OPN expression can be regulated by multiple hormones and cytokines, including the sex steroids progesterone and estrogen [28,51,56-58,94-98]. OPN mediates multiple cellular processes, such as cell-mediated immune responses, inflammation, angiogenesis, cell survival, and tumor metastasis primarily through integrin signaling [3,5,17,99,100]. Integrins are transmembrane glycoprotein receptors composed of non-covalently bound α and β subunits that participate in cell-cell and cell-ECM adhesion, cause cytoskeletal reorganization to stabilize adhesion, and transduce signals through numerous signaling intermediates [7,8]. OPN has an expansive integrin receptor repertoire that includes RGD-mediated binding to $\alpha v \beta 3$ [101,102], $\alpha v \beta 1$ [103], $\alpha v \beta 5$ [103], and $\alpha 8 \beta 1$ [104], as well as alternative binding sequence-mediated interactions with $\alpha 4 \beta 1$ [105], and $\alpha 9 \beta 1$ [106]. OPN binding to these various receptors results in diverse effects including: (1) leukocyte, smooth muscle cell and endothelial cell chemotaxis; (2) endothelial and epithelial cell survival; and (3) fibroblast, macrophage and tumor cell migration [64,66,103,104,107]. Clearly, the ability to bind multiple integrin receptors to produce different cellular outcomes greatly increases OPN's potential role(s) during conceptus development and implantation. Importantly, OPN contains a serine protease cleavage site that when activated generates bioactive OPN fragments [23,108], and two glutamines that support multimerization of the protein [22]. It is notable that OPN is flexible in solution,

allowing for simultaneous binding to more than one integrin receptor [16,109]. Further, OPN can also exist in a polymerized form cross-linked by transglutaminase. Homotypic OPN bonds have high tensile strength, suggesting that self-assembly is involved in cell-cell and cell-matrix interactions [22]. These multimeric complexes may present multiple RGD sequences for simultaneous binding to integrins on multiple surfaces [22,110]. Therefore, OPN has the potential to bind multiple proteins and to participate in assembly of multi-protein complexes that bridge and form the interface between conceptus to uterus during implantation.

OPN expression, regulation and function in the uterus and placenta of gilts

A hallmark of pregnancy in pigs is the protracted peri-implantation period of pregnancy when conceptuses are free within the uterine lumen to elongate from spherical blastocysts to conceptuses with a filamentous morphology (Reviewed in [111]). Pig embryos move from the oviduct into the uterus about 60 to 72 h after onset of estrus, reach the blastocyst stage by Day 5, then shed the zona pellucida and expand to 2–6 mm in diameter by Day 10. At this stage, development of pig embryos diverges from that of rodents or primates. Within a few hours the presumptive placental membranes (trophectoderm and extra-embryonic endoderm) elongate at a rate of 30–45 mm/h from a 10 mm blastocyst to a 150–200 mm long filamentous form, after which further elongation occurs until conceptuses are 800–1,000 mm in length by Day 16 of pregnancy [111]. During this period of rapid elongation, porcine conceptuses secrete estrogen beginning on Days 11 and 12 to signal initiation of pregnancy to the uterus, and by Day 13 begin an extended period of incremental attachment to the uterine LE [17,69]. The attached trophectoderm/chorion-endometrial epithelial bilayer develops microscopic folds, beginning about Day 35 of gestation, and these folds increase the surface area of contact between maternal and fetal capillaries to maximize maternal-to-fetal exchange of nutrients and gases [112].

In pigs, OPN is an excellent candidate for influencing this complex environment of pregnancy, because the OPN gene is located on chromosome 8 under a quantitative trait loci (QTL) peak for prenatal survival and litter size, [113]. The temporal and spatial expression of OPN in the porcine uterus and placenta is complex, with independent and overlapping expression by multiple cell types. Between Days 5 and 9 of the estrous cycle and pregnancy, OPN transcripts are detectable in a small percentage of cells in the sub-epithelial stratum compactum of the endometrial stroma [59]. The morphology and distribution of OPN mRNA- and protein-positive cells in the stratum compactum of the stroma on Day 9 of the estrous cycle and pregnancy suggest that these are immune cells. Certainly Eta-1/OPN, is an established component of the immune system that is secreted by activated T lymphocytes [15]. It is reasonable to speculate that because insemination in pigs is intrauterine, OPN expressing immune cells may protect against pathogens introduced during mating. A similar pattern of distribution of OPN-producing cells is also evident in the allantois of the placenta beginning between Days 20 and 25 of pregnancy, and the number of these cells increases as gestation progresses [58]. The identity of these cells remains to be determined.

OPN expression in uterine LE increases markedly during the peri-implantation period of pigs, but is never observed in uterine LE during the estrous cycle [59]. OPN mRNA is initially induced by conceptus estrogens in discrete regions of the LE juxtaposed to the conceptus just prior to implantation on Day 13, then expands to the entire LE by Day 20 when firm adhesion of conceptus trophectoderm to uterine LE occurs [58]. However, OPN mRNA is not present in pig conceptuses [58,59]. In contrast to mRNA, OPN protein is abundant along the apical surfaces of LE and trophectoderm/chorion, but only in areas of direct contact between the uterus and conceptus [58,59]. Remarkably, OPN mRNA and protein are not present in uterine LE and chorion of areolae where the chorion does not attach to LE, but rather forms a "pocket" of columnar epithelial cells that take up and transport secretions of uterine GE into the placental vasculature by fluid phase pinocytosis [114] (Figure 1). OPN levels remain high at this interface throughout pregnancy [59], as do multiple integrin subunits that potentially form heterodimeric receptors that bind OPN [8,84,90].

All experimental and surgical procedures were in compliance with the Guide for Care and Use of Agricultural Animals in Teaching and Research and approved by the Institutional Animal Care and Use Committee of Texas A&M University.

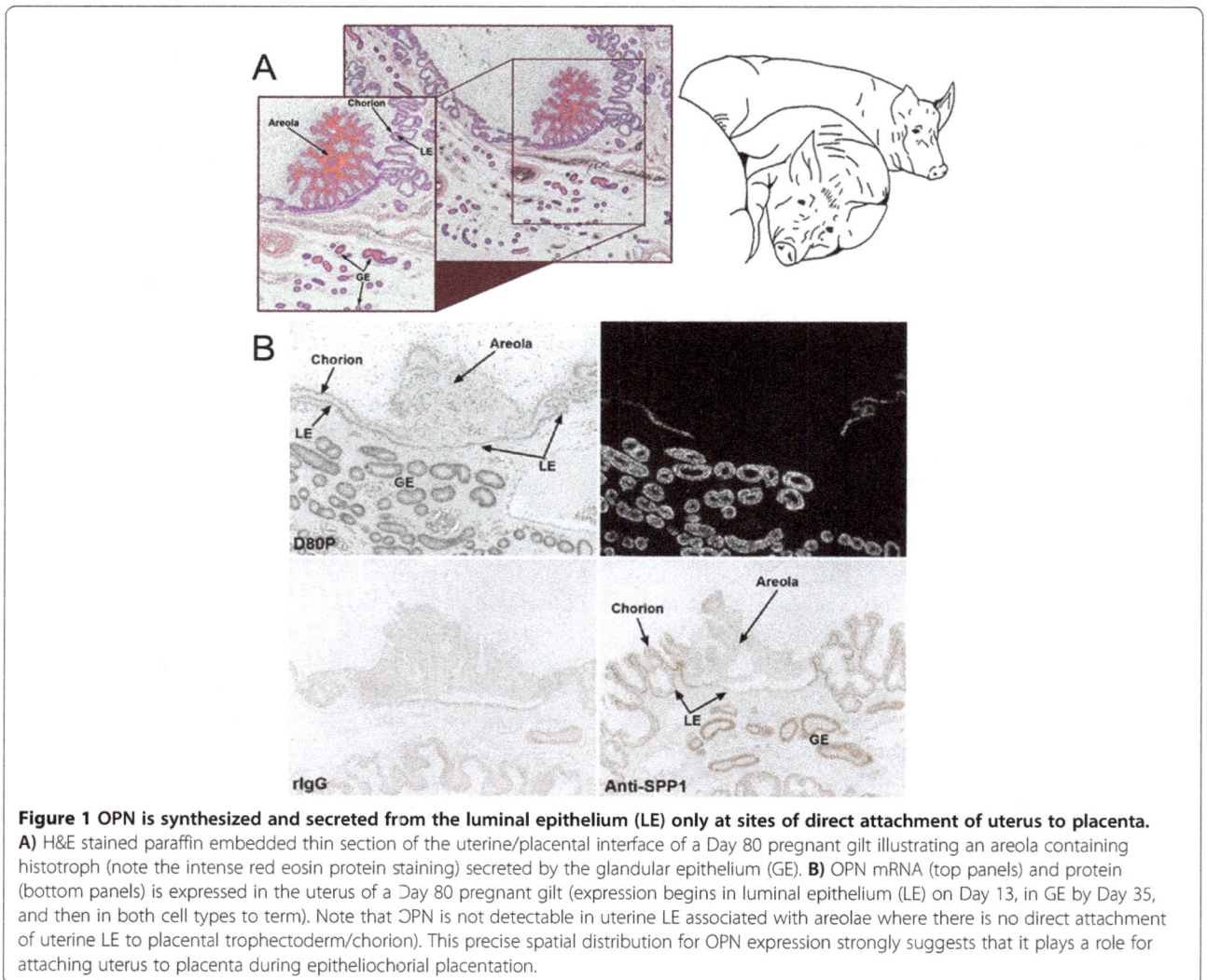

Figure 1 OPN is synthesized and secreted from the luminal epithelium (LE) only at sites of direct attachment of uterus to placenta.
A) H&E stained paraffin embedded thin section of the uterine/placental interface of a Day 80 pregnant gilt illustrating an areola containing histotroph (note the intense red eosin protein staining) secreted by the glandular epithelium (GE). **B)** OPN mRNA (top panels) and protein (bottom panels) is expressed in the uterus of a Day 80 pregnant gilt (expression begins in luminal epithelium (LE) on Day 13, in GE by Day 35, and then in both cell types to term). Note that OPN is not detectable in uterine LE associated with areolae where there is no direct attachment of uterine LE to placental trophectoderm/chorion). This precise spatial distribution for OPN expression strongly suggests that it plays a role for attaching uterus to placenta during epitheliochorial placentation.

Affinity chromatography and immunoprecipitation experiments were performed to test whether the integrin subunits αv, α4, α5, β1, β3, β5, and β6, expressed by porcine trophectoderm cells (pTr2) and porcine uterine epithelial (pUE) cells, directly bind OPN. Detergent extracts of surface-biotinylated pig trophectoderm (pTr2) and uterine epithelial (pUE) cells were incubated with OPN-Sepharose and the proteins that bound to OPN were eluted with EDTA to chelate cations and release bound integrins. To identify these integrins, immunoprecipitation assays were performed using antibodies that successfully immunoprecipitated integrin subunits from pTr2 or pUE cell lysates. OPN directly bound the αvβ6 integrin heterodimer on pTr2 cells and αvβ3 on ULE cells [64]. OPN binding promoted dose- and integrin-dependent attachment of pTr2 and pUE cells, and stimulated haptotactic pTr2 cell migration, meaning that cells migrated directionally along a physical gradient of nonsoluble OPN [64]. Further, immunofluorescence staining revealed that both OPN and αv integrin subunit localized to the apical surface of cells at the interface between uterine LE and conceptus trophectoderm at Day 20 of pregnancy. The αv integrin subunit staining

pattern revealed large aggregates at the junction between trophectoderm and uterine LE, suggesting the formation of OPN-induced *in vivo* focal adhesions at the apical surfaces of both conceptus trophectoderm and uterine LE that facilitate conceptus attachment to the uterus for implantation. The β3 subunit appeared in aggregates on the apical surface of LE cells, but not trophectoderm cells, fitting with affinity chromatography data indicating direct binding of αvβ3 on pUE cells to OPN [64]. Finally, OPN-coated microspheres revealed co-localization of the αv integrin subunit and talin to focal adhesions at the apical domain of pTr2 cells *in vitro* [64]. Collectively, results support that OPN binds integrins to stimulate integrin-mediated focal adhesion assembly, attachment, and cytoskeletal force-driven migration of pTr2 cells to promote conceptus implantation in pigs (Figure 2).

In addition to expression in LE during the peri-implantation period, total uterine OPN mRNA increases 20-fold between Days 25 and 85 of gestation due to induction of OPN expression in uterine GE [59]. The initial significant increase in GE is delayed until between Days 30 and 35 when placental growth and placentation are key events in pregnancy in pigs [5]. OPN expression

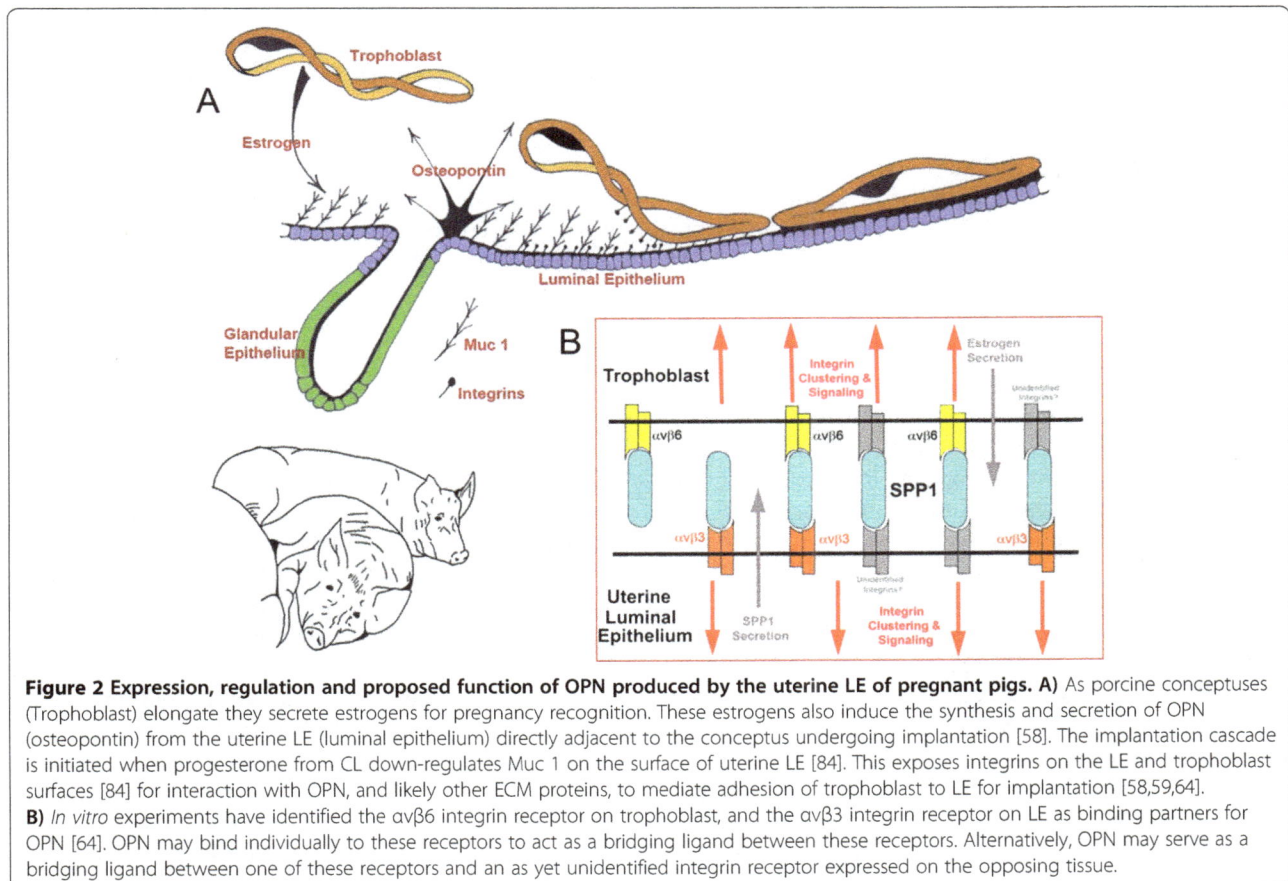

Figure 2 Expression, regulation and proposed function of OPN produced by the uterine LE of pregnant pigs. A) As porcine conceptuses (Trophoblast) elongate they secrete estrogens for pregnancy recognition. These estrogens also induce the synthesis and secretion of OPN (osteopontin) from the uterine LE (luminal epithelium) directly adjacent to the conceptus undergoing implantation [58]. The implantation cascade is initiated when progesterone from CL down-regulates Muc 1 on the surface of uterine LE [84]. This exposes integrins on the LE and trophoblast surfaces [84] for interaction with OPN, and likely other ECM proteins, to mediate adhesion of trophoblast to LE for implantation [58,59,64].
B) *In vitro* experiments have identified the αvβ6 integrin receptor on trophoblast, and the αvβ3 integrin receptor on LE as binding partners for OPN [64]. OPN may bind individually to these receptors to act as a bridging ligand between these receptors. Alternatively, OPN may serve as a bridging ligand between one of these receptors and an as yet unidentified integrin receptor expressed on the opposing tissue.

in GE during later stages of pregnancy is also observed sheep [115], and a microarray study in rats showed that OPN expression increased 60-fold between Day 0 of the estrous cycle and Day 20 of pregnancy, likely within the decidua [116]. Indeed, OPN is expressed by uterine natural killer (uNK) cells of the mouse decidua [44,45]. Secretions of GE in livestock, and the secretions of decidua in rodents and primates, are critical to support implantation, placentation, and fetal growth and development [117,118]. OPN is also expressed in uterine GE of Day 90 of pseudopregnant pigs, suggesting that maintenance of secretion of progesterone by CL is responsible for expression of OPN in GE [58]. Progesterone also regulates OPN expression in the GE of sheep and rabbits [51,54], as well as OPN synthesis by human Ishikawa cells [56].

However, the involvement of progesterone in the regulation of OPN in uterine GE is complex as indicated by recent analysis of long-term progesterone treatment on the expression of OPN in pigs in the absence of ovarian or conceptus factors. In addition to OPN expression, other established progesterone targets including progesterone receptor (PGR) as an index of progesterone's ability to negatively regulate GE gene expression [119], acid phosphate 5, tartrate resistant (ACP5, commonly referred to as uteroferrin) as an index of progesterone's ability to positively regulate early pregnancy GE gene expression [120], and fibroblast growth factor 7 (FGF7, commonly referred to as keratinocyte growth factor) provide an index of progesterone's ability to positively regulate gene expression in uterine GE beyond the peri-implantation period [121]. Pigs were ovariectomized on Day 12 of the estrous cycle when progesterone secretion from CL is high and treated daily with intramuscular injections of progesterone or vehicle for 28 days [122,123]. As anticipated, PGR mRNA decreased, uteroferrin mRNA increased, and FGF7 mRNA increased in uterine GE of pigs injected with progesterone [123]. Surprisingly, long-term progesterone, in the absence of ovarian and/or conceptus factors, did not induce OPN expression in uterine GE [123]. It is currently hypothesized that the hormonal milieu necessary for the production of individual components of histotroph varies, and may require specific servomechanisms, similar to those for sheep and rabbits, which involve sequential exposure of the pregnant uterus to ovarian, conceptus, and/or uterine factors that include progesterone, estrogens and IFNs [124-126]. Recently OPN expression was compared in placental and uterine tissues supplying a normally sized and the smallest fetus carried by hyperprolific Large White and Meishan gilts. Not only were levels of OPN strikingly different between the two breeds of pigs, but OPN was higher in the LE and GE of uteri surrounding smaller sized fetuses, suggesting OPN may be associated with placental efficiency [127].

OPN expression, regulation and function in the uterus and placenta of ewes

Similar to pigs, the conceptuses of sheep remain free-floating within the uterine lumen as they elongate from spherical blastocysts to conceptuses with a filamentous morphology (Reviewed in [88]). Sheep embryos enter the uterus on Day 3, develop to spherical blastocysts and then, after hatching from the zona pellucida, transform from spherical to tubular and filamentous conceptuses between Days 12 and 15 of pregnancy, with extra-embryonic membranes extending into the contralateral uterine horn between Days 16 and 20. During this period of rapid elongation, the mononuclear trophoblast cells of ovine conceptuses secrete interferon tau between Days 10 and 21 of pregnancy, and implantation begins on Day 16 as trophectoderm attaches to the uterine LE [70,88]. The ovine placenta eventually organizes into discrete regions called placentomes that are composed of highly branched placental chorioallantoic villi termed cotyledons which grow rapidly and interdigitate with maternal aglandular endometrial crypts termed caruncles. Approximately 90% of the blood from the uterine artery flows into the placentomes for nutrient transfer from the maternal uterine circulation to the fetus and exchange of gasses between these tissue compartments [128].

The temporal and spatial expression of OPN in the uteri and placentae of sheep is similar to that previously described for the pig, except: 1) unlike in the pig, OPN is not expressed by uterine LE; 2) induction of OPN in uterine GE occurs earlier than in the pig during the peri-implantation period, and expression in the GE is regulated by progesterone; 3) OPN is a prominent component of the stratum compactum stroma; and 4) although large focal adhesions assemble during the peri-implantation period of pigs, they are not observed at the uterine-placental interface until later stages of pregnancy in sheep.

OPN mRNA and protein are present in a small population of cells scattered within the stratum compactum stroma immediately beneath the endometrial LE during the early stages of the estrous cycle and pregnancy in sheep [18]. OPN-producing cells are also present in the allantois of the ovine placenta beginning between Days 20 and 25 of pregnancy and increase in number as gestation progresses [17]. As hypothesized for pigs, these are presumed to be immune cells because Eta-1/OPN is a prominent player in the immune system [15]. In contrast to pigs, in which the OPN-expressing endometrial cells are readily evident in the stratum compactum stroma throughout pregnancy, these cells are difficult to discern in the sheep due to an increase in expression of OPN by stromal cells between Days 20 and 25 gestation [129]. In pregnant mice and primates, OPN in decidualized stroma is considered to be a gene marker for decidualization [130,131]. Decidualization involves transformation of spindle-like fibroblasts into polygonal epithelial-like cells that are hypothesized to limit conceptus trophoblast invasion through the uterine wall during invasive implantation [118]. Although Mossman [132] and Kellas [133] described decidual cell characteristics in the placentomal crypts of sheep and antelope, their reports were largely ignored, and decidualization was not thought to occur in species with central and noninvasive implantation characteristic of domestic animals. However, endometrial stromal cells do increase in size and become polyhedral in shape in pregnant ewes following conceptus attachment, and the classical decidualization markers desmin and α-smooth muscle actin are expressed in these cells, suggesting that OPN expression in this stromal compartment is part of a uterine decidualization-like response to the conceptus during ovine pregnancy [129]. In contrast, no morphological changes in uterine stroma, nor induction of OPN mRNA and protein, or desmin protein, were detected during porcine pregnancy [129]. One of the primary roles of decidua in invasive implanting species is to restrain conceptus trophoblast invasion to a circumscribed region of the endometrium. Both pigs and sheep have noninvasive implantation, but the extent of conceptus invasion into the endometrium differs between these two species. Pig conceptuses undergo true epitheliochorial placentation in which uterine LE remains morphologically intact throughout pregnancy and the conceptus trophectoderm simply attaches to the apical surface of uterine LE surface without contacting uterine stromal cells [134]. Synepitheliochorial placentation in sheep involves extensive erosion of the LE due to formation of syncytia with binucleate cells of the trophectoderm. After Day 19 of pregnancy, conceptus tissue is opposed to, but does not penetrate ovine uterine stroma [135]. Although speculative, differences in stromal expression of OPN between these species suggest that the extent of decidualization is correlated positively with degree of conceptus invasiveness.

In contrast to pigs, OPN is not synthesized by sheep uterine LE, but is nonetheless a component of histotroph secreted from the endometrial GE into the uterine lumen of pregnant ewes as early as Day13. It is not secreted by uterine GE of cyclic ewes [18,29]. OPN mRNA is detected in some uterine glands by Day 13, and is present in all glands by Day 19 of gestation [18]. Progesterone induces expression of OPN in the endometrial GE, and induction is associated with a loss of PGR in uterine GE. Analysis of uterine flushings from pregnant ewes has identified a 45 kDa fragment of OPN with greater binding affinity for αvβ3 integrin receptor than native 70 kDa [29,51,52,108]. Comparison of the spatial distribution of OPN mRNA and protein by *in situ* hybridization and immunofluorescence analyses of cyclic and pregnant ovine uterine sections has provided significant insight into the physiology of uterine OPN during pregnancy. OPN mRNA increases in the endometrial GE during the peri-implantation period;

however, it is not present in LE or conceptus trophectoderm [18]. In contrast, immunoreactive OPN protein is present at the apical surfaces of endometrial LE and GE, and on trophectoderm where the integrin subunits αv, α4, α5, β1, β3, and β5 are expressed constitutively on the apical surfaces of trophectoderm and endometrial LE and could potentially assemble into several heterodimers that could serve as receptors for OPN including αvβ3, αvβ1, αvβ5, α4β1, and α5β1 heterodimers which [29,52]. These results strongly suggest that OPN is a component of histotroph secreted from GE into the uterine lumen of pregnant ewes in response to progesterone, and that OPN binds integrin receptors expressed on endometrial LE and conceptus trophectoderm.

Affinity chromatography and immunoprecipitation experiments, similar to those described previously for pigs, determined whether αv, α4, α5, β1, β3, β5, and β6 integrins expressed by ovine trophectoderm cells (oTr1) directly bind OPN. Successful immunoprecipitation of labeled oTr1 integrins occurred with antibodies to αv and β3 integrin subunits, as well as an antibody to the integrin αvβ3 heterodimer. Antibody to the αv integrin subunit also precipitated a β chain, presumed to be the β3 integrin subunit, as an antibody to the β3 integrin subunit precipitated an α chain at the same relative size as the bands precipitated by an antibody to the αvβ3 heterodimer. Thus, the αvβ3 integrin on oTr1 cells binds OPN [66]. OPN binding to the αvβ3 integrin receptor induced *in vitro* focal adhesion assembly (see Figure 3), a prerequisite for adhesion and migration of trophectoderm, through activation of: 1) P70S6K via crosstalk between FRAP1/MTOR and MAPK pathways; 2) MTOR, PI3K, MAPK3/MAPK1 (Erk1/2) and MAPK14 (p38) signaling to stimulate trophectoderm cell migration; and 3) focal adhesion assembly and myosin II motor activity to induce migration of trophectoderm cells [66]. Collectively, results indicate that OPN binds αvβ3 integrin receptor to activate cell signaling pathways that act in concert to mediate adhesion, migration and cytoskeletal remodeling of trophectoderm cells essential for expansion and elongation of conceptuses and their attachment to uterine LE for implantation (Figure 4).

Focal adhesions, the hallmark of activated integrins, are prominent structures of cells grown in culture; however, they are rarely observed *in vivo*. It is noteworthy that large aggregations of focal adhesion-associated proteins that have been interpreted to be three dimensional focal adhesions are present at the uterine-placental interface of sheep [63]. By day 40 of pregnancy in sheep, the punctate apical surface staining of integrin receptor subunits identified in peri-implantation uterine LE and conceptus trophectoderm [52] is replaced by scattered large aggregates of αv, α4, β1, and β5 subunits in interplacentomal LE and trophectoderm/chorion cells. Integrin aggregates are observed only in gravid uterine horns of

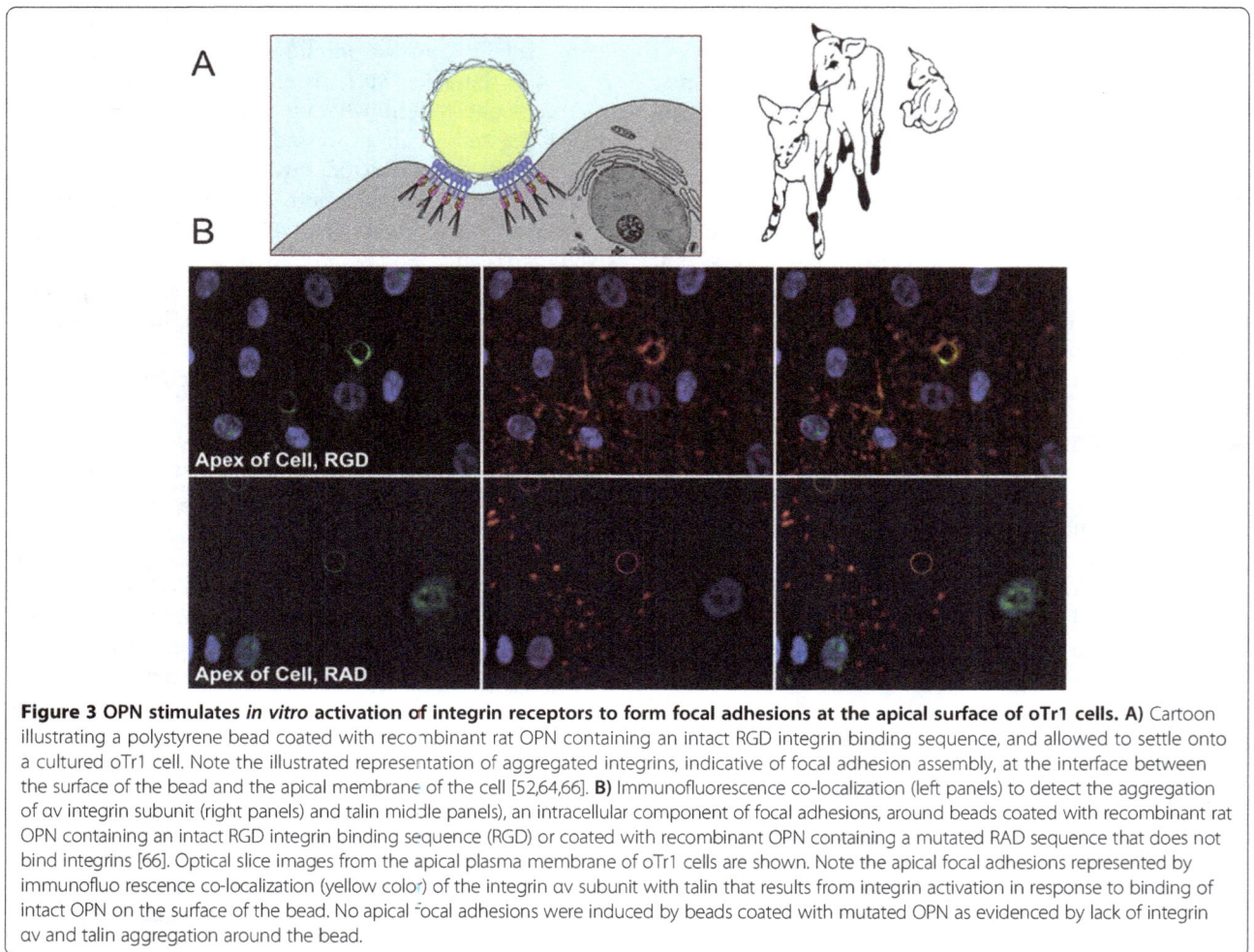

Figure 3 OPN stimulates *in vitro* activation of integrin receptors to form focal adhesions at the apical surface of oTr1 cells. **A)** Cartoon illustrating a polystyrene bead coated with recombinant rat OPN containing an intact RGD integrin binding sequence, and allowed to settle onto a cultured oTr1 cell. Note the illustrated representation of aggregated integrins, indicative of focal adhesion assembly, at the interface between the surface of the bead and the apical membrane of the cell [52,64,66]. **B)** Immunofluorescence co-localization (left panels) to detect the aggregation of αv integrin subunit (right panels) and talin middle panels), an intracellular component of focal adhesions, around beads coated with recombinant rat OPN containing an intact RGD integrin binding sequence (RGD) or coated with recombinant OPN containing a mutated RAD sequence that does not bind integrins [66]. Optical slice images from the apical plasma membrane of oTr1 cells are shown. Note the apical focal adhesions represented by immunofluorescence co-localization (yellow color) of the integrin αv subunit with talin that results from integrin activation in response to binding of intact OPN on the surface of the bead. No apical focal adhesions were induced by beads coated with mutated OPN as evidenced by lack of integrin αv and talin aggregation around the bead.

unilaterally pregnant sheep, demonstrating a requirement for trophectoderm attachment to LE, and aggregates increase in number and size through Day 120 of pregnancy [63]. Interestingly, no accumulation of β3 was observed even though ITGB3 is a prominent component of the uterine-placental interface during the peri-implantation period in sheep [52]. In some regions of the interplacentomal interface, greater subunit aggregation was seen on the uterine side, in other regions it was predominant on the placental side; whereas in some others, both uterine and placental epithelia exhibited prominent focal adhesions. However, by Day 120 of pregnancy, extensive focal adhesions were seen along most of the uterine-placental interface [63]. The placentomes, which provide hematotrophic support to the fetus and placenta, exhibited diffuse immunoreactivity for these integrins compared with interplacentomal regions perhaps due to extensive folding at this interplacentomal interface [63]. These results suggest that focal adhesion assembly at the uterine-placental interface reflects dynamic adaptation to increasing forces caused by the growing conceptus. Cooperative binding of multiple integrins to OPN deposited at the uterine-placental interface may form an adhesive mosaic to maintain a tight connection and increased tensile strength and signaling activity between uterine and placental surfaces along regions of epitheliochorial placentation in sheep.

Steady-state levels of OPN mRNA in total ovine endometrium remain constant between Days 20 and 40, increase 40-fold between Days 40 and 100, and remain maximal thereafter [18]. The major source of this OPN is uterine GE which undergoe hyperplasia through Day 50 followed by hypertrophy and maximal production of histotroph after Day 60 [115]. Additionally, immunofluorescence microscopy demonstrated that the secreted 45-kDa OPN cleavage fragment is exclusively, continuously, and abundantly present along the apical surface of uterine LE, on trophectoderm, and along the entire uterine-placental interface of both interplacentomal and placentomal regions through Day 120 of the 147 day ovine pregnancy [115]. These findings definitively localize OPN as a secretory product of the GE to regions of intimate contact between conceptus and uterus, where OPN may influence

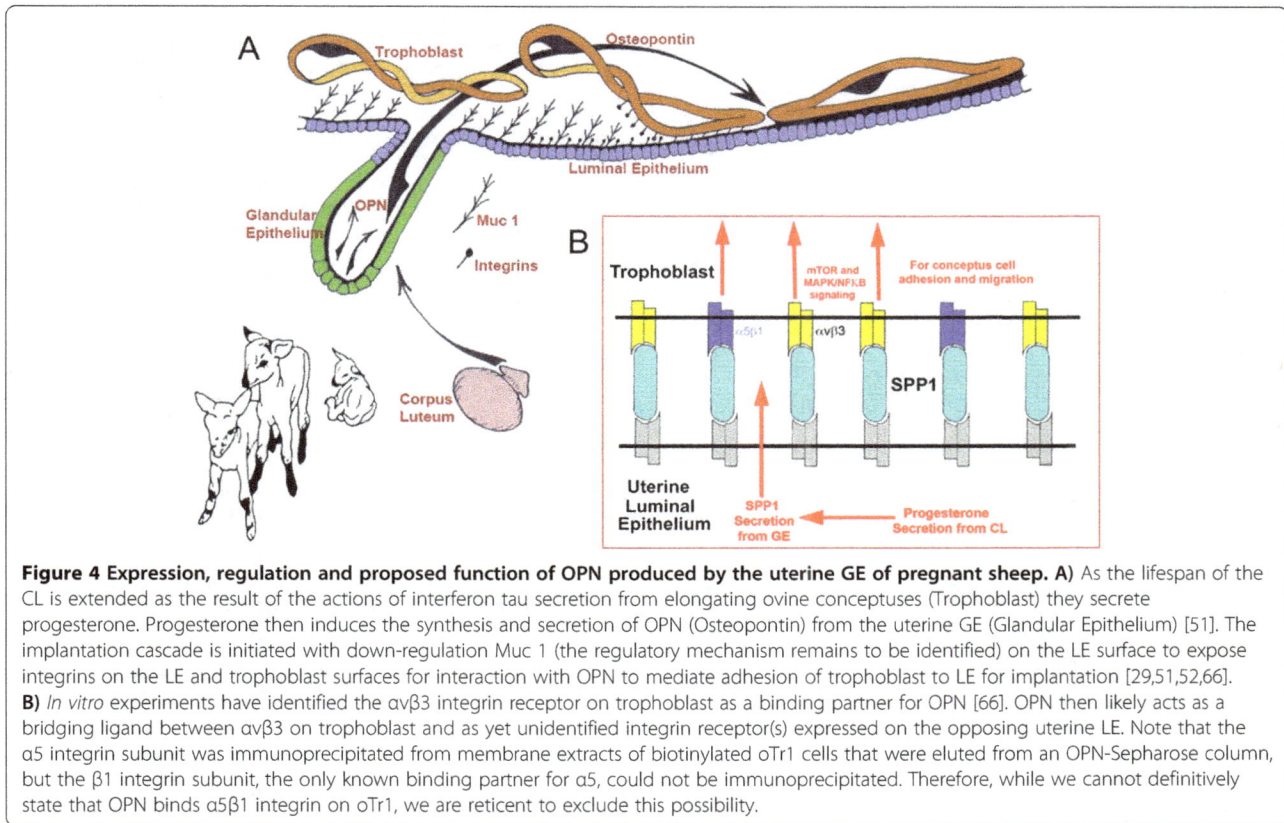

Figure 4 Expression, regulation and proposed function of OPN produced by the uterine GE of pregnant sheep. A) As the lifespan of the CL is extended as the result of the actions of interferon tau secretion from elongating ovine conceptuses (Trophoblast) they secrete progesterone. Progesterone then induces the synthesis and secretion of OPN (Osteopontin) from the uterine GE (Glandular Epithelium) [51]. The implantation cascade is initiated with down-regulation Muc 1 (the regulatory mechanism remains to be identified) on the LE surface to expose integrins on the LE and trophoblast surfaces for interaction with OPN to mediate adhesion of trophoblast to LE for implantation [29,51,52,66]. **B)** *In vitro* experiments have identified the αvβ3 integrin receptor on trophoblast as a binding partner for OPN [66]. OPN then likely acts as a bridging ligand between αvβ3 on trophoblast and as yet unidentified integrin receptor(s) expressed on the opposing uterine LE. Note that the α5 integrin subunit was immunoprecipitated from membrane extracts of biotinylated oTr1 cells that were eluted from an OPN-Sepharose column, but the β1 integrin subunit, the only known binding partner for α5, could not be immunoprecipitated. Therefore, while we cannot definitively state that OPN binds α5β1 integrin on oTr1, we are reticent to exclude this possibility.

fetal/placental development and growth, and mediate communication between placental and uterine tissues to support pregnancy to term.

Increases in OPN from GE are likely influenced by uterine exposure to progesterone, interferon-tau, and placental lactogen which constitute a "servomechanism" that activates and maintains endometrial remodeling, secretory function and uterine growth during gestation. Sequential treatment of ovariectomized ewes with progesterone, interferon tau, placental lactogen, and growth hormone results in GE development similar to that observed during normal pregnancy [126]. Administration of progesterone alone in these experiments induced expression of OPN in GE, and intrauterine infusion of interferon tau and placental lactogen to progesterone-treated ovariectomized ewes increased OPN mRNA levels above those for ewes treated with progesterone alone [126]. An attractive hypothesis for OPN expression in GE is that progesterone interacts with its receptor in GE to down-regulate the progesterone receptor. This removes a progesterone "block" to OPN synthesis, and subsequent increases of OPN expression by GE are augmented by stimulatory effects of placental lactogen. Current studies focus on defining the role(s) of OPN secreted from the uterine GE during later stages of pregnancy.

Conclusions

Research conducted with pigs and sheep has significantly advanced understanding of the role(s) of OPN during implantation through exploitation of 1) the prolonged peri-implantation period of pregnancy when elongating conceptuses are free within the uterine lumen requiring extensive paracrine signaling between conceptus and endometrium, and 2) the protracted and incremental attachment cascade of trophectoderm to uterine LE during implantation. Although OPN is synthesized in different cell types (LE in pigs, GE in sheep) and is regulated by different hormones (conceptus estrogens in pigs, progesterone in sheep), nonetheless OPN protein localizes to the interface between the uterus and trophectoderm where it is well placed to serve as a bifunctional bridging ligand between integrins, expressed by uterine LE and conceptus trophectoderm, to mediate attachment for implantation. It is noteworthy that OPN has been reported to be a prominent component of the uterine-placental environment of other species including primates and rodents, and therefore knowledge gained about the physiology of OPN in sheep and pigs may have significant relevance to human pregnancy. Our understanding of events that underlie successful implantation and placentation across species has been and will likely continue to be advanced by studies of pigs and sheep as biomedical research models.

Competing interests
The authors declare that they have no competing interests.

Authors' contributions
All data illustrated in the Figures presented in this review were produced through experiments designed and implemented by the authors, GAJ, RCB and FWB, and members of their laboratories through a collaborative effort. All authors read and approved the final manuscript.

Acknowledgements
Research supported by USDA-NRICGP 98-35203-6337 to F.W.B. and R.C.B., NRSA-DHHS/NIH 1-F32-HDO 8501 O1A1 to G.A.J., and USDA-NRI 2006-35203-17199 to G.A.J.

Author details
[1]Department of Veterinary Integrative Biosciences, Texas A&M University, College Station, TX 77843-4458, USA. [2]Department of Animal Science, Texas A&M University, College Station, TX 77843, USA.

References
1. Roberts RM, Smith GW, Bazer FW, Cibelli J, Seidel GE Jr, Bauman DE, Reynolds LP, Ireland JJ: **Research priorities. Farm animal research in crisis.** *Science* 2009, **324**:468–469.
2. Hynes RO: **Integrins: a family of cell surface receptors.** *Cell* 1987, **48**:549–554.
3. Denhardt DT, Guo X: **Osteopontin: a protein with diverse functions.** *FASEB J* 1993, **7**:1475–1482.
4. Senger DR, Perruzzi CA, Papadopoulos-Sergiou A, Van de Water L: **Adhesive properties of osteopontin: regulation by a naturally occurring thrombin-cleavage in close proximity to the GRGDS cell-binding domain.** *Mol Biol Cell* 1994, **5**:565–574.
5. Butler WT, Ridall AL, McKee MD: **Osteopontin.** In *Principles of Bone Biology.* New York: Academic Press, Inc; 1996:167–181.
6. Sodek J, Ganss B, McKee MD: **Osteopontin.** *Crit Rev Oral Biol Med* 2000, **11**:279–303.
7. Giancotti FG, Ruoslahti E: **Integrin signaling.** *Science* 1999, **285**:1028–1032.
8. Burghardt RC, Johnson GA, Jaeger LA, Ka H, Garlow JE, Spencer TE, Bazer FW: **Integrins and extracellular matrix proteins at the maternal-fetal interface in domestic animals.** *Cells Tissues Organs* 2002, **172**:202–217.
9. Vogel V: **Mechanotransduction involving multimodular proteins: converting force into biochemical signals.** *Annu Rev Biophys Biomol Struct* 2006, **35**:459–488.
10. Geiger B, Spatz JP, Bershadsky AD: **Environmental sensing through focal adhesions.** *Nat Rev Mol Cell Biol* 2009, **10**:21–33. doi: 10.1038/nrm2593.
11. Kling D, Fingerle J, Harlan JM: **Inhibition of leukocyte extravasation with a monoclonal antibody to CD18 during formation of experimental intimal thickening in rabbit carotid arteries.** *Arterioscler Thromb* 1992, **12**:997–1007.
12. Aplin JD, Seif MW, Graham RA, Hey NA, Behzad F, Campbell S: **The endometrial cell surface and implantation. Expression of the polymorphic mucin MUC-1 and adhesion molecules during the endometrial cycle.** *Ann N Y Acad Sci* 1994, **30**:103–121.
13. Senger DR, Wirth DF, Hynes RO: **Transformed mammalian cells secrete specific proteins and phosphoproteins.** *Cell* 1979, **16**:885–893.
14. Franzen A, Heinegard D: **Isolation and characterization of two sialoproteins present only in bone calcified matrix.** *Biochem J* 1985, **232**:715–724.
15. Patarca R, Freeman GJ, Singh RP, Wei FY, Durfee T, Blattner F, Ragnier DC, Kozak CA, Mock BA, Morse HC III, Jerrells TR, Cantor H: **Structural and functional studies of the early T lymphocyte activation 1 (Eta-1) gene. Definition of a novel T cell-dependent response associated with genetic resistance to bacterial infection.** *J Exp Med* 1989, **170**:145–161.
16. Fisher LW, Torchia DA, Fohr B, Young MF, Fedarko NS: **Flexible structures of SIBLING proteins, bone sialoprotein and osteopontin.** *Biochem Biophys Res Commun* 2001, **280**:460–465.
17. Johnson GA, Burghardt RC, Bazer FW, Spencer TE: **OPN: roles in implantation and placentation.** *Biol Reprod* 2003, **69**:1458–1471.
18. Johnson GA, Spencer TE, Burghardt RC, Bazer FW: **Ovine osteopontin. I. Cloning and expression of mRNA in the uterus during the peri-implantation period.** *Biol Reprod* 1999, **61**:884–891.
19. Prince CW, Oosawa T, Butler WT, Tomana M, Bhown AS, Bhown M, Schrohenloher RE: **Isolation, characterization and biosynthesis of a phosphorylated glycoprotein from rat bone.** *J Biol Chem* 1987, **262**:2900–2907.
20. Nagata T, Todescan R, Goldberg HA, Zhang Q, Sodek J: **Sulphation of secreted phosphoprotein I (SPP-1, osteopontin) is associated with mineralized tissue formation.** *Biochem Biophys Res Commun* 1989, **165**:234–240.
21. Sorensen ES, Hojrup P, Petersen TE: **Post-translational modifications of bovine osteopontin: identification of twenty-eight phosphorylation and three o-glycosylation sites.** *Protein Sci* 1995, **4**:2040–2049.
22. Kaartinen MT, Pirhonen A, Linnala-Kankkunen A, Maenpaa PH: **Cross-linking of osteopontin by tissue transglutaminase increases its collagen binding properties.** *J Biol Chem* 1999, **274**:1729–1735.
23. Agnihotri R, Crawford HC, Haro H, Matrisian LM, Havrda MC, Liaw L: **Osteopontin, a novel substrate for matrix metalloproteinase-3 (stromelysin-1) and matrix metalloproteinase-7 (matrilysin).** *J Biol Chem* 2001, **276**:28261–28267.
24. Senger DR, Perruzzi CA, Papadopoulos A, Tenen DG: **Purification of a human milk protein closely similar to tumor-secreted phosphoproteins and osteopontin.** *Biochim Biophys Acta* 1989, **996**:43–48.
25. Young MF, Kerr JM, Termine JD, Wewer UM, Wang MG, McBride OW, Fisher LW: **cDNA cloning, mRNA distribution and heterogeneity, chromosomal location, and RFLP analysis of human osteopontin (OPN).** *Genomics* 1990, **7**:491–502.
26. Brown LF, Berse B, Van de Water L, Papadopoulos-Sergiou A, Perruzzi CA, Manseau EJ, Dvorak HF, Senger DR: **Expression and distribution of osteopontin in human tissues: widespread association with luminal epithelial surfaces.** *Mol Biol Cell* 1992, **3**:1169–1180.
27. Kohri K, Nomura S, Kitamura Y, Nagata T, Yoshioka K, Iguchi M, Yamate T, Umekawa T, Suzuki Y, Sinohara H, Kurita T: **Structure and expression of the mRNA encoding urinary stone protein (osteopontin).** *J Biol Chem* 1993, **268**:15180–15184.
28. Daiter E, Omigbodun A, Wang S, Walinsky D, Strauss JF III, Hoyer JR, Coutifaris C: **Cell differentiation and endogenous cyclic adenosine 3′,5′-monophosphate regulate osteopontin expression in human trophoblasts.** *Endocrinology* 1996, **137**:1785–1790.
29. Johnson GA, Burghardt RC, Spencer TE, Newton GR, Ott TL, Bazer FW: **Ovine osteopontin II. Osteopontin and alpha(v)beta(3) integrin expression in the uterus and conceptus during the peri-implantation period.** *Biol Reprod* 1999, **61**:892–899.
30. Luedtke CC, McKee MD, Cyr DG, Gregory M, Kaartinen MT, Mui J, Hermo L: **Osteopontin expression and regulation in the testis, efferent ducts, and epididymis of rats during postnatal development through to adulthood.** *Biol Reprod* 2002, **66**:1437–1448.
31. Banerjee A, Burghardt R, Johnson G, White F, Ramaiah S: **The temporal expression of osteopontin (SPP-1) in the rodent model of alcoholic steatohepatitis: a potential biomarker.** *Toxicol Pathol* 2006, **34**:373–384.
32. Kato A, Okura T, Hamada C, Miyoshi S, Katayama H, Higaki J, Ito R: **Cell stress induces upregulation of osteopontin via the ERK pathway in type II alveolar epithelial cells.** *PLoS One* 2014, **9**(6):e100106. doi:10.1371/journal.pone.0100106. eCollection 2014.
33. Ashkar S, Weber GF, Panoutsakopoulou V, Sanchirico ME, Jansson M, Zawaideh S, Rittling SR, Denhardt DT, Glimcher MJ, Cantor H: **Eta-1 (osteopontin): an early component of type-1 (cell-mediated) immunity.** *Science* 2000, **287**:860–864.
34. Malyankar UM, Scatena M, Suchland KL, Yun TJ, Clark EA, Giachelli CM: **Osteoprotegerin is an alpha v beta 3-induced, NF-kappa B dependent survival factor for endothelial cells.** *J Biol Chem* 2000, **275**:20959–20962.
35. Hotte SJ, Winquist EW, Stitt L, Wilson SM, Chambers AF: **Plasma osteopontin: associations with survival and metastasis to bone in men with hormone-refractory prostate carcinoma.** *Cancer* 2002, **95**:506–512.
36. Flores M, Norgard M, Heinegard D, Reinholt FP, Andersson G: **RGD-directed attachment of isolated rat osteoclasts to osteopontin, bone sialoprotein, and fibronectin.** *Exp Cell Res* 1992, **201**:526–530.
37. Butler WT: **The nature and significance of osteopontin.** *Connect Tissue Res* 1989, **23**:123–136.
38. Nabel G, Fresno M, Chessman A, Cantor H: **Use of cloned populations of mouse lymphocytes to analyze cellular differentiation.** *Cell* 1981, **23**:19–28.
39. Hwang S, Lopez CA, Heck DE, Gardner CR, Laskin DL, Laskin JD, Denhardt DT: **Osteopontin inhibits induction of nitric oxide synthase gene expression by inflammatory mediators in mouse kidney epithelial cells.** *J Biol Chem* 1994, **269**:711–715.

40. Leibson HJ, Marrack P, Kappler JW: **B cell helper factors. I. Requirement for both interleukin 2 and another 40,000 mol wt factor.** *J Exp Med* 1981, 154:1681–1693.

41. Hruska KA, Rolnick F, Huskey M, Alvarez U, Cheresh D: **Engagement of the osteoclast integrin αvβ3 by osteopontin stimulates phosphatidylinositol 3-hydroxy kinase activity.** *Endocrinology* 1995, 136:2984–2992.

42. McKee MD, Nanci A, Khan SR: **Ultrastructural immunodetection of osteopontin and osteocalcin as major matrix components of renal calculi.** *J Bone Miner Res* 1995, 10:1913–1929.

43. Nomura S, Wills AJ, Edwards DR, Heath JK, Hogan BLM: **Developmental expression of 2ar (osteopontin) and SPARC (osteonectin) RNA as revealed by in situ hybridization.** *J Cell Biol* 1988, 106:441–450.

44. White FJ, Burghardt RC, Croy BA, Johnson GA: **Osteopontin is expressed by endometrial macrophages and decidual natural killer cells during mouse pregnancy [abstract].** *Biol Reprod* 2005, 73(Suppl 1):155.

45. Herington JL, Bany BM: **The conceptus increases secreted phosphoprotein 1 gene expression in the mouse uterus during the progression of decidualization mainly due to its effects on uterine natural killer cells.** *Reproduction* 2007, 133:1213–1221.

46. Lessey BA, Castelbaum AJ, Buck CA, Lei Y, Yowell CW, Sun J: **Further characterization of endometrial integrins during the menstrual cycle and in pregnancy.** *Fertil Steril* 1994, 62:497–506.

47. Lessey BA, Ilesanmi AO, Lessey MA, Riben M, Harris JE, Chwalisz K: **Luminal and glandular endometrial epithelium express integrins differentially throughout the menstrual cycle: implications for implantation, contraception and infertility.** *Am J Reprod Immunol* 1996, 35:195–204.

48. Lessey BA, Yeh I, Castelbaum AJ, Fritz MA, Ilesanmi AO, Korzeniowski P, Sun J, Chwalisz K: **Endometrial progesterone receptors and markers of uterine receptivity in the window of implantation.** *Fertil Steril* 1996, 65:477–483.

49. Lessey BA, Damjanovich L, Coutifaris C, Castelbaum A, Albelda SM, Buck CA: **Integrin adhesion molecules in the human endometrium. Correlation with the normal and abnormal menstrual cycle.** *J Clin Invest* 1992, 90:188–195.

50. Lessey BA, Castelbaum AJ, Sawin SW, Sun J: **Integrins as markers of uterine receptivity in women with primary unexplained infertility.** *Fertil Steril* 1995, 63:535–542.

51. Johnson GA, Spencer TE, Burghardt RC, Taylor KM, Gray CA, Bazer FW: **Progesterone modulation of osteopontin gene expression in the ovine uterus.** *Biol Reprod* 2000, 62:1315–1321.

52. Johnson GA, Bazer FW, Jaeger LA, Ka H, Garlow JE, Pfarrer C, Spencer TE, Burghardt RC: **Muc-1, integrin and osteopontin expression during the implantation cascade in sheep.** *Biol Reprod* 2001, 65:820–828.

53. Gray CA, Burghardt RC, Johnson GA, Bazer FW, Spencer TE: **Evidence that absence of endometrial gland secretions in uterine gland knockout ewes compromises conceptus survival and elongation.** *Reproduction* 2002, 124:289–300.

54. Illera MJ, Cullinan E, Gui Y, Yuan L, Beyler SA, Lessey BA: **Blockade of the αvβ3 integrin adversely affects implantation in the mouse.** *Biol Reprod* 2000, 62:1285–1290.

55. Illera MJ, Lorenzo PL, Gui YT, Beyler SA, Apparao KBC, Lessey BA: **A role for αvβ3 integrin during implantation in the rabbit model.** *Biol Reprod* 2003, 68:766–771.

56. Apparao KB, Murray MJ, Fritz MA, Meyer WR, Chambers AF, Truong PR, Lessey BA: **Osteopontin and its receptor alphavbeta(3) integrin are coexpressed in the human endometrium during the menstrual cycle but regulated differently.** *J Clin Endocrinol Metab* 2001, 86:4991–5000.

57. Apparao KBC, Illera MJ, Beyler SA, Olson GE, Osteen KG, Corjay MH, Boggess K, Lessey BA: **Regulated expression of osteopontin in the peri-implantation rabbit uterus.** *Biol Reprod* 2003, 68:1484–1490.

58. White FJ, Ross JW, Joyce MM, Geisert RD, Burghardt RC, Johnson GA: **Steroid regulation of cell specific secreted phosphoprotein 1 (osteopontin) expression in the pregnant porcine uterus.** *Biol Reprod* 2005, 73:1294–1301.

59. Garlow JE, Ka H, Johnson GA, Burghardt RC, Jaeger LA, Bazer FW: **Analysis of osteopontin at the maternal-placental interface in pigs.** *Biol Reprod* 2002, 66:718–725.

60. Carson DD, Lagow E, Thathiah A, Al-Shami R, Farach-Carson MC, Vernon M, Yuan L, Fritz MA, Lessey B: **Changes in gene expression during the early to mid-luteal (receptive phase) transition in human endometrium detected by high-density microarray screening.** *Mol Hum Reprod* 2002, 8:871–879.

61. Kao LC, Tulac S, Lobo S, Imani B, Yang JP, Germeyer A, Osteen K, Taylor RN, Lessey BA, Giudice LC: **Global gene profiling in human endometrium during the window of implantation.** *Endocrinology* 2002, 143:2119–2138.

62. Mirkin S, Arslan M, Churikov D, Corica A, Diaz JI, Williams S, Bocca S, Oehninger S: **In search of candidate genes critically expressed in the human endometrium during the window of implantation.** *Hum Reprod* 2005, 20:2104–2117.

63. Burghardt RC, Burghardt JR, Taylor JD II, Reeder AT, Nguyen BT, Spencer TE, Johnson GA: **Enhanced focal adhesion assembly reflects increased mechanosensation and mechanotransduction along the maternal/conceptus interface during pregnancy in sheep.** *Reproduction* 2009, 137:583–593.

64. Erikson DW, Burghardt RC, Bayless KJ, Johnson GA: **Secreted phosphoprotein 1 (SPP1, osteopontin) binds to integrin alphavbeta6 on porcine trophectoderm cells and integrin alphavbeta3 on uterine luminal epithelial cells, and promotes trophectoderm cell adhesion and migration.** *Biol Reprod* 2009, 81:814–825.

65. Massuto DA, Kneese EC, Johnson GA, Hooper NH, Burghardt RC, Ing NH, Jaeger LA: **Transforming growth factor beta (TGFB) signaling is activated during porcine implantation: Proposed role for latency associated peptide-integrins at the conceptus-maternal interface.** *Reproduction* 2009, 139:465–478.

66. Kim J, Erikson DW, Burghardt RC, Spencer TE, Wu G, Bayless KJ, Johnson GA, Bazer FW: **Secreted phosphoprotein 1 binds integrins to initiate multiple cell signaling pathways, including FRAP1/mTOR, to support attachment and force-generated migration of trophectoderm cells.** *Matrix Biol* 2010, 29:369–382.

67. Kang YJ, Forbes K, Carver J, Aplin JD: **The role of the osteopontin-integrin αvβ3 interaction at implantation: functional analysis using three different in vitro models.** *Hum Reprod* 2014, 29:739–749.

68. Frank M, Bazer FW, Thatcher WW, Wilcox CJ: **A study of prostaglandin F2alpha as the luteolysin in swine: III. Effects of estradiol valerate on prostaglandin F, progestins, estrone and estradiol concentrations in the utero-ovarian vein of nonpregnant gilts.** *Prostaglandins* 1977, 14:1183–1196.

69. Bazer FW, Thatcher WW: **Theory of maternal recognition of pregnancy in swine based on estrogen controlled endocrine versus exocrine secretion of prostaglandin F2α by the uterine endometrium.** *Prostaglandins* 1977, 14:397–400.

70. Godkin JD, Bazer FW, Thatcher WW, Roberts RM: **Proteins released by cultured day 15–16 conceptuses prolong luteal maintenance when introduced into the uterine lumen of cyclic ewes.** *J Reprod Fertil* 1984, 71:57–64.

71. Vallet JL, Bazer FW, Fliss MFV, Thatcher WW: **Effect of ovine conceptus secretory proteins and purified ovine trophoblast protein-1 on interoestrous interval and plasma concentrations of prostaglandins F2α and E2 and of 13,14 dihydro-15-keto prostaglandin F2α in cyclic ewes.** *J Reprod Fertil* 1988, 84:493–504.

72. Ashworth CJ, Bazer FW: **Changes in ovine conceptus and endometrial function following asynchronous embryo transfer or administration of progesterone.** *Biol Reprod* 1989, 40:425–433.

73. Gray CA, Taylor KM, Ramsey WS, Hill JR, Bazer FW, Bartol FF, Spencer TE: **Endometrial glands are required for preimplantation conceptus elongation and survival.** *Biol Reprod* 2001, 64:1608–1613.

74. Burton GJ, Watson AL, Hempstock J, Skepper JN, Jauniaux E: **Uterine glands provide histotrophic nutrition for the human fetus during the first trimester of pregnancy.** *J Clin Endocrinol Metab* 2002, 87:2954–2959.

75. Glasser SR, Mulholland J: **Receptivity is a polarity dependent special function of hormonally regulated uterine epithelial cells.** *Microsc Res Tech* 1993, 25:106–120.

76. Denker HW: **Implantation: a cell biological paradox.** *J Exp Zool* 1993, 266:541–558.

77. Bazer FW, First NL: **Pregnancy and parturition.** *J Anim Sci* 1983, 57(Suppl 2):425–460.

78. Johnson GA, Bazer FW, Burghardt RC, Spencer TE, Wu G, Bayless KJ: **Conceptus-uterus interactions in pigs: endometrial gene expression in response to estrogens and interferons from conceptuses.** *Soc Reprod Fertil* 2009, 66(Suppl):321–332.

79. Flint APF, Saunders PTK, Ziecik AJ: **Blastocyst-endometrium interactions and their significanc in embryonic mortality.** In *Control of Pig Reproduction*. Edited by Cole DJA, Foxcroft GR. London: Butterworth Scientific; 1982:253–275.

80. Jainudeen MR, Hafez ESE: **Reproductive failure in females.** In *Reproduction in Farm Animals*. Edited by Hafez ESE, Lea F, Hafez ESE, Lea, Febiger. Philadelphia: Wiley-Blackwell; 1987:399–422.

81. Bazer FW, Johnson GA, Spencer TE: **Growth and Development: (1) Mammalian conceptus peri-implantation period; and (2) Mammalian pre-implantation embryo.** In *Encyclopedia of Animal Science*. Edited by Bell WGPAW. New York: Marcel Dekker; 2005:555–558.

82. Cross JC, Werb Z, Fisher SJ: **Implantation and the placenta: key pieces of the development puzzle.** *Science* 1994, 266:1508–518.

83. Carson DD, Bagchi I, Dey SK, Enders AC, Fazleabas AT, Lessey BA, Yoshinaga K: **Embryo implantation.** *Dev Biol* 2000, 223:217–237.

84. Bowen JA, Bazer FW, Burghardt RC: **Spatial and temporal analyses of integrin and Muc-1 expression in porcine uterine epithelium and trophectoderm in vivo.** *Biol Reprod* 1996, 55:1098–1106.

85. Aplin JD, Meseguer M, Simon C, Ortiz ME, Croxatto H, Jones CJ: **MUC1, glycans and the cell-surface barrier to embryo implantation.** *Biochem Soc Trans* 2001, 29:153–156.

86. Kimber SJ, Illingworth IM, Glasser SR: **Expression of carbohydrate antigens in the rat uterus during early pregnancy and after ovariectomy and steroid replacement.** *J Reprod Fertil* 1995, 103:75–87.

87. Kimber SJ, Spanswick C: **Blastocyst implantation: the adhesion cascade.** *Semin Cell Dev Biol* 2000, 11:77–92.

88. Spencer TE, Johnson GA, Bazer FW, Burghardt RC: **Implantation mechanisms: insights from the sheep.** *Reproduction* 2004, 128:656–668.

89. Ruoslahti E, Pierschbacher MD: **New perspectives in cell adhesion: RGD and integrins.** *Science* 1987, 238:491–497.

90. Burghardt RC, Bowen JA, Newton GR, Bazer FW: **Extracellular matrix and the implantation cascade in pigs.** *J Reprod Fertil* 1997, 52(Suppl):151–164.

91. Lessey BA: **Adhesion molecules and implantation** *J Reprod Immunol* 2002, 55:101–112.

92. Joyce MM, Gonzalez JF, Lewis S, Woldesenbet S, Burghardt RC, Newton GR, Johnson GA: **Caprine uterine and placental osteopontin expression is distinct among epitheliochorial implanting species.** *Placenta* 2005, 26:160–170.

93. White FJ, Burghardt RC, Hu J, Joyce MM, Spencer TE, Johnson GA: **Secreted phosphoprotein 1 (osteopontin) is expressed by stromal macrophages in cyclic and pregnant endometrium of mice, but is induced by estrogen in luminal epithelium during conceptus attachment for implantation.** *Reproduction* 2006, 132:919–929.

94. Craig AM, Smith JH, Denhardt DT: **OPN, a transformation-associated cell adhesion phosphoprotein, is induced by 12-O-tetradecanoylphorbol 13-acetate in mouse epidermis.** *J Biol Chem* 1989, 264:9682–9689.

95. Craig AM, Denhardt DT: **The murine gene encoding secreted phosphoprotein 1 (OPN): promoter structure, activity, and induction in vivo by estrogen and progesterone.** *Gene* 1991, 100:163–171.

96. Singh K, Balligand JL, Fischer TA, Smith TW, Kelly RA: **Glucocorticoids increase OPN expression in cardiac myocytes and microvascular endothelial cells. Role in regulation of inducible nitric oxide synthase.** *J Biol Chem* 1995, 270:28471–28478.

97. Omigbodun A, Ziolkiewicz P, Tessler C, Hoyer JR, Coutifaris C: **Progesterone regulates OPN expression in human trophoblasts: a model of paracrine control in the placenta?** *Endocrinology* 1997, 138:4308–4315.

98. Safran JB, Butler WT, Farach-Carson MC: **Modulation of OPN post-translational state by 1, 25-(OH)2-vitamin D3. Dependence on Ca2+ in-flux.** *J Biol Chem* 1998, 273:29935–29941.

99. Denhardt DT, Noda M, O'Regan AW, Pavlin D, Berman JS: **OPN as a means to cope with environmental insults: regulation of inflammation, tissue remodeling, and cell survival.** *J Clin Invest* 2001, 107:1055–1061.

100. Giachelli CM, Steitz S: **OPN: a versatile regulator of inflammation and biomineralization.** *Matrix Biol* 2000, 19:615–622.

101. Hu DD, Hoyer JR, Smith JW: **Ca2+ suppresses cell adhesion to OPN by attenuating binding affinity for integrin alpha v beta 3.** *J Biol Chem* 1995, 270:9917–9925.

102. Xuan JW, Hota C, Shigeyama Y, D'Errico JA, Somerman MJ, Chambers AF: **Site-directed mutagenesis of the arginine-glycine-aspartic acid sequence in OPN destroys cell adhesion and migration functions.** *J Cell Biochem* 1995, 57:680–690.

103. Hu DD, Lin EC, Kovach NL, Hoyer JR, Smith JW: **A biochemical characterization of the binding of OPN to integrins alpha v beta 1 and alpha v beta 5.** *J Biol Chem* 1995, 270:26232–26238.

104. Denda S, Reichardt LF, Muller U: **Identification of OPN as a novel ligand for the integrin alpha8 beta1 and potential roles for this integrin-ligand interaction in kidney morphogenesis.** *Mol Biol Cell* 1998, 9:1425–1435.

105. Bayless KJ, Meininger GA, Scholtz JM, Davis GE: **OPN is a ligand for the alpha4beta1 integrin.** *J Cell Sci* 1998, 111(Pt 9):1165–1174.

106. Smith LL, Giachelli CM: **Structural requirements for alpha 9 beta 1-mediated adhesion and migration to thrombin-cleaved OPN.** *Exp Cell Res* 1998, 242:351–360.

107. Bayless KJ, Davis GE: **Identification of dual alpha 4beta1 integrin binding sites within a 38 amino acid domain in the N-terminal thrombin fragment of human OPN.** *J Biol Chem* 2001, 276:13483–13489.

108. Senger DR, Perruzzi CA: **Cell migration promoted by a potent GRGDS-containing thrombin-cleavage fragment of OPN.** *Biochim Biophys Acta* 1996, 1314:13–24.

109. Weber GF, Ashkar S, Glimcher MJ, Cantor H: **Receptor-ligand interaction between CD44 and OPN (Eta-1).** *Science* 1996, 271:509–512.

110. Goldsmith HL, Labrosse JM, McIntosh FA, Maenpaa PH, Kaartinen MT, McKee MD: **Homotypic interactions of soluble and immobilized OPN.** *Ann Biomed Eng* 2002, 30:840–850.

111. Bazer FW, Johnson GA: **Pig blastocyst-uterine interactions.** *Differentiation* 2014, 87:52–65.

112. Dantzer V: **Scanning electron microscopy of exposed surfaces of the porcine placenta.** *Acta Anat* 1984, 118:96–106.

113. King AH, Jiang Z, Gibson JP, Haley CS, Archibald AL: **Mapping quantitative trait loci affecting female reproductive traits on porcine chromosome 8.** *Biol Reprod* 2003, 68:21722179.

114. Renegar RH, Bazer FW, Roberts RM: **Placental transport and distribution of uteroferrin in the fetal pig.** *Biol Reprod* 1982, 27:1247–1260.

115. Johnson GA, Burghardt RC, Joyce MM, Spencer TE, Bazer FW, Gray CA, Pfarrer C: **Osteopontin is synthesized by uterine glands and a 45-kDa cleavage fragment is localized at the uterine-placental interface through-out ovine pregnancy.** *Biol Reprod* 2003, 69:92–98.

116. Girotti M, Zingg HH: **Gene expression profiling of rat uterus at different stages of parturition.** *Endocrinology* 2003, 144:2254–2265.

117. Roberts RM, Bazer FW: **The functions of uterine secretions.** *J Reprod Fertil* 1988, 82:875–892.

118. Irwin JC, Giudice LC: **Decidua.** In *Encyclopedia of Reproduction*. Edited by Knobil E, Neill JD. New York: Academic Press; 1999:823–835.

119. Ka H, Al-Ramadan S, Erikson DW, Johnson GA, Burghardt RC, Spencer TE, Jaeger LA, Bazer FW: **Regulation of fibroblast growth factor 7 expression in the pig uterine endometrium by progesterone and estradiol.** *Biol Reprod* 2007, 77:172–180.

120. Schlosnagle DC, Bazer FW, Tsibris JCM, Roberts RM: **An iron-containin phosphatase induced by progesterone in the uterine fluids of pigs.** *J Biol Chem* 1974, 249:7574–7579.

121. Ka H, Jaeger LA, Johnson GA, Spencer TE, Bazer FW: **Keratinocyte growth factor expression is up-regulated by estrogen in porcine uterine endometrium and it functions in trophectodermal cell proliferation and differentiation.** *Endocrinology* 2001, 142:2303–2310.

122. Bailey DW, Dunlap KA, Frank JW, Erikson DW, White BG, Bazer FW, Burghardt RC, Johnson GA: **Effects of long-term progesterone on developmental and functional aspects of porcine uterine epithelia: progesterone alone does not support glandular development of pregnancy.** *Reproduction* 2010, 140:583–594.

123. Bailey DW, Dunlap KL, Erikson DW, Patel A, Bazer FW, Burghardt RC, Johnson GA: **Effects of long-term progesterone exposure on porcine uterine gene expression: progesterone alone does not induce secreted phosphoprotein 1 (osteopontin) in glandular epithelium.** *Reproduction* 2010, 140:595–604.

124. Chilton BS, Mani SK, Bullock DW: **Servomechanism of prolactin and progesterone in regulating uterine gene expression.** *Mol Endocrinol* 1988, 2:1169–1175.

125. Young KH, Kraeling RR, Bazer FW: **Effect of pregnancy and exogenous ovarian steroids on endometrial prolactin receptor ontogeny and uterine secretory response in pigs.** *Biol Reprod* 1990, 43:592–599.

126. Spencer TE, Gray CA, Johnson GA, Taylor KM, Gertler A, Gootwine E, Ott TL, Bazer FW: **Effects of recombinant ovine interferon tau, placental lactogen, and growth hormone on the ovine uterus.** *Biol Reprod* 1999, 61:1409–1418.

127. Hernandez SC, Hogg CO, Billon Y, Sanchez MP, Bidanel JP, Haley CS, Archibald AL, Ashworth CJ: **Secreted phosphoprotein 1 expression in endometrium and placental tissues of hyperprolific large white and meishan gilts.** *Biol Reprod* 2013, 88:120–126.

128. Caton D, Pendergast JF, Bazer FW: **Uterine blood flow: periodic fluctuations of its rate during pregnancy.** *Am J Physiol* 1983, 245:R850–R852.

129. Johnson GA, Burghardt RC, Joyce MM, Spencer TE, Bazer FW, Pfarrer C, Gray CA: **Osteopontin expression in uterine stroma indicates a decidualization-like differentiation during ovine pregnancy.** *Biol Reprod* 2003, **68**:1951–1958.
130. Waterhouse P, Parhar RS, Guo X, Lala PK, Denhardt DT: **Regulated temporal and spatial expression of the calcium-binding proteins calcyclin and OPN (osteopontin) in mouse tissues during pregnancy.** *Mol Reprod Dev* 1992, **32**:315–323.
131. Fazleabas AT, Bell SC, Fleming S, Sun J, Lessey BA: **Distribution of integrins and the extracellular matrix proteins in the baboon endometrium during the menstrual cycle and early pregnancy.** *Biol Reprod* 1997, **56**:348–356.
132. Mossman HW: **Comparative morphogenesis of the foetal membranes and accessory uterine structures.** *Contrib Embryol* 1937, **26**:129–246.
133. Kellas LM: **The placenta and foetal membranes of the antelope Ourebia ourebi (Zimmermann).** *Acta Anat* 1966, **64**:390–445.
134. Bjorkman N: **Fine structure of the fetal-maternal area of exchange in the epitheliochorial and endotheliochorial types of placentation.** *Acat Anat* 1973, **61**(suppl):11–22.
135. Boshier DP: **A histological and histochemical examination of implantation and early placentome formation in sheep.** *J Reprod Fertil* 1969, **19**:51–61.

15

Transplacental induction of fatty acid oxidation in term fetal pigs by the peroxisome proliferator-activated receptor alpha agonist clofibrate

Xi Lin[*], Sheila Jacobi and Jack Odle

Abstract

Background: To induce peroxisomal proliferator-activated receptor α (PPARα) expression and increase milk fat utilization in pigs at birth, the effect of maternal feeding of the PPARα agonist, clofibrate (2-(4-chlorophenoxy)-2-methyl-propanoic acid, ethyl ester), on fatty acid oxidation was examined at full-term delivery (0 h) and 24 h after delivery in this study. Each group of pigs (n = 10) was delivered from pregnant sows fed a commercial diet with or without 0.8% clofibrate for the last 7 d of gestation. Blood samples were collected from the utero-ovarian artery of the sows and the umbilical cords of the pigs as they were removed from the sows by C-section on day 113 of gestation.

Results: HPLC analysis identified that clofibric acid was present in the plasma of the clofibrate-fed sow (~4.2 µg/mL) and its offspring (~1.5 µg/mL). Furthermore, the maternal-fed clofibrate had no impact on the liver weight of the pigs at 0 h and 24 h, but hepatic fatty acid oxidation examined in fresh homogenates showed that clofibrate increased ($P < 0.01$) ^{14}C-accumulation in CO_2 and acid soluble products 2.9-fold from [1-^{14}C]-oleic acid and 1.6-fold from [1-^{14}C]-lignoceric acid respectively. Correspondingly, clofibrate increased fetal hepatic carnitine palmitoyltransferase (CPT) and acyl-CoA oxidase (ACO) activities by 36% and 42% over controls ($P < 0.036$). The mRNA abundance of CPT I was 20-fold higher in pigs exposed to clofibrate ($P < 0.0001$) but no differences were detected for ACO and PPARα mRNA between the two groups.

Conclusion: These data demonstrate that dietary clofibrate is absorbed by the sow, crosses the placental membrane, and enters fetal circulation to induce hepatic fatty acid oxidation by increasing the CPT and ACO activities of the newborn.

Keywords: Clofibrate, Fatty acid oxidation, Pigs, Placenta transfer

Background

High postnatal mortality has been recognized as a critical problem by the modern swine industry since the 1990's. Although genetic improvement has increased the number of pigs born alive per litter, the death rate of newborn pigs during the postnatal period has increased slightly in recent years, from 13.2% in 2006 [1] to 14.8% in 2010 [2]. The epidemiology of postnatal mortality is complex, but data clearly show that inadequate energy (starvation) ranks among the leading causes, especially in the first three days after birth. The immediate postnatal period poses the greatest challenge to the energy balance of neonates, which must rapidly switch from carbohydrates supplied by the mother in utero [3] to predominantly lipids supplied via milk. Sow milk fat concentration is about 6% at birth and increases to 10% within 24 h. This is 67–70% higher than human colostrum (3.5–6%) and constitutes up to 60% of the energy in the milk, indicating that the milk fat is the primary energy source for newborn pigs after birth. Thus, it is critical for newborn pigs to use milk fat efficiently at birth to survive. However, results from previous studies show that over 85% of the fatty acids taken up by newborn pig hepatocytes are re-esterified

* Correspondence: lin_xi@ncsu.edu
Laboratory of Developmental Nutrition, Department of Animal Sciences, North Carolina State University, Box 7621, Raleigh, NC 27695, USA

but not oxidized [4]. In addition, 80% of the acetyl-CoA produced from fatty acid β-oxidation in mitochondria and/or peroxisomes is converted to acetate rather than ketone bodies [5]. These results indicate that newborn pigs have a limited capacity to oxidize milk fat for energy production at birth.

Hepatic fatty acid β-oxidation pathways are controlled mainly by the key enzymes carnitine palmitoyltransferase I (CPT I) in mitochondria and acyl-CoA oxidase (ACO) in peroxisomes. The activities of these key enzymes are transcriptionally regulated by the peroxisomal proliferator-activated receptor α (PPARα; [6]). This central transcription factor, PPARα, can be activated by its natural ligands such as eicosanoids and long-chain fatty acids or a host of pharmaceutical agonists including fibrates, a class of amphipathic carboxylic acids. Clofibrate, a potent pharmaceutical PPARα agonist of the fibrate class, stimulates hepatic mitochondrial and peroxisomal fatty acid β-oxidation by inducing PPARα target genes such as CPT I and ACO mRNA expressions and protein activities. The induction of CPT I and ACO by clofibrate via activation of PPARα has been documented in rodent species [7], laying hens [8], dairy cows [9] and pigs [10,11]. Data from our previous studies showed that the total CPT and ACO activity increase 2- and 3-fold respectively in the livers of pigs fed clofibrate for 2 weeks. Feeding clofibrate also increases hepatic peroxisomal and mitochondrial β-oxidation of [1-^{14}C]-palmitate by 60% and 186%, respectively, in addition to the increased CPT I and ACO specific activities [10]. Moreover, the increase in hepatic fatty acid oxidation is not accompanied by a significant hyperplasia or hepatomegaly [12] as observed in rodent species. These results indicate that the fatty acid oxidative capacity of pigs could be promoted via induction of PPARα without substantial peroxisome proliferation.

Recently we have demonstrated that supplementation of clofibrate could improve *in vivo* fatty acid oxidation in neonatal pigs [13]. However, increasing the fatty acid oxidative capacity of pigs at birth appears to be the key to improving energy utilization and increasing survivability. The question is whether the activities of CPT I and ACO can be increased by activation of PPARα prenatally. Whether PPARα agonist can be delivered to the newborn pigs via maternal diet is not known. Research concerning the impact of maternal feeding of clofibrate during pregnancy on fetal fatty acid oxidative capacity at birth or/and development after birth is very limited. Maternal dietary clofibrate induced peroxisomal proliferation in the liver and intestine tissues, and induced enterocyte peroxisomal catalase and peroxisomal bifunctional enzyme activities of fetuses in rats [14]. Maternal clofibrate also amplified fetal liver endoplasmic reticulum and peroxisomes, and increased the concentrations of peroxisomal membrane protein 70, the specific activity of

dihydroxyacetone phosphate acyltransferase and catalase in the livers of fetal mice [15]. The relative expression of cytochrome P4504A mRNA was increased in the maternal liver and fetal rat tissues [16]. Furthermore, the increase in relative mRNA of ACO, CPT I, medium- and long-chain acyl-CoA dehydrogenases in the liver were also observed in fetal rats from the clofibrate-treated maternal rats [17]. These findings demonstrate that clofibrate is capable of crossing the placenta and increasing peroxisome proliferation and modulating specific gene expression. Therefore, we hypothesize that similar placenta transfer may occur in pigs, although the physiology of extra embryonic membrane attachment in swine are different from rodents and other species. The placental transfer of clofibrate has not been studied directly by measuring clofibrate in the circulation system of the pregnant animals and their fetuses, and the effect of the induction in enzyme activities and mRNA expression on fatty acid oxidative capacity is not known.

To test our hypothesis, we investigated the effect of maternal supplementation of PPARα agonist clofibrate on the development of oxidative capacity of newborn pigs at birth and 24 h after birth. We confirmed that PPARα agonist clofibrate transfers across porcine placental tissues by measuring maternal and fetal plasma clofibrate concentrations and increases hepatic oxidative capacity by measuring hepatic fatty acid oxidation in the newborns using [1-^{14}C]-fatty acid substrates.

Methods

Animals, treatments and sampling

All procedures were approved by the Institutional Animal Care and Use Committee of North Carolina State University (IACUC number: 07-001-A). Twenty newborn pigs from either control (n = 10, body weight = 1.37 ± 0.047 kg) or clofibrate fed (n = 10, body weight = 1.20 ± 0.024 kg) sows were used in this study. Pregnant multiparous crossbred sows were housed at the North Carolina State University Swine Education Unit and were fed a standard gestation diet (3,265 Kcal ME/kg) with or without supplementation of 0.8% colfibrate (w/w) from day 105 of gestation until day 113. The clofibrate was diluted into 10 mL of ethanol and pre-mixed with ~50 g of feed each day. Sows were fed the premixes (clofibrate or vehicle) together with ~1/3 of their daily feed allotment first, to ensure complete consumption, and then the remaining 2/3 allotment was provided. The sows were given 1.75 kg diets per day in total and water was supplied ad libitum. Term fetuses were delivered by caesarian section on d 113 of gestation and plasma was collected simultaneously from the utero-ovarian artery of swine and from the umbilical vein of each fetal pig.

All newborn pigs were weighed and euthanized by AVMA-approved electrocution at time 0 (term fetus, n = 6) or 24 h (n = 4) after delivery. The pigs sampled at 24 h were housed at 35°C in a specialized nursery facility [18] and remained un-fed. Liver samples were immediately homogenized in a buffer (220 mmol/L mannitol, 70 mmol/L sucrose, 2 mmol/L HEPES, and 0.1 mmol/L EDTA; pH 7.2 at 0°C) using a 7 mL glass Pyrex hand homogenizer with 3 complete top to bottom strokes. Fatty acid oxidation in the fresh homogenate was measured immediately using [1-^{14}C]-fatty acids. Homogenate protein was determined using the biuret method [19]. Samples were also immediately frozen in liquid nitrogen and stored at –80°C for enzyme and mRNA assays.

Plasma assay

Clofibrate (2-(4-Chlorophenoxy)-2-methylpropionic acid ethyl ester), clofibric acid (2-(4-chlorophenoxy)-2-methylpropionic acid), 4-chlorophenylacetic acid and its metabolites in the plasma from the sows and term fetuses were extracted using solid-phase extraction (SPE) procedures as described by Du et al. [20]. The extraction was conducted using SPEC.C18 extraction cartridges (Ansys Technologies, CA, and USA). The clofibrate, clofibric acid, and other lipophilic components were eluted with a solvent mixture of acetonitrile : water : formic acid (79% : 20% : 1%, v/v/v) and were analyzed using a Waters HPLC Empower system (Milford, MA. USA). The separation of clofibrate, clofibric acid, and their metabolites was performed on a BDS Hypersil C18 (5 μm, 150 mm × 46 mm) analytical column with BDS C18 (5 μm, 4 mm × 3.0 mm) guard column from Phenomenex (Torrance, CA. USA). The isocratic HPLC system was used and the pump flow rate was set at 1 mL/min. The sample injection volume was 20 μL and the compounds were detected at 230 nm using a photodiode array detector (Waters 996).

Fatty acid oxidation

Hepatic fatty acid oxidations were measured in fresh liver homogenates from the pigs at 0 or 24 h without suckling after delivery using [1-^{14}C]-oleic acid (C18:1, the most enriched fatty acid in pig milk) and [1-^{14}C]-lignoceric acid (C24:0, primary oxidized in peroxisomes) as substrates. The measurements followed the same procedure as described previously by Lin et al. [5]. Specifically, liver homogenates (~40 mg) were incubated in 25 mL Erlenmeyer flasks with a final of volume of 2 mL of the reaction medium with or without addition of antimycin A (50 μmol/L) and rotenone (10 μmol/L) described as previously [13] for determining peroxisomal β-oxidation by inhibition of the electron transport chain of oxidative phosphorylation. The reaction was initiated by adding 2 μmol of either [1-^{14}C]-C18:1 (4.2 MBq/mmol) or [1-^{14}C]-C24:0 (1 MBq/mmol) and terminated by adding 0.5 mL

HCLO$_4$. The ^{14}CO$_2$ collected in ethanolamine and the ^{14}C-acid soluble products (ASP) processed from the acid-killed medium were quantified using liquid scintillation counter (Beckman LS 6000IC. Fullerton, CA. USA). The rate of total β-oxidation and the rate of peroxisomal and mitochondrial β-oxidation were calculated as described by Yu et al. [11].

Enzyme and mRNA analysis

Hepatic ACO activity was determined in liver homogenates using a spectrophotofluorometric assay as described by Walusimbi-kisitu and Harrison [21] with slight modifications. The homogenates were prepared in an ice cold buffer containing 250 mmol/L sucrose, 1 mmol/L EDTA and 1% ethanol. After preparation, the homogenate (0.6 ± 0.035 mg protein) was incubated in a dark room at 37°C in 0.5 mL (final vol.) of a medium with or without 35 μmol/L palmitoyl-CoA for 20 min. The incubations were stopped by adding 2 mL of borate buffer (0.1 mol/L, pH 10). The medium contained 60 mmol/L Tris–HCl, 50 μmol/L FAD, 170 μmol/L CoA, 1 μmol/L scopoletin and 6% of BSA. 200 μL of the incubated medium was transferred into a 96 well plate and measured in a BioTek reader (Synergy HT) with emission at 460 and excitation at 360 nm (BioTek Instruments, Inc., Winooski, VT). The standard curve was generated using H$_2$O$_2$ (30%, w/w) and 150 IU peroxidase under the same incubation conditions and measurements.

CPT activity was measured in liver homogenates using a radio-enzymatic method as described previously by Bremer et al. [22]. The homogenate (65 ± 3.8 mg protein) was pre-incubated with 80 μmol/L palmitoyl-CoA in a medium containing 75 mmol/L KCl, 50 mmol/L HEPES, 0.2 mmol/L EGTA, and 1% of fatty-acid-free BSA in the presence or absence of 2.5 μmol/L malonyl-CoA, and the incubation was conducted at 30°C in a final volume of 1 mL for 3 min [5]. The measurement of the enzyme activity was initiated with 1 μmol of (l-methyl-^{3}H)-carnitine (54 KBq/μmol) and terminated after 6 min by the addition of 2 mL of 6% of HClO$_4$ (vol/vol). The radioactivity in palmitoyl-carnitine, generated during the incubation was extracted with water-saturated butanol and quantified using liquid scintillation spectrometry (Beckman LS 6000IC, Fullerton, CA. USA).

Assay of mRNA was conducted using the methods described by Lin et al. [5]. Total RNA was extracted from the liver samples using guanidine isothiocynate and phenol (TRIzol Reagent, Sigma Chemical, St. Louis, MO. USA). The extracted RNA was quantified using a NanoDrop spectrophotometer (NanoDrop Technologies, Wilmington, DE. USA), and the integrity of the RNA was confirmed using gel electrophoresis with SYBR Safe TM DNA gel stain from Invitrogen Life Technologies (Carlsbad, CA. USA). The RNA (10 μg/50 μL) then was treated with TuboDNase (Ambion, Austin, TX) according to the manufacturer's instructions for removal of genomic DNA and transcribed using the iScriptTM Select cDNA Synthesis Kit provided with oligo (dT) primer mix. (Bio-Rad Laboratories, Hercules, CA). The mRNA abundance of PPARα, CPT I (L isoform) and ACO were determined using the MyiQ Single Color Real-Time PCR Detection System (Bio-Rad Laboratories, CA. USA). The determination was performed in triplicate with primers designed from pig-specific sequences available via GenBank and ordered from Sigma Genosys (St Louis, MO, USA). Amplification efficiencies were verified to be similar for the endogenous control GAPDH and the measured genes. Reactions contained cDNA with 0.4 μmol/L each of reverse and forward primers. The assay conditions and data calculations were the same as described previously [5].

Chemicals

Clofibrate was purchased from Cayman Chemical Company (Ann Arbor, MI. USA). [1-^{14}C]-oleic acid (C18:1) and [1-^{14}C]-lignoceric acid (C24:0) were purchased from American Radiolabeled Chemicals, Inc. (St. Louis, MO. USA). 2-(4-Chlorophenoxy)-2-methylpropionic acid ethyl ester, 2-(4-chlorophenoxy)-2-methylpropionic acid, 4-chlorophenylacetic acid, peroxidase and all other chemicals were obtained from Sigma-Aldrich (St. Louis, MO. USA).

Statistics

Data from fatty acid oxidation measurements were analyzed according to a split-plot design with the main plots in a completely randomized design using the SAS GLM procedure. The clofibrate effect was assigned to the main plot and pig postnatal age and fatty acid chain length effects were treated as the subplots. Least squares means ± SEM for variables are presented in tables and figures. Data from enzyme activity and mRNA enrichment assays were analyzed as a 2 × 2 factorial design using the GLM procedure. Least square means for treatment (clofibrate vs. control) and the postnatal age (0 vs. 24 h) effects were calculated. Differences between least squares means were determined using a Tukey test and considered significant when $P \leq 0.05$.

Results

Plasma assay

Chromatograms from HPLC analysis confirmed that clofibric acid was present in the plasma of both the clofibrate-fed sow and their newborn pigs (Figure 1). In addition to clofibric acid, a clofibrate conjugate metabolite was detected in the plasma of the clofibrate-fed sow but not in the plasma of newborn pigs. An unidentified peak after the reagent peak was observed also in the chromatogram from the plasma analysis of the clofibrate-fed sow and the peak was tiny in the plasma of newborn pigs

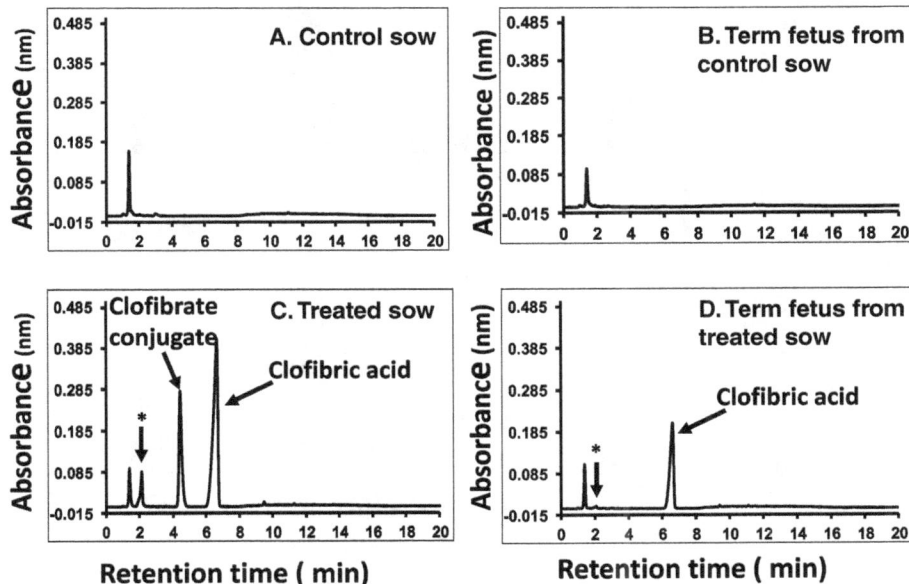

Figure 1 Chromatogram of plasma samples from HPLC analysis. A. control sow, B. newborn pigs from control sow, C. clofibrate-treated sow and D. term pigs from clofibrate-treated sow. See Methods for HPLC conditions and sample injection volumes. Compounds were identified by injection of available standards purchased from Sigma-Aldrich (St. Louis, MO) under the same HPLC conditions. *Unidentified peak.

from the clofibrate-fed sow. No peaks of clofibric acid or its metabolites were detected in the plasma of control sows or their offspring.

Fatty acid oxidation

Results from total fatty acid oxidation in vitro (Figure 2) showed that clofibrate and postnatal age significantly stimulated hepatic β-oxidation of fatty acids ($P < 0.0001$), but the stimulatory effects of clofibrate and postnatal age for C18:1 oxidation were different from C24:0 oxidation. The ^{14}C accumulations for C18:1 on average in CO_2, ASP and $CO_2 + ASP$ (μmol/h/g protein) measured at 0 h and 24 h were 1.9, 2.2, and 2.1 fold higher, respectively, from clofibrate exposed pigs (2.32, 13.36 and 15.69) than control pigs (1.24, 6.17 and 7.41). Similarly, the ^{14}C accumulations in CO_2, ASP and $CO_2 + ASP$ (μmol/h/g protein)

for C18:1 from both clofibrate treatment and control were on average 331%, 49% and 71% greater respectively in the livers of 24 h old pigs (2.89, 11.69 and 14.58) than the pigs at delivery (0.67, 7.83 and 8.51). In contrast with C18:1, the ^{14}C accumulation for C24:0 on average in $CO_2 + ASP$ (μmol/h/g protein) measured at 0 h and 24 h was 1.8 fold higher from clofibrate exposed pigs (3.32) than control pigs (1.81), but no differences were detected in the ^{14}C accumulation in CO_2 and ASP (Figure 2, A & B). In addition, the ^{14}C accumulation in CO_2 from C24:0 was increased on average by 61% in the livers of 24 h old fasted pigs compared to the pigs at delivery (0.32), but there was no postnatal age effect on the ^{14}C accumulations in ASP and $CO_2 + ASP$ (Figure 2, B & C). There was no interaction between clofibrate and pig postnatal age ($P > 0.05$).

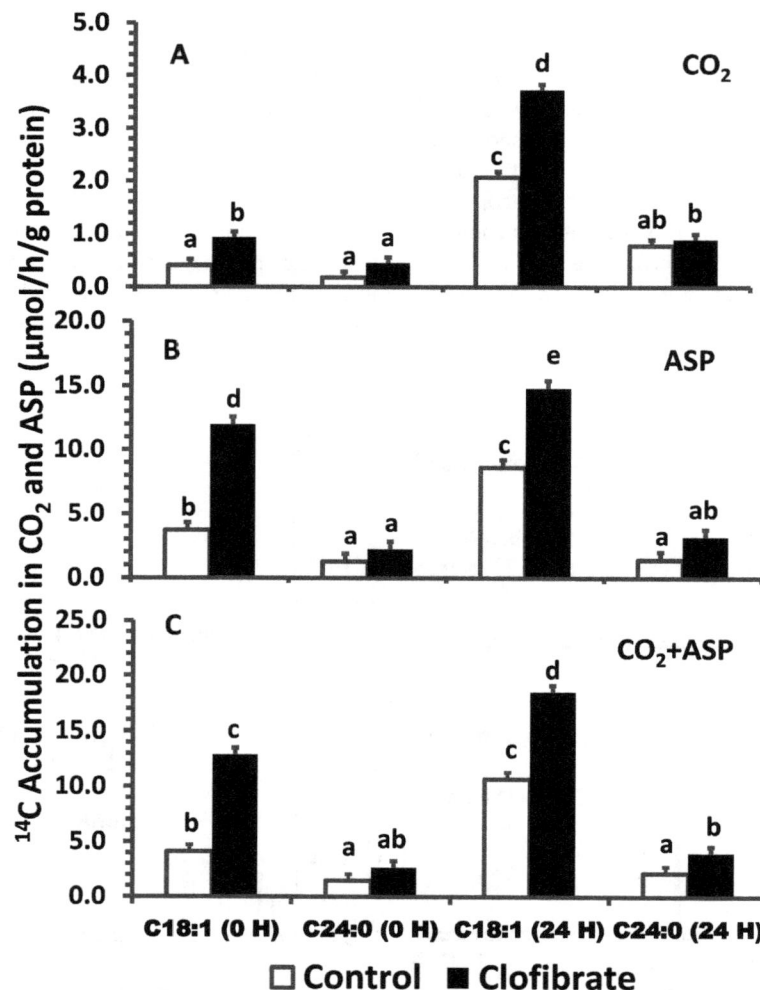

Figure 2 The effect of maternal clofibrate on total (mitochondrial and peroxisomal) fatty acid oxidation in newborn pigs at 0 h and 24 h after delivery by C-section. Total ^{14}C labeled carbon accumulated in CO_2 (**A**), acid soluble products (ASP, **B**) and $CO_2 + ASP$ (**C**). Values are least square means ± SEM (n = 6 for newborn pigs at 0 h and n = 4 for pigs at 24 h). [abcde] Bars within a panel lacking a common superscript differ ($P < 0.05$).

The mitochondrial oxidative flux of C18:1 to CO_2 and ASP (Figure 3, A & B) as well as the total (CO_2 + ASP, Figure 3, C) was affected by the treatment of clofibrate and the postnatal age ($P < 0.0001$). The ^{14}C-accumulations in CO_2, ASP and CO_2 + ASP (μmol/h/g protein) in the livers of the pigs from sow received clofibrate (2.3, 13.1 and 15.5 respectively) were on average 2.1-fold higher than that from controls (1.2, 6.2 and 7.4 respectively). Similarly, the ^{14}C-accumulations in CO_2, ASP and CO_2 + ASP also were increased on average by 3.1, 0.49 and 0.71 fold respectively in the pigs at 24 h after delivery (2.9, 11.7 and 14.6) as compared to the pigs at birth (0.7, 7.8 and 8.5). However, the stimulatory effect by clofibrate on the ^{14}C-accumulations in ASP and CO_2 + ASP tended to attenuate with the increase in postnatal age. There was no effect of age on the ^{14}C accumulation in ASP in the pigs exposed to clofibrate (Figure 3B; $P = 0.079$). In contrast to C18:1, the mitochondrial oxidative flux of C24:0

to ASP and CO_2 + ASP (μmol/h/g protein) remained similar in the livers of the pigs exposed to clofibrate (0.3, 0.83) and the controls (0.4, 0.9), although the flux to CO_2 tended to be higher in the liver from the clofibrate-exposed pigs than control pigs. There was no significant change in the oxidative flux of C24:0 as the postnatal age increased ($P > 0.05$).

No substantial ^{14}C-accumulation in CO_2 (<0.25, μmol/h/g protein) was detected in peroxisomes (Figure 4, A). Clofibrate exposure and postnatal age had no impact on the negligible peroxisomal oxidative flux to CO_2 from C18:1, but significantly increased the flux to ASP and CO_2 + ASP ($P < 0.0001$; Figure 4, B & C). The ^{14}C accumulation in the ASP and CO_2 + ASP (μmol/h/g protein) from C18:1 was on average 2 fold greater from clofibrate-treated (5.4, 5.5) than control (2.7, 2.7) pigs, and 56% higher from the 24 h-old pigs than that from pigs at delivery (1.3, 1.7). In comparison to C18:1,

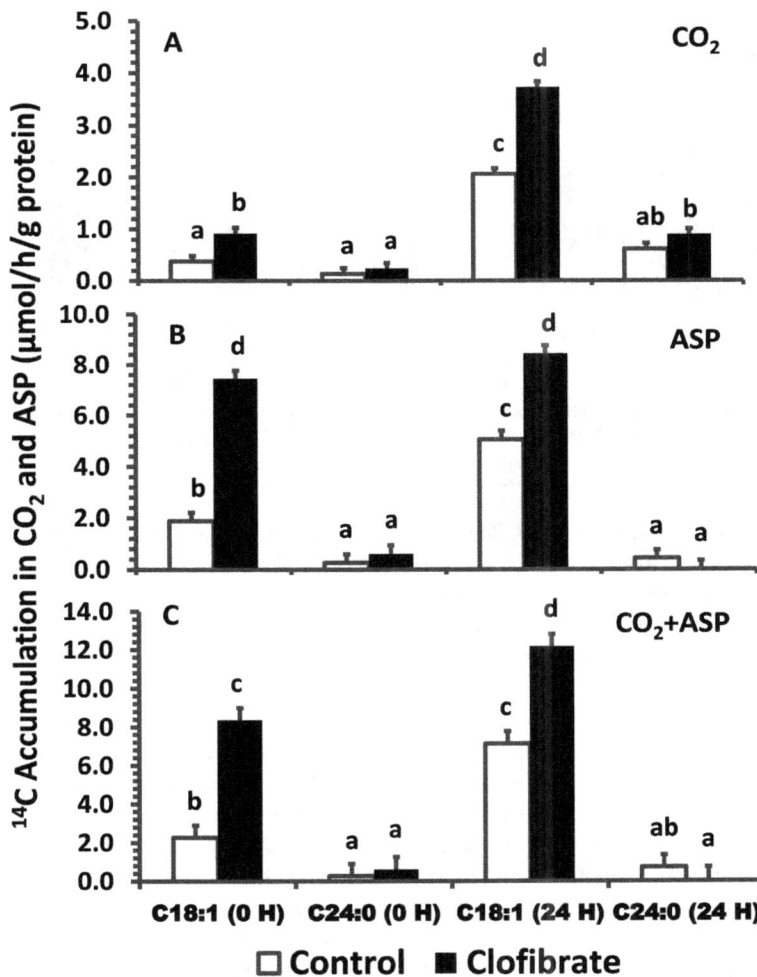

Figure 3 The effect of maternal clofibrate on mitochondrial fatty acid oxidation in newborn pigs at 0 h and 24 h after delivery by C-section.
Total ^{14}C labeled carbon accumulated in CO_2 (**A**), acid soluble products (ASP, **B**) and CO_2 + ASP (**C**). Values are least square means ± SEM (n = 6 for newborn pigs at 0 h and n = 4 for pigs at 24 h). [abcd] Bars within a panel lacking a common superscript differ ($P < 0.05$).

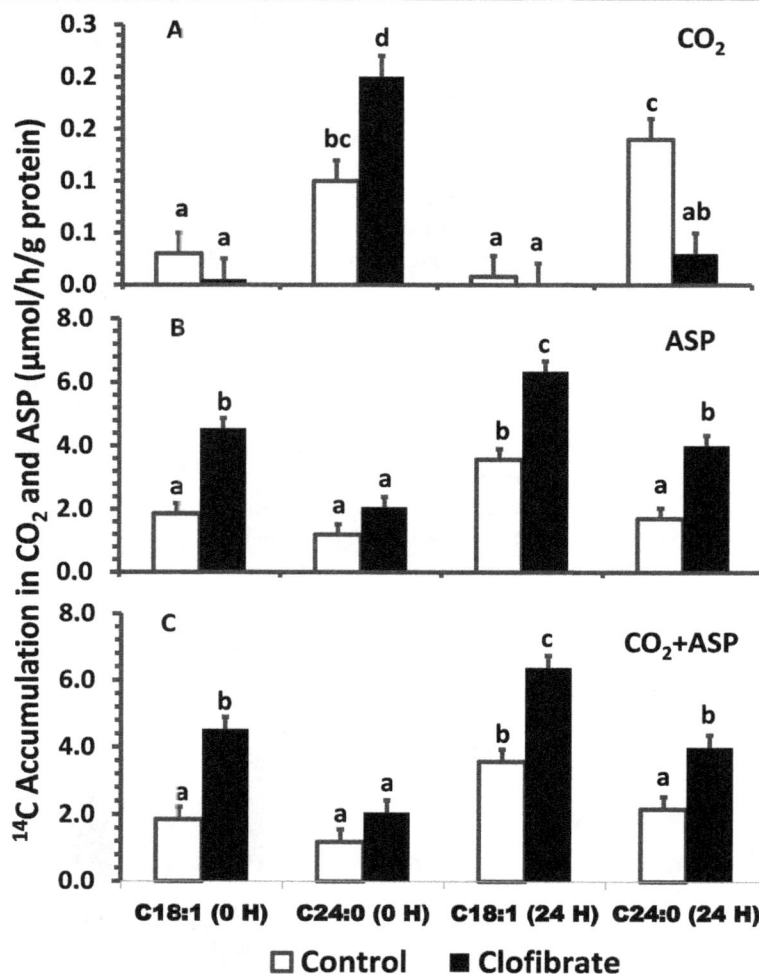

Figure 4 The effect of maternal clofibrate or peroxisomal fatty acid oxidation in newborn pigs at 0 h and 24 h after delivery by C-section. Total ^{14}C labeled carbon accumulated in CO_2 **(A)**, acid soluble products (ASP, **B**) and CO_2 + ASP **(C)**. Values are least square means ± SEM (n = 6 for newborn pigs at 0 h and n = 4 for pigs at 24 h). [abcd] Bars within a panel lacking a common superscript differ ($P < 0.05$).

clofibrate increased the ^{14}C accumulation from C24:0 peroxisomal oxidation in CO_2 in the liver of pigs at delivery but decreased in the liver of 24 h-old pigs. Thus there was no significant difference ($P = 0.93$) in the average flux to CO_2 (µmol/h/g protein) between clofibrate-treated (0.15) and control pigs (0.14). There were no effects of clofibrate on the ^{14}C accumulation in ASP and CO_2 + ASP from C24:0 in pigs at delivery, but the ^{14}C accumulations was 84% greater from clofibrate-treated (4.0) than control in 24 h-old pigs (2.17). Postpartum age also had no influence on C24:0 oxidation in pigs at delivery (P > 0.05), but the ^{14}C accumulation in ASP and CO_2 + ASP was 94% higher from the clofibrate-treated pigs measured at 24 h (4.0) than at delivery (2.06).

Clofibrate and postnatal age had no effect on the distribution of oxidative flux C18:1 and C24:0 between mitochondria and peroxisome (Figure 5, P > 0.05). However, the peroxisomal proportion of total fatty acid oxidation

was on average 2.3 fold higher from C24:0 (85.5) than C18:1 (37.7%).

Enzyme activity and mRNA expression

Hepatic CPT specific activity (Figure 6A) measured with and without malonyl-CoA was 69% and 43% higher, respectively, in the clofibrate-exposed pigs ($P < 0.018$). There was no change in the activities with postnatal age ($P = 0.2$). The total and inhibited enzyme activity (µmol/h/g protein) were 31.2, 52.4 on average for the pigs at delivery and 39.3, 64 on average for the 24 h-old pigs. Hepatic ACO activity was increased by 2.3 fold in the clofibrate-exposed pigs ($P < 0.036$; Figure 6B). The enzyme activity also increased by 1.4 fold in 24-h-old pigs, but the increase had no influence on the degree of clofibrate stimulation ($P > 0.05$).

The relative mRNA expression of CPT I was increased by 19-fold in liver of the clofibrate-exposed pigs ($P < 0.0001$),

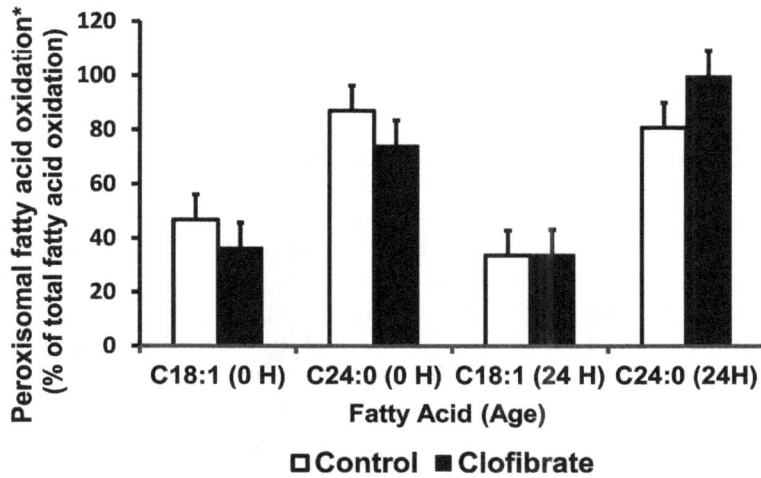

Figure 5 Percentage of peroxisomal fatty acid oxidation in total fatty acid oxidation. Values are least square means ± SEM (n = 6 for newborn pigs at 0 h and n = 4 for pigs at 24 h). *Significantly different from fatty acid (P < 0.05).

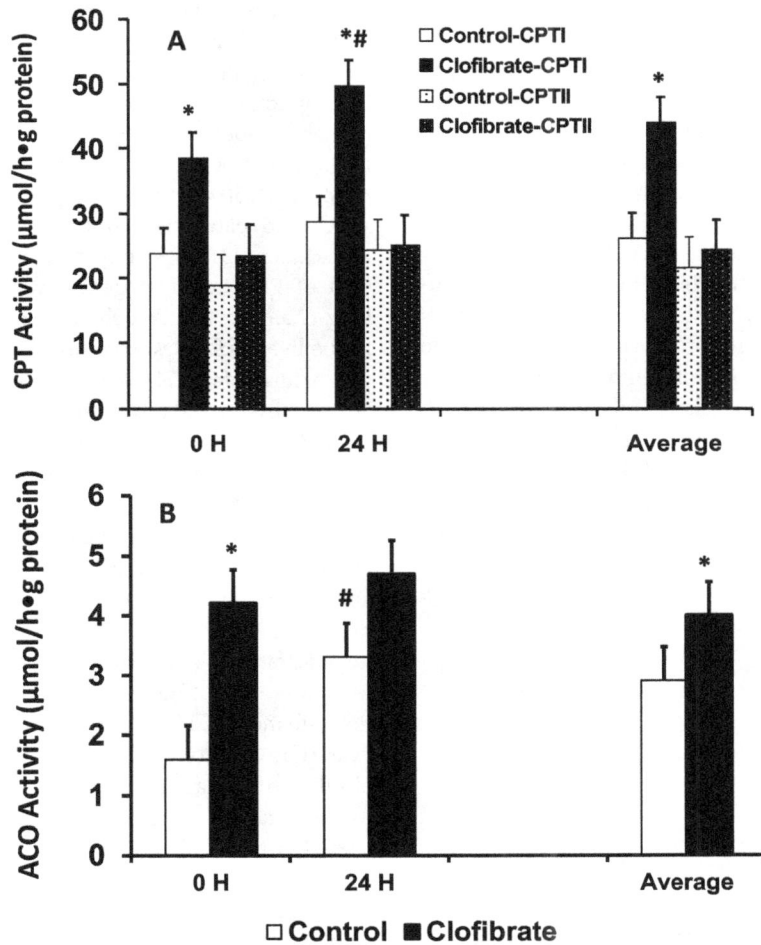

Figure 6 The effect of maternal clofibrate on enzyme specific activity in newborn pigs at birth and 24 h after delivery by C-section. Hepatic activity of carnitine palmitoyltransferase (CPT, **A)** and acyl-CoA oxidase (ACO, **B)**. Values are least square means ± SEM (n = 6 for pigs at 0 h and n = 4 for pigs at 24 h). *Significantly different from clofibrate and # significantly different from postnatal age (P < 0.05).

but no differences were detected for the relative mRNA abundance of either ACO or PPARα ($P > 0.05$, Figure 7).

Discussion

Previous research in rodents has shown peroxisome proliferation was observed in the liver of fetuses over 13 days and/or newborns from clofibrate treated dams during gestation [15,23]. The peroxisomal membrane protein 70 and the marker enzymes dihydroxyacetone phosphate acyltransferase and catalase specific activities were significantly increased in the fetal liver of mice at 19 d gestation [15]; indirectly suggesting that clofibrate and/or its metabolites could cross the placenta barrier to enter the circulation system of the fetus in rodent species. However, there was no direct evidence to demonstrate if clofibrate, clofibrate metabolites, or both actually cross the placental barrier. Additionally, the morphology and function of placenta vary greatly among mammals [24]. The type of placenta in swine has been described as a diffuse, folded, and epitheliochorial placenta type that is different from that of rats, which has a discoidal, labyrinthine, and hemotrichorial placenta type [25]. Considering the disparity between the morphology and mechanisms for an extra embryonic membrane attachment in swine, the placental transfer of clofibrate could be different from the rodent species. The aim of this study is to determine if clofibrate or its metabolites traverse porcine placental tissues to evaluate the effect of clofibrate on lipid metabolism in the offspring of pregnant swine fed clofibrate. Our result clearly established that clofibrate is absorbed and hydrolyzed into clofibric acid, and the clofibric acid crosses the porcine placenta with no chemical or structural modifications. The results are also in line with the previous observation that the biofunctional activity of clofibrate is due to clofibric acid [26]. In addition of the analytical results from plasma analyses,

the in vitro metabolism measurements strongly support that clofibric acid delivered via the maternal diet entered fetal circulation and induced fetal hepatic fatty acid oxidation. Feeding clofibrate to neonatal pigs for 14 days and to young pigs for 28 days caused a mild increase in the liver weight as a percentage of body weight (3.3%, [11]; 3.8%, [27]). The relative weight of the liver (g/kg body weight) measured in this study showed no difference between the pigs from the clofibrate treated group (26.9 ± 0.09) and control group (27.9 ± 0.15), suggesting that exposure to the pharmacological PPARα agonist clofibrate by sows for 9 days during late pregnancy would not cause a morphological change in the liver of the offspring. Although we did not perform a histological examination, clofibrate has no substantial effect on the peroxisome proliferation in this species at the ages of 7 days and 8 weeks [28]. This is similar to humans but significantly different from the rodent species in which activation of PPARα by clofibrate initiates hepatocyte proliferation and induces the well-known hepatocellular carcinoma [29,30].

Evidence from rodent species demonstrated that maternal supplementation of clofibrate orally or by injection stimulates peroxisomal proliferation, decreases oxygen uptake, and alters lipid metabolism in the liver and intestine of fetuses or neonates. Maternal clofibrate induces cytochrome P4504A mRNA expression [16] and increases ACO and catalase specific activities in the fetal liver and kidneys [31]. Recently, the high relative mRNA expression of PPARα target genes ACO, CPT I, medium- and long-chain acyl-CoA dehydrogenases were observed also in the liver of fetuses from pregnant rats fed with clofibrate and oxidized fat [17]. However, as PPARα is a crucial regulator of lipid metabolism, the role of its activation of target genes in fatty acid oxidative metabolism in the fetal and/or neonatal liver from the pregnant animals receiving clofibrate has not been determined in previous studies. In the current study, we examined both peroxisomal and mitochondrial β-oxidation using ^{14}C-labeled long-chain fatty acids C18:1 and C24:0. Results from this study indeed demonstrated that maternal feeding of clofibrate increases hepatic fatty acid oxidation in the pig's liver at birth and 24 h after birth. The increase appears to be associated with the enhanced CPT I and ACO specific activities in the liver. The results achieved in the newborns exposed to clofibrate prenatally via the maternal diet were similar to the increase in vitro and vivo fatty acid oxidation reported in neonatal pigs receiving clofibrate directly postnatally [10,13]. Because the increase of CPT I specific activity was congruent with the great increase in mRNA expression, the stimulation of fatty acid oxidation in the liver, particularly in the mitochondria, resulted from the gene expression potentially due to the activation of PPARα by clofibrate. In contrast with CPT I, however, the expressions

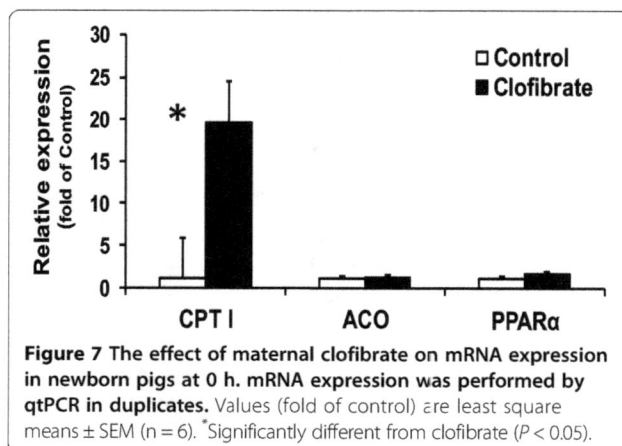

Figure 7 The effect of maternal clofibrate on mRNA expression in newborn pigs at 0 h. mRNA expression was performed by qtPCR in duplicates. Values (fold of control) are least square means ± SEM (n = 6). *Significantly different from clofibrate ($P < 0.05$).

of ACO and PPARα mRNA were not significantly induced in clofibrate-exposed pigs. This suggests that other factors might be involved in the upregulation of ACO specific activity. Data from studies with rodent fetuses indicated that PPARα mRNA expression is associated with the postpartum age and hormonal and/or nutritional status of the mother. The induction of liver PPARα mRNA expression occurs around birth and the expression maintains an elevated level throughout the suckling period [32], while the content of peroxisomes and the activity of peroxisomal enzymes appear to occur in late fetal development and peak dramatically at birth [33]. These results imply that the ACO protein and PPARα mRNA might have reached pinnacle expression at birth. Indeed, the increase in relative PPARα mRNA was not observed in the liver from clofibrate fed rats in a study conducted by Ringseis et al. [34]. They suggested that endogenous free fatty acids might reduce the capability of fibrates to active PPARα, and consequently its metabolic effects, because ACO is the enzyme containing PPAR response element and responds to changes in polyunsaturated fatty acid levels in a PPARα–dependent manner. Furthermore, a different physiological status, such as a fasting state, might provoke a different PPARα response in the liver of fetuses. The induction of peroxisomal oxidation occurs immediately postpartum, is greater in the suckled versus fasted neonatal pigs, and is reliant on the initiation of suckling [35]. Thus, the lower response in mRNA expressions of ACO and PPARα might be due to the developmental stage and the physiological status of the fetuses because the mRNA of PPAR α was also measured in a fasted state in this study.

The effect of the developmental ages of fetuses on oral food intake and gastrointestinal digestion has been described in humans and animals [36,37], but information on the energy metabolism in the fetus is very limited after delivery. Results from earlier studies showed that hepatocytes isolated from term guinea pigs were unable to oxidize fatty acids, but the capability was developed in the first 12 h after birth. The production of ^{14}C measured in CO_2 and ASP from $1-^{14}C$ labeled fatty acids at 6 h was 40–50% of the production at 24 h. At 12 h of age the rate of fatty acid oxidation had already reached the rate at 24 h and did not change during suckling in the first week of life. These data show that the capacity for β-oxidation and ketogenesis develops maximally in this species during the first 6–12 h after birth, and appears to be partly dependent on the development of fatty acid-activating enzymes [38]. Similarly, a low fatty oxidation rate with a high esterification was also observed in hepatocytes isolated from term fetal rabbits, whatever the octanoate concentration in the medium [39]. Consistent with the observations in term guinea pigs and rabbits, the fatty acid oxidation obtained in term pigs at the time

of delivery by C-section was also low even when compared with the rate measured in newborn pigs born naturally [5]. The low fatty acid rate was increased significantly in the first 24 h, in which the rate was increased 2.3 fold in mitochondria and 1.9 fold in peroxisomes. However, the increase in mitochondria apparently was not due to an increase of the key enzyme CPT I activity. The activity measured at 24 h was not significantly different from the activity measured at 0 h after the delivery. A similar phenomenon was also observed in newborn and 24 h fasted pigs [5]. The increase might be associated with an increase in the number of mitochondria observed in neonatal pigs during in the first 12 h [40] and/or a decrease of sensitivity of CPT I to malonyl-CoA inhibition observed in 24-h-old pigs [5]. In contrast with mitochondria, the increase of fatty acid oxidation in peroxisomes was accompanied with a 2.3 fold increase of the ACO. Even so, it is notable that the oxidation rate of C24:0 was not significantly promoted during the first 24 h in peroxisomes, suggesting differences between mitochondria and peroxisomes in the capability of oxidizing fatty acids. Because the very long-chain fatty acid exclusively is activated in micorosomes or/and peroxisomes, the low fatty acid oxidation might be associated with a low activity of very long-chain acyl-CoA synthetase. These results imply the impact of postpartum age on the metabolic pathway varied among subcellular compartments.

Peroxisomal fatty acid oxidation catalyzes chain-length shortening of monounsaturated fatty acid and saturated very long-chain fatty acid. Evidence indicated that C24:0 could only be oxidized initially in peroxisomes after activation by the acyl-CoA sythetase enzyme in endoplasmic reticulum and peroxisomes in brain and liver [41,42]. Peroxisomes at least have two enzyme systems for fatty acid activation: 1 for long-chain fatty acid and 2 for very-long-chain fatty acid [42]. Indeed, there was no substantial ^{14}C accumulation in either CO_2 or ASP or both of CO_2 and ASP from C24:0 in mitochondria. Neither clofibrate-exposure nor age had any influence on the definitely negligible mitochondrial ^{14}C accumulation from C24:0 ($\leq 1\%$). The percentage of C24:0 oxidation in peroxisomes was more than 80% of the total oxidation. Clofibrate had no effect on the relative oxidative capacity of peroxisomes although the efficacy of clofibrate was increased by postpartum age. This result suggests that the very long-chain fatty acid is catabolized exclusively in peroxisomes. In contrast with C24:0, high ^{14}C accumulations in CO_2, ASP and $CO_2 + ASP$ from C18:1 were observed in both mitochondria and peroxisomes, suggesting that C18:1can be oxidized initially in both of the organelles. Similar results were observed also using C18:1 and erucic acid as substrates in our previously study [13].

Conclusions

In conclusion, clofibrate from maternal oral feeding during gestation is absorbed and hydrolyzed to clofibric acid, which can cross the porcine placenta and enter the fetal circulation system. Fetuses that were exposed to clofibric acid via maternal-placental transfer had a higher hepatic capacity to oxidize fatty acid at birth as compared to control fetuses. The high fatty acid oxidative capacity resulted from an increase in key enzyme activities induced by clofibric acid via activation of PPARα and its target genes. The promoted fatty acid oxidation by activation of PPARα and its target genes was attenuated in fasted newborns with postpartum age. Postpartum age also increased fatty acid oxidation. The increase was associated with development and was not influenced by clofibrate supplementation. Results from this study suggest that clofibrate could be examined as an agent to induce precocious development of ACO and possibly improve milk fat oxidation and pig survivability.

Abbreviations

ACO: Acyl-CoA oxidase; ASP: Acid soluble products; CPT: Carnitine palmitoyltransferase; C18:1: Oleic acid; C24:0: Lignoceric acid; PPARα: Peroxisomal proliferator-activated receptor.

Competing interests

This project was supported by National Research Initiative Competitive Grant no. 2007-35206-17897 from the USDA National Institute of Food and Agriculture and by the North Carolina Agriculture Research Service. The author(s) declare that they have no competing interests.

Authors' contributions

XL as the lead and corresponding author was in charge of designing and conducting the experiment. He also collected the samples, performed the biological measurements and completed the statistical analyses. SKJ was involved in sample collection and conducted the molecular analyses. JO was involved in experimental design, sample collection, data analysis and discussion. All authors read and approved the final manuscript.

Authors' information

XL is a Research Associate Professor at the Department of Animal Science at NCSU. His main areas of research are neonatal survival and lipid metabolism primarily focused on the regulation of fatty acid oxidation during neonatal development and epigenetic regulation of fetal development and placenta growth. He is also interested in the role of polyunsaturated fatty acid in the development of intestine in neonates and the role of trace minerals and vitamins in the stressed animals. SKJ is a Senior Research Scientist at the Department of Animal Science at NCSU. Her main area of interest is identifying the molecular mechanisms of nutritional immunology associated with neonatal gastrointestinal health and development. She is interested in the fundamental question of how bioactive nutrients are involved in the developing gastrointestinal tract of the pig and using the pig as a model for infant nutrition. JO is a Williams Neal Reynolds Professor in the Department of Animal Science at NCSU, his research interests are molecular and metabolic regulation of lipid digestion and metabolism; neonatal nutrition; intestinal growth and metabolism in normal and pathophysiological states. His program is focused on using the young pigs as a model for the human infant in nutrition and digestive physiology. He also has teaching responsibilities in the areas of nutrition and biochemistry.

Acknowledgements

We thank Anthony Blikslager, DVM, PhD, DACVS for caesarean section of the sows, and staff at the North Carolina State University Swine Educational Unit, Raleigh for assistance of the animal management.

References

1. NAHMS. Swine 2006 Part III: Reference of Swine Health, Productivity, and General Management in the United States, 2006. 2006. (pdf 688kb 3/08). http://www.aphis.usda.gov.
2. Knauer MT, Hostetler CE. US swine industry productivity analysis, 2005 to 2010. J Swine Health Prod. 2013;21:248–52.
3. Girard J, Pégorier JP. An overview of early post-partum nutrition and metabolism. Biochem Soc Trans. 1998;26:69–74. Review.
4. Pégorier JP, Duée PH, Girard J, Peret J. Metabolic fate of non-esterified fatty acids in isolated hepatocytes from newborn and young pigs. Evidence for a limited capacity for oxidation and increased capacity for esterification. Biochem J. 1983;212:93–7.
5. Lin X, Shim K, Odle J. Carnitine palmitoyltransferase I control of acetogenesis, the major pathway of fatty acid {beta}-oxidation in liver of neonatal swine. Am J Physiol Regul Integr Comp Physiol. 2010;298:R1435–43.
6. Rakhshandehroo M, Knoch B, Müller M, Kersten S. Peroxisome proliferator-activated receptor alpha target genes. PPAR Res. 2010;2010:612089. doi:10.1155/2010/612089.
7. Brady PS, Marine KA, Brady LJ, Ramsay RR. Co-ordinate induction of hepatic mitochondrial and peroxisomal carnitine acyltransferase synthesis by diet and drugs. Biochem J. 1989;260:93–100.
8. König B, Kluge H, Haase K, Brandsch C, Stangl GI, Eder K. Effects of clofibrate treatment in laying hens. Poult Sci. 2007;86:1187–95.
9. Litherland NB, Bionaz M, Wallace RL, Loor JJ, Drackley JK. Effects of the peroxisome proliferator-activated receptor-alpha agonists clofibrate and fish oil on hepatic fatty acid metabolism in weaned dairy calves. J Dairy Sci. 2010;93:2404–18.
10. Peffer PL, Lin X, Odle J. Hepatic beta-oxidation and carnitine palmitoyltransferase I in neonatal pigs after dietary treatments of clofibric acid, isoproterenol, and medium-chain triglycerides. Am J Physiol Regul Integr Comp Physiol. 2005;288:R1518–24.
11. Yu XX, Drackley JK, Odle J. Rates of mitochondrial and peroxisomal beta-oxidation of palmitate change during postnatal development and food deprivation in liver, kidney and heart of pigs. J Nutr. 1997;127:1814–21.
12. Cheon Y, Nara TY, Band MR, Beever JE, Wallig MA, et al. Induction of overlapping genes by fasting and a peroxisome proliferator in pigs: evidence of functional PPARalpha in nonproliferating species. Am J Physiol Regul Integr Comp Physiol. 2005;288:R1525–35.
13. Bai X, Lin X, Drayton J, Liu Y, Ji C, Odle J. Clofibrate increases long-chain fatty acid oxidation by neonatal pigs. J Nutr. 2014;144:1688–93.
14. Laclide-Drouin H, Masutti JP, Hatier R, Dauça M, Grignon G. Effect of clofibrate on the peroxisomes of the intestine of the rat during foetal development. Ital J Anat Embryol. 1995;1:411–7.
15. Wilson GN, King T, Argyle JC, Garcia RF. Maternal clofibrate administration amplifies fetal peroxisomes. Pediatr Res. 1991;29:256–62.
16. Simpson AE, Brammar WJ, Pratten MK, Cockcroft N, Elcombe CR. Placental transfer of the hypolipidemic drug, clofibrate, induces CYP4A expression in 18.5-day fetal rats. Drug Metab Dispos. 1996;24:547–54.
17. Ringseis R, Gutgesell A, Dathe C, Brandsch C, Eder K. Feeding oxidized fat during pregnancy up-regulates expression of PPARalpha-responsive genes in the liver of rat fetuses. Lipids Health Dis. 2007;6:6.
18. Herfel TM, Jacobi SK, Lin X, Fellner V, Walker DC, et al. Polydextrose enrichment of infant formula demonstrates prebiotic characteristics by altering intestinal microbiota, organic acid concentrations, and cytokine expression in suckling piglets. J Nutr. 2011;141:2139–45.
19. Gornall AG, Bardawill CJ, David MM. Determination of serum proteins by means of the biuret reaction. J Biol Chem. 1949;177:751–66.
20. Du L, Xu Y, Musson DG. Simultaneous determination of clofibrate and its active metabolite clofibric acid in human plasma by reversed-phase high-performance liquid chromatography with ultraviolet absorbance detection. J Chromatogr B Analyt Technol Biomed Life Sci. 2003;794:343–51.
21. Walusimbi-Kisitu M, Harrison EH. Fluorometric assay for rat liver peroxisomal fatty acyl-coenzyme A oxidase activity. J Lipid Res. 1983;24:1077–84.
22. Bremer J, Woldegiorgis G, Schalinske K, Shrago E. Carnitine palmitoyltransferase: activation by palmitoyl-CoA and inactivation by malonyl-CoA. Biochim Biophys Acta. 1985;833:9–16.

23. Stefanini S, Mauriello A, Farrace MG, Cibelli A, Ceru MP. Proliferative response of foetal liver peroxisomes to clofibrate treatment of pregnant rats. A quantitative evaluation. Biol Cell. 1989;67:299–305.

24. Beck F. Comparative placental morphology and function. Environ Health Perspect. 1976;18:5–12.

25. Ridderstråle Y, Persson E, Dantzer V, Leiser R. Carbonic anhydrase activity in different placenta types: a comparative study of pig, horse, cow, mink, rat, and human. Microsc Res Tech. 1977;38:115–24.

26. Cayen MN, Ferdinandi ES, Greselin E, Robinson WT, Dvornik D. Clofibrate and clofibric acid. Comparison of the metabolic disposition in rats and dogs. J Pharmacol. 1977;200:33–43.

27. Luci S, Kluge H, Hirche F, Eder K. Clofibrate increases hepatic triiodothyronine (T3)- and thyroxine (T4)-glucuronosyltransferase activities and lowers plasma T3 and T4 concentrations in pigs. Drug Metab Dispos. 2006;34:1887–92.

28. Luci S, Giemsa B, Hause G, Kluge H, Eder K. Clofibrate treatment in pigs: effects on parameters critical with respect to peroxisome proliferator-induced hepatocarcinogenesis in rodents. BMC Pharmacol. 2007;7:6. 2007. doi:10.1186/1471-2210-7-6.

29. Gonzalez FJ. Regulation of hepatocyte nuclear factor 4 alpha-mediated transcription. Drug Metab Pharmacokinet. 2008;23:2–7. Review.

30. Peters JM, Shah YM, Gonzalez FJ. The role of peroxisome proliferator-activated receptors in carcinogenesis and chemoprevention. Nat Rev Cancer. 2012;12:181–95.

31. Sartori C, Stefanini S, Cimini A, Di Giulio A, Cerù MP. Liver peroxisomes in newborns from clofibrate-treated rats. II. A biochemical study of the recovery period. Biol Cell. 1992;74:315–24.

32. Panadero M, Bocos C, Herrera E. Relationship between lipoprotein lipase and peroxisome proliferator-activated receptor-alpha expression in rat liver during development. J Physiol Biochem. 2006;62:189–98.

33. Brun S, Carmona MC, Mampel T, Viñas O, Giralt M, et al. Uncoupling protein-3 gene expression in skeletal muscle during development is regulated by nutritional factors that alter circulating non-esterified fatty acids. FEBS Lett. 1999;453:205–9.

34. Ringseis R, Eder K. Influence of pharmacological PPARalpha activators on carnitine homeostasis in proliferating and non-proliferating species. Pharmacol Res. 2009;60:179–84.

35. Yu XX, Drackley JK, Odle J. Food deprivation changes peroxisomal beta-oxidation activity but not catalase activity during postnatal development in pig tissues. J Nutr. 1998;128:1114–21.

36. Buddington RK, Sangild PT, Hance B, Huang EY, Black DD. Prenatal gastrointestinal development in the pig and responses after preterm birth. J Anim Sci. 2012;4:290–8.

37. Rasch S, Sangild PT, Gregersen H, Schmidt M, Omari T, et al. The preterm piglet - a model in the study of oesophageal development in preterm neonates. Acta Paediatr. 2010;99:201–8.

38. Shipp DA, Parameswaran M, Arinze IJ. Development of fatty acid oxidation in neonatal guinea-pig liver. Biochem J. 1982;208:723–30.

39. Pégorier JP, Duée PH, Clouet P, Kohl C, Herbin C, et al. Octanoate metabolism in isolated hepatocytes and mitochondria from fetal, newborn and adult rabbit. Evidence for a high capacity for octanoate esterification in term fetal liver. Eur J Biochem. 1989;184:681–6.

40. Mersmann HJ, Goodman J, Houk JM, Anderson S. Studies on the biochemistry of mitochondria and cell morphology in the neonatal swine hepatocyte. J Cell Biol. 1972;53:335–47.

41. Bhushan A, Singh RP, Singh I. Characterization of rat brain microsomal acyl-coenzyme A ligases: different enzymes for the synthesis of palmitoyl-coenzyme A and lignoceroyl-coenzyme A. Arch Biochem Biophys. 1986;246:374–80.

42. Wanders RJ, van Roermund CW, van Wijland MJ, Schutgens RB, Heikoop J, et al. Paroxysmal fatty acid beta-oxidation in relation to the accumulation of very long chain fatty acids in cultured skin fibroblasts from patients with Zellweger syndrome and other peroxisomal disorders. J Clin Invest. 1987;80:1778–83.

16

Comparative analysis of proteomic profiles between endometrial caruncular and intercaruncular areas in ewes during the peri-implantation period

Yang Wang[1†], Chao Wang[1†], Zhuocheng Hou[2], Kai Miao[1], Haichao Zhao[1], Rui Wang[1], Min Guo[1], Zhonghong Wu[3], Jianhui Tian[1] and Lei An[1*]

Abstract

The endometrium of sheep consists of plenty of raised aglandular areas called caruncular (C), and intensely glandular intercaruncular areas (IC). In order to better understand the endometrium involved mechanisms of implantation, we used LC-MS/MS technique to profile the proteome of ovine endometrial C areas and IC areas separately during the peri-implantation period, and then compared the proteomic profiles between these two areas. We successfully detected 1740 and 1813 proteins in C areas and IC areas respectively. By comparing the proteome of these two areas, we found 170 differentially expressed proteins (DEPs) ($P < 0.05$), functional bioinformatics analysis showed these DEPs were mainly involved in growth and remodeling of endometrial tissue, cell adhesion and protein transport, and so on. Our study, for the first time, provided a proteomic reference for elucidating the differences between C and IC areas, as an integrated function unit respectively, during the peri-implantation period. The results could help us to better understand the implantation in the ewes. In addition, we established a relatively detailed protein database of ovine endometrium, which provide a unique reference for further studies.

Keywords: Caruncular areas, Endometrium, Ewe, Implantation, Intercaruncular areas

Background

Implantation, the sign and initial phase of pregnancy, is a process lading to attachment of developing conceptus to the maternal endometrium, and resulting in the establishment of placental structure. During the peri-implantation period, a synchronized and accurate crosstalk between conceptus and maternal endometrium, which is referred as maternal-fetal dialogue, must be established to support pregnancy [1]. It has been demonstrated that both the development of an embryo to the implantation-competent stage, as well as the transformation of the uterus into a receptive stage, are required for successful implantation [2].

Ovine has been extensively used as model to research maternal–fetal dialogue during implantation [3]. As an early sensor of embryos [4], endometrial functions are regulated primarily by progesterone (P4) from the corpus luteum, as well as cytokines and hormones secreted from the trophectoderm/chorion, including interferon tau (IFNT), to enter into a receptive status during early pregnancy [5]. In order to better understand the implantation process, detailed and comprehensive profiling of endometrium is necessary. However, the structure of endometrium in ruminants differs from other mammalian species Ovine uterine wall can be functionally divided into the endometrium and the myometrium. The normal adult ovine endometrium consists of LE, glandular epithelium (GE), several types of stroma (stratum compactum and stratum spongiosum), blood vessels and immune cells. In sheep, the endometrium has two distinct areas – aglandular caruncular (C) and glandular intercaruncular (IC). The C areas have LE and compact stroma and are the

* Correspondence: anleim@cau.edu.cn
†Equal contributors
[1]Ministry of Agriculture Key Laboratory of Animal Genetics, Breeding and Reproduction, National engineering laboratory for animal breeding, College of Animal Sciences and Technology, China Agricultural University, No.2 Yuanmingyuan Xi Lu, Haidian, Beijing 100193, China
Full list of author information is available at the end of the article

sites of superficial implantation and placentation [6], while the IC area, which is suffused with glandular epithelial cells [7,8], is mainly responsible for the synthesis and secretion histotroph, including enzymes, cytokines, growth factors, ions, hormones, glucose, transport proteins, and adhesion molecules to support early conceptus survival, development, implantation and placentation [2,9]. These two areas play different roles in implantation process, and both are essential for the establishment of pregnancy. Considering the significant structural and functional differences between C and IC areas, a comprehensive comparison between those two distinct endometrial areas could facilitate the understanding of endometrium involved implantations in ruminants.

Although implantation is of prime importance for establishment of pregnancy, the underlying mechanisms responsible for this complex physiological process, are still unclear. The high-throughput or 'omic' approaches, including RNA sequencing, microarray and proteome, has been applied to profile the expression patterns of genes or proteins for the endometrium during peri-implantation period. Walker et al. investigated the endometrial receptivity and maternal immunoregulation at day 17 of pregnancy in cattle using microarray [2]. Mansouri-Attia et al. recently profiled transcriptome of bovine endometrium during the peri-implantation period, after artificial insemination (AI) compared with the estrous cyclic endometrium using microarray, and found many factors and pathways essential to implantation in the C and IC areas [9]. Similarly, transcriptome of endometrium during early pregnancy have also been profiled in ewes [10], humans [11], and mice [12]. However, so far, most of these related studies were based on transcriptomic analysis. Compared to transcriptomic analysis, the proteomic approach has advantages of allowing people directly investigate functional molecules. Mullen et al. used label-free liquid chromatography-tandem mass spectrometry (LC-MS/MS) shotgun proteomics approach to characterize the uterine proteome at preimplantation stage in high fertility cattle [13]. Koch et al. first utilized LC-MS/MS proteomic technology to obtain the signature profile of proteins in the uterine lumen of ewes during early pregnancy [14]. However, in most of those studies, C and IC areas were not analyzed separately or compared mutually.

In the present study, we used LC-MS/MS technique to profile the proteome of endometrial C and IC areas separately at Day 17 of pregnancy, and established a relative detailed protein database of ovine endometrium during the time window of implantation. This database would provide a reference for future studies on implantation. By comparing the proteomic profiles between C areas and IC areas, we revealed the functional differences between these two areas during the critical period, which provided new information for studying the underlying

molecular mechanisms responsible for structural and functional differentiation of these two areas. Finally, to our knowledge, this is the first report of endometrial proteome of C and IC areas, as an integrated tissue layer respectively, during peri-implantation in ewes.

Materials and methods
Animals and treatments
The experiments were performed were accordance with the Guide for the Care and Use of Agricultural Animals in Agricultural Research and Teaching, and all procedures were approved by the Institutional Animal Care and Use Committee at China Agricultural University (Beijing, China). Forty-six Chinese Small Tail Han ewes with normal estrus cycle were selected in our study. All ewes were fed and managed under a unified and optimized condition of environment and nutrition.

Forty-six cyclic ewes were estrus synchronized by using progesterone-impregnated (0.3 g) vaginal implants with controlled intra-vaginal drug release (CIDR-B™, Pfizer Animal Health, Auckland, New Zealand) for 13 d. Then each of these 46 ewes received 15 mg of prostaglandin F2α (Lutalyse, Pfizer, New York, NY, USA) intramuscularly 2 d before the progesterone vaginal implant was removed. Twenty-four h after removing the progesterone vaginal implant, three artificial insemination (AI) was performed within a 12-h interval. The day of the progesterone withdrawal was considered as day 0.

Endometrial tissue recovery
All the ewes were slaughtered at Day 17 of pregnancy. Their uteri were flushed with PBS and collected. Ewes with the presence of normal elongated conceptus with a length of 25 cm or more attached to the endometrium were assigned pregnant and were sampled for proteomic analysis. Samples of the endometrial C and IC areas were taken and processed as described by Mansouri-Attia. The ipsilateral uterine horn was longitudinally opened by scissors. C areas were first carefully cut out then the IC areas were collected [4,9].

Experimental design
Given the significant structural and functional differences associated with C and IC areas, these two distinct endometrial zones need to analyze separately for a more comprehensive understanding of implantation in sheep. The global proteomic analysis of C and IC areas would provide new insights into the mechanisms responsible for implantation. Thus, endometrial C and IC areas were sampled and analyzed separately in the present study. For LC-MS/MS analysis, endometrial samples from ewes were divided into three pools for biological replicates, and each pooled sample was divided into two equal aliquots and processed as technical replicates (as shown in

Figure 1). Data for each pool were obtained by averaging results from the two technical replicates.

Protein extraction

All samples were ground to powder in liquid nitrogen and stored overnight at −20°C after adding a five-fold volume of chilled acetone containing 10% trichloroacetic acid (TCA) and 10 mmol/L dithiothreitol (DTT). Then the samples were centrifuged at 4°C, 16,000 × g for 20 min and the supernatant was discarded. The precipitates were added in 1 mL chilled acetone containing 10 mmol/L DTT before storing at −20°C for 30 min, then centrifuged at 4°C, 20,000 × g for 30 min. Centrifugation was repeated several times until the supernatant was colorless. The pellets were air-dried, then dissolved in lysis buffer containing 1 mmol/L phenylmethanesulfonyl fluoride (PMSF), 2 mmol/L ethylenediaminetetraacetic acid (EDTA) and 10 mmol/L DTT and sonicated at 200 Watts for 15 min before being centrifuged at 30,000 × g at room temperature for 30 min. The final protein concentration of supernatant was detected by using the Bradford method.

Peptide digestion

Each sample taken 50 µg protein, then all isopycnic samples were formed by adding 8 mol/L urea solution. The samples were incubated with 10 mmol/L DTT at 56°C for 1 h to reduce disulfide bonds,and then added to 55 mmol/L iodoacetamide (IAM) in a dark room for 45 min to block cysteine bonding. Subsequently, each sample was diluted 8-fold with using 50 mmol/L ammonium

bicarbonate and digested with Trypsin Gold at a protein: trypsin ratio of 20:1 at 37°C for 16 h. Final desalting using a Strata X C18 column (Phenomenex), the samples were vacuum dried. Peptides generated from digestion were directly loaded for LC-MS/MS analysis.

LC-ESI-MS/MS analysis with LTQ-orbitrap collision induced dissociation (CID)

Each sample was resuspended in buffer A (2% acetonitrile (ACN), 0.1% formic acid (FA)) and centrifuged at 20,000 × g for 10 min. The final peptide concentration for each sample was about 0.5 µg/mL. The digested samples were fractionated using a Shimadzu LC-20 AD nano-high performance liquid chromatography (nano-HPLC) system. Sample loading (10 µL) was achieved using a 2-cm C18 trapping column in line with auto-sampler, and the peptides were eluted onto a resolving 10-cm analytical C18 column prepared in-house. The samples were loaded at a flow rate of 15 µL/ min for 4 min, and then a 91- min gradient from 2% to 35% buffer B (98% ACN, 0.1% FA) was run at a flow rate of 400 nL/ min, followed by a 5- min linear gradient to 80% buffer B that was maintained for 8 min before finally returning to 2% buffer B within 2 min. The peptides were subjected to nanoelectrospray ionization and then detected by MS/MS in an LTQ Orbitrap Velos (Thermo Fisher Scientific, Bremen, Germany) connected online to a HPLC system. The whole peptides were detected in the Orbitrap analyzer at a resolution of 60,000. Peptides were selected for MS/MS using the CID operating mode with a normalized collision energy setting of 35%, and ion fragments were detected in the LTQ. One MS scan followed by ten MS/MS scans was applied for the ten most abundant precursor ions above a threshold ion count of 5,000 in the MS survey scan. Dynamic exclusion was applied to increase dynamic range and maximize peptide identifications, and the parameters were set as follows: repeat counts = 2; repeat duration = 30 s; and exclusion duration = 120 s. The electrospray voltage was 1.5 kV. Automatic gain control (AGC) was used to prevent overfilling of the ion trap; 1×10^4 ions were accumulated in the ion trap for generation of CID spectra. MS survey scans from m/z 350 to 2,000 Da.

Proteomic analysis

Mass spectra were analyzed by using the MaxQuant software (version 1.1.1.36). As the genomic data of sheep is incomplete, we generated a reference protein database by integrating the following databases and sequences of cow proteins and current known sheep proteins and removed duplicate proteins, including GenBank nr (20110403), Uniprot cow proteins (20110503), sheep proteins [http://www.livestockgenomics.csiro.au/sheep/] and cow proteins [http://genomes.arc.georgetown.edu/drupal/bovine/]. The

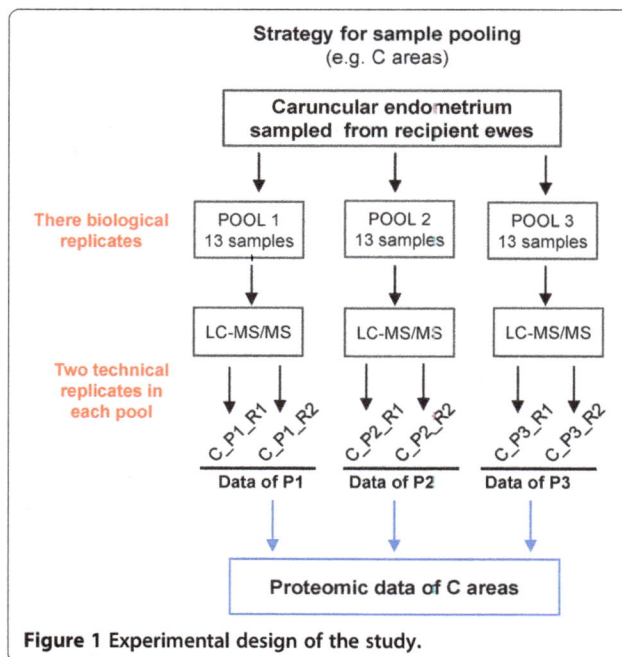

Figure 1 Experimental design of the study.

MS/MS data were searched against the reference protein database using the search engine embedded in MaxQuant. Up to two missed cleavages were allowed. The first search was set to 20 ppm, and the MS/MS tolerance for CID was set to 0.5 Da. To warrant the reliability and stability of our detection platform, following criteria for inclusion/ exclusion of peptides and proteins were used, as previous studies [15,16]: The false discovery rate (FDR) was set to 0.01 for peptide and protein identifications. Proteins were considered identified when at least two peptides were identified and at least one of which was uniquely assignable to the corresponding sequence. Contents of the protein table were filtered to eliminate identifications from the reverse database and common contaminants. In the case of identified peptides that were all shared between two proteins, these were combined and reported as one protein group. Contents of the protein table were filtered to eliminate identifications from the reverse database and common contaminants. The minimum peptide length was set to 6 amino acids. A minimum of two peptides with one being unique was required for protein identification. To perform label-free quantification analysis, the MaxQuant software suite containing an algorithm based on the extracted ion currents (XICs) of the peptides was used. Xcalibur 2.1 (Thermo Scientific) was used as quality control program to check the quality of chromatographs.

Data analysis

In data analysis, all proteins were mapped to the Ensembl Bos taurus gene ID. The expression quantity of each protein normalized on the basis of the numbers of peptides by using MaxQuant software (version 1.1.1.36). Only peptides which is corresponding to unique proteins could use to protein quantization in the comparison between peptides and proteins. In the comparison of proteomic profiles between C and IC areas, the measured value of each biological replicate was achieved by averaging every two technical replicates of a biological replicate. Then the Student's t-test was used to detect the significance of the differentially expressed proteins (DEPs) according to the measured value of every three biological replicates in each group, and $P < 0.05$ was considered significant.

We used DAVID (The Database for Annotation, Visualization and Integrated Discovery) version 6.7 platform [http://david.abcc.ncifcrf.gov/] annotate biological themes for DEPs between C and IC areas. This platform often used to analyze high-throughput data [17,18].

To assess the similarities of the different replicates, and to obtain a visual understanding of the relationship between the different areas, hierarchical clustering was carried out using CLUSTER 3.0 data analysis tool based on the clusters of protein expression profile of different technical and biological replicates.

Results

Summary of the endometrial proteome in C and IC areas

Among 46 recipient ewes, there were 39 ewes successful pregnant. Then we collected endometrium samples of these pregnant ewes at Day 17 of pregnancy (Figure 2). By LC-MS/MS proteomic analysis, we successfully detected 7459 and 7933 peptides in C areas and IC areas. After further protein identification, we identified 1740 and 1813 proteins in C areas and IC areas, relatively. Hierarchical clustering was performed based on the overall similarity of protein expression patterns of different areas and replicates. Results showed a striking separation of C and IC area samples into major opposing branches, implying that the endometrial proteomes of two areas are very distinct from each other. In addition, technical replicates in each pool were tightly clustered in the same branch, confirming the reliability of our detection system (Figure 3). The precision of quantitation between the technical replicates was evaluated by Pearson's correlation coefficient as previous study [15]. As shown in Additional file 1: Table S1 and Additional file 2: Figure S1, we found an average PCC of 0.9871 for the protein level and 0.9401 for the peptide level. According to the criteria proposed in Waanders et al.'s study (0.87 and 0.98 for peptide and protein respectively) [16], the correlation values between technical replicates should be satisfying.

Comparison of proteomic profile between C areas and IC areas

By comparing proteomic profile between C areas and IC areas, we found 170 DEPs ($P < 0.05$). Among these DEPs, 60 proteins were up-regulated in C areas, and 110 proteins were up-regulated in IC areas. The most increased protein (fold change > 5) in C areas was GLANS (N-acetylgalactosamine-6-sulfatase precursor, 5.0-fold). In

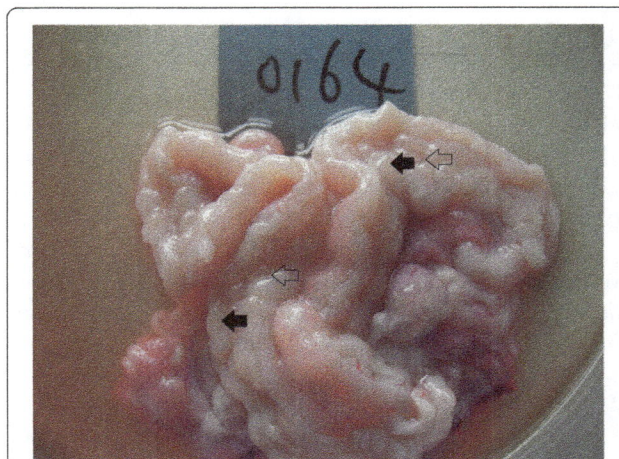

Figure 2 The endometrium of Chinese Small Tail Han ewes.
Black arrows pointed to C areas, white arrows pointed to IC areas.

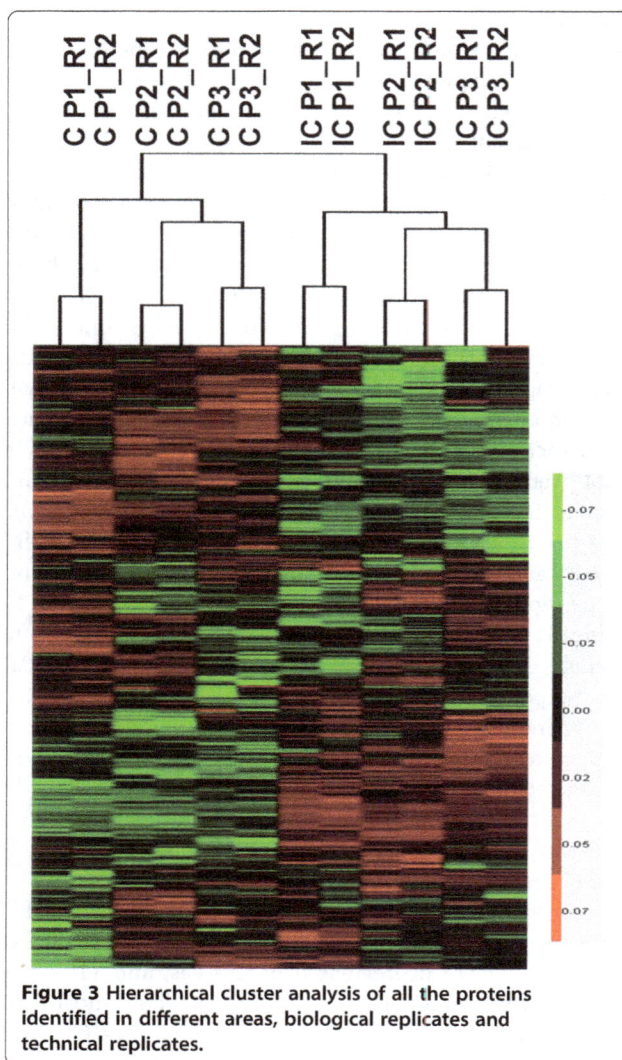

Figure 3 Hierarchical cluster analysis of all the proteins identified in different areas, biological replicates and technical replicates.

enriched Kyoto Encyclopedia of Genes and Genomes (KEGG) pathways ($P < 0.05$), including "Focal adhesion", "Regulation of actin cytoskeleton" and "ECM-receptor interaction" (Table 1).

To better understand the structural and functional differences between these two areas, we divided all the DEPs into 2 groups: 60 DEPs that were up-regulated in C areas and 110 DEPs that were up-regulated in IC areas. These two groups of DEPs were then analyzed using DAVID platform respectively. For DEPs up-regulated in C areas, we identified 12 significant enriched GO terms ($P < 0.01$), based on the major category of "biological process". "primary metabolic processes", "cellular metabolic processes", "cellular protein metabolic process" and "proteolysis involved in cellular protein catabolic process" were the most represented processes (Figure 6). KEGG pathway analysis just identified 1 significant pathway ($P < 0.05$), which is "Valine, leucine and isoleucine degradation".

For DEPs that were up-regulated in IC areas, we found 26 significant enriched GO categories ($P < 0.01$) based on the major category of "biological process", including "cellular process", "cellular component organization", "actin cytoskeleton organization", "multicellular organismal development" and "cell adhesion" (Figure 7). Our analysis also identified 7 significant enriched KEGG pathways ($P < 0.05$), which were major involved in cell adhesion and cell migration. These pathways included "Focal adhesion", "ECM-receptor interaction" and "Tight junction" (Table 2).

Besides, we also used DAVID Bioinformatics Resources 6.7 to analyze the specifically expressed proteins in IC areas. Finally, we found these proteins were significant enriched ($P < 0.01$) in two GO categories, "organic acid metabolic process" and "regulation of cell communication".

Discussion

The endometrium of sheep consists of C and IC areas, both of which are essential for the establishment of implantation and maintenance of pregnancy. In order to study the different contribution of C areas and IC areas to the implantation, we used LC-MS/MS technique to profile the proteome of these two areas separately at Day 17 of pregnancy. By comparing the proteomic profiles between C areas and IC areas, we found many biologically meaningful DEPs. Based on these DEPs, bioinformatics analysis revealed the different biological functions between these two areas on protein level. These important functional differences were mainly associated with growth and remodeling of endometrial tissue, cell adhesion and protein transport. Our results provided data references and materials for further studies. It should be noted that these two regions contain varying amounts of different cell types, including LE, GE, stroma, blood and lymph vessels, and immune cells, and the gene or protein

IC areas, the highest increased proteins (fold change > 5) in C areas included PIP4K2C (Phosphatidylinositol 5-phosphate 4-kinase type-2 gamma, 16.2-fold), PLIN4 (Uncharacterized protein, 6.6-fold), EML4 (echinoderm microtubule-associated protein-like 4, 6.6-fold) and ITGA1 (integrin, alpha 1, 6.1-fold). In addition, we detected 3 proteins specifically expressed in the C areas, and 22 proteins specifically expressed in the IC areas (Figure 4).

To gain insight into different biological functions involved in implantation associated with C and IC areas, the 170 DEPs were analyzed using Functional Annotation Tool of DAVID Bioinformatics Resources 6.7. Finally, we identified 43 significant enriched Gene Ontology (GO) categories ($P < 0.01$) based on the major category of "biological process", including "primary metabolic processes", "cellular metabolic process", "protein metabolic process", "cell adhesion" and "multicellular organismal development" (Figure 5). The analysis also identified 9 significant

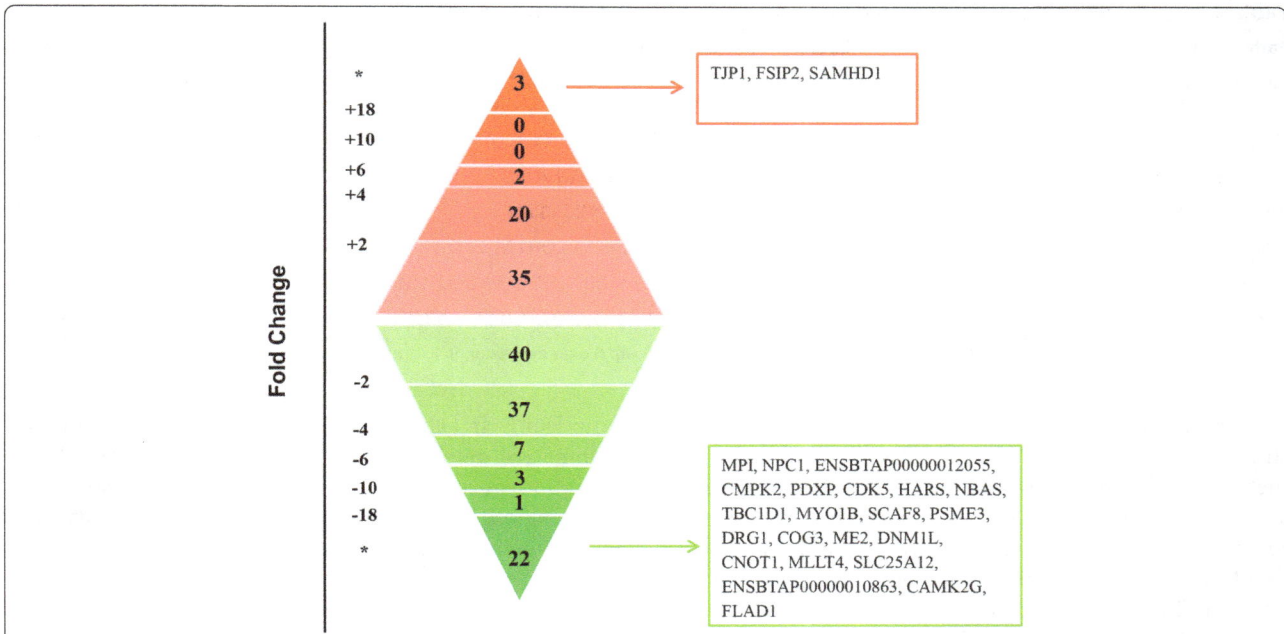

Figure 4 Distribution of differentially expressed proteins with different fold change in the comparison between C and IC areas. "*" means these proteins specifically expressed in C or IC areas.

expression pattern shows temporal and spatial changes in endometrium [19], Therefore, previous study have profiled the endometrial transcriptome of individual cell populations using laser capture microdissection [20]. However, in the most of the studies concerning the endometrium during establishment of pregnancy in ruminants, endometrium was sampled as an integrated tissue layer for studying the endometrium involved mechanisms of pregnancy establishment or maternal-fetal interaction [2,4,9,21,22], In addition, due to the relative large requirements of sample amount for LC-MS/MS analysis, profiling the proteome of a single cell type in endometrium is

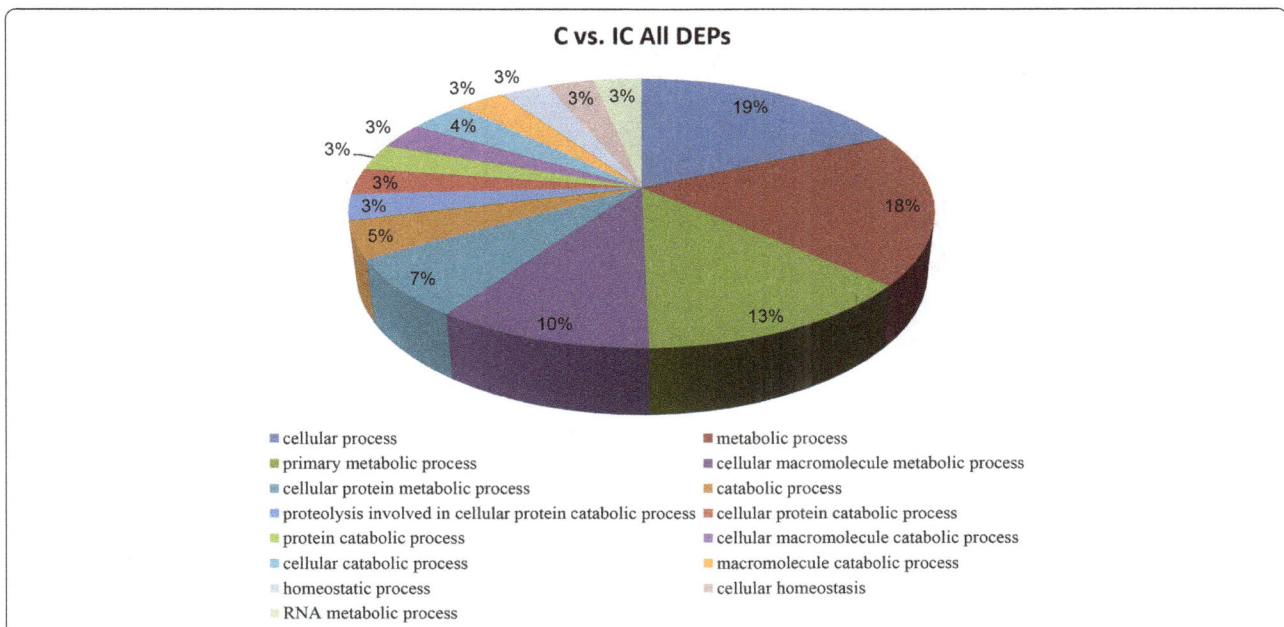

Figure 5 Function annotation clustering results with differentially expressed proteins between C and IC areas based on GO biological process.

Table 1 Significantly enriched KEGG pathways for DEPs between C and IC areas

Pathway name	Count	Differentially expressed proteins	P-value
Focal adhesion	14	ACTG1,ACTN1,COL1A2,COL6A1,COL6A2,FLNA,ITGA1,LAMA4,LAMC1, MYLK,MYL9,COL6A3,TLN1,VCL	6.3E-6
ECM-receptor interaction	8	COL1A2,COL6A1,COL6A2,HSPG2,ITGA1,LAMA4,,LAMC1,COL6A3	2.0E-4
Pyruvate metabolism	4	**ALDH7A1**,**DLD**,LDHA,ME2	2.2E-2
Adherens junction	5	ACTG1,ACTN1,MLT4,**TJP1**,VCL	2.4E-2
Valine, leucine and isoleucine degradation	4	**HMGCL**,**ACAA1**,**ALDH7A1**,**DLD**	3.5E-2
Regulation of actin cytoskeleton	8	ACTG1,ACTN1,GSN,ITGA1,MYLK,MYL9,PIP4K2C,VCL	3.5E-2
Tight junction	6	ACTG1,ACTN1,MLT4,MYH11,MYL9,**TJP1**	4.9E-2

Proteins with higher expression level in C areas were shown in boldface; proteins with higher expression in IC areas were shown in lightface.

impractical. Therefore, in our study, a wildly accepted strategy was used to establish the sample pools of endometrium with different cell types. The collection of tissue samples were performed strictly according to the method reported in the previous studies [4,9]. We did not provide the detailed proteomic difference between the distinct cell types in C and IC areas, however, this limitation did not compromise the significance of our study.

Growth and remodeling of endometrial tissue

During implantation, ovine endometrium always undergoes extensive and sufficient structural modification and tissue growth, in order to accommodate and support rapid conceptus development and growth [3]. The series of biological processes were referred to as endometrial

remodeling. It has been demonstrated that appropriate endometrial remodeling is essential for successful implantation during early pregnancy [3]. Given structural and functional differences between C and IC areas, they have different ways in remodeling.

As C areas are major responsible for conceptus attachment and placentation, the substantial remodeling in these areas are crucial for successful implantation. In the present study, GO category "proteolysis" (PSMB3, NEDD8, UCHL3 and USP34) was significantly over-represented in C areas. Among these proteins, PSMB3 is a member of proteasome. The proteasome is an ATP-dependent, multisubunit, multi-catalytic protease complex that is involved in recognizing and degrading ubiquitinated proteins in most non-lysosomal pathway of protein degradation in

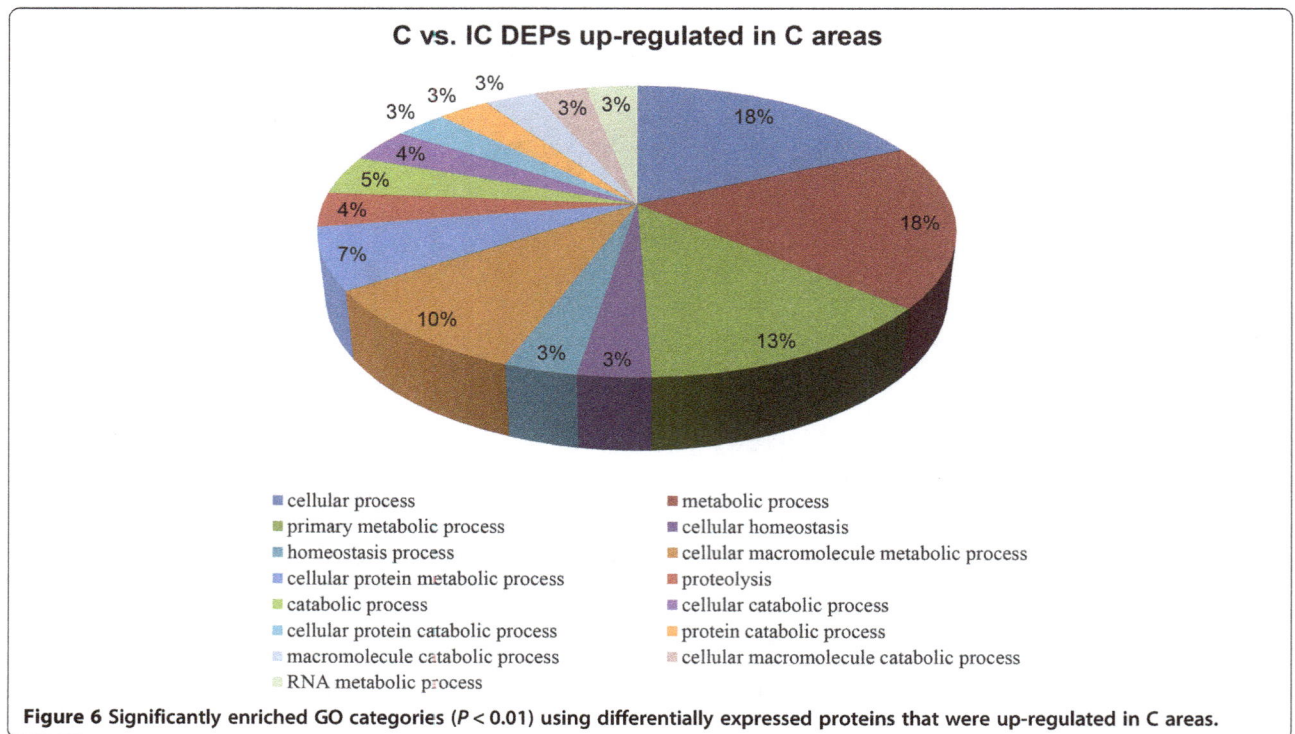

Figure 6 Significantly enriched GO categories (P < 0.01) using differentially expressed proteins that were up-regulated in C areas.

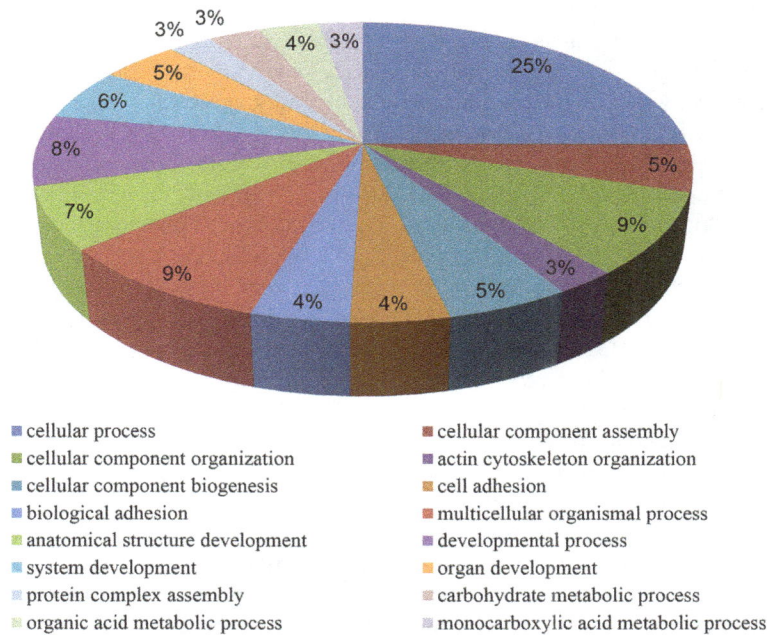

Figure 7 Significantly enriched GO categories ($P < 0.01$) using differentially expressed proteins that were up-regulated in IC areas.

eukaryotic cells [23]. It has been reported that proteasome participate in tissue remodeling of the conceptus and endometrium during pregnancy in human and mouse [24]. Up to now, however, only few studies have reported the specific functions of proteasome in ovine endometrium during early pregnancy. Our results indicated that proteasome are likely to participate in the tissue remodeling of endometrial C areas through degrading proteins during implantation. NEDD8, UCHL3 and USP4 are highly associated with ubiquitination. Many previous studies have demonstrated that ubiquitination-related proteins are expressed in cells of uterus in many species, and these proteins play vital roles in endometrial remodeling and placentation during pregnancy [25]. The up-regulation of above mentioned proteins in C areas indicated that C

areas undergo extensive and substantial tissue remodeling through ubiquitin-proteasome system (UPS) based proteolysis during the peri-implantation period. Besides, we found the expression of TAGLN was drastically increased in C areas (fold-change = 2.15). TAGLN is constitutively expressed in pregnant uterus, and previously regarded as a biomarker for arterial vessel remodeling in uterine tissue [14]. The up-regulation of TAGLN in C areas implied that vascular remodeling may play crucial roles in endometrial remodeling preparing for implantation. As C areas are key tissues for placentation, the increased vascular remodeling in C areas would also be essential for ensuring normal placental function.

Unlike C areas, there are many endometrial glands localized in IC areas. During the peri-implantation period,

Table 2 Significantly enriched KEGG pathways ($P < 0.05$) using proteins with higher expression levels in IC areas

Pathway name	Count	Differentially expressed proteins	P-value
Focal adhesion	14	ACTG1,ACTN1,COL1A2,COL6A1,COL6A2,FLNA,ITGA1,LAMA4,LAMC1, MYLK,MYL9,COL6A3,TLN1,VCL	7.1E-8
ECM-receptor interaction	8	COL1A2,COL6A1,COL6A2,HSPG2,ITGA1,LAMA4,,LAMC1,COL6A3	1.8E-5
Regulation of actin cytoskeleton	8	ACTG1,ACTN1,GSN,ITGA1,MYLK,MYL9,PIP4K2C,VCL	5.2E-3
Arrhythmogenic right ventricular cardiomyopathy (ARVC)	4	ACTG1,ACTN1,DES,ITGA1	3.7E-2
Leukocyte transendothelial migration	5	ACTG1,ACTN1,MLT4,MYL9,VCL	3.7E-2
Adherens junction	4	ACTG1,ACTN1,MLT4,VCL	4.0E-2
Tight junction	5	ACTG1,ACTN1,MLT4,MYH11,MYL9	4.9E-2

these endometrial glands of the IC areas experience substantially growth in length and width, to ensure normal synthesis of nutrition to support the development of conceptus [3]. Thus, we can conclude that IC areas also underwent a considerable remodeling process in the form of proliferation and morphogenesis of the intercaruncular endometrial glands. In our study, many DEPs that were up-regulated in the IC areas were significantly enriched in GO categories and KEGG pathways functionally associated with endometrial gland morphogenesis. Most of these proteins, including ACTA2, ACTG1, ACTN1, MYLK and MYL9, are the structural constituents of cytoskeleton, which is essential for morphologic changes of tissues and cells [26]. In these proteins, ACTA2, ACTG1 and ACTN1 belong to actin family of proteins. It has been reported that actin plays important roles in cell motility, structure, and integrity [27]. MYLK and MYL9 belong to myosin, which is responsible for muscular movement. Myosin may participate in cell motility and contraction in endometrial glands. The up-regulation of all above proteins in IC areas indicated that endometrial glands of IC areas undergo extensive changes in cell morphology during the peri-implantation period, which may prepare for increased nutrient production. In addition, many proteins specifically expressed in IC areas were enriched in GO categories related to cell proliferation, including CAMK2G, CDK5, CMPK2 and HARS. CDK5 belongs to cyclin dependent kinases, which is involved in regulating cell proliferation and differentiation, as well as apoptosis. These specifically expressed proteins made us to speculate that IC areas underwent an increased cell proliferation compared with C areas during the peri-implantation period. Together our findings in IC areas, is presumable that both cytoskeleton remodeling and cell proliferation play crucial roles in endometrial gland morphogenesis.

Based on all above results, we provided proteomics evidence that both C areas and IC areas were characterized by a considerable remodeling process, which was a preparation for successful implantation. However, these two areas participated in endometrial remodeling through different patterns. The tissue remodeling in C areas was mainly through proteasome-dependent proteolysis and vascularization. Unlike C areas, IC areas underwent endometrial gland morphogenesis through accelerating cytoskeleton remodeling and cell proliferation.

Cell adhesion

Cell adhesion is one of the most important physiological processes involved in implantation. Abnormal cell adhesion may lead to an early pregnancy failure. In the present study, many DEPs were functionally associated with cell adhesion.

We found two proteins related to cell adhesion, TJP1 and POSTN, were specifically expressed in C areas. TJP1

belongs to tight junction proteins. Tight junctions are always located on the plasma membrane, and maintain cell polarity, cell-cell contact, and cell adhesion [28]. POSTN, an important extracellular matrix (ECM) protein, plays an important role in facilitating cell adhesion and migration. It has been reported that POSTN may induce PI3K/AKT signaling to stimulate attachment and migration of ovine trophectoderm (oTr) cells during early pregnancy in ewes. Besides, POSTN may influence placentation and placental functions [27]. Therefore, our results indicated that cell adhesion, as an very important and specialized physiological process in the C areas during the peri-implantation period, were ensured by specifically expression of related proteins.

We also found many DEPs up-regulated in the IC areas were significantly enriched in GO categories and KEGG pathways related to cell adhesion. GO category "cell junction" (TLN1, TNS1 and VCL) was overrepresented in the IC areas. These three proteins are important cytoskeleton proteins. They play crucial roles in integrin-mediated cell adhesion, migration and proliferation by activating integrin signaling pathway [29,30]. Besides, KEGG pathways "ECM-receptor interaction" and "focal adhesion" (COL1A2, COL6A2, LAMA4, LAMC1 and ITGA1) were over-represented in IC areas. Among these proteins, ITGA1 was dramatically increased in IC areas (fold-change = 6.07). ITGA is one kind of integrins. Integrins are a family of cell adhesion molecules that have now been largely accepted as biomarkers of uterine receptivity [31]. During implantation, integrins are responsible for mediating adhesion between maternal and embryonic epithelium [32,33]. COL1A2, COL6A2, LAMA4, LAMC1 belong to ECM. ECM participates in cell adhesion mainly through binding to cell adhesion molecules. These results indicated that contrast with traditional knowledge, IC areas may also played important roles in cell adhesion through many biological processes and pathways, such as cell junction, ECM-receptor interaction and focal adhesion.

Taken together, we found both C areas and IC areas participate in cell adhesion during the peri-implantation period. Considering that C areas are the major sites for conceptus attachment, most previous studies related to cell adhesion during implantation focused on these areas. However, our results indicated that IC areas also played important roles in cell adhesion. One possible explanation was that endometrial glands of IC areas may synthesize a lot of proteins related to cell adhesion, such as cytoskeletal proteins, integrins, collagens and laminins, and then secrete them into C areas or uterine lumen to participate in cell adhesion. Thus, we inferred that C areas mediate cell adhesion during implantation in collaboration with IC areas.

Protein transport

In ruminants, the growth and development of the conceptus depend largely on endometrial glands in IC areas. During early pregnancy, these glands synthesize, secrete or transport various enzymes, growth factors, cytokines, hormones, transport proteins and adhesion molecules, collectively referred as histotroph. Gray et al. reported that secretions of endometrial glands were required for peri-implantation conceptus survival and development, and uterine gland knockout ewes (UGKO) were unable to support pregnancy up to Day 25 of pregnancy [34]. Besides, many studies demonstrated that some secretions of endometrial glands are the primary regulators of fetal-maternal dialogues.

In the present study, GO category "protein transport" (HSPA8, SYPL1 and TEMD10) was significantly overrepresented in IC areas. HSPA8 functions as an ATPase during transport of membrane proteins through the cell. SYPL1 is associated with GLUT4-cotaining vesicles [35]. GLUT4 has been found in the human placenta, which is mainly responsible for controlling uptake of glucose into cells [36]. Therefore, the reduced expression of GLUT4 in endometrial cells may lead to metabolic defects, which has adverse effect on implantation. TMED10 is type I membrane protein, which is localized to the plasma membrane and golgi cisternae and is involved in vesicular protein trafficking.

Together these results, we found protein transport is an important physiological process associated with IC areas. Compared to C areas, protein transport played a more important role in IC areas. Abnormal protein transport may have adverse effects on secretion function of endometrial glands, leading to implantation failure.

Conclusion

We established a relatively comprehensive and detailed protein database of ovine endometrium by using LC-MS/MS technique, providing data references for further researches. By comparing proteomic profiles between C and IC areas, we detected 170 DEPs. Subsequent bioinformatics analysis indicated these DEPs were mainly involved in a lot of important physiological processes, including growth and remodeling of endometrial tissue, cell adhesion and protein transport. Based on these results, we provided proteomics evidence that both C areas and IC areas were characterized by a considerable remodeling process, which was a preparation for successful implantation. In addition, we found that C areas could mediate cell adhesion during implantation in collaboration with IC areas. Lastly, results showed that protein transport was an important physiological process associated with endometrial glands, which played a more important role in IC areas relatively.

Additional files

> **Additional file 1: Table S1.** Summary of Pearson's correlation coefficient (R) between technical replicates in each pool of different groups.
>
> **Additional file 2: Figure S1.** Pearson correlation coefficient (PCC) of protein level between technique replicates in each pool of different groups. The reliability of protein quantitation between the technical replicates was evaluated by PCC.

Abbreviations

C: Caruncles; IC: Intercaruncular areas; DEPs: Differentially expressed proteins; P4: Progesterone; IFNT: Interferon tau; LC-MS/MS: Liquid chromatography-tandem mass spectrometry; CIDR: Controlled intra-vaginal drug release; AI: Artificial insemination; TCA: Trichloroacetic acid; DTT: Dithiothreitol; PMSF: Phenylmethanesulfonyl fluoride; EDTA: Ethylenediaminetetraacetic acid; IAM: Iodoacetamide; ACN: Acetonitrile; FA: Formic acid; nano-HPLC: Nano-high performance liquid chromatography; UPS: Ubiquitin-proteasome system; oTr: Ovine trophectoderm; UGKO: Uterine gland knockout.

Competing interests

The authors declare that they have no competing interests.

Authors' contributions

YW and CW analyzed the data and drafted the manuscript. LA designed the study and revised the manuscript. ZCH preformed the statistical analysis. KM, HCZ, RW and MG participated in the collection of samples. JHT and ZHW participated in the design of the study. All authors have read and approved the final manuscript.

Acknowledgements

We thank our laboratory members for their helpful comments on the manuscript. This work was supported by grants from the National Key Technology R&D Program (Nos. 2011BAD19B01, 2011BAD19B03, 2011BAD19B04) and the National High-Tech R&D Program (No. 2011AA100303, 2013AA102506).

Author details

[1]Ministry of Agriculture Key Laboratory of Animal Genetics, Breeding and Reproduction, National engineering laboratory for animal breeding, College of Animal Sciences and Technology, China Agricultural University, No.2 Yuanmingyuan Xi Lu, Haidian, Beijing 100193, China. [2]National Engineering Laboratory for Animal Breeding and MOA Key Laboratory of Animal Genetics and Breeding, China Agricultural University, Beijing 100193, China. [3]State Key Laboratory of Animal Nutrition, College of Animal Sciences and Technology, China Agricultural University, No.2 Yuanmingyuan Xi Lu, Haidian, Beijing 100193, China.

References

1. Bazer FW, Song G, Kim J, Dunlap KA, Satterfield MC, Johnson GA, Burghardt RC, Wu G: Uterine biology in pigs and sheep. *J Anim Sci Biotechnol* 2012, **3**(1):23.
2. Walker CG, Meier S, Littlejohn MD, Lehnert K, Roche JR, Mitchell MD: Modulation of the maternal immune system by the pre-implantation embryo. *BMC Genomics* 2010, **11**:474.
3. Lee KY, DeMayo FJ: Animal models of implantation. *Reproduction* 2004, **128**(6):679–695.
4. Mansouri-Attia N, Sandra O, Aubert J, Degrelle S, Everts RE, Giraud-Delville C, Heyman Y, Galio L, Hue I, Yang X, *et al*: Endometrium as an early sensor of in vitro embryo manipulation technologies. *Proc Natl Acad Sci U S A* 2009, **106**(14):5687–5692.
5. Satterfield MC, Song G, Kochan KJ, Riggs PK, Simmons RM, Elsik CG, Adelson DL, Bazer FW, Zhou H, Spencer TE: Discovery of candidate genes and pathways in the endometrium regulating ovine blastocyst growth and conceptus elongation. *Physiol Genomics* 2009, **39**(2):85–99.
6. Spencer TE, Johnson GA, Bazer FW, Burghardt RC: Implantation mechanisms: insights from the sheep. *Reproduction* 2004, **128**(6):657–668.
7. Igwebuike UM: A review of uterine structural modifications that influence conceptus implantation and development in sheep and goats. *Anim Reprod Sci* 2009, **112**(1–2):1–7.

8. Forde N, Lonergan P: Transcriptomic analysis of the bovine endometrium: what is required to establish uterine receptivity to implantation in cattle? *J Reprod Dev* 2012, **58**(2):189–195.

9. Mansouri-Attia N, Aubert J, Reinaud P, Giraud-Delville C, Taghouti G, Galio L, Everts RE, Degrelle S, Richard C, Hue I, *et al*: Gene expression profiles of bovine caruncular and intercaruncular endometrium at implantation. *Physiol Genomics* 2009, **39**(1):14–27.

10. Gray CA, Abbey CA, Beremand PD, Choi Y, Farmer JL, Adelson DL, Thomas TL, Bazer FW, Spencer TE: Identification of endometrial genes regulated by early pregnancy, progesterone, and interferon tau in the ovine uterus. *Biol Reprod* 2006, **74**(2):383–394.

11. Savaris RF, Hamilton AE, Lessey BA, Giudice LC: Endometrial gene expression in early pregnancy: lessons from human ectopic pregnancy. *Reprod Sci* 2008, **15**(8):797–816.

12. Bethin KE, Nagai Y, Sladek R, Asada M, Sadovsky Y, Hudson TJ, Muglia LJ: Microarray analysis of uterine gene expression in mouse and human pregnancy. *Mol Endocrinol* 2003, **17**(8):1454–1469.

13. Mullen MP, Elia G, Hilliard M, Parr MH, Diskin MG, Evans AC, Crowe MA: Proteomic characterization of histotroph during the preimplantation phase of the estrous cycle in cattle. *J Proteome Res* 2012, **11**(5):3004–3018.

14. Koch JM, Ramadoss J, Magness RR: Proteomic profile of uterine luminal fluid from early pregnant ewes. *J Proteome Res* 2010, **9**(8):3878–3885.

15. Geiger T, Wehner A, Schaab C, Cox J, Mann M: Comparative proteomic analysis of eleven common cell lines reveals ubiquitous but varying expression of most proteins. *Mol Cell Proteomics* 2012, **11**(3):M111–M14050.

16. Waanders LF, Chwalek K, Monetti M, Kumar C, Lammert E, Mann M: Quantitative proteomic analysis of single pancreatic islets. *Proc Natl Acad Sci U S A* 2009, **106**(45):18902–18907.

17. Huang DW, Sherman BT, Tan Q, Kir J, Liu D, Bryant D, Guo Y, Stephens R, Baseler MW, Lane HC, *et al*: DAVID bioinformatics resources: expanded annotation database and novel algorithms to better extract biology from large gene lists. *Nucleic Acids Res* 2007, **35**:W169–W175. Web Server issue.

18. Huang DW, Sherman BT, Lempicki RA: Systematic and integrative analysis of large gene lists using DAVID bioinformatics resources. *Nat Protoc* 2009, **4**(1):44–57.

19. Spencer TE, Sandra O, Wolf E: Genes involved in conceptus-endometrial interactions in ruminants: insights from reductionism and thoughts on holistic approaches. *Reproduction* 2008, **135**(2):165–179.

20. Niklaus AL, Pollard JW: Mining the mouse transcriptome of receptive endometrium reveals distinct molecular signatures for the luminal and glandular epithelium. *Endocrinology* 2006, **147**(7):3375–3390.

21. Bauersachs S, Ulbrich SE, Gross K, Schmidt SE, Meyer HH, Wenigerkind H, Vermehren M, Sinowatz F, Blum H, Wolf E: Embryo-induced transcriptome changes in bovine endometrium reveal species-specific and common molecular markers of uterine receptivity. *Reproduction* 2006, **132**(2):319–331.

22. Klein C, Bauersachs S, Ulbrich SE, Einspanier R, Meyer HH, Schmidt SE, Reichenbach HD, Vermehren M, Sinowatz F, Blum H, *et al*: Monozygotic twin model reveals novel embryo-induced transcriptome changes of bovine endometrium in the preattachment period. *Biol Reprod* 2006, **74**(2):253–264.

23. Voges D, Zwickl P, Baumeister W: The 26S proteasome: a molecular machine designed for controlled proteolysis. *Annu Rev Biochem* 1999, **68**:1015–1068.

24. Bebington C, Doherty FJ, Fleming SD: The possible biological and reproductive functions of ubiquitin. *Hum Reprod Update* 2001, **7**(1):102–111.

25. Bebington C, Doherty FJ, Fleming SD: Ubiquitin and ubiquitin-protein conjugates are present in human cytotrophoblast throughout gestation. *Early Pregnancy* 2000, **4**(4):240–252.

26. Chae JI, Kim J, Lee SG, Jeon YJ, Kim DW, Soh Y, Seo KS, Lee HK, Choi NJ, Ryu J, *et al*: Proteomic analysis of pregnancy-related proteins from pig uterus endometrium during pregnancy. *Proteome Sci* 2011, **9**:41.

27. Ahn HW, Farmer JL, Bazer FW, Spencer TE: Progesterone and interferon tau-regulated genes in the ovine uterine endometrium: identification of periostin as a potential mediator of conceptus elongation. *Reproduction* 2009, **138**(5):813–825.

28. Satterfield MC, Dunlap KA, Hayashi K, Burghardt RC, Spencer TE, Bazer FW: Tight and adherens junctions in the ovine uterus: differential regulation by pregnancy and progesterone. *Endocrinology* 2007, **148**(8):3922–3931.

29. Roberts GC, Critchley DR: Structural and biophysical properties of the integrin-associated cytoskeletal protein talin. *Biophys Rev* 2009, **1**(2):61–69.

30. Humphries JD, Wang P, Streuli C, Geiger B, Humphries MJ, Ballestrem C: Vinculin controls focal adhesion formation by direct interactions with talin and actin. *J Cell Biol* 2007, **179**(5):1043–1057.

31. MacIntyre DM, Lim HC, Ryan K, Kimmins S, Small JA, MacLaren LA: Implantation-associated changes in bovine uterine expression of integrins and extracellular matrix. *Biol Reprod* 2002, **66**(5):1430–1436.

32. Lessey BA: Endometrial integrins and the establishment of uterine receptivity. *Hum Reprod* 1998, **13**(3):247–258. 259–261.

33. Johnson GA, Bazer FW, Jaeger LA, Ka H, Garlow JE, Pfarrer C, Spencer TE, Burghardt RC: Muc-1, integrin, and osteopontin expression during the implantation cascade in sheep. *Biol Reprod* 2001, **65**(3):820–828.

34. Gray CA, Taylor KM, Ramsey WS, Hill JR, Bazer FW, Bartol FF, Spencer TE: Endometrial glands are required for preimplantation conceptus elongation and survival. *Biol Reprod* 2001, **64**(6):1608–1613.

35. Yokouchi K, Mizoguchi Y, Watanabe T, Iwamoto E, Sugimoto Y, Takasuga A: Identification of a 3.7-Mb region for a marbling QTL on bovine chromosome 4 by identical-by-descent and association analysis. *Anim Genet* 2009, **40**(6):945–951.

36. Hay WJ: Placental-fetal glucose exchange and fetal glucose metabolism. *Trans Am Clin Climatol Assoc* 2006, **117**:321–339. 339–340.

17

Emerging evidence of the physiological role of hypoxia in mammary development and lactation

Yong Shao and Feng-Qi Zhao[*]

Abstract

Hypoxia is a physiological or pathological condition of a deficiency of oxygen supply in the body as a whole or within a tissue. During hypoxia, tissues undergo a series of physiological responses to defend themselves against a low oxygen supply, including increased angiogenesis, erythropoiesis, and glucose uptake. The effects of hypoxia are mainly mediated by hypoxia-inducible factor 1 (HIF-1), which is a heterodimeric transcription factor consisting of α and β subunits. HIF-1β is constantly expressed, whereas HIF-1α is degraded under normal oxygen conditions. Hypoxia stabilizes HIF-1α and the HIF complex, and HIF then translocates into the nucleus to initiate the expression of target genes. Hypoxia has been extensively studied for its role in promoting tumor progression, and emerging evidence also indicates that hypoxia may play important roles in physiological processes, including mammary development and lactation. The mammary gland exhibits an increasing metabolic rate from pregnancy to lactation to support mammary growth, lactogenesis, and lactation. This process requires increasing amounts of oxygen consumption and results in localized chronic hypoxia as confirmed by the binding of the hypoxia marker pimonidazole HCl in mouse mammary gland. We hypothesized that this hypoxic condition promotes mammary development and lactation, a hypothesis that is supported by the following several lines of evidence: i) Mice with an HIF-1α deletion selective for the mammary gland have impaired mammary differentiation and lipid secretion, resulting in lactation failure and striking changes in milk compositions; ii) We recently observed that hypoxia significantly induces HIF-1α-dependent glucose uptake and GLUT1 expression in mammary epithelial cells, which may be responsible for the dramatic increases in glucose uptake and GLUT1 expression in the mammary gland during the transition period from late pregnancy to early lactation; and iii) Hypoxia and HIF-1α increase the phosphorylation of signal transducers and activators of transcription 5a (STAT5a) in mammary epithelial cells, whereas STAT5 phosphorylation plays important roles in the regulation of milk protein gene expression and mammary development. Based on these observations, hypoxia effects emerge as a new frontier for studying the regulation of mammary development and lactation.

Keywords: Glucose transporter, Hypoxia, Hypoxia inducible factor, Lactation, Mammary development, Metabolism

Introduction

Oxygen is critical for cellular aerobic metabolism in many higher organisms, including mammals, as it is the final electron acceptor in the electron transport chain of oxidative phosphorylation in mitochondria (Figure 1). Aerobic metabolism is 19 times more efficient in energy production than anaerobic metabolism: a molecule of glucose, the major energy source in most mammalian cells, produces up to 38 molecules of ATP in aerobic metabolism though only 2 in anaerobic metabolism [1]. In addition to efficient energy production, aerobic metabolism produces the end product H_2O, whereas anaerobic metabolism produces lactate, which is normally removed in the liver with the requirement of oxygen in mammals. Therefore, an oxygen supply and the maintenance of oxygen homeostasis are essential in aerobic organisms.

Tissues in the body are exposed to different levels of oxygen, which are affected by the oxygen supply and tissue metabolic rate and range from near 159 mm Hg in the upper airway [the O_2 partial pressure in the atmosphere at sea level (pO_2); 21% of atmospheric air] to as low as 5 mm Hg (~1%) in the retina [2]. In most tissues, the oxygen tension is between 10 and 45 mm Hg (3-6%) [3,4]. Hypoxia is a deficiency of oxygen supply in the entire body (such as in high elevation exposure) or locally within

* Correspondence: fzhao@uvm.edu
Laboratory of Lactation and Metabolic Physiology, Department of Animal Science, University of Vermont, Burlington, Vermont 05405, USA

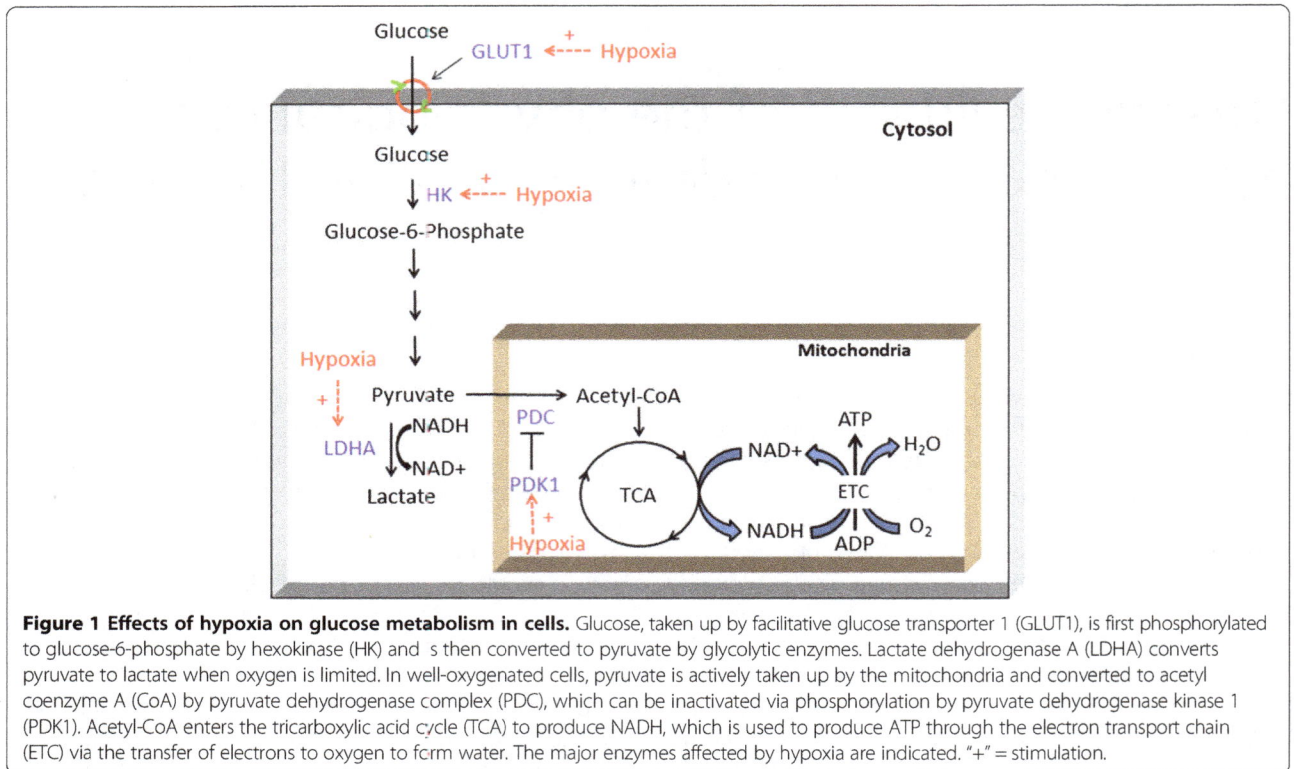

Figure 1 Effects of hypoxia on glucose metabolism in cells. Glucose, taken up by facilitative glucose transporter 1 (GLUT1), is first phosphorylated to glucose-6-phosphate by hexokinase (HK) and s then converted to pyruvate by glycolytic enzymes. Lactate dehydrogenase A (LDHA) converts pyruvate to lactate when oxygen is limited. In well-oxygenated cells, pyruvate is actively taken up by the mitochondria and converted to acetyl coenzyme A (CoA) by pyruvate dehydrogenase complex (PDC), which can be inactivated via phosphorylation by pyruvate dehydrogenase kinase 1 (PDK1). Acetyl-CoA enters the tricarboxylic acid cycle (TCA) to produce NADH, which is used to produce ATP through the electron transport chain (ETC) via the transfer of electrons to oxygen to form water. The major enzymes affected by hypoxia are indicated. "+" = stimulation.

a tissue when the oxygen delivery or availability is reduced due to different causes (such as tissue injury). The precise hypoxic pO_2 is different between organs, but a venous pO_2 of below 6% O_2 can induce a hypoxic response in most tissues, and 0.5-1% induces the maximal effects [5].

The body and its tissues have specific mechanisms to sense oxygen levels and make the necessary adaptations for survival under hypoxic conditions. Global oxygen is sensed by central chemoreceptors located on ventro-lateral surface of the medulla oblongata and peripheral chemoreceptors in the aortic and carotid bodies [6-8]. These chemoreceptors control the respiration and heart rates to adjust the oxygen supply in the entire body. In localized hypoxia and chronic hypoxia, the tissues and cells can also sense decreases in oxygen tension through transcription factor complexes known as hypoxia-inducible factors (HIFs) to restore homeostasis [2,9].

Cellular hypoxia signaling and HIFs

The effects of hypoxia in many tissues and cells are primarily mediated by HIF-1 [10]. HIF-1 is a heterodimer protein consisting of an HIF-1α subunit and an HIF-1β subunit (Figure 2). Both subunits are similar in structure and contain an N-terminal basic helix-loop-helix (bHLH) domain, a Per-ARNT-Sim (PAS) domain, and a C-terminal domain [11]. The PAS domain facilitates the heterodimerization of HIF-1α and HIF-1β [12], the bHLH domain mediates the binding of HIF-1 to the consensus hypoxia

response element (HRE) 5′-RCGTG-3′(R = A or G) in target gene promoters or enhancers [9], and the C-terminus of the proteins contains a transactivation domain and recruits transcriptional co-factors. HIF-1β is constitutively expressed in cells, whereas HIF-1α is under tight regulation by oxygen levels [13]. Under normoxic conditions, specific proline residues of HIF-1α are hydroxylated by prolyl hydroxylase domain proteins (PHDs) [14] (Figure 2), and hydoxylated HIF-1α is bound by the Von Hippel-Lindau (vHL) protein, a component of the E3 ubiquitin ligase complex, leading to HIF-1α ubiquitination and subsequent proteasomal degradation [15-17] (Figure 2). Thus, the transcriptional activity of HIF-1 is primarily determined by the cellular level of HIF-1α protein.

In addition to HIF-1α degradation, normoxia also regulates the transcriptional activity of HIF-1α, where a C-terminal asparagine in the transactivation domain of HIF-1α is hydroxylated by factor inhibiting HIF-1 (FIH). This hydroxylation of HIF-1α blocks its interaction with the transcriptional coactivator p300 and CREB binding protein (CBP) [18,19], thus disrupting the proper assembly if HIF-1 at the HRE.

Cells sense oxygen availability through PHDs and FIH, both of which are 2-oxoglutarate-dependent dioxygenases that require oxygen as a substrate. One oxygen atom is used to hydroxylate a proline or asparagine of HIF-1α, and the other oxygen atom is added to the co-substrate α-ketoglutarate, converting it to carbon dioxide

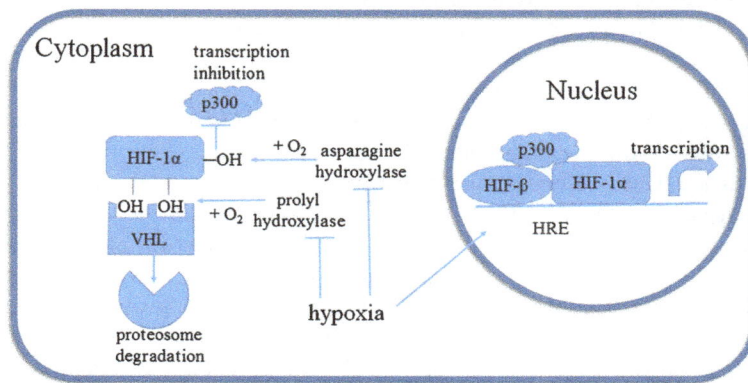

Figure 2 The HIF-1α protein is stabilized under hypoxia and translocates to the nucleus to activate gene transcription. Two proline residues of HIF-1α are hydroxylated by prolyl hydroxylase under normoxia. Hydroxylated HIF-1α is bound by the Von Hippel-Lindau (vHL) protein, leading to its ubiquitination and subsequent degradation by the proteasome. In addition, a C-terminal asparagine in the transactivation domain of HIF-1α is hydroxylated by factor inhibiting HIF-1 (FIH) under normoxia. The hydroxylation of HIF-1α blocks its interaction with the transcriptional coactivator p300 and CREB binding protein (CBP), thereby disrupting the proper assembly of HIF-1 at the hypoxia response element (HRE) of its target genes. Without oxygen, the proline and asparagine residues of HIF-1α cannot be hydroxylated; therefore, HIF-1α protein is stabilized, translocates to the nucleus, forms a heterodimer with HIF-β and activates target gene transcription.

and succinate [17,20]. Therefore, under hypoxic conditions, the activities of PHDs and FIH are repressed, and HIF-1α is stabilized and able to translocate to the nucleus, heterodimerize with HIF-1β, and activates target gene transcription.

In addition to being regulated by oxygen stress, HIF-1α has been shown to be regulated by other stimuli, including nitric oxide, reactive oxygen species, nutrient stress, and glycolytic intermediates (e.g., pyruvate and lactate) [10,21-23]. Indeed, glycolytic intermediates may act as competitors of the PHD substrate 2-oxoglutarate, preventing PHD hydroxylation of HIF-1α.

HIF isoforms in addition to HIF-1α and HIF-1β have been identified [2,24]. There are currently three structurally related HIF-α isoforms that share a similar functional mechanism [2]. In contrast to the ubiquitous expression of HIF-1α, the expression of HIF-2α is commonly found in tissues involved in oxygen delivery, such as endothelial cells and cardiac myocytes [25]. The various isoforms may also have selective transcriptional targets [2]. Additionally, HIF-3α lacks the transactivation domain and thus may play a dominant-negative role [26]. HIF-1α deletion in mice is embryonic lethal, with cardiac and vascular defects and decreased erythropoiesis [27-29]; in contrast, HIF-2α deletion is either embryonic lethal, or death occurs shortly after birth due to respiratory failure or mitochondrial dysfunction [30].

Physiological responses of tissues to hypoxia

All nucleated cells have the ability of physiological alterations to adapt to transient or chronic hypoxia. These changes include enhancing glucose uptake, metabolic switching from oxidative phosphorylation to anaerobic glycolysis to reduce oxygen consumption, and increasing oxygen delivery by promoting angiogenesis and erythropoiesis.

Glucose uptake

Glucose utilization in most mammalian cells is controlled by its uptake. Glucose transport across the plasma membrane of most mammalian cells is mediated by a family of facilitative glucose transporters (GLUTs), which includes 14 structurally related members designated GLUT1-12, GLUT14, and HMIT (H^+/myo-inositol co-transporter) [31]. Each transporter has a tissue-specific distribution and distinct kinetic properties and exhibits differential regulation by ambient glucose, oxygen levels, and hormones [31]. Glucose transport is stimulated by hypoxia in many cell types, and this response is largely mediated by enhancing GLUT1 and GLUT3 transcription through HIF-1α activation [32-34]. An HRE identified in an enhancer region located approximately 2.7 kb upstream of the transcription start site in mouse and rat GLUT1 genes is reported to convey the response to hypoxia [35,36].

In addition to the direct regulation of GLUT transcription by HIF-1α, hypoxia may also stimulate glucose uptake by secondary mechanisms, such as by inhibiting oxidative phosphorylation in mitochondria. The exposure to hypoxia results in the inhibition of oxidative phosphorylation, which, without the reduction of oxygen tension, in turn stabilizes GLUT1 mRNA and up-regulates GLUT1 expression through enhancer elements that are independent of HIF binding [35,36]. A serum-responsive element (SRE) located approximately 100 nucleotides upstream of an HRE in the mouse GLUT1 gene conveys the responses to mitochondrial inhibitors [36]. In the rat GLUT1 gene, a

666-bp sequence at 6 kb upstream of the transcription start site responds to the inhibition of oxidative phosphorylation [35]. Furthermore, the induction of GLUT1 expression by low oxygen availability is rapid, whereas the half-life of GLUT1 mRNA is not affected. In contrast, the up-regulation of GLUT1 mRNA by the inhibition of oxidative phosphorylation is a delayed response, and the GLUT1 mRNA half-life increases significantly [35].

Metabolism

In aerobic metabolism, glucose is first converted to two pyruvates through glycolysis in the cytoplasm; pyruvate dehydrogenase complex (PDC) in the mitochondria then catalyzes the conversion of pyruvate to acetyl-CoA, which enters the tricarboxylic acid (TCA) cycle and produces ATP via oxidative phosphorylation (Figure 1). However, when the availability of O_2 becomes limited, cells switch from oxidative metabolism to anaerobic glycolysis by up-regulating the expression of several key glycolytic genes, including hexokinases 1 and 2 (HKs), lactate dehydrogenase A (LDHA), and pyruvate dehydrogenase kinase 1 (PDK1) [2,4,13]. The up-regulation of glycolytic gene expression is also mediated by HIF-1 via direct binding to promoters or enhancers [37-40].

In addition to preventing substrates from entering oxidative phosphorylation, hypoxia also regulates mitochondrial activity: it regulates cytochrome c oxidase (COX) by switching isoform COX4-1 to COX4-2, which can use the limited oxygen more efficiently [41]. Moreover, hypoxia inhibits mitochondrial biogenesis by decreasing c-Myc activity [42] and induces mitochondrial autophagy by up-regulating BNIP3 [43].

Angiogenesis and erythropoiesis

Another strategy against hypoxia is to increase oxygen delivery, which is performed in two ways. i) Hypoxia induces angiogenesis, the growth of new blood vessels, to deliver more oxygen and nutrients to hypoxic tissues, a process that is primarily mediated via the up-regulation of vascular endothelial growth factor (VEGF) expression by HIF [44]. Hypoxia-induced angiogenesis plays important roles in wound healing, inflammation, and pregnancy. ii) In systemic hypoxia, genes related to red blood cell production and oxygen transport are up-regulated [45]. Erythropoietin (EPO) was the first gene shown to be regulated by hypoxia, and research on the underlying mechanism led to the discovery of HIF [46,47]. HIF increases the expression of EPO in the kidney and liver, stimulating the production of red blood cells in the bone marrow. Hypoxia also induces other proteins involved in iron uptake and utilization to support the production of hemoglobin [47].

In addition to the above changes, hypoxia can also induce changes in gene expression to maintain a more alkaline intracellular pH to overcome the hostile acidic extracellular environment resulted from lactate accumulation [2].

Pathological implications of hypoxia in cancer and other diseases

Many studies on the effects of hypoxia have been conducted in cancer cells [48]. Indeed, hypoxia is a characteristic feature of malignant solid cancers, such as breast cancer. Cancer cells can grow and proliferate without external growth signals and are insensitive to growth suppressors and resistant to cell death [49]. Thus, highly aggressive and rapidly growing tumors are exposed to hypoxia as a consequence of an inadequate blood supply, which in turn plays a pivotal role in promoting tumor progression and the resistance to therapy. During tumor growth, hypoxia provides an array of effects to favor tumor survival under the hypoxic condition. These effects include gene expression changes to suppress apoptosis and support autophagy [50,51], genotype selection of cells with diminished apoptotic potential [52], increased GLUT expression and glucose uptake [53], anabolic metabolism switching [54], and the loss of genomic stability through the down-regulation of DNA repair [55] and possibly increased reactive oxygen species (ROS) production [56]. In addition, hypoxia can also suppress immune responses [57] and enhance tumor angiogenesis, vasculogenesis, the epithelial-mesenchymal transition (EMT), invasiveness, and metastasis [2,58]. These effects may also play roles in the resistance to cancer therapy [59]. Due to its multiple contributions to tumor progression, hypoxia has been suggested to be a negative prognostic factor for patient outcomes [60], and many studies have shown that endogenous hypoxia markers (e.g., HIF-1α, GLUT1, CA IX) are correlated with poorer outcomes and more aggressive malignancies [60,61]. Thus, it is not surprising that hypoxia has been a high priority target for cancer therapy [59].

In addition to cancer, hypoxia is implicated in other human diseases, such as ischemic heart disease (IHD), stroke, kidney disease, chronic lung disease, and inflammatory disorders [62]. Furthermore, cellular hypoxia may be a key factor in increasing glucose uptake by adipocytes, contributing to adipose tissue dysfunction in obesity [63,64].

Involvement of hypoxia in physiological processes

Studies have also demonstrated that HIF-1α is required for embryonic development and the development and growth of several murine tissues, implying a role for hypoxia in these physiological processes. Most mammalian embryos develop at oxygen concentrations of 1% - 5%, thus hypoxia and HIF play important roles in embryogenesis and placenta development by regulating gene expression, cell behaviour, and cell fate [65]. HIF-1α null mice die by embryonic day (E) 10.5 due to an impaired circulatory system [13]. HIF-1α deletion in the cartilaginous growth plate of developing bone leads to death of cells in the interior of the growth plate, indicating that hypoxia plays an important role in chondrogenesis [66]. Hypoxia promotes angiogenesis and osteogenesis in bone by elevating the VEGF levels induced by HIF-1α in osteoblasts, as HIF-1α overexpression results in extremely dense, heavily vascularized long bones and a high level of VEGF, whereas the lack of HIF-1α in osteoblasts leads to significantly thinner and less vascularized long bones [67]. Hypoxia inhibits adipogenesis and the conditional knock-out of HIF-1α in mouse embryonic fibroblasts impairs the hypoxia-mediated inhibition of adipogenesis [68]. In HIF-1α-deficient mice, hematopoietic stem cells (HSCs) in bone marrow lose cell cycle quiescence, and HSC numbers decrease [69]. B lymphocyte development is abnormal in HIF-1α-deficient chimeric mice, inducing autoimmunity [70].

Below, we introduce the emerging evidence of the involvement of hypoxia in mammary development and lactation.

Mammary development and lactation

The development of the mammary gland can be divided into three stages, embryo, puberty, and pregnancy, and the major development occurs during postnatal stages. In mice, two milk lines form between the fore and hind limbs of an E 10.5 embryo, and five pairs of placodes arise along each of the two milk lines at E 11.5. The placodes then invaginate into the underlying mesenchyme to form small bulb-shaped buds. The mammary epithelial cells of the buds proliferate from E 15.5 to form sprouts that penetrate to the underlying fat pad and form the rudimentary ductal tree and teats [71,72]. Canonical Wnt signaling plays an important role during embryonic mammary development: Wnt 6, Wnt 10a, and Wnt 10b are expressed during this period, and the inhibition of Wnt signaling results in impaired placode formation [72,73]. From birth to puberty, the mammary gland first undergoes isometric growth (with the same rate as the body), followed by allometric growth (2-4X faster than body fat deposition). Robust duct branching then begins at the onset of puberty. Terminal end buds (TEBs) form from the tip of the rudimentary ductal tree of the mouse mammary gland and drive pubertal mammary development. TEBs are highly proliferative, penetrating further into the fat pad, and the bifurcations of TEBs form the primary ducts. TEBs continue to invade into the fat pad until the primary ducts reach the border of the fat pad, and the secondary ducts then branch laterally from the primary ducts until the fat pad is filled with extensive ducts [74]. In ruminants,

the terminal ductal lobular units (TDLUs) are the characteristic structures of postpubertal mammary development. During this stage, estrogen and growth hormone regulate duct branching, as the knock-out of estrogen receptor (ER) α or growth hormone receptor was found to impair duct development during puberty [75,76]. During the early stage of pregnancy, short tertiary side-branches form along the ductal system developed during puberty, a process that is regulated by progesterone, progesterone receptor, and downstream Wnt 4 signaling [77,78]. The mammary epithelial cells then proliferate rapidly to form a lobuloalveolar structure within the ductal branches. Prolactin is important for alveolar morphogenesis, as the knock-out of prolactin receptor or STAT5a, a major prolactin downstream-signaling molecule, results in lobuloalveolar defects [79,80]. During late pregnancy, the mammary epithelial cells differentiate, and lactogenesis occurs under the synergistic effects of prolactin, glucocorticoids, and insulin [81]. Milk protein genes and lipogenic genes are expressed, and lipid droplets form in the epithelial cells. At the onset of lactation, the peak blood levels of lactogenic hormones and withdrawal of progesterone lead to copious milk secretion [81]. Extensive angiogenesis to supply nutrients is associated with all stages of mammary development [82]; in particular, extensive capillaries form a basket-like architecture to surround the alveoli during lobuloalveolar development. The vascular density doubles from day 18 of pregnancy to day 5 of lactation in mice [83]. In consistent with vascular development, the expression of VEGF and VEGF receptor in the rodent mammary gland increases dramatically during pregnancy and lactation and decreases during involution [84].

Oxygen uptake in the mammary gland during mammary development and lactation

The mammary gland has high metabolic rates during development and lactation and is thus considered to be a benign, highly regulated tumor [85]. Although extensive studies have been performed in the pathology of hypoxia in breast cancer, limited attention has been given to oxygen utilization in normal mammary development and lactation, and reports on mammary oxygen uptake have been largely limited to only a few early studies [86-88]. The average O_2 uptake of a lactating mammary gland is 0.51-0.73 μmol/min/g in goats and 2.06 in rats [86-88]. Mammary O_2 uptake is steadily increased during late pregnancy and reaches the highest levels in early lactation [89]. The O_2 uptake in lactating goats is twice that in the preparturient goat, and there is a correlation between mammary O_2 uptake and milk secretion [87,88]. Starvation results in the virtual cessation of milk production, with a 75% reduction in mammary oxygen uptake [86], whereas growth hormone administration increases mammary oxygen uptake [90]. In 10-wk virgin mice, the pO_2 level in the mammary fat pad is, on average, 13.0 mm Hg (~2%), which is considerably lower than in muscle (29.1 mm Hg, ~4%) [91]. However, a recent study using phosphonated trityl probes in mice reported an average mammary gland tissue pO_2 of 52 mmHg [92], which is consistent with the pO_2 value reported in normal breast tissue [93]. Nevertheless, it is likely that the mammary gland develops chronic hypoxia during the rapid mammary development in late pregnancy and in early lactation because the oxygen consumption increases in these periods to meet the increased metabolic rates.

To examine possible hypoxic conditions in the mammary gland during mammary development, we recently injected the hypoxia marker pimonidazole HCl into mice from the virgin state to the early lactation state. Pimonidazole HCl is a chemical that forms adducts with thiol groups in proteins, peptides, and amino acids in hypoxic cells (http://www.hypoxyprobe.com/), and these pimonidazole adducts can be detected with specific antibodies. Immunohistochemical staining of the mouse mammary glands from different stages showed a hypoxic mammary gland in all examined stages, but the staining was stronger during late pregnancy and early lactation (Shao and Zhao, unpublished preliminary observations), indicating a more hypoxic condition in these stages.

Emerging evidence of a role of hypoxia in mammary development and lactation
Consistent with the hypoxic condition in the mammary gland during development and lactation, there is emerging evidence that supports possible physiological roles of hypoxia in these processes.

1) Selective deletion of HIF-1α in the mouse mammary gland results in impaired mammary development and lactation
The most compelling evidence of hypoxia's role in mammary development and lactation is from a targeted gene-knockout study by Seagroves et al. (2003) [94]. Because the whole-genome knockout of the *HIF-1α* gene is embryonic lethal, Seagroves et al. specifically removed the *HIF-1α* gene from the mammary epithelial cells (MECs) using mouse mammary tumor virus (MMTV) promoter-directed cre-lox technology. No morphological defects were observed in the $HIF-1\alpha^{-/-}$ glands until day 15 of pregnancy when the mammary gland is well into secretory differentiation. By day 15 of pregnancy, striking histological differences became obvious between the wild-type and $HIF-1\alpha^{-/-}$ glands. In particular, the $HIF-1\alpha^{-/-}$ glands had smaller alveoli with no protein or lipid droplets due to impaired mammary differentiation. In addition, the expression of milk proteins β-casein and α-lactalbumin was decreased by over 50% in these glands.

Surprisingly, the $HIF-1\alpha^{-/-}$ glands showed no abnormality in microvessel pattering and vascular density, despite the important role of hypoxia in angiogenesis [94]. However, this observation could be explained by *HIF-1α* only being deleted in MECs and not in other cell types, such as endothelial and stroma cells, and these cells could still respond to the hypoxia conditions and retain intact angiogenesis.

Furthermore, consistent alveolar defects and trapped large lipid droplets were observed in the $HIF-1\alpha^{-/-}$ glands during lactation [94]. The glands weighed ~50% of the wild-type glands at mid-lactation; additionally, the $HIF-1\alpha^{-/-}$ animals produced less milk than the wild-type controls, and their milk was more viscous and contained significantly elevated fat and ion contents.

These data clearly indicate that, although the expression of HIF-1α in MECs is not required for early mammary development (mainly ductal morphogenesis), it is essential for mammary secretory differentiation, milk production, and lipid secretion, implying a role for hypoxia in these physiological processes.

2) Hypoxia increases glucose uptake and GLUT1 expression in MECs
As in all mammalian cells, glucose is an important source of energy and NADH and also serves as a substrate for lipid, protein, and nucleotide syntheses in MECs. Glucose is also the major and an essential precursor of lactose synthesis in the lactating MECs. Mammary glucose uptake increases gradually from late pregnancy and peaks at early lactation [87,89], and mammary glucose transport activity increases approximately 40-fold from the virgin state to mid-lactation state in mice [95]. Glucose uptake in the mammary gland is mediated by facilitative glucose transporters (GLUTs) [31,96]. Mammary cells mainly express GLUT1, GLUT8, and GLUT 12, with GLUT1 being the predominant isoform [97], and there is a dramatic increase in mammary GLUT expression from late pregnancy to early lactation [97], concomitantly with mammary glucose uptake. We originally hypothesized that the lactogenic hormones (prolactin, glucocorticoids, insulin, and estrogen) are responsible for stimulating GLUT expression during lactogenesis. However, our recent study challenged this hypothesis because these hormones failed to stimulate GLUT expression in bovine mammary explants and primary MECs, even though they were able to dramatically stimulate the expression of milk protein and lipogenic genes [98].

We, thus, hypothesized that the mechanism underlying the increase in GLUT expression in MECs during the transition period from pregnancy to lactation involves hypoxia signaling through hypoxia inducible factor-1α (HIF-1α). To test this hypothesis, we recently studied the effects of hypoxia on GLUT expression in bovine MECs.

Hypoxia (below 5% O_2) significantly stimulated glucose uptake and GLUT1 mRNA and protein expression in bovine MECs yet decreased GLUT8 mRNA expression in these cells (Shao and Zhao, unpublished observations). A robust induction of HIF-1α protein was observed in the bovine MECs, consistent with the observation in mouse MECs [94]. Furthermore, an siRNA against HIF-1α completely abolished the up-regulation of GLUT1 by hypoxia but had no effect on GLUT8 expression (Shao and Zhao, unpublished observations).

Consistent with our study, the expression of GLUT1 in mouse MECs is HIF-1α-dependent in a stage-dependent manner *in vivo*. GLUT1 expression decreases by 60% in *HIF-1α$^{-/-}$* glands by day 16 of pregnancy, whereas no difference was observed at mid-lactation [94]. In addition, it has been shown that prolonged hypoxia stimulates GLUT1 expression in bovine endothelial cells [99,100].

Taken together, the above evidence clearly show that MECs and mammary endothelial cells are responsive to hypoxia (as high as 5% O_2) through HIF-1α.

3) Possible interactions of HIF-1α with other signaling pathways in the mammary gland

The signal transducer and activator of transcription 5 (STAT5) is essential for mammary gland differentiation and lactation [101,102] and is mainly activated by prolactin and its receptor in the mammary gland. After binding prolactin, the prolactin receptor phosphorylates and activates Janus kinase (JAK) 2. JAK2 then phosphorylates STAT5, and STAT5 dimerizes and translocates to the nucleus to stimulate target genes' expression [103,104]. In addition to the prolactin pathway, STAT5 can also be activated by epidermal growth factor (EGF), growth hormone, insulin growth factor (IGF), estrogen, and progesterone signaling pathways in the mammary gland [105]. STAT5 controls the population of luminal progenitor cells that will differentiate to alveolar cells [106,107]. During lactogenesis and lactation, the prolactin-STAT5 pathway controls the expression of milk protein genes and lipogenic genes [81,108]. STAT5-null mice have impaired mammary alveologenesis due to a reduction in the mammary luminal progenitor cell population and exhibit impaired milk protein gene expression [109]. It has been reported that hypoxia and HIF-1α can induce STAT5 phosphorylation and enhance its DNA-binding activity in mammary epithelial cells and breast cancer cells [110,111]. Thus, hypoxia may be involved in mammary development and lactation by regulating STAT5 activity.

Notch signaling responds to extrinsic or intrinsic developmental cues and regulates multiple cellular processes, such as stem cell maintenance, cell fate specification, and differentiation [112]. Upon ligand binding, the Notch protein is cleaved by presenilin/γ-secretase to release the active intracellular domain (ICD), which translocates to the nucleus to regulate the transcription of target genes [113]. In the mammary gland, Notch represses mammary stem cells expansion in the basal cell compartment [114] and promotes luminal cell fate specification and prevents myoepithelial cell lineage during pregnancy [114,115]. Hypoxia prevents differentiation in various stem and precursor cells, and research has shown that HIF-1α interacts with the Notch ICD to activate Notch target genes, inhibiting the differentiation of myogenic and neutral precursor cells [116]. In addition, FIH negatively regulates Notch by hydroxylating two asparagine residues of ICD [117]. Therefore, hypoxia and Notch signaling may cross-talk to regulate cell differentiation in the mammary gland.

Wnt proteins are involved in multiple events during embryogenesis and adult tissue development, with effects on cell fate specification, differentiation, mitogenic stimulation, and stem cell self-renewal [118]. The Wnt signaling pathways include the canonical pathway involving β-catenin and noncanonical pathways; the canonical pathway has been investigated intensively and best characterized. Wnt proteins bind to the cell surface receptor Frizzled in conjunction with low-density lipoprotein receptor-related proteins (LRPs), which transduce the signal to intracellular proteins, leading to the stabilization of β-catenin. β-Catenin then translocates to the nucleus and interacts with transcription factor lymphoid enhancer-binding factor 1/T cell-specific transcription factor (LEF/TCF) to affect target gene transcription [118]. Wnt signaling is essential for mammary rudiment formation in embryos, as the lack of TCF or the overexpression of the Wnt inhibitor Dkk-1 impairs the formation of mammary placodes [73,119,120]. Wnt signals drive ductal branching during pubertal mammary development, because the deletion of LRP reduces duct branching and the overexpression of LRP increases duct branching in the virgin mouse

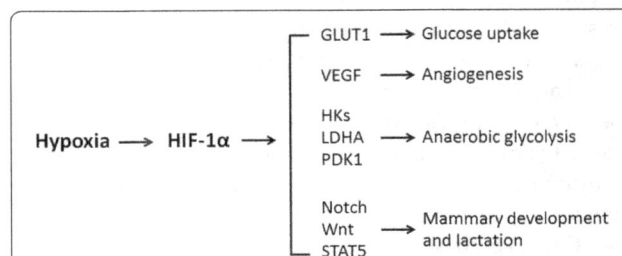

Figure 3 Schematic diagram of proposed effects of hypoxia on the mammary gland. Hypoxia stabilizes the hypoxia-inducible factor-1α (HIF-1α), which stimulates: i) glucose uptake through enhancing the expression of facilitative glucose transporter 1 (GLUT1), ii) angiogenesis through enhancing the expression of vascular endothelial growth factor (VEGF), iii) anaerobic glycolysis through up-regulating the expression of several key glycolytic genes, including hexokinases 1 and 2 (HKs), lactate dehydrogenase A (LDHA), and pyruvate dehydrogenase kinase 1 (PDK1), and iv) mammary development, differentiation and lactation through regulation of Notch, Wnt and STAT5 signaling pathways.

mammary gland [121-123]. Wnt-4 works as a paracrine factor downstream of progesterone signaling during pregnancy to stimulate lobular alveolar development [77,124]. During mammary development, both ductal and alveolar epithelial cells originate from mammary stem cells, and Wnt signaling serves as a rate-limiting self-renewal signal to maintain mammary stem cells [125]. Most stem cells reside in hypoxic niches, and research has shown that Wnt signaling is modulated by hypoxia in embryonic stem cells and neural stem cells by enhancing β-catenin activation and increasing LEF/TCF proteins [126]. Such interactions may also exist in the mammary gland.

Conclusion
Increased oxygen and energy requirements, nutrient stress, extensive angiogenesis, and metabolic switching to glycolysis [127] in the mammary gland during the transition period from pregnancy to lactation make hypoxia a possible regulator of these processes. The possible effects of hypoxia on the mammary gland include, but not limit to the stimulations of: i) glucose uptake through enhancing the expression of GLUT1, ii) angiogenesis through enhancing the expression of VEGF, iii) anaerobic glycolysis through up-regulating the expression of several key glycolytic genes, including HKs, LDHA, and PDK1, and iv) mammary development, differentiation and lactation through regulation of Notch, Wnt and STAT5 signaling pathways (Figure 3).

However, based on above evidence, it is important to note that, although HIF-1α is clearly required in normal mammary development and lactation, there is a possibility that hypoxia per se may not be the major or only stimulus of HIF-1α activity in the mammary gland. As mentioned above, in addition to being regulated by hypoxia, HIF-1α stability has been shown to be regulated by other stimuli, including nitric oxide, reactive oxygen species, nutrient stress, and glycolytic intermediates [10,21-23]. Furthermore, several studies have shown that hypoxia causes mammary epithelial disorganization and induces a cancer cell-like phenotype in human MECs [128-130]. Thus, to link an effect of hypoxia to mammary development and lactation, it is essential to quantitatively measure oxygen tension changes in the mammary gland from pregnancy to lactation. Our hypothesis is that a small degree of the hypoxic condition is involved in mammary development and lactation from late pregnancy to early lactation, whereas more severe hypoxia is involved in breast cancer development.

Abbreviations
FIH: Factor inhibiting HIF-1; GLUT: Facilitative glucose transporter; HIF: Hypoxia-inducible factor; HRE: Hypoxia response element; LDHA: Lactate dehydrogenase A; MECs: Mammary epithelial cells; PDC: Pyruvate dehydrogenase complex; PDK1: Pyruvate dehydrogenase kinases 1; PHD: Prolyl hydroxylase domain protein; SRE: Serum responsive element; STAT5: Signal transducer and activator of transcription 5; TCA: Tricarboxylic acid cycle.

Competing interests
The authors declare that they have no competing interests.

Authors' contributions
YS and FQZ drafted the manuscript, and FQZ revised the manuscript. Both authors read and approved the final manuscript.

Acknowledgements
This research was supported by National Research Initiative Competitive grant 2007-35206-18037 from the USDA National Institute of Food and Agriculture (to FQZ).

References
1. Rich PR: The molecular machinery of Keilin's respiratory chain. *Biochem Soc Trans* 2003, 31:1095–1105.
2. Cassavaugh J, Lounsbury KM: Hypoxia-mediated biological control. *J Cell Biochem* 2011, 112:735–744.
3. Braun RD, Lanzen JL, Snyder SA, Dewhirst MW: Comparison of tumor and normal tissue oxygen tension measurements using OxyLite or microelectrodes in rodents. *Am J Physiol Heart Circ Physiol* 2001, 280:H2533–2544.
4. Aragones J, Fraisl P, Baes M, Carmeliet P: Oxygen sensors at the crossroad of metabolism. *Cell Metab* 2009, 9:11–22.
5. Stroka DM, Burkhardt T, Desbaillets I, Wenger RH, Neil DA, Bauer C, Gassmann M, Candinas D: HIF-1 is expressed in normoxic tissue and displays an organ-specific regulation under systemic hypoxia. *FASEB J* 2001, 15:2445–2453.
6. Prabhakar NR: O2 sensing at the mammalian carotid body: why multiple O2 sensors and multiple transmitters? *Exp Physiol* 2006, 91:17–23.
7. Lahiri S, Roy A, Baby SM, Hoshi T, Semenza GL, Prabhakar NR: Oxygen sensing in the body. *Prog Biophy Mol Biol* 2006, 91:249–286.
8. Prabhakar NR: Sensing hypoxia: physiology, genetics and epigenetics. *J Physiol* 2013, 591:2245–2257.
9. Xia X, Lemieux ME, Li W, Carroll JS, Brown M, Liu XS, Kung AL: Integrative analysis of HIF binding and transactivation reveals its role in maintaining histone methylation homeostasis. *Proc Natl Acad Sci USA* 2009, 106:4260–4265.
10. Prabhakar NR, Semenza GL: Adaptive and maladaptive cardiorespiratory responses to continuous and intermittent hypoxia mediated by hypoxia-inducible factors 1 and 2. *Physiol Rev* 2012, 92:967–1003.
11. Wang GL, Jiang BH, Rue EA, Semenza GL: Hypoxia-inducible factor 1 is a basic-helix-loop-helix-PAS heterodimer regulated by cellular O2 tension. *Proc Natl Acad Sci USA* 1995, 92:5510–5514.
12. Jiang BH, Rue E, Wang GL, Roe R, Semenza GL: Dimerization, DNA binding, and transactivation properties of hypoxia-inducible factor 1. *J Biol Chem* 1996, 271:17771–17778.
13. Semenza GL: Hypoxia-inducible factors in physiology and medicine. *Cell* 2012, 148:399–408.
14. Epstein AC, Gleadle JM, McNeill LA, Hewitson KS, O'Rourke J, Mole DR, Mukherji M, Metzen E, Wilson MI, Dhanda A, *et al*: C. elegans EGL-9 and mammalian homologs define a family of dioxygenases that regulate HIF by prolyl hydroxylation. *Cell* 2001, 107:43–54.
15. Cockman ME, Masson N, Mole DR, Jaakkola P, Chang GW, Clifford SC, Maher ER, Pugh CW, Ratcliffe PJ, Maxwell PH: Hypoxia inducible factor-alpha binding and ubiquitylation by the von Hippel-Lindau tumor suppressor protein. *J Biol Chem* 2000, 275:25733–25741.
16. Maxwell PH, Wiesener MS, Chang GW, Clifford SC, Vaux EC, Cockman ME, Wykoff CC, Pugh CW, Maher ER, Ratcliffe PJ: The tumour suppressor protein VHL targets hypoxia-inducible factors for oxygen-dependent proteolysis. *Nature* 1999, 399:271–275.
17. Schofield CJ, Ratcliffe PJ: Oxygen sensing by HIF hydroxylases. *Nat Rev Mol Cell Biol* 2004, 5:343–354.
18. Mahon PC, Hirota K, Semenza GL: FIH-1: a novel protein that interacts with HIF-1alpha and VHL to mediate repression of HIF-1 transcriptional activity. *Genes Dev* 2001, 15:2675–2686.
19. Lando D, Peet DJ, Gorman JJ, Whelan DA, Whitelaw ML, Bruick RK: FIH-1 is an asparaginyl hydroxylase enzyme that regulates the transcriptional activity of hypoxia-inducible factor. *Genes Dev* 2002, 16:1466–1471.

20. Kaelin WG Jr, Ratcliffe PJ: Oxygen sensing by metazoans: the central role of the HIF hydroxylase pathway. *Mol Cell* 2008, 30:393–402.

21. Lu H, Dalgard CL, Mohyeldin A, McFate T, Tait AS, Verma A: Reversible inactivation of HIF-1 prolyl hydroxylases allows cell metabolism to control basal HIF-1. *J Biol Chem* 2005, 280:41928–41939.

22. Lu H, Forbes RA, Verma A: Hypoxia-inducible factor 1 activation by aerobic glycolysis implicates the Warburg effect in carcinogenesis. *J Biol Chem* 2002, 277:23111–23115.

23. Hirota K, Semenza GL: Regulation of hypoxia-inducible factor 1 by prolyl and asparaginyl hydroxylases. *Biochem Biophys Res Commun* 2005, 338:610–616.

24. Loboda A, Jozkowicz A, Dulak J: HIF-1 and HIF-2 transcription factors–similar but not identical. *Mol Cells* 2010, 29:435–442.

25. Wiesener MS, Jurgensen JS, Rosenberger C, Scholze CK, Horstrup JH, Warnecke C, Mandriota S, Bechmann I, Frei UA, Pugh CW, et al: Widespread hypoxia-inducible expression of HIF-2alpha in distinct cell populations of different organs. *FASEB J* 2003, 17:271–273.

26. Maynard MA, Evans AJ, Hosomi T, Hara S, Jewett MA, Ohh M: Human HIF-3alpha4 is a dominant-negative regulator of HIF-1 and is down-regulated in renal cell carcinoma. *FASEB J* 2005, 19:1396–1406.

27. Ryan HE, Lo J, Johnson RS: HIF-1 alpha is required for solid tumor formation and embryonic vascularization. *EMBO J* 1998, 17:3005–3015.

28. Compernolle V, Brusselmans K, Franco D, Moorman A, Dewerchin M, Collen D, Carmeliet P: Cardia bifida, defective heart development and abnormal neural crest migration in embryos lacking hypoxia-inducible factor-1alpha. *Cardiovasc Res* 2003, 60:569–579.

29. Yoon D, Pastore YD, Divoky V, Liu E, Mlodnicka AE, Rainey K, Ponka P, Semenza GL, Schumacher A, Prchal JT: Hypoxia-inducible factor-1 deficiency results in dysregulated erythropoiesis signaling and iron homeostasis in mouse development. *J Biol Chem* 2006, 281:25703–25711.

30. Scortegagna M, Ding K, Oktay Y, Gaur A, Thurmond F, Yan LJ, Marck BT, Matsumoto AM, Shelton JM, Richardson JA, et al: Multiple organ pathology, metabolic abnormalities and impaired homeostasis of reactive oxygen species in Epas1-/- mice. *Nat Genet* 2003, 35:331–340.

31. Zhao FQ, Keating AF: Functional properties and genomics of glucose transporters. *Curr Genomics* 2007, 8:113–128.

32. Zhang JZ, Behrooz A, Ismail-Beigi F: Regulation of glucose transport by hypoxia. *Am J Kidney Dis* 1999, 34:189–202.

33. Baumann MU, Zamudio S, Illsley NP: Hypoxic upregulation of glucose transporters in BeWo choriocarcinoma cells is mediated by hypoxia-inducible factor-1. *Am J Physiol Cell Physiol* 2007, 293:C477–485.

34. Ren BF, Deng LF, Wang J, Zhu YP, Wei L, Zhou Q: Hypoxia regulation of facilitated glucose transporter-1 and glucose transporter-3 in mouse chondrocytes mediated by HIF-1alpha. *Joint Bone Spine* 2008, 75:176–181.

35. Behrooz A, Ismail-Beigi F: Dual control of glut1 glucose transporter gene expression by hypoxia and by inhibition of oxidative phosphorylation. *J Biol Chem* 1997, 272:5555–5562.

36. Ebert BL, Firth JD, Ratcliffe PJ: Hypoxia and mitochondrial inhibitors regulate expression of glucose transporter-1 via distinct Cis-acting sequences. *J Biol Chem* 1995, 270:29083–29089.

37. Semenza GL, Jiang BH, Leung SW, Passantino R, Concordet JP, Maire P, Giallongo A: Hypoxia response elements in the aldolase A, enolase 1, and lactate dehydrogenase A gene promoters contain essential binding sites for hypoxia-inducible factor 1. *J Biol Chem* 1996, 271:32529–32537.

38. Semenza GL: Transcriptional regulation by hypoxia-inducible factor 1 molecular mechanisms of oxygen homeostasis. *Trends Cardiovasc Med* 1996, 6:151–157.

39. Kim JW, Tchernyshyov I, Semenza GL, Dang CV: HIF-1-mediated expression of pyruvate dehydrogenase kinase: a metabolic switch required for cellular adaptation to hypoxia. *Cell Metab* 2006, 3:177–185.

40. Stellingwerff T, Spriet LL, Watt MJ, Kimber NE, Hargreaves M, Hawley JA, Burke LM: Decreased PDH activation and glycogenolysis during exercise following fat adaptation with carbohydrate restoration. *Am J Physiol Endocrinol Metab* 2006, 290:E380–388.

41. Fukuda R, Zhang H, Kim JW, Shimoda L, Dang CV, Semenza GL: HIF-1 regulates cytochrome oxidase subunits to optimize efficiency of respiration in hypoxic cells. *Cell* 2007, 129:111–122.

42. Zhang H, Gao P, Fukuda R, Kumar G, Krishnamachary B, Zeller KI, Dang CV, Semenza GL: HIF-1 inhibits mitochondrial biogenesis and cellular respiration in VHL-deficient renal cell carcinoma by repression of C-MYC activity. *Cancer Cell* 2007, 11:407–420.

43. Zhang H, Bosch-Marce M, Shimoda LA, Tan YS, Baek JH, Wesley JB, Gonzalez FJ, Semenza GL: Mitochondrial autophagy is an HIF-1-dependent adaptive metabolic response to hypoxia. *J Biol Chem* 2008, 283:10892–10903.

44. Forsythe JA, Jiang BH, Iyer NV, Agani F, Leung SW, Koos RD, Semenza GL: Activation of vascular endothelial growth factor gene transcription by hypoxia-inducible factor 1. *Mol Cell Biol* 1996, 16:4604–4613.

45. Wang GL, Semenza GL: Molecular basis of hypoxia-induced erythropoietin expression. *Curr Opin Hematol* 1996, 3:156–162.

46. Semenza GL, Nejfelt MK, Chi SM, Antonarakis SE: Hypoxia-inducible nuclear factors bind to an enhancer element located 3′ to the human erythropoietin gene. *Proc Natl Acad Sci USA* 1991, 88:5680–5684.

47. Haase VH: Regulation of erythropoiesis by hypoxia-inducible factors. *Blood Rev* 2013, 27:41–53.

48. Kunz M, Ibrahim SM: Molecular responses to hypoxia in tumor cells. *Mol Cancer* 2003, 2:23.

49. Hanahan D, Weinberg RA: Hallmarks of cancer: the next generation. *Cell* 2011, 144:646–674.

50. Erler JT, Cawthorne CJ, Williams KJ, Koritzinsky M, Wouters BG, Wilson C, Miller C, Demonacos C, Stratford IJ, Dive C: Hypoxia-mediated down-regulation of Bid and Bax in tumors occurs via hypoxia-inducible factor 1-dependent and -independent mechanisms and contributes to drug resistance. *Mol Cell Biol* 2004, 24:2875–2889.

51. Rouschop KM, van den Beucken T, Dubois L, Niessen H, Bussink J, Savelkouls K, Keulers T, Mujcic H, Landuyt W, Voncken JW, et al: The unfolded protein response protects human tumor cells during hypoxia through regulation of the autophagy genes MAP1LC3B and ATG5. *J Clin Invest* 2010, 120:127–141.

52. Graeber TG, Osmanian C, Jacks T, Housman DE, Koch CJ, Lowe SW, Giaccia AJ: Hypoxia-mediated selection of cells with diminished apoptotic potential in solid tumours. *Nature* 1996, 379:88–91.

53. Airley RE, Mobasheri A: Hypoxic regulation of glucose transport, anaerobic metabolism and angiogenesis in cancer: novel pathways and targets for anticancer therapeutics. *Chemotherapy* 2007, 53:233–256.

54. Cairns RA, Harris IS, Mak TW: Regulation of cancer cell metabolism. *Nat Rev Cancer* 2011, 11:85–95.

55. Bristow RG, Hill RP: Hypoxia and metabolism. Hypoxia, DNA repair and genetic instability. *Nat Rev Cancer* 2008, 8:180–192.

56. Guzy RD, Hoyos B, Robin E, Chen H, Liu L, Mansfield KD, Simon MC, Hammerling U, Schumacker PT: Mitochondrial complex III is required for hypoxia-induced ROS production and cellular oxygen sensing. *Cell Metab* 2005, 1:401–408.

57. Yotnda P, Wu D, Swanson AM: Hypoxic tumors and their effect on immune cells and cancer therapy. *Methods Mol Biol* 2010, 651:1–29.

58. Pennacchietti S, Michieli P, Galluzzo M, Mazzone M, Giordano S, Comoglio PM: Hypoxia promotes invasive growth by transcriptional activation of the met protooncogene. *Cancer Cell* 2003, 3:347–361.

59. Wilson WR, Hay MP: Targeting hypoxia in cancer therapy. *Nat Rev Cancer* 2011, 11:393–410.

60. Vaupel P, Mayer A: Hypoxia in cancer: significance and impact on clinical outcome. *Cancer Metastasis Rev* 2007, 26:225–239.

61. Young RJ, Moller A: Immunohistochemical detection of tumour hypoxia. *Methods Mol Biol* 2010, 611:151–159.

62. Semenza GL: Oxygen sensing, homeostasis, and disease. *N Engl J Med* 2011, 365:537–547.

63. Wood IS, de Heredia FP, Wang B, Trayhurn P: Cellular hypoxia and adipose tissue dysfunction in obesity. *Proc Nutr Soc* 2009, 68:370–377.

64. Wood IS, Wang B, Lorente-Cebrian S, Trayhurn P: Hypoxia increases expression of selective facilitative glucose transporters (GLUT) and 2-deoxy-D-glucose uptake in human adipocytes. *Biochem Biophys Res Commun* 2007, 361:468–473.

65. Dunwoodie SL: The role of hypoxia in development of the Mammalian embryo. *Dev Cell* 2009, 17:755–773.

66. Schipani E, Ryan HE, Didrickson S, Kobayashi T, Knight M, Johnson RS: Hypoxia in cartilage: HIF-1alpha is essential for chondrocyte growth arrest and survival. *Genes Dev* 2001, 15:2865–2876.

67. Wang Y, Wan C, Deng L, Liu X, Cao X, Gilbert SR, Bouxsein ML, Faugere MC, Guldberg RE, Gerstenfeld LC, et al: The hypoxia-inducible factor alpha pathway couples angiogenesis to osteogenesis during skeletal development. *J Clin Invest* 2007, 117:1616–1626.

68. Yun Z, Maecker HL, Johnson RS, Giaccia AJ: **Inhibition of PPAR gamma 2 gene expression by the HIF-1-regulated gene DEC1/Stra13: a mechanism for regulation of adipogenesis by hypoxia.** *Dev Cell* 2002, **2**:331–341.

69. Takubo K, Goda N, Yamada W, Iriuchishima H, Ikeda E, Kubota Y, Shima H, Johnson RS, Hirao A, Suematsu M, Suda T: **Regulation of the HIF-1alpha level is essential for hematopoietic stem cells.** *Cell Stem Cell* 2010, **7**:391–402.

70. Kojima H, Gu H, Nomura S, Caldwell CC, Kobata T, Carmeliet P, Semenza GL, Sitkovsky MV: **Abnormal B lymphocyte development and autoimmunity in hypoxia-inducible factor 1alpha -deficient chimeric mice.** *Proc Natl Acad Sci USA* 2002, **99**:2170–2174.

71. Hens JR, Wysolmerski JJ: **Key stages of mammary gland development: molecular mechanisms involved in the formation of the embryonic mammary gland.** *Breast Cancer Res* 2005, **7**:220–224.

72. Robinson GW: **Cooperation of signalling pathways in embryonic mammary gland development.** *Nat Rev Genet* 2007, **8**:963–972.

73. Chu EY, Hens J, Andl T, Kairo A, Yamaguchi TP, Brisken C, Glick A, Wysolmerski JJ, Millar SE: **Canonical WNT signaling promotes mammary placode development and is essential for initiation of mammary gland morphogenesis.** *Development* 2004, **131**:4819–4829.

74. Howlin J, McBryan J, Martin F: **Pubertal mammary gland development: insights from mouse models.** *J Mammary Gland Biol Neoplasia* 2006, **11**:283–297.

75. Mueller SO, Clark JA, Myers PH, Korach KS: **Mammary gland development in adult mice requires epithelial and stromal estrogen receptor alpha.** *Endocrinology* 2002, **143**:2357–2365.

76. Gallego MI, Binart N, Robinson GW, Okagaki R, Coschigano KT, Perry J, Kopchick JJ, Oka T, Kelly PA, Hennighausen L: **Prolactin, growth hormone, and epidermal growth factor activate Stat5 in different compartments of mammary tissue and exert different and overlapping developmental effects.** *Dev Biol* 2001, **229**:163–175.

77. Brisken C, Heineman A, Chavarria T, Elenbaas B, Tan J, Dey SK, McMahon JA, McMahon AP, Weinberg RA: **Essential function of Wnt-4 in mammary gland development downstream of progesterone signaling.** *Genes Dev* 2000, **14**:650–654.

78. Brisken C, Park S, Vass T, Lydon JP, O'Malley BW, Weinberg RA: **A paracrine role for the epithelial progesterone receptor in mammary gland development.** *Proc Natl Acad Sci USA* 1998, **95**:5076–5081.

79. Ormandy CJ, Camus A, Barra J, Damotte D, Lucas B, Buteau H, Edery M, Brousse N, Babinet C, Binart N, Kelly PA: **Null mutation of the prolactin receptor gene produces multiple reproductive defects in the mouse.** *Genes Dev* 1997, **11**:167–178.

80. Liu X, Robinson GW, Wagner KU, Garrett L, Wynshaw-Boris A, Hennighausen L: **Stat5a is mandatory for adult mammary gland development and lactogenesis.** *Genes Dev* 1997, **11**:179–186.

81. Neville MC, McFadden TB, Forsyth I: **Hormonal regulation of mammary differentiation and milk secretion.** *J Mammary Gland Biol Neoplasia* 2002, **7**:49–66.

82. Djonov V, Andres AC, Ziemiecki A: **Vascular remodelling during the normal and malignant life cycle of the mammary gland.** *Microsc Res Tech* 2001, **52**:182–189.

83. Matsumoto M, Nishinakagawa H, Kurohmaru M, Hayashi Y, Otsuka J: **Pregnancy and lactation affect the microvasculature of the mammary gland in mice.** *J Vet Med Sci* 1992, **54**:937–943.

84. Pepper MS, Baetens D, Mandriota SJ, Di Sanza C, Oikemus S, Lane TF, Soriano JV, Montesano R, Iruela-Arispe ML: **Regulation of VEGF and VEGF receptor expression in the rodent mammary gland during pregnancy, lactation, and involution.** *Dev Dyn* 2000, **218**:507–524.

85. Williamson DH, Evans RD, Wood SC: **Tumor growth and lipid metabolism during lactation in the rat.** In *Advances in Enzyme Regulation. Volume 27.* Edited by Weber G. Oxford and New York: Pergamon Press; 1988:93–104.

86. Williamson DH, Lund P, Evans RD: **Substrate selection and oxygen uptake by the lactating mammary gland.** *Proc Nutr Soc* 1995, **54**:165–175.

87. Linzell JL: **Mammary-gland blood flow and oxygen, glucose and volatile fatty acid uptake in the conscious goat.** *J Physiol* 1960, **153**:492–509.

88. Reynolds M: **Mammary respiration in lactating goats.** *Am J Physiol* 1967, **212**:707–710.

89. Davis AJ, Fleet IR, Goode JA, Hamon MH, Walker FM, Peaker M: **Changes in mammary function at the onset of lactation in the goat: correlation with hormonal changes.** *J Physiol* 1979, **288**:33–44.

90. Davis SR, Collier RJ, McNamara JP, Head HH, Croom WJ, Wilcox CJ: **Effects of thyroxine and growth hormone treatment of dairy cows on mammary uptake of glucose, oxygen and other milk fat precursors.** *J Anim Sci* 1988, **66**:80–89.

91. Matsumoto A, Matsumoto S, Sowers AL, Koscielniak JW, Trigg NJ, Kuppusamy P, Mitchell JB, Subramanian S, Krishna MC, Matsumoto K: **Absolute oxygen tension (pO(2)) in murine fatty and muscle tissue as determined by EPR.** *Magn Reson Med* 2005, **54**:1530–1535.

92. Dhimitruka I, Bobko AA, Eubank TD, Komarov DA, Khramtsov VV: **Phosphonated trityl probes for concurrent in vivo tissue oxygen and pH monitoring using electron paramagnetic resonance-based techniques.** *J Am Chem Soc* 2013, **135**:5904–5910.

93. Tatum JL, Kelloff GJ, Gillies RJ, Arbeit JM, Brown JM, Chao KS, Chapman JD, Eckelman WC, Fyles AW, Giaccia AJ, et al: **Hypoxia: importance in tumor biology, noninvasive measurement by imaging, and value of its measurement in the management of cancer therapy.** *Int J Radiat Biol* 2006, **82**:699–757.

94. Seagroves TN, Hadsell D, McManaman J, Palmer C, Liao D, McNulty W, Welm B, Wagner KU, Neville M, Johnson RS: **HIF1alpha is a critical regulator of secretory differentiation and activation, but not vascular expansion, in the mouse mammary gland.** *Development* 2003, **130**:1713–1724.

95. Prosser CG, Topper YJ: **Changes in the rate of carrier-mediated glucose transport by mouse mammary epithelial cells during ontogeny: hormone dependence delineated in vitro.** *Endocrinology* 1986, **119**:91–96.

96. Mueckler M, Thorens B: **The SLC2 (GLUT) family of membrane transporters.** *Mol Aspects Med* 2013, **34**:121–138.

97. Zhao FQ, Keating AF: **Expression and regulation of glucose transporters in the bovine mammary gland.** *J Dairy Sci* 2007, **90**(Suppl 1):E76–86.

98. Shao Y, Wall EH, McFadden TB, Misra Y, Qian X, Blauwiekel R, Kerr D, Zhao FQ: **Lactogenic hormones stimulate expression of lipogenic genes but not glucose transporters in bovine mammary gland.** *Domest Anim Endocrinol* 2013, **44**:57–69.

99. Loike JD, Cao L, Brett J, Ogawa S, Silverstein SC, Stern D: **Hypoxia induces glucose transporter expression in endothelial cells.** *Am J Physiol* 1992, **263**:C326–333.

100. Takagi H, King GL, Aiello LP: **Hypoxia upregulates glucose transport activity through an adenosine-mediated increase of GLUT1 expression in retinal capillary endothelial cells.** *Diabetes* 1998, **47**:1480–1488.

101. Hennighausen L, Robinson GW, Wagner KU, Liu X: **Developing a mammary gland is a stat affair.** *J Mammary Gland Biol Neoplasia* 1997, **2**:365–372.

102. Wagner KU, Rui H: **Jak2/Stat5 signaling in mammogenesis, breast cancer initiation and progression.** *J Mammary Gland Biol Neoplasia* 2008, **13**:93–103.

103. Liu X, Robinson GW, Gouilleux F, Groner B, Hennighausen L: **Cloning and expression of Stat5 and an additional homologue (Stat5b) involved in prolactin signal transduction in mouse mammary tissue.** *Proc Natl Acad Sci USA* 1995, **92**:8831–8835.

104. Gouilleux F, Wakao H, Mundt M, Groner B: **Prolactin induces phosphorylation of Tyr694 of Stat5 (MGF), a prerequisite for DNA binding and induction of transcription.** *EMBO J* 1994, **13**:4361–4369.

105. Furth PA, Nakles RE, Millman S, Diaz-Cruz ES, Cabrera MC: **Signal transducer and activator of transcription 5 as a key signaling pathway in normal mammary gland developmental biology and breast cancer.** *Breast Cancer Res* 2011, **13**:220.

106. Yamaji D, Na R, Feuermann Y, Pechhold S, Chen W, Robinson GW, Hennighausen L: **Development of mammary luminal progenitor cells is controlled by the transcription factor STAT5A.** *Genes Dev* 2009, **23**:2382–2387.

107. Vafaizadeh V, Klemmt P, Brendel C, Weber K, Doebele C, Britt K, Grez M, Fehse B, Desrivieres S, Groner B: **Mammary epithelial reconstitution with gene-modified stem cells assigns roles to Stat5 in luminal alveolar cell fate decisions, differentiation, involution, and mammary tumor formation.** *Stem Cells* 2010, **28**:928–938.

108. Anderson SM, Rudolph MC, McManaman JL, Neville MC: **Key stages in mammary gland development. Secretory activation in the mammary gland: it's not just about milk protein synthesis!** *Breast Cancer Res* 2007, **9**:204.

109. Barash I: **Stat5 in the mammary gland: controlling normal development and cancer.** *J Cell Physiol* 2006, **209**:305–313.

110. Joung YH, Park JH, Park T, Lee CS, Kim OH, Ye SK, Yang UM, Lee KJ, Yang YM: Hypoxia activates signal transducers and activators of transcription 5 (STAT5) and increases its binding activity to the GAS element in mammary epithelial cells. *Exp Mol Med* 2003, **35**:350–357.

111. Lee MY, Joung YH, Lim EJ, Park JH, Ye SK, Park T, Zhang Z, Park DK, Lee KJ, Yang YM: Phosphorylation and activation of STAT proteins by hypoxia in breast cancer cells. *Breast* 2006, **15**:187–195.

112. Chiba S: Notch signaling in stem cell systems. *Stem Cells* 2006, **24**:2437–2447.

113. LaVoie MJ, Selkoe DJ: The notch ligands, jagged and delta, are sequentially processed by alpha-secretase and presenilin/gamma-secretase and release signaling fragments. *J Biol Chem* 2003, **278**:34427–34437.

114. Bouras T, Pal B, Vaillant F, Harburg G, Asselin-Labat ML, Oakes SR, Lindeman GJ, Visvader JE: Notch signaling regulates mammary stem cell function and luminal cell-fate commitment. *Cell Stem Cell* 2008, **3**:429–441.

115. Buono KD, Robinson GW, Martin C, Shi S, Stanley P, Tanigaki K, Honjo T, Hennighausen L: The canonical Notch/RBP-J signaling pathway controls the balance of cell lineages in mammary epithelium during pregnancy. *Dev Biol* 2006, **293**:565–580.

116. Gustafsson MV, Zheng X, Pereira T, Gradin K, Jin S, Lundkvist J, Ruas JL, Poellinger L, Lendahl U, Bondesson M: Hypoxia requires notch signaling to maintain the undifferentiated cell state. *Dev Cell* 2005, **9**:617–628.

117. Zheng X, Linke S, Dias JM, Gradin K, Wallis TP, Hamilton BR, Gustafsson M, Ruas JL, Wilkins S, Bilton RL, *et al*: Interaction with factor inhibiting HIF-1 defines an additional mode of cross-coupling between the Notch and hypoxia signaling pathways. *Proc Natl Acad Sci USA* 2008, **105**:3368–3373.

118. Logan CY, Nusse R: The Wnt signaling pathway in development and disease. *Annu Rev Cell Dev Biol* 2004, **20**:781–810.

119. Boras-Granic K, Chang H, Grosschedl R, Hamel PA: Lef1 is required for the transition of Wnt signaling from mesenchymal to epithelial cells in the mouse embryonic mammary gland. *Dev Biol* 2006, **295**:219–231.

120. Brennan KR, Brown AM: Wnt proteins in mammary development and cancer. *J Mammary Gland Biol Neoplasia* 2004, **9**:119–131.

121. Lindvall C, Evans NC, Zylstra CR, Li Y, Alexander CM, Williams BO: The Wnt signaling receptor Lrp5 is required for mammary ductal stem cell activity and Wnt1-induced tumorigenesis. *J Biol Chem* 2006, **281**:35081–35087.

122. Lindvall C, Zylstra CR, Evans N, West RA, Dykema K, Furge KA, Williams BO: The Wnt co-receptor Lrp6 is required for normal mouse mammary gland development. *PloS ONE* 2009, **4**:e5813.

123. Zhang J, Li Y, Liu Q, Lu W, Bu G: Wnt signaling activation and mammary gland hyperplasia in MMTV-LRP6 transgenic mice: implication for breast cancer tumorigenesis. *Oncogene* 2010, **29**:539–549.

124. Oakes SR, Hilton HN, Ormandy CJ: The alveolar switch: coordinating the proliferative cues and cell fate decisions that drive the formation of lobuloalveoli from ductal epithelium. *Breast Cancer Res* 2006, **8**:207.

125. Zeng YA, Nusse R: Wnt proteins are self-renewal factors for mammary stem cells and promote their long-term expansion in culture. *Cell Stem Cell* 2010, **6**:568–577.

126. Mazumdar J, O'Brien WT, Johnson RS, LaManna JC, Chavez JC, Klein PS, Simon MC: O2 regulates stem cells through Wnt/beta-catenin signalling. *Nat Cell Biol* 2010, **12**:1007–1013.

127. Mazurek S, Weisse G, Wust G, Schafer-Schwebel A, Eigenbrodt E, Friis RR: Energy metabolism in the involuting mammary gland. *In Vivo* 1999, **13**:467–477.

128. Whelan KA, Reginato MJ: Surviving without oxygen: hypoxia regulation of mammary morphogenesis and anoikis. *Cell Cycle* 2011, **10**:2287–2294.

129. Whelan KA, Caldwell SA, Shahriari KS, Jackson SR, Franchetti LD, Johannes GJ, Reginato MJ: Hypoxia suppression of Bim and Bmf blocks anoikis and luminal clearing during mammary morphogenesis. *Mol Biol Cell* 2010, **21**:3829–3837.

130. Vaapil M, Helczynska K, Villadsen R, Petersen OW, Johansson E, Beckman S, Larsson C, Pahlman S, Jogi A: Hypoxic conditions induce a cancer-like phenotype in human breast epithelial cells. *PLoS ONE* 2012, **7**:e46543.

18

An efficient method for the sanitary vitrification of bovine oocytes in straws

Yanhua Zhou[1], Xiangwei Fu[1], Guangbin Zhou[3], Baoyu Jia[1], Yi Fang[1], Yunpeng Hou[2] and Shien Zhu[1*]

Abstract

Background: At present, vitrification has been widely applied to humans, mice and farm animals. To improve the efficiency of vitrification in straw, bovine oocytes were used to test a new two-step vitrification method in this study.

Results: When *in vitro* matured oocytes were exposed to 20% ethylene glycol (EG20) for 5 min and 40% ethylene glycol (EG40) for 30 s, followed by treatment with 30% glycerol (Gly30), Gly40 or Gly50, a volume expansion was observed in Gly30 and Gly40 but not Gly50. This indicates that the intracellular osmotic pressure after a 30 s differs between EG40 and ranged between Gly40 (approximately 5.6 mol/L) and Gly50 (approximately 7.0 mol/L). Since oocytes are in EG40 just for only a short period of time (30 s) and at a lower temperature (4°C), we hypothesize that the main function of this step in to induce dehydration. Based on these results, we omitted the EG40 step, before oocytes were pretreated in EG20 for 5 min, exposed to pre-cooled (4°C) Gly50, for 30 s, and then dipped into liquid nitrogen. After warming, 81.1% of the oocytes survived, and the surviving oocytes developed into cleavage stage embryos (63.5%) or blastocysts (20.0%) after parthenogenetic activation.

Conclusions: These results demonstrate that in a two-step vitrification procedure, the permeability effect in the second step is not necessary. It is possible that the second step is only required to provide adequate osmotic pressure to condense the intracellular concentration of CPAs to a level required for successful vitrification.

Keywords: Bovine, Cryopreservation, Oocytes, Straw, Vitrification

Background

Vitrification is the rapid cooling of cells in liquid medium in the absence of ice crystal formation. Vitrification can be achieved when the intracellular concentration of cryoprotective agents (CPAs) is higher than 6 mol/L [1]. The benefits of a two-step vitrification method are that it allows establishment of a relatively complete equilibrium while reducing exposure of the oocyte to potential toxic effects of CPAs. Previously, oocytes or embryos were first exposed to non-vitrifying solutions containing permeating CPAs [2,3]. Next, the oocytes were exposed for a short time (45–60 s) to a vitrifying solution (VS) containing high concentrations of penetrating (4.8–6.4 mol/L) and non-penetrating (0.5–

0.75 mol/L) CPAs before being plunged into liquid nitrogen (LN2) [2-4].

Since the first successful vitrification of mouse embryos by Rall and Fahy in 1985 [1], this method has been used widely for oocyte and embryo cryopreservation. Numerous research articles have focused on CPA permeability and the rate at which it enters cells [5,6]. Other studies have investigated incubation times in both the pretreatment and vitrification solutions and found that the temperature used during the handling procedure is also important for successful vitrification [7-9]. The open-pulled straw (OPS) method originally described by Vajta and colleagues, allows for faster heat transfer between the solution and the environment, achieving cooling/warming rates on the order of 20,000°C/min [10]. In 1999, when Le Gal and Massip compared three approaches (standard 0.25-mL straw, OPS, and Microdrop) for cooling a vitrification solution containing bovine oocytes, the highest cleavage rate was achieved with the traditional straw [11]. Dinnyés and colleagues [12]

---

* Correspondence: zhushien@cau.edu.cn
[1]National Engineering Laboratory for Animal Breeding, Key Laboratory of Animal genetics, Breeding and Reproduction, Ministry of Agriculture, College of Animal Science and Technology, China Agricultural University, Beijing 100193, P.R. China
Full list of author information is available at the end of the article

described the use of solid-surface vitrification (SSV). In 2004, an early report using cryotops for bovine oocyte vitrification was published [13]. These variations make the vitrification method seem difficult to master which has limited the application of this technology in the field of reproductive biology.

Cells react to changes in extracellular osmolarity by altering their volume. Cells exposed to hypotonic or hypertonic solutions initially react either by swelling (hypotonic solutions) or shrinking (hypertonic solutions) due to water exchange but later recover as permeant solutes equilibrate across the cell membrane [4,5,14,15]. Vanderzwalmen et al. [3,4] estimated the final intracellular concentration of cryoprotectant (ICCP) after incubation in vitrification solutions by exposing cells to sucrose solutions with defined molarities. The ICCP was calculated from the sucrose concentration that produced no change in cell volume, i.e., when intra- and extracellular osmolarities were equivalent [4].

In 1977, Whittingham successfully cryopreserved mouse oocytes [16]. Bovine oocytes were also vitrified and remained viable for offspring production after in vitro fertilization and embryo transplantation [17,18]. vitrified buffalo oocytes with 51.1% glycerol via the straw method, obtaining a maturation rate of 23.5% after thawing [18]. When glycerol was used with EG, which increased permeability of the cell membrane during oocyte vitrification, and maturation rates of 30 s exposure groups did not differ from those of controls [19]. Additionally, the OPS (open pulled straw) method results in a better survival rate during cryopreservation than the straw method [20]. However, unlike other methods, the straw method is safer for oocyte vitrification because the oocytes are free of bacterial contamination due to a lack of direct contact with liquid nitrogen.

In our experiments glycerol was used as an extracellular measure for ICCP. In the first part of this study, bovine oocytes were used to test changes in intracellular cryoprotectant concentration during a widely used two-step vitrification method. Oocytes were pretreated with 20% EG (EG20) for 5 min, transferred to pre-cooled (4°C) 40% EG (EG40) for 30 s, then treated with pre-cooled glycerol either at 30% (Gly30), 40% (Gly40) or 50% (Gly50) concentration. The intracellular EG molarity was then determined from the extracellular glycerol molarity. In the second part of the experiment, oocytes were pretreated with EG20 for 5 min, transferred directly to pre-cooled (4°C) 50% glycerol (Gly50) for 30 s, and then plunged directly into liquid nitrogen for cryopreservation in an insemination straw. Vitrified-warmed oocytes were parthenogenetically activated and cultured in vitro to assess viability.

In this study, experiments were designed to improve the efficacy of vitrification in straws. To optimize the ideal CPA treatment for this two-step vitrification method, different cryoprotectants (EG and Gly) were used in each step, which differs from methods reported previously. It has been reported that the permeability of glycerol is relatively low [14]. The present experiments examined whether CPA permeability during the second step is a key factor for vitrification. We investigated the possibility that the second equilibration step provides a high osmotic pressure increase intracellular CPA to a level required for successful vitrification.

Methods

The Institution Animal Care and Use Committee at China Agricultural University (Beijing, China) reviewed and approved the protocols used in this study. All chemicals and media were purchased from Sigma Chemical Co. (St. Louis, MO, USA) unless otherwise indicated.

Solution preparation

Modified phosphate-buffered saline (mPBS) was prepared by adding 10% (v/v) fetal bovine serum (FBS, Gibco), 0.3% (w/v) BSA and 50 mg/mL gentamycin to Dulbecco's phosphate-buffered saline (DPBS, Gibco).

EG20 was prepared by adding 20% (v/v) ethylene glycol to mPBS;

EG40 was prepared by adding 40% (v/v) ethylene glycol to mPBS;

Gly30 was prepared by adding 30% (v/v) glycerol to mPBS;

Gly40 was prepared by adding 40% (v/v) glycerol to mPBS;

Gly50 was prepared by adding 50% (v/v) glycerol to mPBS;

Dilution medium was 0.5 mol/L sucrose in mPBS.

Oocyte collection and *in vitro* maturation

Bovine (*Bos taurus*, 3 to 6 yr of age) ovaries were transported from the abattoir to the laboratory in a physiological saline solution at 26°C to 30°C within 2 h of slaughter. Antral follicles (2 mm to 8 mm in diameter) were manually aspirated using an 18-gauge needle attached to a 10 mL syringe. Oocytes with at least four layers of compact cumulus cells (COCs) were selected for in vitro maturation (IVM). Oocytes were washed three times in HEPES-buffered TCM-199 medium and then washed twice in $NaHCO_3$-buffered TCM-199. Fifty COCs were transferred to 0.75 mL maturation medium (M199 with 10 mg/mL oFSH [Ovagen, Auckland, New Zealand], 10 mg/mL oLH [Ovagen], 1 mg/mL estradiol [Ovagen] and 10% fetal bovine serum [FBS; Gibco]) in 4-well plates (Nunclon). The COCs were cultured for 22 h at 38.5°C in a humidified atmosphere with 5% CO_2. Oocytes were denuded after 22 h maturation by repeated pipetting with a 200 μL pipette for approximately 1 min in 38°C 0.1% w/v. hyaluronidase. Cumulus-free oocytes with the first polar body were selected and randomly allocated to experimental groups.

Oocyte volume

Oocyte volumes were determined using established methods [14] with modifications. Oocytes were fixed in a 5 µL mPBS drop using a holding needle (outer diameter 50 µm; inner diameter 30 µm) attached to an Olympus inverted microscope, 200 µL of EG20 was then flushed onto the drop. The drop of EG20 (~200 µL) was aspirated 5 min later with a transferpettor. In the second step, 500 µL of pre-cooled (4°C) EG40 (experiment 1-a) or Gly50 (experiment 1-e) was flushed into the drop. For Experiments 1-b, c and d, 200 µL pre-cooled solution (EG40) was flushed into the drop. EG40 was aspirated off the oocyte after 30 s, and then 500 µL of pre-cooled (4°C) glycerol solution (either of Gly30, Gly40 or Gly50) was flushed into the drop.

The entire procedure was video recorded using a CCD camera on an inverted microscope. Screenshots of the video recording were taken at the desired times. Cross-sectional areas of the oocytes were calculated using EZ-C1 3.00 Free Viewer software. The relative change in volume was determined according to a previously pubulished method [10]. Briefly, the oocyte area relative to that in isotonic mPBS medium was calculated and converted into a relative volume (considered as 1 V). The volume was assumed to change proportionally, and the equation $V = S^{3/2}$ was used, where S is relative cross-sectional area and V is the relative volume. For each treatment, 5 oocytes were examined.

Oocyte vitrification and warming

Oocytes were vitrified by a two-step method as previously reported, with modifications [9,21,22]. Briefly, oocytes were placed in EG20 for 5 min at 25°C. The oocytes were then transferred to pre-cooled Gly50 for 30 s, pipetted into sections of an insemination straw (250 µL, IMV, L'Aigle, France), as shown in Figure 1. and then straws were sealed with seal powder and plunged into liquid nitrogen. Two oocytes were loaded into each straw.

After one week of storage in liquid nitrogen, the straws were plunged into 25°C water for 10 s. As the crystallized sucrose solutions in the straw melted, the straws were removed from the water and quickly wiped dry. The straws were then held at the sealed end and shaken three times by hand to mix the vitrification solution and the sucrose solution. Subsequently, the seals of the straw were removed and the oocytes were expelled from the straw into a dry culture dish. Oocytes were put into fresh 0.5 mol/L sucrose for 5 min and then washed in two other mPBS dishes for 5 min each.

After a 30 min recovery in mPBS, the oocytes were assessed for survival. Surviving oocytes were those with regular, spherical shapes that were not lysed, shrunken, swollen or blackened. The surviving oocytes were parthenogenetically activated and cultured in vitro.

Parthenogenetic activation

Oocytes were washed three times in HEPES-buffered TCM-199 with 10% FBS (H199) and then activated as follows: (1) incubation for 5 min in 7% ethanol in IVM medium at room temperature and (2) cultured for 4 h in 2 mmol/L 6-DMAP in culture medium. Fifteen oocytes were transferred to 60 µL Charles Rosenkran's 1 medium [23] with BSA (3 mg/mL, Sigma A3311) covered with mineral oil (Sigma M8410) in a 35 mm × 35 mm Nunclon dish and cultured in an incubator (38.5°C with 5% CO_2 in air) for up to 48 h before determining the rates of activation and cleavage. Cleaved embryos were cultured for an additional 5 d in Charles Rosenkran's 1 medium with 5% FBS.

Experimental design

Experiment 1. In this section oocytes were randomly allocated to five experimental groups, and each experiment was repeated five times.

(a) Oocytes were incubated in EG20 for 5 min followed by addition of pre-cooled (4°C) EG40. Oocytes → EG20 (25°C, 5 min) → EG40 (4°C, 4 min).

(b) Oocytes were incubated in EG20 for 5 min, pre-cooled (4°C) EG40 for 30 s, and then incubated in pre-cooled (4°C) Gly30 for 3 min. Oocytes → EG20 (25°C, 5 min) → EG40 (4°C, 30 s) → Gly30 (4°C, 3 min).

(c) Oocytes were incubated in EG20 for 5 min, pre-cooled (4°C) EG40 for 30 s, and then pre-cooled (4°C) Gly40 for 3 min. Oocytes → EG20 (25°C, 5 min) → EG40 (4°C, 30 s) → Gly40 (4°C, 3 min).

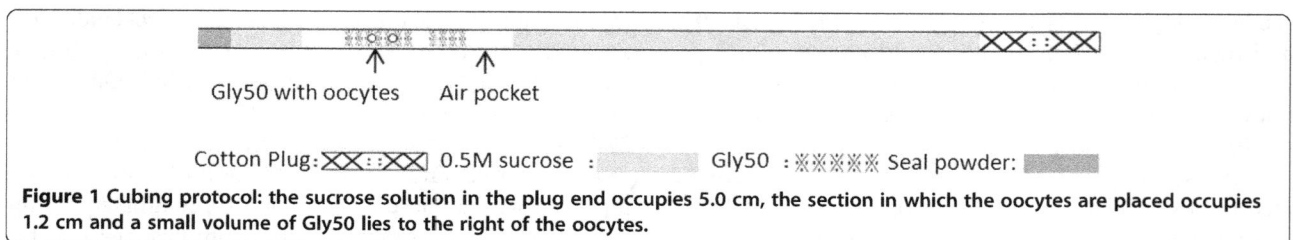

Gly50 with oocytes   Air pocket

Cotton Plug: ▨▨▨   0.5M sucrose :   Gly50 : ※※※※※ Seal powder: ▨▨▨

Figure 1 Cubing protocol: the sucrose solution in the plug end occupies 5.0 cm, the section in which the oocytes are placed occupies 1.2 cm and a small volume of Gly50 lies to the right of the oocytes.

(d) Oocytes were incubated in EG20 for 5 min, pre-cooled (4°C) EG40 for 30 s, and then pre-cooled (4°C) Gly50 for 3 min. Oocytes → EG20 (25°C, 5 min) → EG40 (4°C, 30 s) → Gly50 (4°C, 3 min).

(e) Oocytes were incubated in EG20 for 5 min and then pre-cooled (4°C) Gly50 for 4 min. Oocytes → EG20 (25°C, 5 min) → Gly50 (4°C, 4 min).

The volume changes of oocytes during all of these procedures were analyzed and used to generate a curve diagram over time.

Experiment 2. Development of cryopreserved oocytes after parthenogenetic activation.

Cumulus-free oocytes with the first polar body and normal morphology were selected and allocated randomly to the following experimental groups:

(1) Control group: Oocytes without CPA treatment or vitrification were cultured after parthenogenetic activation.

(2) Toxicity group: Oocytes were exposed to the same solutions as the vitrification group but were not plunged into liquid nitrogen. These oocytes were diluted and parthenogenetically activated according to the procedure used for the vitrification group.

(3) Vitrification group: based on the results of experiment 1, oocytes were pre-treated in EG20 (25°C) for 5 min and then transferred to Gly50 (4°C) for 30 s before being plunged into liquid nitrogen.

Statistical analysis

Embryos development experiments were repeated three times. The percentage data were subjected to arcsine transformation before statistical analysis. The data were analyzed by one-way ANOVA combined with the LSD test. $P < 0.05$ was considered statistically significant.

Results

Experiment 1. Oocytes volume changes for five different protocols:

(a) Oocytes → EG20 (25°C, 5 min) → EG40 (4°C, 4 min). When exposed to EG20, the oocytes shrank to 0.48 V (◇ in Figure 2a) in 20 s and then swelled slowly to 0.80 V after 5 min of exposure. When the oocytes were flushed with pre-cooled EG40, they shrank to 0.52 V (△ in Figure 2) and then gradually swelled again.

(b) Oocytes → EG20 (25°C, 5 min) → EG40 (4°C, 30 s) → Gly30 (4°C, 3 min). At the end of the 30 s exposure to EG40, the oocytes shrank to 0.59 V (◇ in Figure 2b). Subsequently, pre-cooled Gly30 was flushed over the oocytes and an expansion in volume to 0.65 V was observed

(from ◇ to △, as shown in Figure 2b), indicating a higher intracellular osmotic pressure as compared to the extracellular pressure. After a 25-s exposure to Gly30, the oocytes began to gradually shrink in volume.

(c) Oocytes → EG20 (25°C, 5 min) → EG40 (4°C, 30 s) → Gly40 (4°C, 3 min). As shown in Figure 2c (from ◇ to △), the oocytes swelled from 0.53 V to 0.59 V, indicating that the intracellular osmotic pressure after treatment was higher than the extracellular osmotic pressure generated by Gly40.

(d) Oocytes → EG20 (25°C, 5 min) → EG40 (4°C, 30 s) → Gly50 (4°C, 3 min). As shown in Figure 2d, when the oocytes were flushed with pre-cooled Gly50 followed by a 5 min treatment with EG20 and a 30 s treatment with EG40, no expansion was observed. After immersion in Gly50, the oocytes began to shrink, and within 60 s, the oocytes gradually reached a minimum volume (from ◇ to △ in Figure 2d). This result indicated that the intracellular osmotic pressure was not higher than the extracellular osmotic pressure exerted by Gly50.

(e) Oocytes → EG20 (25°C, 5 min) → Gly50 (4°C, 4 min). In this experiment, oocytes were pretreated with EG20 and then flushed with pre-cooled Gly50. Volume changes are shown in Figure 2e. The oocytes shrank quickly in Gly50 after EG20 pretreatment. The oocytes reached a minimum volume (△ part in Figure 2e) within approximately 100 s.

Experiment 2. Development of cryopreserved oocytes after parthenogenetic activation

As shown in Table 1, after vitrification, warming and parthenogenetic activation, surviving bovine oocytes develop to the cleavage stage embryos and blastocysts. After this vitrification protocol, 81.1% oocytes survived and 63.5% of them cleaved after parthenogenetic activation. Finally, we observed a 20.0% blastocyst rate. There was no difference in rates of blastocyst development between the control and toxicity groups (38.6% *vs* 36.0%, $P > 0.05$). However, after oocyte vitrification, the rates of blastocyst development decreased ($P < 0.05$).

Discussion

There are few reports that analyze vitrification of bovine metaphase oocytes by the straw method. In the present study, we achieved cleavage (63.5%) and blastocyst development (20.0%) after parthenogenetic activation of vitrified-warmed bovine oocytes similar to that from oocytes vitrified by the open-pulled straw method (57.0% cleavage and 23.0% blastocyst development, respectively)

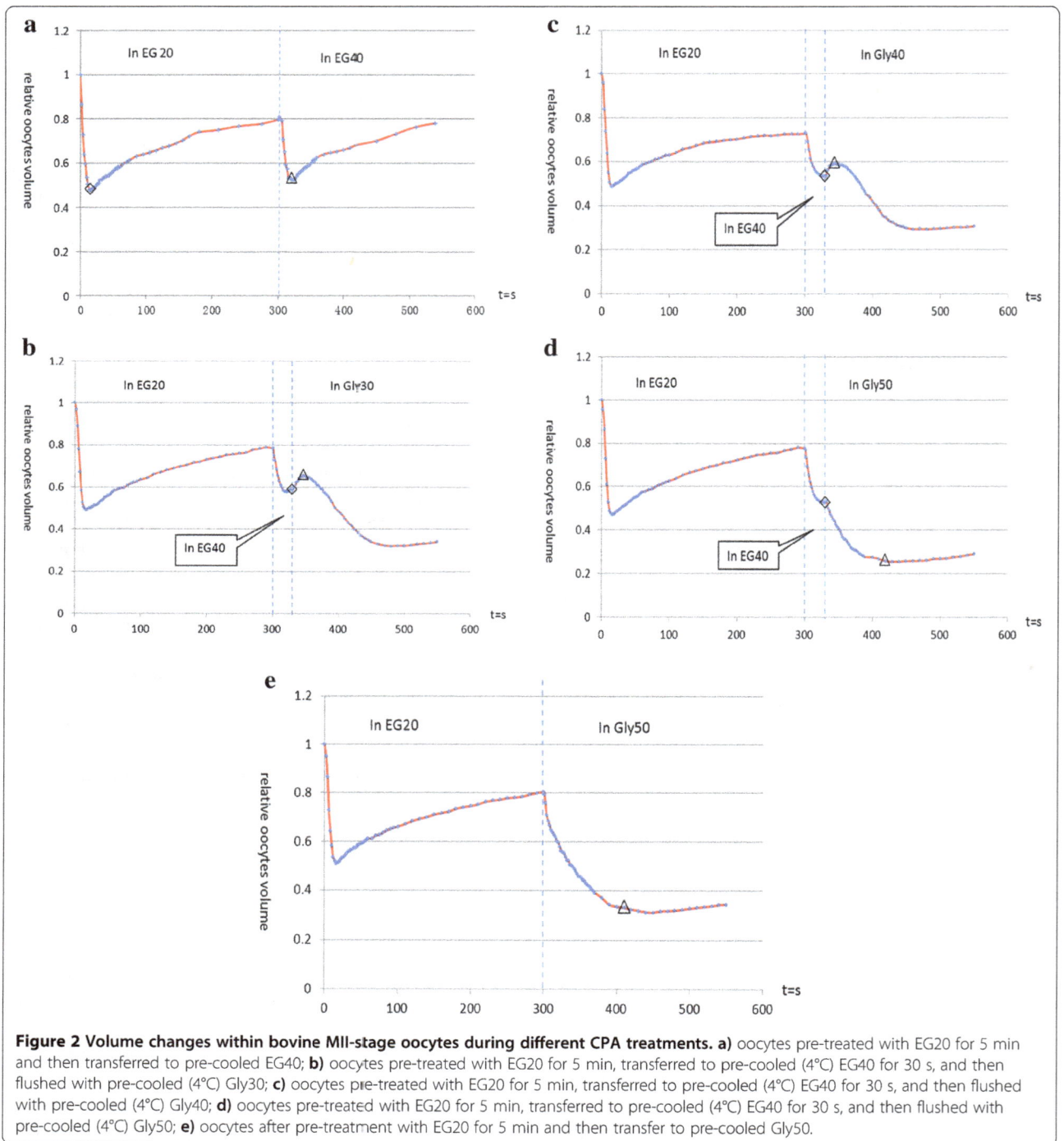

Figure 2 Volume changes within bovine MII-stage oocytes during different CPA treatments. a) oocytes pre-treated with EG20 for 5 min and then transferred to pre-cooled EG40; **b)** oocytes pre-treated with EG20 for 5 min, transferred to pre-cooled (4°C) EG40 for 30 s, and then flushed with pre-cooled (4°C) Gly30; **c)** oocytes pre-treated with EG20 for 5 min, transferred to pre-cooled (4°C) EG40 for 30 s, and then flushed with pre-cooled (4°C) Gly40; **d)** oocytes pre-treated with EG20 for 5 min, transferred to pre-cooled (4°C) EG40 for 30 s, and then flushed with pre-cooled (4°C) Gly50; **e)** oocytes after pre-treatment with EG20 for 5 min and then transfer to pre-cooled Gly50.

[24]. However, the straw method has an advantage for bovine oocyte vitrification, because oocytes do not directly contact the liquid nitrogen and thus potential bacterial contamination is avoided.

Most two-step vitrification methods are similar, with some differences in exposure time or CPA combination. In fact, exposure times greatly influence the outcome of the vitrification method [2,7-9,22]. It has been reported that when bovine blastocysts are exposed to EFS40 from 1 to 3 min, the survival rate drops from 77.0% to 7.0% [8]. Campos-Chillon tested a range of pretreatment times (from 1 to 3 min) in 3.5 mol/L ethylene glycol on bovine morulae and found that a 1-min exposure was ideal [7]. Fujihira found a significant relationship ($P <$ 0.05) between the rate of development of morphologically normal oocytes after vitrification and equilibration

Table 1 Effects of oocyte vitrification on embryo development after parthenogenetic activation

Group	No. oocytes	Survival rate	Cleavage rate	Blastocyst rate
Control	132	100% ± 0 (132/132)[a]	84.21% ± 2% (111/132)[a]	38.62% ± 0.75% (51/132)[a]
Toxicity	103	93.22% ± 1.56% (96/103)[b]	76.74% ± 2.27% (79/96)[b]	35.93% ± 1.47% (37/96)[a]
Vitrified	105	81.08% ± 2.86% (89/105)[c]	63.49% ± 2.1% (66/105)[c]	19.96% ± 1.06% (21/105)[b]

Different superscripts (a, b, c) in the same column represent a significant difference ($P < 0.05$). Percentage data are presented as mean ± SEM.

time in pigs [25]. Others used much longer exposure times (10-20 min) for oocytes [19] or embryos [21,26] before vitrification. Despite these variations, the ultimate goal of these procedures was to determine proper intracellular concentrations of CPA for successful vitrification of oocytes.

In the present study, bovine oocytes exposed to EG20 for 5 min and then transferred to EG40 shrank to 0.52 V before gradually regaining volume. Furthermore, mouse [5] and bovine [15] oocytes exhibited ideal osmotic responses when their volumes were analyzed using the Boyle-van't Hoff relationship. Here, cells stopped shrinking at 0.52 V, which can be inferred as the state of equilibrium between intra- and extracellular osmotic pressures resulting from exposure to hypotonic (swelling) or hypertonic (shrinking) solutions [4]. In subsequent experiments, oocytes were sequentially exposed to EG20 (5 min) and EG40 (30 s) followed by treatment either with Gly30 and Gly40 or Gly50. Volume expansion was observed in Gly30 and Gly40, suggesting that the intracellular osmotic pressure was higher than that produced by Gly40. However, volume expansion did not occur when the oocytes were flushed with Gly50, suggesting that the extracellular osmotic pressure was higher in Gly50 than the intracellular osmotic pressure.

The permeability of CPAs is strongly decreased at low temperatures [15,27]; therefore, during a short exposure of oocytes (30 s) to pre-cooled (4°C) EG40 in the second equilibrium step, the intra- and extracellular EG levels changed minimally. As a result, we omitted the EG40 step and modified the second equilibrium step by exposing oocytes to EG20 for 5 min and to pre-cooled Gly50 for 30 s before plunging into liquid nitrogen. As shown in Figure 2e, the oocytes shrank to a minimum volume in approximately 100 s, and this shrinking would result in concentration of intracellular EG during this period. It has been reported by Jin [28] that vitrification solutions with higher osmotic pressure could facilitate intracellular vitrification, yielding better results. In Experiment 2, oocytes were vitrified after treatment with pre-cooled (4°C) Gly50 for 30 s. After warming, 81.1% of the oocytes survived, and the surviving oocytes developed into cleavage stage embryos (63.5%) and blastocysts (20.0%) after parthenogenetic activation. As long as the CPA concentration higher than EFS30 (which corresponds to PB1 medium containing 30% (v/v ethylene glycol, 21% (w/v) Ficoll 70, and 0.35 mol/L sucrose) oocytes will

not devitrify during warming [29]. The high survival of oocytes here which indicates that the intra- and extracellular CPAs were vitrified during both the freezing and warming procedures.

In this study, we did not compare oocyte survival and development in the absence and presence of Gly50 (after EG20 (5 min) > EG40 (30 s)). We believed that treatment with Gly50 would yield better results, as it has been reported that solutions containing 40% ethylene glycol remain transparent when plunged into liquid nitrogen but crystallize during warming [29]. Although we can not rule out the possibility that EG leaves the oocytes under high osmotic pressure, the concentration of EG remaining is sufficient to achieve vitrification according to our results in Table 1.

Conclusion

Results from these experiments provide clear evidence that during the two-step equilibrium before vitrification, if proper pretreatment (EG20 for 5 min) was taken, the permeability of CPA into oocytes is unnecessary during the second equilibration step. What is efficient is a vitrification solution (Gly50) with high osmotic pressure only during the second equilibrium period to concentrate the intracellular CPAs adequately to facilitate intracellular vitrification.

Abbreviations
EG20: 20% v/v ethylene glycol in mPBS; EG40: 40% v/v ethylene glycol in mPBS; Gly30: 30% v/v glycerol in mPBS; Gly40: 40% v/v glycerol in mPBS; Gly50: 50% v/v glycerol in mPBS; EFS30: PB1 medium containing 30% v/v ethylene glycol, 21% (w/v) Ficoll 70, and 0.35 mol/L sucrose; CPA: Cryoprotective agents; ICCP: Intracellular concentration of cryoprotectant; VS: Vitrification solution; V: Volume; S: Sucrose; Min: Minute; Sec: Second.

Competing interests
The authors declare that they have no competing interests.

Authors' contributions
ZYH designed the study, conducted the experiments and analyses and wrote the manuscript. ZSE collaborated in the design of the study and analysis and oversaw the work of laboratory staff. FXW collaborated in the design of the study and the analysis and helped write the manuscript. ZGB collaborated in the design of the study and reviewed the manuscript. JBY selected the assays to measure cross-sectional area and assisted in the interpretation of results. FY helped with the culture of oocytes and early embryos. HYP collaborated in the interpretation of results and writing of the manuscript. All authors read and approved the final manuscript.

Authors' information
Yanhua Zhou is Ph.D candidate of College of Animal Science and Technology, China Agricultural University, Shien Zhu is professor of College of Animal Science and Technology, China Agricultural University. Xiangwei

Fu is associate professor of College of Animal Science and Technology, China Agricultural University, Yunpeng Hou is associate professor of State Key Laboratory for Agrobiotechnology, College of Biological Sciences, China Agricultural University, Guangbin Zhou is professor of College of Animal Science and Technology, Sichuan Agricultural University (Chengdu Campus). Baoyu Jia and Yi Fang are Ph.D candidate of College of Animal Science and Technology, China Agricultural University.

Acknowledgments

This work was supported by the National "863" Project Foundation of China (No. 2011AA100303) and the National Science and Technology Support Projects of China (No. 2011BAD19B01). We thank Nature Publishing Group Language Editing (NPGLE) for proof-reading the manuscript.

Author details

[1]National Engineering Laboratory for Animal Breeding, Key Laboratory of Animal genetics, Breeding and Reproduction, Ministry of Agriculture, College of Animal Science and Technology, China Agricultural University, Beijing 100193, P.R. China. [2]State Key Laboratory for Agrobiotechnology, College of Biological Sciences, China Agricultural University, Beijing 100193, P.R. China. [3]Institute of Animal Genetics and Breeding, College of Animal Science and Technology, Sichuan Agricultural University (Chengdu Campus), Wenjiang 611130, P.R. China.

References

1. Rall WF, Fahy GM: **Ice-free cryopreservation of mouse embryos at -196°C by vitrification.** Nat Commun 1985, 313:573–575.
2. Kuwayama M, Vajta G, Kato O, Leibo SP: **Highly efficient vitrification method for cryopreservation of human oocytes.** Reprod Biomed Online 2005, 11:300–308.
3. Vanderzwalmen P, Ectors F, Grobet L, Prapas Y, Panagiotidis Y, Vanderzwalmen S, Stecher A, Frias P, Liebermann J, Zech NH: **Aseptic vitrification of blastocysts from infertile patients, egg donors and after IVM.** Reprod Biomed Online 2009, 19:700–707.
4. Vanderzwalmen P, Connan D, Grobet L, Wirleitner B, Remy B, Vanderzwalmen S, Zech N, Ectors FJ: **Lower intracellular concentration of cryoprotectants after vitrification than after slow freezing despite exposure to higher concentration of cryoprotectant solutions.** Hum Reprod 2013, 28:2101–2110.
5. Liang W, Jun L, Guangbin Z, Yunpeng H, Junjie L, Shien Z: **Quantitative investigations on the effects of exposure durations to the combined cryoprotective agents on mouse oocyte vitrification procedures.** Biol Reprod 2011, 85:884–894.
6. Edashige K, Ohta S, Tanaka M, Kuwano T, Valdez DM, Hara T, Jin B, Takahashi S, Seki S, Koshimoto C: **The role of aquaporin 3 in the movement of water and cryoprotectants in mouse morulae.** Biol Reprod 2007, 77:365–375.
7. Campos-Chillon LF, Walker DJ, De La Torre-Sanchez JF, Seidel GE Jr: **In vitro assessment of a direct transfer vitrification procedure for bovine embryos.** Theriogenology 2006, 65:1200–1214.
8. Tachikawa S, Otoi T, Kondo S, Machida T, Kasai M: **Successful vitrification of bovine blastocysts, derived by in vitro maturation and fertilization.** Mol Reprod Dev 1993, 34:266–271.
9. Vajta G, Holm P, Greve T, Callesen H: **Factors affecting survival rates of in vitro produced bovine embryos after vitrification and direct in-straw rehydration.** Anim Reprod Sci 1996, 45:191–200.
10. Vajta G, Holm P, Kuwayama M, Booth PJ, Jacobsen H, Greve T, Callesen H: **Open pulled straw (OPS) vitrification: a new way to reduce cryoinjuries of bovine ova and embryos.** Mol Reprod Dev 1998, 51:53–58.
11. Le Gal F, Massip A: **Cryopreservation of cattle oocytes: effects of meiotic stage, cycloheximide treatment, and vitrification procedure.** Cryobiology 1999, 38:290–300.
12. Dinnyés A, Dai Y, Jiang S, Yang X: **High developmental rates of vitrified bovine oocytes following parthenogenetic activation, in vitro fertilization, and somatic cell nuclear transfer.** Biol Reprod 2000, 63:513–518.
13. KATO O, NAGAI T: **High survival rate of bovine oocytes matured in vitro following vitrification.** J Reprod Dev 2004, 50:685–596.
14. Pedro PB, Yokoyama E, Zhu SE, Yoshida N, Valdez DM Jr, Tanaka M, Edashige K, Kasai M: **Permeability of mouse oocytes and embryos at various developmental stages to five cryoprotectants.** J Reprod Dev 2005, 512:235–246.
15. Wang X, Al Naib A, Sun D, Lonergan P: **Membrane permeability characteristics of bovine oocytes and development of a step-wise cryoprotectant adding and diluting protocol.** Cryobiology 2010, 61:58–65.
16. Whittingham DG: **Fertilization in vitro and development to term of unfertilized mouse oocytes previously stored at—196 C.** J Reprod Fertil 1977, 49:89–94.
17. Fuku E, Kojima T, Shioya Y, Marcus GJ, Downey BR: **In vitro fertilization and development of frozen-thawed bovine oocytes.** Cryobiology 1992, 29:485–492.
18. Wani NA, Misra AK, Maurya SN: **Maturation rates of vitrified-thawed immature buffalo (Bubalus bubalis) oocytes: effect of different types of cryoprotectants.** Anim Reprod Sci 2004, 84:327–335.
19. Yamada C, Caetano HVA, Simões R, Nicacio AC, Feitosa WB, Assumpção MEOD, Visintin JA: **Immature bovine oocyte cryopreservation: comparison of different associations with ethylene glycol, glycerol and dimethylsulfoxide.** Anim Reprod Sci 2007, 99:384–388.
20. Sharma GT, Dubey PK, Chandra V: **Morphological changes, DNA damage and developmental competence of in vitro matured, vitrified-thawed buffalo (Bubalus bubalis) oocytes: a comparative study of two cryoprotectants and two cryodevices.** Cryobiology 2010, 60:315–321.
21. Saha S, Otoi T, Takagi M, Boediono A, Sumantri C, Suzuki T: **Normal calves obtained after direct transfer of vitrified bovine embryos using ethylene glycol, trehalose, and polyvinylpyrrolidone.** Cryobiology 1996, 33:291–299.
22. Zhu SE, Kasai M, Otoge H, Sakurai T, Machida T: **Cryopreservation of expanded mouse blastocysts by vitrification in ethylene glycol-based solutions.** J Reprod Fertil 1993, 98:139–145.
23. Rosenkrans CF, Zeng GQ, McNamara GT, Schoff PK, First NL: **Development of bovine embryos in vitro as affected by energy substrates.** Biol Reprod 1993, 49:459–462.
24. Hou Y, Dai Y, Zhu S, Zhu H, Wu T, Gong G, Wang H, Wang L, Liu Y, Li R: **Bovine oocytes vitrified by the open pulled straw method and used for somatic cell cloning supported development to term.** Theriogenology 2005, 64:1381–1391.
25. Fujihira T, Nagai H, Fukui Y: **Relationship between equilibration times and the presence of cumulus cells, and effect of taxol treatment for vitrification of in vitro matured porcine oocytes.** Cryobiology 2005, 51:339–343.
26. Van Wagtendonk-de Leeuw AD, Den Daas J, Rall WF: **Field trial to compare pregnancy rates of bovine embryo cryopreservation methods: vitrification and one-step dilution versus slow freezing and three-step dilution.** Theriogenology 1997, 48:1071–1084.
27. Leibo SP, Mazur P, Jackowski SC: **Factors affecting survival of mouse embryos during freezing and thawing.** Exp Cell Res 1974, 89:79–88.
28. Jin B, Mochida K, Ogura A, Koshimoto C, Matsukawa K, Kasai M, Edashige K: **Equilibrium vitrification of mouse embryos at various developmental stages.** Mol Reprod Dev 2012, 79:785–794.
29. Kasai M, Komi JH, Takakamo A, Tsudera H, Sakurai T, Machida T: **A simple method for mouse embryo cryopreservation in a low toxicity vitrification solution, without appreciable loss of viability.** J Reprod Fertil 1990, 89:91–97.

19

Mammary stem cells: expansion and animal productivity

Ratan K Choudhary

Abstract

Identification and characterization of mammary stem cells and progenitor cells from dairy animals is important in the understanding of mammogenesis, tissue turnover, lactation persistency and regenerative therapy. It has been realized by many investigators that altered lactation, long dry periods (non-milking period between two consecutive lactation cycles), abrupt cessation of lactation (common in water buffaloes) and disease conditions like mastitis, greatly reduce milk yield thus render huge financial losses within the dairy sector. Cellular manipulation of specialized cell types within the mammary gland, called mammary stem cells (MaSCs)/progenitor cells, might provide potential solutions to these problems and may improve milk production. In addition, MaSCs/progenitor cells could be used in regenerative therapy against tissue damage caused by mastitis. This review discusses methods of MaSC/progenitor cell manipulation and their mechanisms in bovine and caprine animals. Author believes that intervention of MaSCs/progenitor cells could lessen the huge financial losses to the dairy industry globally.

Keywords: Hormones, Mammary stem cell, Manipulation, Milk production, Ruminant, Xanthosine

Introduction

The ultimate goal of mammary gland and mammary stem cell biologists in dairy science is to enhance milk production in lactating dairy animals. Milk production is affected by the number of secretory cells in the mammary alveolar epithelium and the secretory activity per cell. The differentiation status of mammary epithelial cells determines their secretory activity. Poorly differentiated mammary epithelial cells are often non-secretory, whereas intermediate and fully differentiated cells are often secretory in nature [1,2]. Classification of these cells into poor-, intermediate- and fully-differentiated stages, are based on observation of cellular morphology at high magnification for the presence of secretory vacuoles, lipid droplets, nuclear location, cytoplasmic area and cell shape. Although hormones, like estrogen, progesterone and prolactin influence cytological differentiation of these cells but their regeneration depends upon the activity of mammary progenitor cells. Mammary progenitor cells are trans-amplifying cells [3] and are the progeny of mammary stem cells (MaSCs). Reports indicate that MaSCs are multipotent, giving rise to luminal and basal/myoepithelial cell types [4,5]. However, a recent report has indicated MaSCs as lineage-restricted unipotent stem cells in the mouse [6]. This suggests that re-evaluation of MaSCs is required to understand the biology of their cell regulation. For milk-producing dairy animals, more in-depth analysis of the characterization and regulation of MaSCs and progenitor cells is needed before we can understand how to influence cell turnover for increased milk production and tissue homeostasis of the mammary gland. Manipulation of mammary gland development and milk production can be achieved using management of photoperiod, frequent milking, machine milking and bovine somatotrophic (bST) hormone [7,8]. However, manipulation of MaSCs and progenitor cells for increasing milk production is novel and promising, and was first hypothesized by Capuco et al. [9,10]. This review deals with relatively recent studies performed towards expansion of MaSCs for determining the impact on milk production. Readers are encouraged to take note of two recent comprehensive review articles on MaSCs in animals of veterinary importance, including a comparative study of post-natal mammary gland development and mammary stem cells in murine and bovine animals [11,12].

Correspondence: vetdrrkc@gmail.com
School of Animal Biotechnology, Guru Angad Dev Veterinary and Animal Science University, Ludhiana, Punjab 141004, India

Review

Mammary stem cells, their identification and characterization

MaSCs/progenitor cells

MaSCs are multipotent adult stem cells giving rise to cells of luminal and myoepithelial cell origins. Conventionally, MaSCs are epithelial in origin. In addition to epithelial cells, mammary tissue also comprises cells of mesenchymal origin, including fad pad and connective tissues. Transplantation of dispersed cells into cleared mammary fat pad and clonal expansion of transplanted cell into functional mammary gland, have become gold standard methods to assess the self-renewal property of MaSCs and support the existence MaSC multipotency [13]. In addition, researchers have reported MaSCs as bi-potent [14] and lineage-restricted unipotent stem cells [6]. Indeed, the precise identification and subsequent characterization of MaSCs are conflicting [15] and need to be re-evaluated in the context of their dynamics [16]. Identifying different MaSC subtypes will allow precise targets to be found for optimal manipulation of increased milk production.

One of the main roles of adult stem cells is to proliferate, ensuring organ growth and maintaining tissue homeostasis of the resident organ. During proliferation, stem cells divide symmetrically and when maintaining tissue homeostasis, they divide asymmetrically. Symmetrical division of a stem cell involves mitotic division of the cell into two daughter stem cells or terminally differentiated cells. During asymmetrical division, the stem cell produces one daughter stem cell and one differentiated cell, both cells possessing dissimilar phenotypes. Although, adult stem cells have an unlimited proliferation capacity but divide infrequently *in situ*. Progenitor cells, the progeny of stem cells, have a more limited proliferation capacity in comparison with stem cells, but divide more frequently. Lineage restricted progenitor cells have a tremendous proliferation capacity and are responsible for the generation of differentiated cells to ensure ductal growth, alveolar development and ultimately milk production. The regeneration capacity of MaSCs is evaluated *in vivo* using a transplantation assay in the mammary fat pad of mice that are devoid of mammary epithelium [17,18]. Likewise, the regeneration capacity of progenitor cells is tested *in vitro* by the colony formation assay [19-21].

Identification of MaSCs/progenitor cells

Various methods for identification of MaSCs have been performed in different species, as reviewed recently [22,23]. Among these various methods for enriching the MaSC population, utilization of cell surface marker expression is the most common. This method has been used to successfully identify MaSCs in various species including human [24,25], murine [18,26] and bovine [27].

BrdU (bromodeoxyuridine) label-retaining epithelial cells (LRECs) are stem cells identified in various organs, including murine and bovine mammary glands [28,29]. LRECs do not express estrogen and progesterone receptors (ER⁻ and PR⁻ cells), similar to mammary stem cells identified by multiparameter cell sorting in mice [30]. Detailed investigation of LRECs from heifer mammary glands has demonstrated their transcriptome profile that was harvested from the basal layer (hypothesized location of MaSCs) and embedded layer of mammary epithelium layers [31]. Basal layer LRECs were enriched with stem cell transcripts, and therefore were characteristic of stem cells. Likewise, LRECs from the embedded layer were enriched with a few stem cell transcripts, indicative of progenitor cell characteristics. However, this method of MaSC and progenitor cell isolation is challenging because it pushes the limits of research to identify, isolate and profile the gene signature of the harvested cells. Furthermore, identification of BrdU-LRECs with anti-BrdU antibody itself is challenging because anti-BrdU antibody only binds with BrdU antigens when the DNA is single stranded. To expose BrdU antigens in mammary cryosections, antigen retrieval using harsh chemicals, like alkali, acids or heat, is imperative. This likely destroys the morphology of the cells, as well as their nucleic acids and proteins. Additionally, the heat generated using a laser beam for microdissection will degrade RNA quality of tissue sections on glass slides [32,33]. The scant amount of nucleic acid isolated from microdissected cells was barely sufficient to perform global gene expression analysis. A novel method that permits the identification of BrdU-LRECs without compromising RNA quality [34] is reported for the laser microdissection of LRECs and non-LRECs (control cells) to enable transcriptome profiling of bovine MaSCs and progenitor cells [31]. Unfortunately, this method does not permit *in vitro* or *in vivo* analysis of the microdissected cells because the cells apoptose during harvesting. Interestingly, this method does permit study of the stem cell niche, because the cells are harvested from specific *in situ* locations.

Characterization of bovine and caprine MaSCs/progenitor cells

Several studies have been performed to identify bovine MaSCs and progenitor cells. Initial investigations were based on staining and morphological characteristics, namely the intensity of staining, size and shape of the cell and nucleus, nucleus to cytoplasmic ratio, and presence of cell organelles. Light stained cells were suggestive of MaSCs, which were pleomorphic and occurred singularly or in pairs [35]. Paired light stained cells were suggestive of the proliferation potential of these cells, which was later confirmed by Ki-67 expression. In prepubertal bovine mammary glands, approximately 10% of the epithelial cells displayed light staining, of which 50% were proliferating (Ki-67 positive).

Multiparameter cell sorting using a cocktail of antibodies appeared to be the most common method to identify MaSCs and progenitor cells in human, mice and bovine tissues. Expression of cluster of differentiation (CD) molecules, like CD24 (heat stable antigen) and CD49f (integrin alpha 6) on Lin- sorted cells, revealed features of bovine MaSCs (CD24^{med}, CD49f^{pos}), basal bipotent progenitors (CD24^{neg}, CD49f^{pos}), luminal unipotent progenitors (CD24^{high}, CD49f^{neg}), and luminal unipotent cells (CD24^{med}, CD49f^{neg}) [27].

Stem cell antigen 1 (Sca-1) appears to be a controversial marker for MaSCs. Sca-1 is a glycosyl phosphatidylinositol (GPI)-anchored cell surface protein present in the lipid raft of the cell membrane and regulates many signaling events [36]. For identification of putative bovine MaSCs in one study, Sca-1 sorted cells appeared to be located in the stroma and elicited hematopoietic transcriptomic characteristics [37]. However, MaSCs are epithelial in origin and should be localized within the epithelial compartment of the bovine mammary gland. A combination of Sca-1 marker with a panel of existing MaSC markers should enable an enriched stem cell population to be distinguished with respect to the unipotent, bipotent and truly differentiated cells. A recent study of murine mammary glands indicated that the differential gene expression profile of sorted and non-sorted cells using Sca-1, CD24 and CD49f, identified two types of luminal cells (Sca-1^{pos} and Sca-1^{neg} cells, both CD24^{high}), basal cells (Sca1^{neg} CD24^{low} CD49f^{high}) and myoepithelial cells (Sca1^{neg} CD24^{low} CD49f^{low}) [26]. Basal cells with high CD49f expression were considered as putative MaSCs located in the basal layer.

Although fluorescent activated cell sorting (FACS) can be used to identify MaSCs and progenitor cells in various species, it has failed to provide information about the stem cell niche. This is because the preparation of a single cell suspension of mammary tissue involves enzymatic digestion of tissues and therefore disrupts all cellular and extracellular attachment of prospective stem cells. An alternative approach, BrdU-label retention method, successfully identified LRECs as enriched population of MaSC and progenitor cells [31]. LRECs had low expression of estrogen receptor (ESR1) and high expression of aldehyde dehydrogenase 3B1 (ALDH3B1) in the basal LRECs. Higher expression of nuclear receptor subfamily 5, group A, member 2 (NR5A2), a pluripotency transcription factor [38] and little to no expression of XIST, X-chromosome inactivation factor [39] in basal LREC is consistent with stem cell characteristics [31]. In the same study, embedded LRECs appeared to be more committed progenitor cells, evidenced by down-regulation of cell survival and proliferation factors IGF2, HSPB6,

NR5A2, and nestin. Nestin is a mammary stem cells marker [40].

The discovery of pluripotency factors, including OCT4, SOX2 and NANOG, as new markers for MaSCs is novel [41,42]. Furthermore, the presence of MaSCs in milk and the fact that milk is a cheap and non-invasive source of MaSCs is of considerable interest [43]. However, use of pluripotency factors as additional markers of MaSC and progenitor cells remains to be validated. Understanding MaSC plasticity and the interactions between stem cells, progenitor cells, differentiated cells and stroma, is important to comprehend their biology and regulation within the gland. This will allow for the development of an effective strategy for improving milk production and livestock management.

The first demonstration of different cell types within goat mammary tissue reported the existence of luminal and myoepithelial cells [44], which were based on expression of cytokeratins (CK). Further, analyses suggested there were certain cells that were undifferentiated (observed by loss of CK expression) which remained in the alveoli of the lactating goat mammary gland. These undifferentiated cells with loss of CK, including the luminal epithelial cell differentiation marker, CK18, which indicated the presence of mammary stem or progenitor cells in the goat mammary tissue. Convincingly, the presence of caprine MaSCs and progenitor cells was demonstrated based upon in vivo transplantation of sorted cells in NOD/SCID mice and in vitro by the colony formation assay [14].

Methods of MaSCs/progenitor cell expansion

The idea of MaSC expansion to increase cell turn over, enhance tissue regeneration and secretory activity of mammary epithelial cells was proposed by Capuco et al. [10,45,46]. During early postnatal life, increased activity of MaSCs and progenitor cells is responsible for ductal growth [46,47], which later during established lactation declines [10,48]. Stem and progenitor cells have three different fates; 1) they divide symmetrically and increase their numbers, 2) they divide asymmetrically and maintain their numbers and 3) they differentiate into terminally differentiated cells and die via programmed cell death or apoptosis (Figure 1). Endogenous factors, including the estrogen, progesterone and growth hormones, as well as exogenous compounds including bST, xanthosine and inosine, have been shown to expand MaSC and progenitor cell numbers and mammary epithelial cell populations [7,9,49-53]. It appears that these factors and compounds affect cell kinetics and enhance proliferation (Figure 2). Another instance that influences the rate of stem and progenitor cell activity is the dynamics of mammary gland physiology. An increased progenitor cell population during pregnancy indicates a role of progesterone in influencing

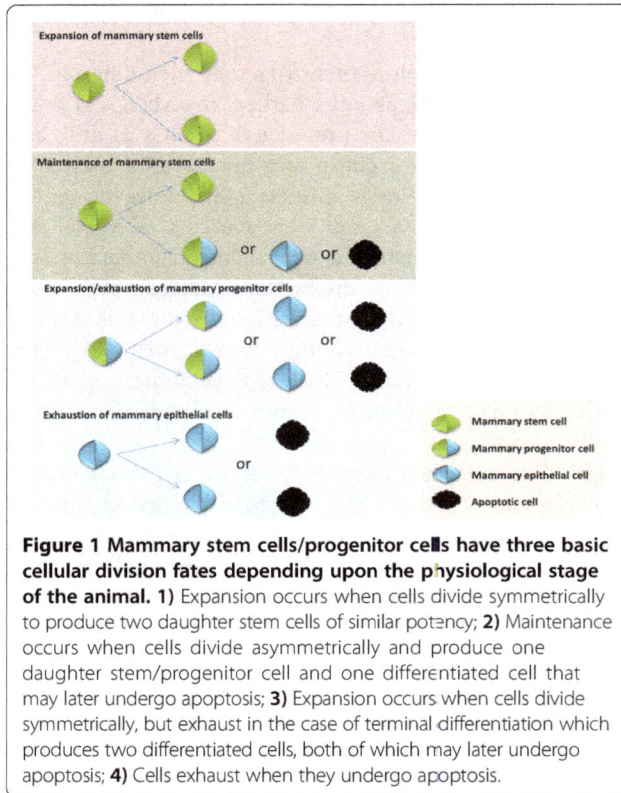

Figure 1 Mammary stem cells/progenitor cells have three basic cellular division fates depending upon the physiological stage of the animal. 1) Expansion occurs when cells divide symmetrically to produce two daughter stem cells of similar potency; **2)** Maintenance occurs when cells divide asymmetrically and produce one daughter stem/progenitor cell and one differentiated cell that may later undergo apoptosis; **3)** Expansion occurs when cells divide symmetrically, but exhaust in the case of terminal differentiation which produces two differentiated cells, both of which may later undergo apoptosis; **4)** Cells exhaust when they undergo apoptosis.

stem and progenitor cell activity. In mice, parity- based (pregnancy associated) progenitor cells, also termed 'parity induced MaSCs (PI-MaSCs) are reported to be located in the terminal duct of the alveolar unit, and are thought to originate from cells that skipped apoptosis during the last pregnancy [54]. The presence of PI-MaSCs has been confirmed in multiparous mice (absent in nulliparous) as multipotent stem cells by a transplantation study [55]. However, this was later refuted by generation of a mammosphere from tissue explants obtained from nulliparous mice [56]. Multiparous animals have a greater number of PI-MaSCs in the luminal epithelium than nulliparous mice [57], which is consistent with the previous study [58]. This suggests that expression of novel markers of bovine MaSCs and progenitor cells, including NUP153, NR5A2 and HNF4A [31], were significantly increased in multiparous lactating animals (peak lactation) than in nulliparous (heifer) animals. This is consistent with the idea that multiparous animals have greater numbers of PI-MaSCs than nulliparous animals. However, it remains unknown whether MaSCs and PI-MaSCs are similar or different. Taken together, these studies indicate that the physiology of the animals affects the number and activity of MaSCs/progenitor cells.

Use of nucleosides xanthosine and inosine

Xanthosine and inosine are purine nucleosides that act as precursors of *de novo* biosynthesis of guanine ribonucleotide. Sherley and colleagues [59,60] demonstrated that p53 mediates asymmetric division of rat hepatic stem cells and hair follicle stem cells. This action is mediated via down-regulation of inosine-5′-monophosphate dehydrogenase (IMPDH), the rate-limiting enzyme for guanine ribonucleotide synthesis. Addition of xanthosine or inosine into the system circumvents the IMPDH-mediated step and thus increases guanine concentration in the cell, thereby promoting symmetric division of stem cells and their

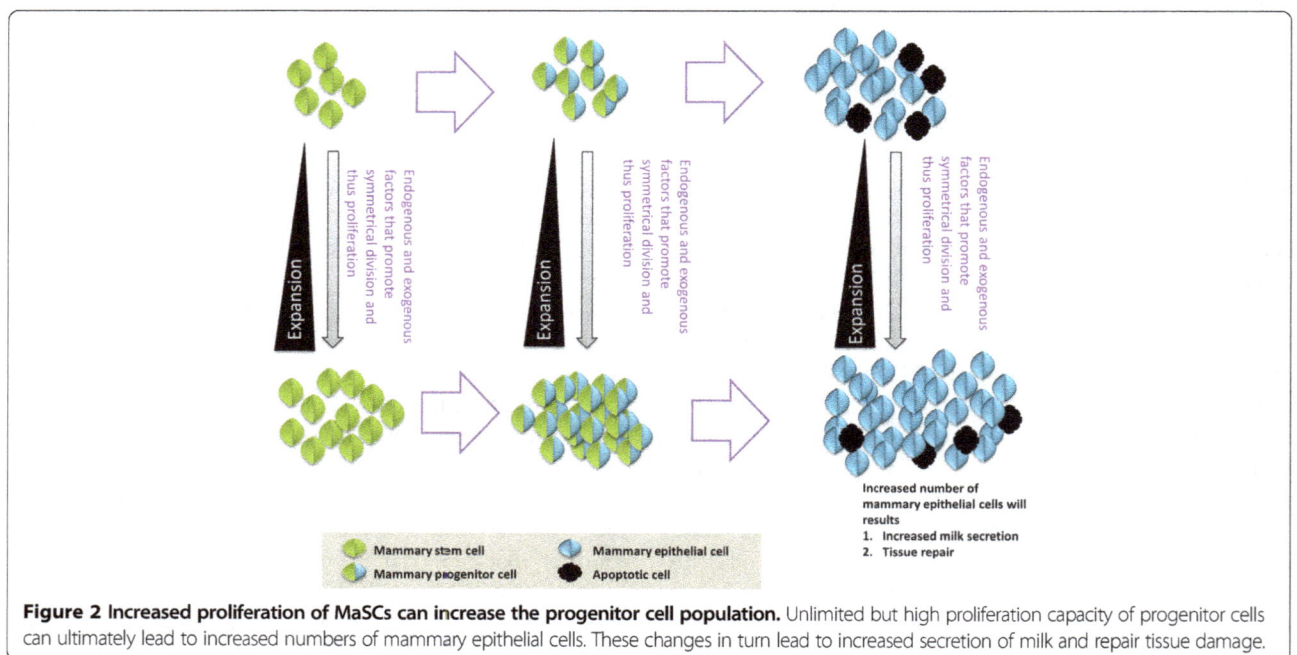

Figure 2 Increased proliferation of MaSCs can increase the progenitor cell population. Unlimited but high proliferation capacity of progenitor cells can ultimately lead to increased numbers of mammary epithelial cells. These changes in turn lead to increased secretion of milk and repair tissue damage.

expansion [53]. Xanthosine has been used successfully to increase stem cells, including LRECs, within the mammary gland of heifers *in vivo* [9]. Capuco and colleagues [52] further provided evidence that xanthosine increases mammary epithelial cell proliferation and the putative stem cell population in primary cultures of lactating bovine mammary epithelial cells. Details of the transcriptomic changes induced by xanthosine and the molecular mechanisms of how xanthosine alters cell proliferation, awaits further elucidation.

Use of growth hormone and estrogen

The scarcity of MaSCs and a universal marker to identify their pure population has hampered their study. Investigators have realized that during puberty, there is ductal growth of the mammary tree in the presence of estrogen and growth hormone. Therefore, estrogen and growth hormone might be responsible for the proliferation of MaSCs/progenitor cells present in the ducts [61]. Growth of Sca-1^{pos} cells in the presence of estrogen and growth hormone resulted in a greater number of Sca-1^{pos} cells in culture, evidenced by growth of the mammospheres and differentiation potential [62]. This study suggests that MaSC and progenitor cell populations could be increased when cells are grown in the presence of estrogen and growth hormone.

Use of progesterone and progestin

Progesterone is a hormone, which maintains pregnancy of the animal. One study provided the evidence of natural progesterone triggers mammary alveologenesis and expansion of MaSCs (CD24^{pos}CD49f^{high}) in mice [49], which was consistent with the finding that progesterone increases DNA replication and progenitor cell population in the breast [63]. These studies indicate that progesterone certainly has a role in the regulation of MaSCs/progenitor cells. In the mammary system, progesterone acts on MaSCs in a paracrine fashion [64]. Immunohistochemical analysis of bovine mammary tissue revealed PR expression in the nuclei of mammary epithelial cells, stromal and endothelial cells in heifers and lactating animals [65]. Mammary epithelial cells of the basal layer, the hypothesized location of MaSCs, usually lacks PR expression [28,65]. WNT4 and RANKL pathways mediate the mitogenic effect of progesterone [49,66]. Increased expression of RANKL in luminal cells and RANK (the receptor of RANKL) in basal cells, are the likely effectors of progesterone in basal MaSCs. Progesterone in combination with estrogen resulted in higher cell proliferation in the mammary gland than estrogen alone [67].

Xanthosine, inosine and lactation persistency

Xanthosine has been shown to increase mammary epithelial cell proliferation [9,52]. Changes in mammary epithelial cell dynamics during lactation affects milk production. For instance, increased secretory activity of the epithelium is reported to be responsible for increased milk production from early lactation to peak lactation in cows. However a decline in milk production from peak lactation to late lactation is due to a decline in epithelial cell number [7]. Although secretory activity per cell did not change significantly from peak to late lactation, the number of secretory cells declined due to increased apoptosis in non-pregnant cows, which was responsible for the declining milk production. In pregnant and lactating cows, the effect of declining milk production was more pronounced owing to concomitant demands of nutrients for pregnancy and lactation. A reduction in milk yield was evident in continuously milked dairy cows [68-71], unlike that of goats where continuous milking did not adversely affect milk production [2,72]. In goats, continuously milked glands had a greater number of fully differentiated cells (maximum secretory activity per cell) but with fewer alveoli and thus a reduced number of mammary epithelial cells [2]. The rate of proliferation and epithelial cell differentiation also varies depending on the parity of animals. Primiparous goats are more persistent owing to higher cell proliferation and cell survival after parturition than multiparous animals [73]. If xanthosine increases mammary epithelial cell proliferation, than it would diminish cell apoptosis during late lactation. A diminished rate of cell apoptosis during late lactation will likely lead to increased availability of secretory cells (flatten the milking curve), thereby maintaining the milk production for an extended period. In other words, xanthosine treatment could be used to extend persistency of lactation.

Inosine, a compound similar to xanthosine, has been successfully used to increase milk production in transgenic goats [51]. Transgenic goats are poor milk producers owing to accelerated cell death [74] and intramammary administration of inosine during early lactation has been shown to increase milk production from day one to peak lactation period (50 days). This study was based on the experiment that demonstrated xanthosine increased MaSCs and progenitor cells in heifers [9]. This study further tested the hypothesis that stimulating MaSCs using inosine could induce the cascade of cell proliferation in transgenic goats and prevent premature cessation of lactation. Clearly, the study indicates the role of inosine in increasing MaSC numbers in transgenic goats. It has been well- documented that increased concentration of guanine ribonucleotides in stem cells favor symmetric mitotic cell division [53]. An increased number of epithelial cells, due to increased MaSCs, could have resulted increased secretory cells. An increased number of mammary secretory cells have produced more milk for extended time. It is imperative here to validate this result, that inosine really, increases MaSC number. Additionally, dosage, frequency and time of inosine administration to goats and other dairy animals need to be evaluated.

Management of the dry period to ensure more milk production during next lactation

The non-lactating period between two consecutive lactation cycles is called the dry period. The dry period is critical to regenerate mammary epithelial cells. It is the time to replace the senescent cells that have lost production ability, with new epithelial cells that can be used to produce milk during next lactation. It can be hypothesize that during the dry period, progenitor cell activity is high which is responsible for increased cell turn over. Usually animals during the dry period are pregnant and therefore influenced by progesterone hormone, because hormone maintains pregnancy. Progesterone harbingers the mammogenic effects that are manifested by increased cell turnover and MaSC/progenitor cell activity. Usually, the length of the dry period in the cow is 50–60 days. A reduction in the non-productive dry period length is indirectly associated with increased milk production due to the reduced time of the non-productive period. At the end of the dry period lactation starts. Lactation cycle is divided into various stages- early, peak and late stage- depending upon the amount of milk produced by the animals. Maintaining the peak stage of milk production longer than average milk production is called persistency of lactation and such animals are called persistent. These persistent animals are less exposed to calving-related stress and low milk production potential during the initial lactation cycle. Apart from this, reducing the length of the dry period from 60 to 30 days could be another approach to increase the efficiency of lactation. It has been shown that a shortened dry period (30 days) or omitted dry period in the presence of bST hormone, did not alter milk production of multiparous cows during the next lactation [70]. This result was consistent with the finding that administration of bST in lactating Holstein cow increased mammary epithelial cell renewal, as evidenced by a 3-fold increase in expression of the proliferation marker, Ki-67, in bST-treated animal compared with control animals [7]. In other words, a 60-day dry period could be reduced to 30-day for regeneration of mammary epithelium without deleterious effect on milk production in next lactation.

Conclusions and future directions

This review describes manipulation methods of MaSCs/progenitor cells that could influence future milk production, mainly in dairy animals. Emphasis has been given to cows and goat mammary glands with some imperative missing information supplemented from mice. Proliferation of both MaSCs and progenitor cells with natural hormonal treatments like progesterone, estrogen and growth hormone, and by exogenous administration of xanthosine or inosine, could potentially increase milk production of dairy animals by increasing cell turnover or persistency of lactation. Further investigations are essential to understand the biology of MaSCs/progenitor cells and their role in mammary gland morphogenesis, tissue turnover and homeostasis. Recent reports of the existence of MaSCs in breast milk and pluripotency factors as additional markers of MaSC [43], raises many questions like whether MaSCs possess multi-lineage potential. Why MaSCs are present in the milk? Does MaSCs have any role in infants who drink mother's milk? It would also be useful to determine whether these pluripotency transcription factors are involved in the self-renewal of MaSC? Use of milk as a non-invasive source of MaSCs for their identification and characterization is a novel and promising.

Abbreviations

MaSCs: Mammary stem cells; PI-MaSCs: Parity-induced mammary stem cells; PR: Progesterone receptor; CK: Cytokeratin; BrdU: Bromodeoxyuridine; LREC: Label-retaining epithelial cells; SE: Standard error; ESR1: Estrogen receptor; IGF2: Insulin like growth factor 2; HSPB6: Heat shock protein beta 6; OCT4: Octamer-binding transcription factor 4 (also called as POU5F1); SOX2: SRY (sex determining region Y)-box 2; NANOG: Nanog homeobox; NUP153: Nucleoporin 153; HNF4A: Hepatocyte nuclear factor 4 alpha; RANK: Receptor activator of nuclear factor κ B; bST: Bovine somatotrophin.

Competing interests

It is declared that the author has no competing interests.

Author's contributions

RKC collected the information, drafted and finalized the manuscript.

Author's information

Ratan K. Choudhary, Ph.D., is an Assistant Professor at the School of Animal Biotechnology at Guru Angad Dev Veterinary and Animal Science University, Ludhiana, Punjab, India. The author has been working with bovine mammary stem cells, their molecular characterization and expansion for the last seven years.

References

1. Akers RM, Capuco AV, Keys JE: **Mammary histology and alveolar cell differentiation during late gestation and early lactation in mammary tissue of beef and dairy heifers.** *Livest Sci* 2006, **105**:44–49.
2. Safayi S, Theil PK, Hou L, Engbaek M, Nørgaard JV, Sejrsen K, Nielsen MO: **Continuous lactation effects on mammary remodeling during late gestation and lactation in dairy goats.** *J Dairy Sci* 2010, **93**:203–17.
3. Visvader JE, Lindeman GJ: **Mammary stem cells and mammopoiesis.** *Cancer Res* 2006, **66**:9798–9801.
4. Smith GH: **Experimental mammary epithelial morphogenesis in an in vivo model: evidence for distinct cellular progenitors of the ductal and lobular phenotype.** *Breast Cancer Res Treat* 1996, **39**:21–31.
5. Esmailpour T, Huang T: **Advancement in mammary stem cell research.** *Pathology* 2008, **4**:131–138.
6. Van Keymeulen A, Rocha AS, Ousset M, Beck B, Bouvencourt G, Rock J, Sharma N, Dekoninck S, Blanpain C: **Distinct stem cells contribute to mammary gland development and maintenance.** *Nature* 2011, **479**:189–193.
7. Capuco AV, Wood DL, Baldwin R, Mcleod K, Paape MJ: **Mammary cell number, proliferation, and apoptosis during a bovine lactation: relation to milk production and effect of bST.** *J Dairy Sci* 2001, **84**:2177–87.

8. Wall EH, Auchtung TL, Dahl GE, Ellis SE, McFadden TB: **Exposure to short day photoperiod during the dry period enhances mammary growth in dairy cows.** *J Dairy Sci* 2005, **88**:1994–2003.

9. Capuco AV, Evock-Clover CM, Minuti A, Wood DL: **In vivo expansion of the mammary stem/ progenitor cell population by xanthosine infusion.** *Exp Biol Med (Maywood)* 2009, **234**:475–482.

10. Capuco AV, Ellis SE, Hale SA, Long E, Erdman RA, Zhao X, Paape MJ: **Lactation persistency: Insights from mammary cell proliferation studies.** *J Anim Sci* 2003, **81**:18–31.

11. Borena BM, Bussche L, Burvenich C, Duchateau L, Van de Walle GR: **Mammary stem cell research in veterinary science: an update.** *Stem Cells Dev* 2013, **22**:1743–1751.

12. Capuco AV, Ellis SE: **Comparative Aspects of Mammary Gland Development and Homeostasis.** *Annu Rev Anim Biosci* 2013, **1**:179–202.

13. Kordon EC, Smith GH: **An entire functional mammary gland may comprise the progeny from a single cell.** *Development* 1998, **125**:1921–1930.

14. Prpar S, Martignani E, Dovc P, Baratta M: **Identification of goat mammary stem/progenitor cells.** *Biol Reprod* 2012, **86**:117.

15. Kaimala S, Bisana S, Kumar S: **Mammary gland stem cells: more puzzles than explanations.** *J Biosci* 2012, **37**:349–358.

16. Joshi PA, Khokha R: **The mammary stem cell conundrum: is it unipotent or multipotent?** *Breast Cancer Res* 2012, **14**:305.

17. Sleeman KE, Kendrick H, Ashworth A, Isacke CM, Smalley MJ: **CD24 staining of mouse mammary gland cells defines luminal epithelial, myoepithelial/basal and non-epithelial cells.** *Breast Cancer Res* 2006, **8**:R7.

18. Shackleton M, Vaillant F, Simpson KJ, Stingl J, Smyth GK, Asselin-Labat M-L, Wu L, Lindeman GJ, Visvader JE: **Generation of a functional mammary gland from a single stem cell.** *Nature* 2006, **439**:84–88.

19. Dey D, Saxena M, Paranjape AN, Krishnan V, Giraddi R, Kumar MV, Mukherjee G, Rangarajan A: **Phenotypic and functional characterization of human mammary stem/progenitor cells in long term culture.** *PLoS One* 2009, **4**:e5329.

20. Smalley MJ, Titley J, O'Hare MJ: **Clonal characterization of mouse mammary luminal epithelial and myoepithelial cells separated by fluorescence-activated cell sorting.** *In Vitro Cell Dev Biol Anim* 1998, **34**:711–721.

21. Stingl J, Raouf A, Emerman JT, Eaves CJ: **Epithelial progenitors in the normal human mammary gland.** *J Mammary Gland Biol Neoplasia* 2005, **10**:49–59.

22. Smalley MJ, Kendrick H, Sheridan JM, Regan JL, Prater MD, Lindeman GJ, Watson CJ, Visvader JE, Stingl J: **Isolation of mouse mammary epithelial subpopulations: a comparison of leading methods.** *J Mammary Gland Biol Neoplasia* 2012, **17**:91–97.

23. Choudhary RK, Choudhary S: **Ruminant mammary stem cells: Methods of identification and status.** *Rumin Sci* 2012, **1**:101–108.

24. Eirew P, Stingl J, Raouf A, Turashvili G, Aparicio S, Emerman JT, Eaves CJ: **A method for quantifying normal human mammary epithelial stem cells with in vivo regenerative ability.** *Nat Med* 2008, **14**:1384–1389.

25. Stingl J, Eirew P, Ricketson I, Shackleton M, Vaillant F, Choi D, Li HI, Eaves CJ: **Purification and unique properties of mammary epithelial stem cells.** *Nature* 2006, **439**:993–997.

26. Meier-Abt F, Milani E, Roloff T, Brinkhaus H, Duss S, Meyer DS, Klebba I, Balwierz PJ, van Nimwegen E, Bentires-Alj M: **Parity induces differentiation and reduces Wnt/Notch signaling ratio and proliferation potential of basal stem/progenitor cells isolated from mouse mammary epithelium.** *Breast Cancer Res* 2013, **15**:R36.

27. Rauner G, Barash I: **Cell hierarchy and lineage commitment in the bovine mammary gland.** *PLoS One* 2012, **7**:e30113.

28. Capuco AV: **Identification of putative bovine mammary epithelial stem cells by their retention of labeled DNA strands.** *Exp Biol Med (Maywood)* 2007, **232**:1381–1390.

29. Smith GH: **Label-retaining epithelial cells in mouse mammary gland divide asymmetrically and retain their template DNA strands.** *Development* 2005, **132**:681–687.

30. Sleeman KE, Kendrick H, Robertson D, Isacke CM, Ashworth A, Smalley MJ: **Dissociation of estrogen receptor expression and in vivo stem cell activity in the mammary gland.** *J Cell Biol* 2007, **176**:19–26.

31. Choudhary RK, Li RW, Evock-Clover CM, Capuco AV: **Comparison of the transcriptomes of long-term label retaining-cells and control cells microdissected from mammary epithelium: an initial study to characterize potential stem/progenitor cells.** *Front Oncol* 2013, **3**:21.

32. Vogel A, Horneffer V, Lorenz K, Linz N, Hüttmann G, Gebert A: **Principles of laser microdissection and catapulting of histologic specimens and live cells.** *Methods Cell Biol* 2007, **82**:153–205.

33. Legres LG, Janin A, Masselon C, Bertheau P: **Beyond laser microdissection technology: follow the yellow brick road for cancer research.** *Am J Cancer Res* 2014, **4**:1–28.

34. Choudhary RK, Daniels KM, Evock-Clover CM, Garrett W, Capuco AV: **Technical note: A rapid method for 5-bromo-2'-deoxyuridine (BrdU) immunostaining in bovine mammary cryosections that retains RNA quality.** *J Dairy Sci* 2010, **93**:2574–2579.

35. Ellis S, Capuco AV: **Cell proliferation in bovine mammary epithelium: identification of the primary proliferative cell population.** *Tissue Cell* 2002, **34**:155–163.

36. Epting CL, King FW, Pedersen A, Zaman J, Ritner C, Bernstein HS: **Stem cell antigen-1 localizes to lipid microdomains and associates with insulin degrading enzyme in skeletal myoblasts.** *J Cell Physiol* 2008, **217**:250–260.

37. Motyl T, Bierła JB, Kozłowski M, Gajewska M, Gajkowska B, Koronkiewicz M: **Identification, quantification and transcriptional profile of potential stem cells in bovine mammary gland.** *Livest Sci* 2011, **136**:136–149.

38. Heng J-CD, Feng B, Han J, Jiang J, Kraus P, Ng J-H, Orlov YL, Huss M, Yang L, Lufkin T, Lim B, Ng H-H: **The nuclear receptor Nr5a2 can replace Oct4 in the reprogramming of murine somatic cells to pluripotent cells.** *Cell Stem Cell* 2010, **6**:167–174.

39. Navarro P, Chambers I, Karwacki-Neisius V, Chureau C, Morey C, Rougeulle C, Avner P: **Molecular coupling of Xist regulation and pluripotency.** *Science* 2008, **321**:1693–1695.

40. Li H, Cherukuri P, Li N, Cowling V, Spinella M, Cole M, Godwin AK, Wells W, DiRenzo J: **Nestin is expressed in the basal/myoepithelial layer of the mammary gland and is a selective marker of basal epithelial breast tumors.** *Cancer Res* 2007, **67**:501–510.

41. Simões BM, Piva M, Iriondo O, Comaills V, López-Ruiz JA, Zabalza I, Mieza JA, Acinas O, Vivanco MDM: **Effects of estrogen on the proportion of stem cells in the breast.** *Breast Cancer Res Treat* 2011, **129**:23–35.

42. Lengerke C, Fehm T, Kurth R, Neubauer H, Scheble V, Müller F, Schneider F, Petersen K, Wallwiener D, Kanz L, Fend F, Perner S, Bareiss PM, Staebler A: **Expression of the embryonic stem cell marker SOX2 in early-stage breast carcinoma.** *BMC Cancer* 2011, **11**:42.

43. Hassiotou F, Beltran A, Chetwynd E, Stuebe AM, Twigger A-J, Metzger P, Trengove N, Lai CT, Filgueira L, Blancafort P, Hartmann PE: **Breastmilk is a novel source of stem cells with multilineage differentiation potential.** *Stem Cells* 2012, **30**:2164–2174.

44. Li P, Wilde CJ, Finch LM, Fernig DG, Rudland PS: **Identification of cell types in the developing goat mammary gland.** *Histochem J* 1999, **31**:379–393.

45. Capuco AV, Ellis S: **Bovine mammary progenitor cells: current concepts and future directions.** *J Mammary Gland Biol Neoplasia* 2005, **10**:5–15.

46. Capuco AV, Choudhary RK, Daniels KM, Li RW, Evock-Clover CM: **Bovine mammary stem cells: cell biology meets production agriculture.** *Animal* 2012, **6**:382–393.

47. Rios AC, Fu NY, Lindeman GJ, Visvader JE: **In situ identification of bipotent stem cells in the mammary gland.** *Nature* 2014, **506**:322–327.

48. Boutinaud M, Guinard-Flamenta J, Jammes H: **The number and activity of mammary epithelial cells, determining factors for milk production.** *Reprod Nutr Dev* 2004, **44**:499–508.

49. Joshi PA, Jackson HW, Beristain AG, Di Grappa MA, Mote PA, Clarke CL, Stingl J, Waterhouse PD, Khokha R: **Progesterone induces adult mammary stem cell expansion.** *Nature* 2010, **465**:803–807.

50. Rahal OM, Simmen RCM: **Paracrine-acting adiponectin promotes mammary epithelial differentiation and synergizes with genistein to enhance transcriptional response to estrogen receptor β signaling.** *Endocrinology* 2011, **152**:3409–3421.

51. Baldassarre H, Deslauriers J, Neveu N, Bordignon V: **Detection of endoplasmic reticulum stress markers and production enhancement treatments in transgenic goats expressing recombinant human butyrylcholinesterase.** *Transgenic Res* 2011, **20**:1265–1272.

52. Choudhary RK, Capuco AV: **In vitro expansion of the mammary stem/progenitor cell population by xanthosine treatment.** *BMC Cell Biol* 2012, **13**:14.

53. Lee H-S, Crane GG, Merok JR, Tunstead JR, Hatch NL, Panchalingam K, Powers MJ, Griffith LG, Sherley JL: **Clonal expansion of adult rat hepatic stem cell lines by suppression of asymmetric cell kinetics (SACK).** *Biotechnol Bioeng* 2003, **83**:760–771.

54. Matulka LA, Triplett AA, Wagner K-U: **Parity-induced mammary epithelial cells are multipotent and express cell surface markers associated with stem cells.** *Dev Biol* 2007, **303**:29–44.

55. Boulanger CA, Wagner K-U, Smith GH: **Parity-induced mouse mammary epithelial cells are pluripotent, self-renewing and sensitive to TGF-beta1 expression.** *Oncogene* 2005, **24**:552–560.

56. Booth BW, Boulanger CA, Smith GH: **Alveolar progenitor cells develop in mouse mammary glands independent of pregnancy and lactation.** *J Cell Physiol* 2007, **212**:729–736.

57. Wagner K-U, Smith GH: **Pregnancy and stem cell behavior.** *J Mammary Gland Biol Neoplasia* 2005, **10**:25–36.

58. Choudhary RK, Evock-Clover CM, Capuco AV: **Expression of noval, putative mammary stem markers in prepubertal and lactating bovine mammary glands.** *J Dairy Sci* 2011, **94**:180.

59. Sherley JL: **Guanine nucleotide biosynthesis is regulated by the cellular p53 concentration.** *J Biol Chem* 1991, **266**:24815–28428.

60. Sherley JL: **Asymmetric cell kinetics genes: The key to expansion of adult stem cells in culture.** *Stem Cells* 2002, **20**:561–572.

61. Asselin-Labat M-L, Vaillant F, Sheridan JM, Pal B, Wu D, Simpson ER, Yasuda H, Smyth GK, Martin TJ, Lindeman GJ, Visvader JE: **Control of mammary stem cell function by steroid hormone signalling.** *Nature* 2010, **465**:798–802.

62. Dou X, Zhang B, Liu R, Li J, Shi D, Lu C, Zhu X, Liao L, Du Z, Zhao RC: **Expanding Sca-1(+) mammary stem cell in the presence of oestrogen and growth hormone.** *Clin Transl Oncol* 2012, **14**:444–451.

63. Graham JD, Mote PA, Salagame U, van Dijk JH, Balleine RL, Huschtscha LI, Reddel RR, Clarke CL: **DNA replication licensing and progenitor numbers are increased by progesterone in normal human breast.** *Endocrinology* 2009, **150**:3318–3326.

64. Beleut M, Rajaram RD, Caikovski M, Ayyanan A, Germano D, Choi Y, Schneider P, Brisken C: **Two distinct mechanisms underlie progesterone-induced proliferation in the mammary gland.** *Proc Natl Acad Sci U S A* 2010, **107**:2989–2994.

65. Schams D, Kohlenberg S, Amselgruber W, Berisha B, Pfaffl MW, Sinowatz F: **Expression and localisation of oestrogen and progesterone receptors in the bovine mammary gland during development, function and involution.** *J Endocrinol* 2003, **177**:305–317.

66. Roarty K, Rosen JM: **Wnt and mammary stem cells: hormones cannot fly wingless.** *Curr Opin Pharmacol* 2010, **10**:643–649.

67. Wood CE, Branstetter D, Jacob AP, Cline JM, Register TC, Rohrbach K, Huang L-Y, Borgerink H, Dougall WC: **Progestin effects on cell proliferation pathways in the postmenopausal mammary gland.** *Breast Cancer Res* 2013, **15**:R62.

68. Collier RJ, Annen-Dawson EL, Pezeshki A: **Effects of continuous lactation and short dry periods on mammary function and animal health.** *Animal* 2012, **6**:403–414.

69. Annen EL, Collier RJ, Mcguire MA, Vicini JL, Ballam JM, Lormore MJ: **Effects of Dry Period Length on Milk Yield and Mammary Epithelial Cells *.** *J Dairy Sci* 2004, **87**(June 2003):66–76.

70. Annen EL, Collier RJ, McGuire MA, Vicini JL, Ballam JM, Lormore MJ: **Effect of modified dry period lengths and bovine somatotropin on yield and composition of milk from dairy cows.** *J Dairy Sci* 2004, **87**:3746–3761.

71. Annen EL, Stiening CM, Crooker BA, Fitzgerald AC, Collier RJ: **Effect of continuous milking and prostaglandin E2 on milk production and mammary epithelial cell turnover, ultrastructure, and gene expression.** *J Anim Sci* 2008, **86**:1132–1144.

72. Caja G, Salama AAK, Such X: **Omitting the dry-off period negatively affects colostrum and milk yield in dairy goats.** *J Dairy Sci* 2006, **89**:4220–4228.

73. Safayi S, Theil PK, Elbrønd VS, Hou L, Engbaek M, Nørgaard JV, Sejrsen K, Nielsen MO: **Mammary remodeling in primiparous and multiparous dairy goats during lactation.** *J Dairy Sci* 2010, **93**:1478–1490.

74. Baldassarre H, Schirm M, Deslauriers J, Turcotte C, Bordignon V: **Protein profile and alpha-lactalbumin concentration in the milk of standard and transgenic goats expressing recombinant human butyrylcholinesterase.** *Transgenic Res* 2009, **18**:621–632.

20

Practical starter pig amino acid requirements in relation to immunity, gut health and growth performance

Bob Goodband[1*], Mike Tokach[1], Steve Dritz[2], Joel DeRouchey[1] and Jason Woodworth[1]

Abstract

Immune system activation begins a host of physiological responses. Infectious agents are recognized by monocytes and macrophages which in turn stimulate cytokine production. It is the hormone-like factors called cytokines that orchestrate the immune response. The classic responses observed with immune system activation and cytokine production include: anorexia, fever, lethargy, recruitment of other immune cells, and phagocytosis. While production of immune system components is known to require some amino acids, increases in amino acid requirements are more than offset by the associated decrease in protein accretion and increased muscle protein degradation that also accompanies immune system activation. However, the biggest impact of cytokine production is a decrease in feed intake. Therefore, as feed intake decreases, the energy needed to drive protein synthesis is also decreased. This suggests that diets should still be formulated on a similar calorie:lysine ratio as those formulated for non-immune challenged pigs. The evidence is sparse or equivocal for increasing nutrient requirements during an immune challenge. Nutritionists and swine producers should resist the pressure to alter the diet, limit feed, or add expensive feed additives during an immune challenge. While immune stimulation does not necessitate changes in diet formulation, when pigs are challenged with non-pathogenic diarrhea there are potential advantages on gut health with the increased use of crystalline amino acids rather than intact protein sources (i.e., soybean meal). This is because reducing crude protein decreases the quantity of fermentable protein entering the large intestine, which lowers post weaning diarrhea. It also lowers the requirement for expensive specialty protein sources or other protein sources such as soybean meal that present immunological challenges to the gut. The objective of this review is two-fold. The first is to discuss immunity by nutrition interactions, or lack thereof, and secondly, to review amino acid requirement estimates for nursery pigs.

Keywords: Amino acids, Immunity, Pigs, Requirements

Introduction

World-wide swine production has evolved dramatically in the last decade. Genetic improvements have dramatically increased reproductive traits such as litter size as well as improve growth traits like daily gain and feed efficiency. Multiple site production has made a large impact on herd health and weaned pig flow management. Practical nutrition programs continue evolving to keep pace with these rapid changes and to improve profitability of pork producers. One important concept that has risen from these changes is the interaction of nutrition and immunity or herd health. Feeding pigs based on their immune status or pathogen challenge was once a novel idea based on feeding specifically formulated diets to meet the different amino acid requirements of the immune system. While studies have observed that up-regulation of the immune system may slightly impact amino acid requirements for leukocyte and cytokine production [1], the major driver of a nutrition/immune response interaction resides in the response to an immune challenge of lower feed intake and in some cases poorer feed efficiency [2-4]. Therefore, because of decreased energy (feed) intake, the body will most likely not support excess amino acid supplementation for protein synthesis to combat the effects of immune system activation [5]. In addition, from an enteric pathogen stress point of view, recent studies have observed that

* Correspondence: Goodband@ksu.edu
[1]Department of Animal Sciences and Industry, Kansas State University, Manhattan, KS 66506-0201, USA
Full list of author information is available at the end of the article

practices to minimize post weaning scours, such as restrictive feeding of pigs at weaning or providing high-fiber diets will contribute to the decreased growth performance in the nursery stage [6,7]. Therefore, these data suggest that from a practical feeding standpoint, there is no interaction between immune challenge and diet complexity on pig performance which indicates that relatively high complexity diets containing specialty protein sources are just as valuable for healthy pigs as those faced with an immune challenge [5]. Thus, producers should avoid reformulating diets if environmental conditions are less than ideal. Ultimately, by maintaining a relatively high amount of specialty protein sources and utilizing the proper amino acid ratios with crystalline amino acid supplementation, dietary crude protein can be lowered and excellent growth performance in the nursery can be maintained.

Nutrition by immune system activation interactions

To evaluate immune system activation by nutrition interactions in pigs, Williams et al. [2,3] observed that the efficiency of lysine utilization for protein deposition was similar among pigs with high or low immune system activation. Thus, differences in feed efficiency among challenge groups could be explained by shifts in ratios of lean and fat deposition and proportion for maintenance. This indicates that healthy pigs with relatively low immune system activation have greater need for dietary lysine as a consequence of greater protein deposition compared to those pigs with high immune system activation. Increasing pathogen load stimulated pro-inflammatory cytokine production and endocrine shifts which not only decreased feed intake, but increased muscle catabolism [1]. The high immune stimulated pigs have less protein deposition, hence less energy needed to drive body protein accretion. Decreasing pathogen load and thus lowering immune stimulation resulted in greater feed intake and growth performance [2,3]. More recently, research evaluating PCV2 vaccination under commercial conditions indicated that vaccinated pigs had a greater need for lysine on a grams per day basis [5]. However, in this study when evaluated on a lysine to calorie ratio, the requirement was not different between vaccinated and unvaccinated pigs even though there were large differences in growth performance and mortality rates. From a practical standpoint this would support the idea that a similar calorie:lysine ratio should be maintained regardless of immune system activation status, or in other words diet modifications are not warranted when the immune system is activated.

It is possible that immune system activation will affect the utilization of some amino acids relative to lysine. However data to support this effect is difficult to find. Methionine is probably the most studied amino acid other than lysine in response to an immune challenge.

Naturally diets deficient in essential amino acids like lysine or methionine, will not support cytokine (IL-1) production and will further reduce growth beyond that observed due to decreased feed intake [8]. Again, decreased muscle protein accretion and increased degradation appear to offset shifts in immune-related protein synthesis.

It is important to recognize that viral, bacterial, or mixed pathogen challenges may elicit different immune responses. However, from the chronic mixed challenge model used by Williams et al. [1,2] to the more acute, viral challenge of Shelton [5]; responses in protein deposition and feed intake were similar. However more research in this area is necessary to determine if other types of immune challenge may have differing effects on feed intake.

As it appears that there is little need to adjust diets based on immune system activation in growing-finishing pigs, Dritz et al. [4] evaluated the interactive effects of a lipopolysaccharide (LPS) induced immune challenge and diet complexity on weanling pig performance. In this study the three comparisons consisted of control pigs fed ad libitum, pigs challenged with LPS and fed ad libitum, or non-challenged pigs pair fed to the same feed intake level of the LPS pigs. In addition, there were 3 diet complexity regimens used: a complex diet using high amounts of specialty protein sources (animal blood plasma, fish meal, blood meal, and dried whey), intermediate amounts of these specialty ingredients, and then a very simple diet with minimal amounts of these ingredients. The LPS challenged pigs had increased haptoglobin concentrations indicating the inflammatory cytokine production was increased in immune challenged pigs. Control pigs had increased ADG and were heavier at the end of the study, whereas LPS challenged pigs were the lightest, with pair fed pigs intermediate (Figure 1). There were no immune status × diet complexity interactions observed suggesting that the response to immune challenge is independent from diet complexity. That is, pigs fed the complex diet regimen had the greatest ADG regardless of immune system activation or pair feeding. Pigs administered LPS had poorer performance than those that were pair fed resulting from a combination of reduced ADFI and poor G:F. The intermediate performance of the pair-fed pigs suggests that approximately 2/3 of the reduction in growth was feed intake related, whereas 1/3 was due to poorer G:F. Ultimately this study confirms that the diets fed to pigs in an immune challenged environment should be similar to that of pigs fed in a clean environment.

More recently, the effects of an immune challenge as a result of housing weanling pigs in either "clean" or "dirty" environments has been addressed by Montagne et al. [6]. To potentially reduce the incidence of post weaning diarrhea, one hypothesis is that feeding a diet high in

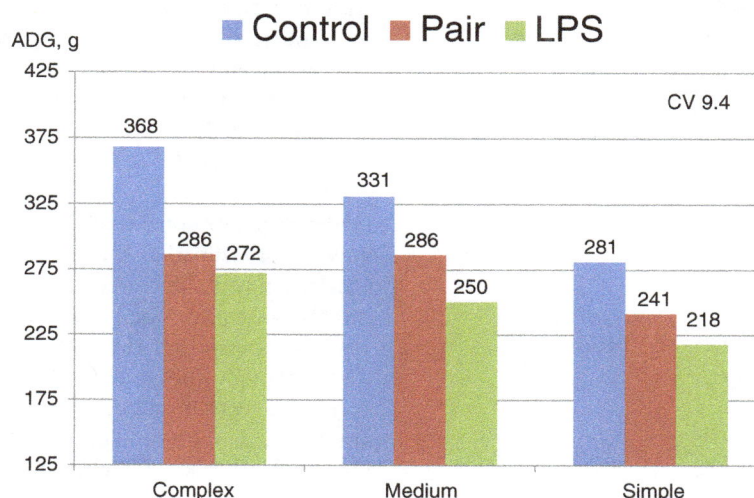

Figure 1 The effects of immune system activation and diet complexity on average daily gain. Dietary treatments include feeding a complex starter diet, a medium complexity diet, and a low complexity diet. Immune activation includes control pigs fed ad libitum, LPS injected pigs fed ad libitum, and control pigs pair-fed to that of the LPS challenged pigs. Main effects of diet complexity and immune system activation are significant (P < 0.01). There was no immune system by diet complexity interactions (Adapted from Dritz et al., 1996 [4]).

fermentable fiber might increase the population of beneficial bacteria. This is thought to alleviate the effects of pathogenic bacteria by competitive exclusion for binding sites within the gut. In this study, weanling pigs were housed in either a clean (washed and disinfected) nursery or one that had not been cleaned after the previous group of pigs [6]. They were either fed a control diet or high fiber diet (d 0 to 14: 3.25 or 4.89% crude fiber and 12.1 or 16.9% total dietary fiber, respectively). There were no environment × diet interactions, and pigs in the dirty environment had poorer G:F than those in the clean environment (Table 1). The addition of fiber to the diet decreased NE intake and tended to decrease ADG and ADFI. Pigs housed in the dirty environment and fed the high fiber diet were 0.50 kg lighter than counterparts fed the control diet after 1 wk. The authors confirmed

that poor sanitary conditions reduced pig growth and increased the incidence of digestive disorders in the first week post-weaning. Feeding a high fiber diet to pigs housed in a dirty environment further decreased growth [6]. Rather than being beneficial, the addition of fiber reduced performance in both the clean and dirty environment.

Recently, other options to reduce the risk of post-weaning diarrhea evaluated restricted vs.ad libitum feeding immediately post weaning [7]. Similar to the previous paper [6], pigs were housed in either a clean or dirty environment but in this study they were either fed ad libitum or a restrictive feeding regimen from day 2 to 7 after weaning. Again, no environment × feeding regimen interactions were observed indicating that the response to each was independent (Table 2). The authors concluded that feed restriction immediately after weaning exacerbated the effects of poor sanitary conditions [7].

Results of these 4 studies indicate that when faced with a disease challenge, weanling pigs need a high quality diet, but not one different than what would be provided to pigs with minimal disease challenge.

In an excellent meta-analysis covering 121 different studies, Pastorelli et al. [9] examined the effects of an immune system challenge on feed intake and growth responses. They examined the performance responses to digestive bacterial infections, sanitary housing conditions, LPS challenge, mycotoxicoses, parasitic infections and respiratory disease. They established the percentage change in growth as a result of poorer G:F or reduced daily feed intake (Figure 2). Digestive bacterial infections had the greatest negative impact on growth responses with

Table 1 Effects of added dietary fiber in either a clean or dirty environment on weanling pig performance (Montagne et al. [6])[1]

Items	Clean		Dirty	
	Control	Fiber	Control	Fiber
d 0 to 14				
ADG, g[2]	128	127	132	91
ADFI, g[2]	228	218	275	241
G:F[3]	.524	.543	.452	.424

[1]Pigs assigned to the good sanitary conditions were housed in cleaned and disinfected rooms; pigs assigned to the poor sanitary conditions were housed in rooms that were not cleaned; the Control and Fiber diets used during d 0 to 14 were 121 and 169 g/kg of total dietary fiber, respectively.
[2]Effect of added dietary fiber, (P < 0.10).
[3]Effect of sanitary condition, (P < 0.01).

Table 2 Effects of restrictive feeding in either a clean or dirty environment on weanling pig performance (Pastorelli et al. [7])[1]

Items	Clean		Dirty	
	Ad libitum	Restricted	Ad libitum	Restricted
d 0 to 11				
ADG, g[2,3]	257	159	173	.95
ADFI, g[3]	336	219	319	225
G:F[2,4]	.753	.729	537	.393
Overall (d 0 to 60)				
ADG, g[5]	511	492	463	394
ADFI, g	875	821	826	705
G:F[2]	.587	.599	562	.555

[1]Pigs assigned to the good sanitary conditions were housed in cleaned and disinfected rooms and received an antibiotic supplementation; pigs assigned to the poor sanitation conditions were housed in rooms that were not cleaned; the ad libitum group corresponded to pigs nourished ad libitum on overall experimental period; the restricted group corresponded to pigs that received, from 2 to 7 d after weaning, respectively, 20, 30, 40, 60, 80, and 90% of the amounts of feed voluntary consumed by ad libitum pigs in both sanitary conditions at each previous day. No feed restriction × sanitary conditions interactions, ($P > 0.10$).
[2]Effect of sanitary condition, ($P < 0.01$).
[3]Effect of feed restriction, ($P < 0.01$).
[4]Effect of feed restriction, ($P < 0.07$).
[5]Effect of sanitary condition, ($P < 0.10$).

approximately 2/3 related to poorer G:F and 1/3 related to poorer feed intake.

Like the response to digestive bacterial infection, unsanitary housing conditions resulted in the majority of the decreased performance as a result of poorer G:F, whereas LPS challenge, mycotoxicosis, and respiratory disease were almost solely feed intake driven. Again, while there may be transient changes in amino acid requirements for maintaining the immune system, decreased feed intake and poorer G:F suggest that highly digestible diets

with ingredients that stimulate feed intake seem to be the best course of action in getting pigs through an immune system challenge. However, the method of supplying these amino acids in properly formulated low-protein, amino acid fortified diets may be one option to reduce post-weaning diarrhea.

Minimizing nutritional challenges to the gut - low-protein, amino acid fortified diets

One method to decrease the dietary challenge imposed on the gastrointestinal system is to lower the dietary crude protein level. However, it is crucial to emphasize that although these diets are lower in crude protein compared with traditional formulations; they still meet amino acid requirements and support excellent pig growth performance. Reducing the crude protein content lowers the requirement for protein ingredients, such as soybean meal, that present immunological challenges to the gut as well as decreases inclusion of expensive specialty protein sources. Presenting the large intestine with a large quantity of undigested nitrogen appears to be a factor in post weaning diarrhea [10]. Lowering the quantity of protein in the diet decreases the ammonia concentration in the small intestine and urea nitrogen and volatile fatty acids in the ileum [11]. It is thought that the decreased nitrogen concentrations are due to reduced protein fermentation by the bacteria [12].

In summary, these studies would suggest that major changes in diet formulations offered to pigs during immune system activation are not warranted. The only considerations would be to ensure that crude protein is not overfed by using the optimum levels of crystalline amino acids which helps prevent large amounts of undigested nitrogen being present in the large intestine and thus contributing to diarrhea. Because changes to dietary amino

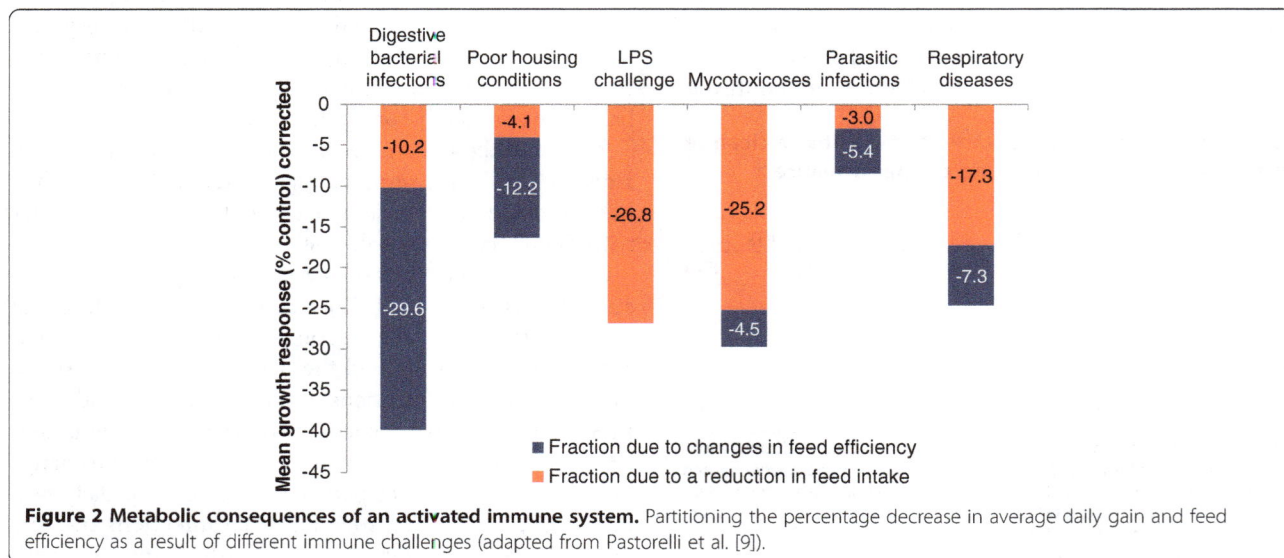

Figure 2 Metabolic consequences of an activated immune system. Partitioning the percentage decrease in average daily gain and feed efficiency as a result of different immune challenges (adapted from Pastorelli et al. [9]).

acid concentrations on a lysine to calorie ratio basis are not needed during immune system activation, the remainder of this review will focus on defining the amino acid levels for weanling pig diets.

Lysine requirements for weanling pigs

Numerous trials have explored the SID lysine requirement of nursery pigs in recent years and requirement estimates have been established (Table 3). The requirement estimate for the 5 to 10 kg pig was found to be between 1.35 and 1.40% standardized ileal digestible (SID) lysine (4.0 to 4.2 g/Mcal ME [13]. This requirement was similar to the estimate found by Dean et al. [14] of 1.40% SID lysine or 18.9 g of SID lysine per kg of gain for 6 to 12 kg pigs.

For 10 to 25 kg pigs, Kendall et al. [15] conducted 5 experiments with 3,628 pigs and found the SID lysine requirement to be 1.30% SID lysine (3.80 g/Mcal ME). This was equivalent to 19 g of SID lysine per kg of gain. Schneider et al. [16] evaluated energy and lysine levels simultaneously in two separate trials with different genotypes. With one genotype, the optimal SID lyine:ME ratio was 3.4 to 3.6 g/Mcal ME, while the optimal ratio was 3.9 to 4.2 g/Mcal ME for the other genotype. However, when expressed relative to gain, the requirement was approximately 19.0 g of SID lysine/kg of gain for both genotypes. In another large field study, Lenehan et al. [17] found the SID lysine requirement for 10 to 20 kg pigs was 1.40%; and when calculated on a g/kg of gain basis, the optimal level was again 19 g of SID lysine/kg of gain.

Although lysine requirements of nursery pigs have increased in recent years and vary with environmental conditions and genotype, when expressed relative to growth rate, empirical studies have consistently found the requirement to be 19 g per kg of gain.

While historically diets for early weaned pigs (4.5 to 5.5 kg) have been formulated to 1.65 or 1.70% total lysine (1.55 to 1.65% SID lysine) or greater, Nemecheck et al. [18] observed that slightly lower dietary lysine levels can be fed in the early nursery phases without negative impact

Table 3 Effects of lysine level fed during each phase on nursery pig performance[a]

Items	Standardized ileal digestible lysine, %											
d 0 to 7	1.35	1.35	1.35	1.35	1.55	1.55	1.55	1.55				
d 7 to 21	1.15	1.15	1.35	1.35	1.15	1.15	1.35	1.35		Probability, $P <$		
d 21 to 35	1.05	1.25	1.05	1.25	1.05	1.25	1.05	1.25	SEM	Phase 1	Phase 2	Phase 3
d 0 to 7												
ADG, g	161	151	152	162	155	163	159	161	19.9	0.69	0.89	0.72
ADFI, g	171	164	157	164	145	150	149	162	15.0	0.37	0.94	0.55
G:F	0.962	0.926	0.965	0.997	1.054	1.089	1.074	0.984	0.059	0.01	0.93	0.63
d 7 to 21												
ADG, g	363	365	366	371	346	333	370	375	15.8	0.41	0.18	0.98
ADFI, g	541	530	512	521	508	506	498	517	18.4	0.16	0.49	0.78
G:F	0.674	0.687	0.716	0.711	0.680	0.660	0.742	0.723	0.016	0.75	0.03	0.43
d 21 to 35												
ADG, g	561	616	579	614	555	573	540	593	35.1	0.20	0.78	0.001
ADFI, g	934	915	943	956	907	883	883	925	34.6	0.37	0.53	0.85
G:F	0.601	0.674	0.614	0.643	0.613	0.649	0.612	0.640	0.031	0.60	0.39	<.0001
d 0 to 35												
ADG, g	402	422	406	426	389	395	395	419	11.3	0.15	0.30	0.03
ADFI, g	745	726	730	747	711	701	696	732	20.5	0.38	0.74	0.65
G:F	0.645	0.692	0.666	0.683	0.658	0.676	0.681	0.688	0.011	0.52	0.07	0.001
BW, kg												
d 0	5.71	5.70	5.73	5.68	5.71	5.75	5.71	5.71	0.05	0.59	0.24	0.43
d 7	6.84	6.76	6.79	6.81	6.80	6.89	6.83	6.83	0.19	0.67	0.91	0.85
d 21	11.93	11.86	11.95	12.00	11.67	11.55	12.01	12.09	0.32	0.54	0.14	0.94
d 35	19.78	20.64	20.05	20.59	19.44	19.57	19.57	20.38	0.36	0.14	0.37	0.04

[a]A total of 320 weanling pigs (PIC 1050 barrows, initially 5.71 ± 0.05 kg and 21 d of age) were used in a 35-d trial with 8 pens per treatment. Phase 1, 2, and 3 diets were fed from d 0 to 7 (SID Lys 1.35 vs 1.55%), 7 to 21 (SID Lys 1.15 vs 1.35%), and 21 to 35 (SID Lys 1.45 vs 1.25%) after weaning, respectively. There were no interactions among the different phases. Nemecheck et al. [19].

on overall ADG or BW, as long as diets during the late nursery period are adequate in lysine (Table 3). In this study, there were a total of 8 dietary treatments arranged in a 2 × 2 × 2 factorial. During phase 1 (d 0 to 7), pigs were fed diets containing either 1.35 or 1.55% SID lysine, followed by either 1.15 or 1.35% SID lysine in phase 2 (d 7 to 21), and 1.05 or 1.25% SID lysine during phase 3 (d 21 to 35). The low dietary lysine concentrations were achieved by reducing both crystalline lysine and a portion of the intact protein sources from the high lysine diets. From d 0 to 7, there were no differences in ADG or ADFI but increasing SID lysine improved G:F. Similar to phase 1, from d 7 to 21, there were no differences in ADG or ADFI between pigs fed the two lysine levels, but increasing SID lysine improved ($P < 0.03$) G:F. During phase 3, feeding the high lysine diet increased ADG and G:F, but had no effect on ADFI. For the overall trial (d 0 to 35), pigs fed the high lysine during phase 3 had the greatest improvement in ADG and G:F. There were no interactions between phases, which indicate that the response to lysine in one phase is not influenced by the lysine level fed in other phases. This allows for formulation of lower lysine (and thus crude protein) diets in early nursery phases and could result in an economical advantage by reducing feed costs while maintaining optimal growth performance.

Until recently, lowering the crude protein level in the diet usually corresponded with reduced growth performance because the minimum requirement for the fourth, fifth, or sixth amino acids (often tryptophan, valine, or isoleucine) or nonessential amino acids that have a role in gut development (arginine, glutamine, or glycine) were not met. However, numerous recent research trials have demonstrated that performance can be maintained when the crude protein level in the diet is reduced by using crystalline amino acids to replace intact protein sources [19,20].

When lowering the crude protein level in the diet, it is critical that we first ensure that diets are not formulated too far below the lysine requirement (Table 4). Assuming a protein deposition of 150 g/d from 20 to 120 kg, adapting equations from Main et al. (2008) [21] and the National Swine Nutrition Guide (van Heugten, 2010 [22]), the equation: g /SID Lysine:Mcal = $0.000146 \times (BW, kg)^2 - 0.0377 \times (BW, kg) + 4.352$; describes the SID Lysine:calorie ratio for barrows while; g/SID Lysine:Mcal = $-0.00000094 \times (BW, kg)^3 + 0.000306 \times (BW, kg)^2 - 0.0435 \times (BW, kg) + 4.414$) describes the g SID Lysine:Mcal ratio for gilts (Table 1). This model is relatively similar to the model recently presented by the NRC [23] with the exception that the proposed model above increases lysine concentrations for late finishing pigs by about 0.05 percentage units. A second option for estimating Lysine requirements uses g Lysine/kg gain. A review of the literature indicates that for nursery pigs (< 20 kg) require approximately 19 g of SID

Table 4 Standardized ileal digestible lysine recommendations as influenced by weight

Pig weight, kg	g/kg of gain	Barrows[1]		Gilts[2]	
		g/Mcal ME	%[3]	g/Mcal ME	%[3]
5	19	4.17	1.40	4.20	1.40
10	19	3.99	1.34	4.01	1.34
15	19	3.82	1.28	3.83	1.28
20	19	3.66	1.22	3.66	1.23
30	20	3.35	1.12	3.36	1.13
40	20	3.08	1.03	3.10	1.04
50	20	2.83	0.95	2.89	0.97
60	20	2.62	0.88	2.70	0.91
70	20	2.43	0.81	2.55	0.85
80	20	2.27	0.76	2.41	0.81
90	20	2.14	0.72	2.29	0.77
100	20	2.04	0.68	2.18	0.73
110	20	1.97	0.66	2.08	0.70
120	20	1.93	0.65	1.98	0.66

[1]Barrow lysine concentrations based on the equation: g/SID Lys:Mcal = $0.000146 \times (BW, kg)^2 - 0.0377 \times (BW, kg) + 4.352$.
[2]Gilt lysine concentrations based on the equation: g/SID Lys:Mcal = $-0.00000094 \times (BW, kg)^3 + 0.000306 \times (BW, kg)^2 - 0.0435 \times (BW, kg) + 4.414$).
[3]Percentages are for diet containing 3,350 kcal ME/kg using NRC [23] nutrient values.

lysine/kg of gain, whereas finishing pigs require approximately 20 g/kg of gain. With this approach, accurate growth and energy intake curves are required to generate a customized Lysine:calorie ratio. As an increasing variety of feed ingredients are used, the range of dietary energy levels has expanded increasing the need for accurate Lysine:calorie ratios in diet formulation. The requirements for the other essential amino acids in relation to lysine must also be considered to allow crude protein to be lowered to minimal levels (Table 5).

Threonine:lysine ratio

Deficiencies of threonine result in relatively small reductions in growth and efficiency as compared to deficiencies of the other major amino acids. However, the large difference between apparent and standardized digestibility values for threonine has caused some confusion when setting requirements on a digestible basis. Compared with other amino acids, threonine digestibility increases the most when moving from an apparent to standardized digestibility basis. Van Milgen and Le Bellego [25] conducted a meta-analysis of 22 different studies and found the optimal threonine:lysine ratio increased from 58% at 15 kg to 65% at 110 kg using a linear-plateau model. Use of curvilinear models resulted in higher requirement estimates. In two separate experiments, Lenehan et al. [26] found an optimal threonine:lysine level of 64 to 66% for 10 to 20 kg pigs. James et al. [27] also found the optimal

Table 5 Suggested minimum standardized ileal digestible amino acid ratios for growing swine[1]

Amino acid	Pig weight range, kg					
	4 to 25	25 to 40	40 to 60	60 to 80	80 to 100	100 to 130
Lysine	100	100	100	100	100	100
Threonine[2]	62	61	61	62	63	64
Methionine[3]	28	28	28	28	28	28
Methionine + cysteine[4]	58	56	56	56	57	58
Tryptophan [5]	18+	18+	18+	18+	18+	18+
Isoleucine[6]	52	52	52	52	52	52
Valine	65	65	65	65	65	65

[1]Adapted from Shannon and Allee, [24] with updates from recent research conducted by the authors and summarized in this paper. Ratios are based on NRC [23] nutrient levels for ingredients. Nutritionists should review their ingredient nutrient values relative to NRC [23] to apply these ratios to their diets.
[2]Threonine:lysine = $0.0000130x^2 - 0.0014229x + 0.6387290$.
[3]Methionine:lysine = $0.0000020x^2 - 0.0000808x + 0.2806061$.
[4]Methionine & Cysteine: lysine = $0.0000113x^2 - 0.0012621x + 0.5785238$.
[5]Tryptopan:lysine ratio appears to be increased when the diet contains large excesses of large neutral amino acids (leucine, isoleucine, valine, phenylalanine, and tyrosine). Improvements in pig growth have been observed with Trp:Lys ratios greater than 18%.
[6]Ratio is at least 60% when high levels of blood meal or cells are included in the diet. Ratio may be lower than 52% when blood cells are not included, but more research is required to verify and to determine the optimal ratio of isoleucine to leucine.

threonine:lysine ratio to be 60 to 65% for 10 to 20 kg pigs. Although Wang et al. [28] did not report a SID threonine: lysine ratio, the growth rate of pigs in their study can be used to estimate the SID lysine requirement (19 g/kg of gain) to calculate SID threonine to be at least 60% of lysine. Based on the above findings, we believe that the threonine requirement can be modeled as a ratio relative to lysine in early growing pig diets ($0.0000130BW^2$ - $0.0014229BW + 0.6387290$), and like NRC 2012 [23] estimates, increases as the pig becomes heavier (Table 5).

TSAA:lysine ratio

Considerable research has been conducted in recent years on the total sulfur amino acid requirement and individual requirements for methionine and cysteine. It is generally assumed that methionine must constitute at least 50% of the TSAA ratio (NRC = 48% on weight basis); however, recent data suggests that methionine may need to be slightly greater (55% on weight basis; 50% on molar basis) than cysteine in the ratio [29].

For nursery pigs, Dean et al., [14] suggested that the requirement for total sulfur amino acids was 10.1 g/kg gain or 54% of lysine for 6 to 12 kg pigs. Gaines et al. [26,30] found a slightly higher ratio of 57 to 61% depending on the response criteria and method of assessing the breakpoint with 8 to 26 kg pigs. Yi et al. [31] found a similar TSAA:lysine ratio of 58% for optimal ADG of 12 to 24 kg pigs. In a series of experiments, Schneider et al. [32] found a similar range of SID TSAA: lysine ratios of 57 to 60% for 10 to 20 kg pigs.

Tryptophan:lysine ratio

Conclusions as to the optimal tryptophan to lysine ratio are difficult to assess for several reasons. Because of the relatively low inclusion rates and small differences in

range of tryptophan levels tested (ex. 14 to 22% of lysine), diet manufacturing can be challenging to ensure the low volume test ingredient additions are thoroughly mixed. Also, tryptophan is a difficult amino acid to analyze and different analytical techniques yield different results adding to the confusion. There is also disagreement in the quantity of tryptophan present in key basal ingredients used in many of the research trials, which can dramatically impact the projected ratios because the basal ingredients such as corn make up such a large proportion of the tryptophan in test diets. Finally, the level of other large neutral amino acids in the diet may influence the response to increasing tryptophan levels. The optimal tryptophan:lysine ratio suggested by most studies ranges from 16 to 20%. Although this range is relatively small, the difference can lead to large changes in diet formulation and cost.

On the low end of the recommended range for nursery pigs, Ma et al. [33] suggested that the SID tryptophan: lysine requirement may be as low as 15% for 11 to 22 kg pigs; however, data from Nemechek et al. [34] demonstrates that 15% SID tryptophan:lysine results in lower ADFI and ADG than a ratio of 20%. Guzik et al. [35] estimated the SID tryptophan requirement for nursery pigs at 0.21, 0.20, and 0.18% of the diet for 5.2 to 7.3 kg, 7.3 to 10.2 kg, and 10.3 to 15.7 kg pigs, respectively. Using the SID lysine levels suggested above, these ratios would all be less than 16% of lysine. Jansman et al. [36] found higher estimates for SID tryptophan for 10 to 20 kg pigs, both as a percentage of the diet (0.22%) and as a ratio to lysine (21.5%). In a review of 33 experiments, Susenbeth [37] summarized that the SID tryptophan:lysine requirement is below 17.4% and likely near 16.0%. Susenbeth also concluded that feeding at 17% would include a safety margin to cover most of biological variations and that the tryptophan:lysine ratio seemed to be unaffected by body weight, growth rate, lysine and protein concentration in the diet, or genetic potential of the animals.

Recently Nitikachana et al. [38,39] conducted a series of tryptophan studies in nursery and finishing pigs designed to determine the requirement relative to lysine on an SID basis. They observed that the ideal ratio was no less than 19 to 20% of lysine, which is much greater than previous estimates when evaluated on an economic basis. Furthermore, Slayer et al. [40] also observed a tryptophan requirement of at least 19% of lysine in finishing pigs fed diets containing 30% dried distillers grains with solubles. What is interesting in the pig's response to tryptophan is that while an optimum "requirement" level can be determined, there is usually a continued, albeit, small improvement in growth performance when feeding levels above the requirement. As a result, from an economic analysis, it is by far safer and, even more economical to be over the estimated requirement than to be below the requirement estimate. This attribute is demonstrated when looking at tryptophan:lysine ratios relative to income over feed cost (IOFC; Figure 3). When the tryptophan:lysine ratio drops below 16% of lysine, profitability decreases dramatically; however, feeding higher ratios in most studies does not decrease IOFC and in some cases increases profitability. When comparing methods to increase the tryptophan:lysine ratio, research suggests that using either added crystalline tryptophan or soybean meal results in similar pig performance [39].

Valine:lysine ratio

Numerous valine trials have been published in the last 10 years. Mavromichalis et al. [43] was one of the first publications to suggest that the valine requirement of

nursery pigs was greater than the level suggested by NRC 1998 [44]. Their data suggested that 10 to 20 kg pigs required 12.5 g of SID valine per kg of gain. Gaines et al. [45] found a similar requirement of 12.3 g of SID valine/kg of gain for 13 to 32 kg pigs. Using the requirement of 19 g of SID lysine per kg of gain for nursery pigs found by several researchers and discussed earlier in this paper, a SID valine:lysine ratio of 66% can be calculated, which is similar to the 65% reported by Gaines et al. [45] for 13 to 32 kg pigs and 65 to 67% reported by Wiltafsky et al. [46] for 8 to 25 kg pigs. A 65% SID valine:lysine ratio was observed by Nemechek et al. [47] using 7 to 12 kg pigs. Nutrient profiles for different ingredients are important when discussing amino acid ratios. For example, a corn-soybean meal based diet formulated to a 65% SID valine:lysine ratio using nutrient values from NRC 1998 [44] will contain 69% SID valine:lysine using values from NRC 2012 [23]. Thus, the ratio used in diet formulation needs to be increased simply due to a change in nutrient profiles.

Isoleucine:lysine ratio

Similar to other amino acids, our understanding of the optimal ratios of isoleucine to lysine has increased greatly in the last 10 years. The main confusion in understanding the optimal isoleucine to lysine ratio is the interaction between isoleucine and other branch chain amino acids, in particular leucine. Excess leucine in the diet increases branch chain keto acid dehydrogenase levels which lead to catabolism of all branch chain amino acids, further leading to increased requirement for isoleucine due to the increased breakdown.

Spray dried blood cells have been used in several isoleucine studies to create a basal diet with a low isoleucine:lysine ratio [48-51]. The problem with this

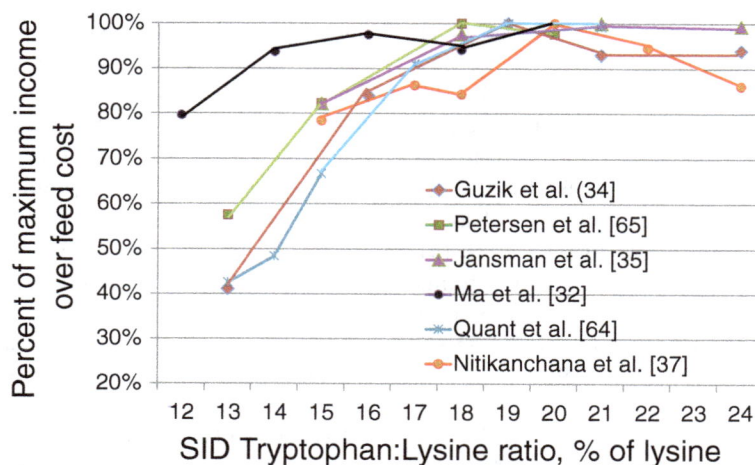

Figure 3 The effects of increasing tryptophan:lysine ratio on income over feed costs. Lines represent the change in income over feed cost with increasing standardized ileal digestible tryptophan:lysine ratio from 6 experiments [33,35,36,38,41,42].

approach is that blood cells contain high leucine levels, which later were determined to increase the isoleucine:lysine recommendation. Subsequently, Fu et al. [52,53], Dean et al. [54], and Wiltafsky et al. [55] demonstrated that the SID isoleucine:lysine requirement was 60% or greater in diets containing blood meal or blood cells and closer to 50% for diets without high levels of blood cells. The requirement of 50% or less for SID isoleucine:lysine when blood cells are not included in the diet was confirmed by Barea et al. [56] for 11 to 23 kg pigs. Lindemann et al. [57] also found the SID isoleucine:lysine requirement to be between 48 and 52% for ADG. Norgaard and Fernandez [58] found that increasing the isoleucine:lysine ratio from 53 to 62% did not influence performance of 9 to 22 kg pigs. Therefore, it appears that the SID isoleucine:lysine is less than 52% for diets that don't contain a protein source such as blood products that provide excess leucine in relation to the isoleucine level.

Caution is advised with all branch chain amino acids such as valine, isoleucine, and leucine, as feeding as little as 5% below the minimum ratio (ex. 45 vs 50% of lysine) will greatly reduce feed intake and daily gain. Another concern is that with low-protein amino acid fortified diets formulated to the 5th and 6th limiting amino acid, leucine can become limiting or very near its requirement estimate at 100% of SID lysine [23].

Nonessential amino acid requirement

Although the order can vary with different dietary ingredient mixtures, typically the first 5 limiting amino acids for most practical diets are lysine, threonine, methionine, tryptophan, and valine. However, formulating diets with high levels of crystalline amino acids to the optimal ratio for the first 5 limiting amino acids often has resulted in poorer performance than diets with high levels of intact protein sources. Kendall et al. [59] found that certain nonessential amino acids (ex. glycine) were required in corn-soybean meal diets with high levels of crystalline lysine and that the nitrogen could not be provided by non-protein nitrogen. In a series of experiments, Powell et al. [60,61] and Southern et al. [62] found that glycine and another amino acid to provide nitrogen were required in diets formulated to the fifth or sixth limiting amino acid in order to maintain feed efficiency.

Another method to ensure that the diet contains enough nonessential amino acids is to place a maximum on the total lysine to total crude protein ratio in diet formulation. The biological basis for a lysine:CP ratio originates from the level of total lysine as a percentage of crude protein in muscle, which ranges from 6.5 to 7.5%. Although an average lysine:CP ratio of 6.8% is often cited, a higher lysine:CP ratio can be used in the diet because the lysine released during normal muscle protein breakdown is conserved and recycled with greater efficiency than other amino acids. Ratliff et al. [63] suggested that the total lysine:CP ratio should not exceed 7.1%. Nemechek et al. [64] found that feed efficiency was only poorer when the total lysine:CP ratio exceeded 7.35%. More research is clearly needed to continue to expand our understanding of nonessential amino acid needs of the pig.

Nonessential amino acids appear to play a particularly important role immediately after weaning due to their high requirement for intestinal growth. For instance, glutamine serves as a primary fuel for the intestinal mucosa. Glutamine and glycine stimulate polyamine synthesis. Arginine is the precursor for polyamines and nitric oxide which is important for regulation of intestinal blood flow and migration of intestinal epithelial cells. Numerous other roles of the nonessential amino acids are reviewed by Wu [65].

Conclusion

The immune system elicits a variety of responses orchestrated by cytokines. Of these responses, anorexia or reduced energy intake is the limiting factor for protein synthesis. While the amino acid requirements may increase with immune system activation, from a practical standpoint, the decrease in muscle accretion will offset most of the changes in requirements. The evidence is sparse or equivocal for increasing nutrient requirements during an immune challenge. However, some ingredients and diet formulation techniques can help the pig counteract some of the normal gut changes that occur at weaning. Low-protein, amino acid fortified diets can limit the amount of fermentable protein presented to the gut and help reduce post-weaning diarrhea. In these cases, proper amino acid fortification and ratios relative to lysine are essential not to limit pig growth. The ultimate goal for nutritionists is to help the pig transition through this phase without incurring excessive diet cost.

Competing interests
The authors declare that they have no competing interests.

Authors' contributions
BG, MT, SD, JD, and JW were all involved in preparing and contributing to the review. All authors read and approved the final manuscript.

Acknowledgement
Contribution no. 14-090-J from the Kansas Agricultural Experiment Station, Manhattan, KS 66506 USA.

Author details
[1]Department of Animal Sciences and Industry, Kansas State University, Manhattan, KS 66506-0201, USA. [2]Department of Diagnostic Medicine/ Pathobiology, College of Veterinary Medicine, Kansas State University, Manhattan, KS 66506-0201, USA.

References

1. Kalsing KC: **Nutritional aspects of leukocytic cytokines.** *J Nutr* 1988, **118**:1436–1446.

2. Williams NH, Stahly TS, Zimmerman DR: **Effect of chronic immune system activation on the rate, efficiency, and composition of growth and lysine needs of pigs fed from 6 to 27 kg.** *J Anim Sci* 1997, **75**:2463–2471.

3. Williams NH, Stahly TS, Zimmerman DR: **Effect of chronic immune system activation on body nitrogen retention, partial efficiency of lysine utilization, and lysine needs of pigs.** *J Anim Sci* 1997, **75**:2472–2480.

4. Dritz SS, Owen KQ, Goodband RD, Nelssen JL, Tokach MD, Chengappa MM, Blecha F: **Influence of lipopolysaccharide-induced immune challenge and diet complexity on growth performance and acute-phase protein production in segregated early-weaned pigs.** *J Anim Sci* 1996, **74**:1620–1628.

5. Shelton NW, Tokach MD, Dritz SS, Goodband RD, Nelssen JL, DeRouchey JM, Usry JL: **Effects of porcine circovirus type 2 vaccine and increasing standardized ileal digestible lysine:ME ratio on growth performance and carcass composition of growing and finishing pigs.** *J Anim Sci* 2012, **90**:361–372.

6. Montagne L, Le Floc'h N, Arturo-Schaan M, Foret R, Urdaci MC, Le Gall M: **Comparative effects of level of dietary fiber and sanitary condition on the growth and health of weanling pigs.** *J Anim Sci* 2012, **90**:2556–2569.

7. Pastorelli H, Le Floc'h N, Merlot E, Meunier-Salaün MC, van Milgen J, Montagne L: **Feed restriction applied after weaning has different effects on pig performance and health depending on the sanitary conditions.** *J Anim Sci* 2012, **90**:4866–4875.

8. Klassing KC, Barns DM: **Decreased amino acid requirements of growing chicks due to immunologic stress.** *J Nutr* 1988, **118**:1158–1164.

9. Pastorelli H, van Milgen J, Lovatto P, Montagne L: **Meta-analysis of feed intake and growth responses of growing pigs after a sanitary challenge.** *Animal* 2012, **6**(6):952–961.

10. Heo JM, Opapeju FO, Pluske JR, Kim JC, Hampson DJ, Nyachoti CM: **Gastrointestinal health and function in weaned pigs: a review of feeding strategies to control post-weaning diarrhoea without using in-feed antimicrobial compounds.** *J Anim Physiol Nutr* 2013, **97**:207–237. doi:10.1111/j.1439-0396.2012.01284.x.

11. Bikker PA, Dirkzwager A, Fledderus J, Trevisi P, le H.Jerou-Luron I, Lalles JP, Awati A: **The effect of dietary protein and fermentable carbohydrates levels on growth performance and intestinal characteristics in newly weaned piglets.** *J Anim Sci* 2006, **84**:3337–3345.

12. Nyachoti CM, Omogbenigun FO, Rademacher M, Blank G: **Performance responses and indicators of gastrointestinal health in early-weaned pigs fed low-protein amino acid-supplemented diets.** *J Anim Sci* 2006, **2006**(84):125–134.

13. Gaines AM, Kendall DC, Allee GL, Tokach MD, Dritz SS, Usry JL: **Evaluation of the true ileal digestible (TID) lysine requirement for 7 to 14 kg pigs.** *J Anim Sci* 2003, **81**(1):139. Abstr.

14. Dean DW, Southern LL, Kerr BJ, Bidner TD: **The lysine and total sulfur amino acid requirements of six- to twelve-kilogram pigs.** *Prof Anim Sci* 2007, **23**:527–535.

15. Kendall DC, Gaines AM, Allee GL, Usry JL: **Commercial validation of the true ileal digestible lysine requirement for eleven- to twenty-seven-kilogram pigs.** *J Anim Sci* 2008, **86**:324–332.

16. Schneider JD, Tokach MD, Dritz SS, Nelssen JL, DeFouchey JM, Goodband RD: **Determining the effect of lysine:calorie ratio on growth performance of ten- to twenty-kilogram of body weight nursery pigs of two different genotypes.** *J Anim Sci* 2010, **88**:137–146.

17. Lenehan NA, Dritz SS, Tokach MD, Goodband RD, Nelssen JL, Usry JL: **Effects of lysine level fed from 10 to 20 kg on growth performance of barrows and gilts.** *J Anim Sci* 2003, **81**(1):46. Abstr.

18. Nemechek JE, Tokach MD, Dritz SS, Goodband RD, DeRouchey JM, Nelssen JL: **Evaluation of SID lysine level, replacement of fish meal with crystalline amino acids, and lysine:CP ratio on growth perfromance of nursery pigs from 6.8 to 11.3 kg.** *Anim Sci* 2011, **89**(E-Suppl. 2):220. Abstr.

19. Nemechek JE, Tokach MD, Dritz SS, Goodband RD, DeRouchey JM, Nelssen JL: **Does lysine level fed in one phase influence performance during another phase in nursery pigs?** *J Anim Sci* 2010, **88**(E - Supp. 3):13. Abstr.

20. Heo JM, Kim JC, Hansen CF, Mullan BP, Hampson DJ, Pluske JR: **Feeding a diet with decreased protein content reduces indices of protein fermentation and the incidence of postweaning diarrhea in weaned pigs challenged with an enterotoxigenic strain of Escherichia coli.** *J Anim Sci* 2009, **87**:2833–2843.

21. Main RG, Dritz SS, Tokach MD, Goodband RD, Nelssen JL: **Determining an optimum lysine:calorie ratio for barrows and gilts in a commercial finishing facility.** *J Anim Sci* 2008, **86**:2190–2207.

22. Van Heugton E: *Growing-finishing swine nutrient recommendations and feeding management.* Ames, IA, USA: National Swine Nutrition Guide, US Pork Center of Excellence; 2010.

23. NRC: *Nutrient requirements of swine: 11th revised edition.* Washington DC: Natl. Acad. Press; 2012.

24. Shannon MC, Allee GL: **Protein and amino acid sources in swine diets.** In *National Swine Nutrition Guide.* Edited by Meisinger DJ. 2010.

25. Van Milgen J, LeBellego L: **A meta-analysis to estimate the optimum threonine to lysine ratio in growing pigs.** *J Anim Sci* 2003, **81**(1):553. Abstr.

26. Lenehan NA, Tokach MD, Dritz SS, Goodband RD, Nelssen JL, Usry JL, DeRouchey JM, Frantz NZ: **The optimal true ileal digestible lysine and threonine requirement for nursery pigs between 10 and 20 kg.** *J Anim Sci* 2004, **82**(1):293. Abstr.

27. James BW, Tokach MD, Goodband RD, Dritz SS, Nelssen JL, Usry JL: **The optimal true ileal digestible threonine requirement for nursery pigs between 11 to 22 kg.** *J Anim Sci* 2003, **81**(1):42. Abstr.

28. Wang X, Qiao SY, Liu M, Ma YX: **Effects of graded levels of true ileal digestible threonine on performance, serum parameters and immune function of 10–25 kg pigs.** *Anim Feed Sci Tech* 2006, **129**:264–278.

29. Gillis AM, Reijmers A, Pluske JR, de Lange CFM: **Influence of dietary methionine to methionine plus cysteine ratios on nitrogen retention in gilts fed purified diets between 40 and 80 kg live body weight.** *Can J Anim Sci* 2007, **87**:87–92.

30. Gaines AM, Yi GF, Ratliff BW, Srichana P, Kendall DC, Allee GL, Knight CD, Perryman KR: **Estimation of the ideal ratio of true ileal digestible sulfur amino acids:lysine in 8- to 26-kg nursery pigs.** *J Anim Sci* 2005, **83**:2527–2534.

31. Yi GF, Gaines AM, Ratliff BW, Srichana P, Allee GL, Perryman KR, Knight CD: **Estimation of the true ileal digestible lysine and sulfur amino acid requirement and comparison of the bioefficacy of 2-hydroxy-4-(methylthio) butanoic acid and DL-methionine in eleven- to twenty-six-kilogram nursery pigs.** *J Anim Sci* 2006, **84**:1709–1721.

32. Schneider JD, Tokach MD, Dritz SS, Goodband RD, Nelssen JL, Usry JL, DeRouchey JM, Hastad CW, Lenehan NA, Frantz NZ, James BW, Lawrence KR, Groesbeck CN, Gottlob RO, Young MG: **2 The optimal true ileal digestible lysine and total sulfur amino acid requirement for nursery pigs between 10 and 20 kg.** *J Anim Sci* 2004, **82**(1):293. 3 Abstr.

33. Ma L, Zhu ZP, Hinson RB, Allee GL, Less JD, Hall DD, Yang H, Holzgraefe DP: **Determination of SID Trp:lysine ratio requirement of 11- to 22-kg pigs fed diets containing 30% DDGS.** *J Anim Sci* 2010, **88**(3):151. Abstr.

34. Nemechek JE, Tokach MD, Dritz SS, Goodband RD, DeRouchey JM, Nelssen JL: **Effects of deleting crystalline amino acids from low-CP, amino acid-fortified diets and dietary valine:lysine ratio of for nursery pigs from 6.8 to 11.3 kg.** *J Anim Sci* 2011A, **89**(E - Suppl. 3):97. Abstr.

35. Guzik AC, Southern LL, Bidner TD, Kerr BJ: **The tryptophan requirement of nursery pigs.** *J Anim Sci* 2002, **80**:2646–2655.

36. Jansman AJM, van Diepen JTM, Melchior D: **The effect of diet composition on tryptophan requirement for young piglets.** *J Anim Sci* 2010, **88**:1017–1027.

37. Susenbeth A: **Optimum tryptophan:lysine ratio in diets for growing pigs: Analysis of literature data.** *Livest Sci* 2006, **101**:32–45.

38. Nitikanchana S, Tokach MD, Dritz SS, DeRouchey JM, Goodband RD, Nemecheck JE, Nelssen JL, Usry J: **Influence of standardized ileal digestible tryptophan:lysine ratio on growth performance of 6- to 10-kg nursery pigs.** *J Anim Sci* 2012, **90**(2):151. Abstr.

39. Nitikanchana S, Tokach MD, Dritz SS, Usry J, Goodband RD, DeRouchey JM, Nelssen JL: **The effects of Sid Trp:Lys ratio and Trp source in diets containing DDGS on growth performance and carcass characteristics of finishing pigs.** *J Anim Sci* 2013, **91**(Suppl. 2):73. Abstr.

40. Slayer JA, Tokach MD, DeRouchey JM, Goodband RD, Dritz SS, Nelssen JL: **Effects of standardized ileal digestible tryptophan:lysine ratio in diets containing 30% dried distillers grains with solubles (DDGS) on finishing pig performance and carcass traits.** *J Anim Sci* 2013, **91**:3244–3252.

41. Quant AD, Lindemann MD, Cromwell GI, Kerr BJ, Payne RL: **Determining the optimum ratio of dietary tryptophan to lysine in growing pigs fed non-US-type ingredients.** *J Anim Sci* 2008, **86**(Suppl. 1):91. Abstr.

42. Peterson GI, Stein HH: **Determination of the lys and trp requirements in 10 to 20 kg pigs.** *J Anim Sci* 2011, **89**(Suppl. 1):62. Abstr.

43. Mavromichalis I, Kerr BJ, Parr TM, Albin DM, Gabert VM, Baker DH: **Valine requirement of nursery pigs.** *J Anim Sci* 2001, **79**:1223–1229.

44. NRC: *Nutrient requirements of swine: 10th revised edition.* DC, Washington: Natl. Acad. Press; 1998.

45. Gaines AM, Kendall DC, Allee GL, Usry JL, Kerr BJ: Estimation of the standardized ileal digestible valine to lysine ratio in 13 to 32 kg pigs. *J Anim Sci* 2010, 89:736–742.

46. Wiltafsky MK, Schmidtlein B, Roth FX: Estimates of the optimum dietary ratio of standardized ileal digestible valine to lysine for eight to twenty-five kilograms of body weight pigs. *J Anim Sci* 2009, 87:2544–2553.

47. Nemechek JE, Tokach MD, Dritz SS, Goodband RD, DeRouche JM, Nelssen JL: Effects of deleting crystalline amno acids from low-CP, amino acid-fortified diets and dietary valine:lysine ratio for nursery pigs from 6.8 to 11.3 kg. *J Anim Sci* 2011, 97(E-Suppl. 2):97. Abstr.

48. Parr TM, Kerr BJ, Baker DH: Isoleucine requirement of growing (25 to 45 kg) pigs. *J Anim Sci* 2003, 81:745–752.

49. Fu SX, Fent RW, Srichana P, Allee GL, Usry JL: Effects of protein source on true ileal digestible (TID) isoleucine:lysine ratio in late-finishing barrows. *J Anim Sci* 2005, 83(2):149. Abstr.

50. Parr TM, Kerr BJ, Baker DH: Isoleucine requirement for late-finishing (87 to 100 kg) pigs. *J Anim Sci* 2004, 82:1334–1338.

51. Kerr BJ, Kidd MT, Cuaron JA, Bryant KL, Parr TM, Maxwell CV, Campbell JM: Isoleucine requirements and ratios in starting (7 to 11 kg) pigs. *J Anim Sci* 2004, 82:2333–2342.

52. Fu SX, Kendall DC, Fent RW, Allee GL, Usry JL: True ileal digestible (TID) isoleucine:lysine ratio of late-finishing barrows fed corn-blood cell or corn-amino acid diets. *J Anim Sci* 2005, 83(2):148. Abstr.

53. Fu SX, Fent RW, Allee GL, Usry JL: Branched chain amino acid interactions and isoleucine imbalance in late-finishing pigs. *J Anim Sci* 2006a, 84(1):371. Abstr.

54. Dean DW, Southern LL, Kerr BJ, Bidner TD: Isoleucine requirement of 80- to 120-kilogram barrows fed corn-soybean meal or corn-blood cell diets. *J Anim Sci* 2005, 83:2543–2553.

55. Wiltafsky MK, Bartelt J, Relandeau C, Roth FX: Estimation of the optimum ratio of standardized ileal digestible isoleucine to lysine for eight- to twenty-five-kilogram pigs in diets containing spray-dried blood cells or corn gluten feed as a protein source. *J Anim Sci* 2009, 87:2554–2564.

56. Barea R, Brossard L, Le Floc'h N, Primot Y, van Milgen J: The standardized ileal digestible isoleucine-to-lysine requirement ratio may be less than fifty percent in eleven- to twenty-three-kilogram piglets. *J Anim Sci* 2009, 87:4022–4031.

57. Lindemann MD, Quant AD, Cho JH, Kerr BJ, Cromwell GL, Htoo JK: Determining the optimium ratio of standardized ileal digestible (SID) isoleucine to lysine for growing pigs fed wheat-barley based diets. *J Anim Sci* 2010, 88(E-Suppl. 3):43. Abstr.

58. Nørgaard JV, Fernández JA: Isoleucine and valine supplementation of crude protein-reduced diets for pigs aged 5–8 weeks. *Anim Feed Sci Tech* 2009, 154:248–253.

59. Kendall DC, Fent RW, Usry JL, Allee GL: The essentiality of nonessential amino acids in low protein diet formulations for 11 to 30 kg barrows. *J Anim Sci* 2004, 82(2):125. Abstr.

60. Powell S, Bidner T, Southern L, Payne R: Growth performance of 20- to 50-kilogram pigs fed low crude protein diets supplemented with Glycine and L-Arginine. *J Anim Sci* 2009a, 87(E-Suppl. 3):153. Abstr.

61. Powell S, Greely J, Bidner T, Southern L, Payne R: Growth performance of 20- to 50-kilogram pigs fed low crude protein diets supplemented with L-Histidine, L-Cystine, and Glycine. *J Anim Sci* 2009b, 87(E-Suppl. 3):152. Abstr.

62. Southern LL, Ross ML, Waguespack AM, Powell S, Bidner TD, Payne RL: Developing low protein amino acid supplemented diets for swine. *J Anim Sci* 2010, 88(E-Suppl. 3):42. Abstr.

63. Ratliff BW, Gaines AM, Srichana P, Allee GL, Usry JL: Evaluation of high synthetic amino acid inclusion and supplemental arginine in starter diets. *J Anim Sci* 2005, 83(Suppl. 2):69. Abstr.

64. Nemechek JE, Tokach MD, Dritz SS, Goodband RD, DeRouchey JM, Nelssen JL, Usry JL: Evaluation of SID lysine level, the replacement of fish meal with crystalline amino acids, and lysine:CP ratio on growth performance of nursery pigs from 6.8 to 11.3 kg. *J Anim Sci* 2011b, 89(Suppl. 2):220. Abstr.

65. Wu G: *Importance of intestinal amino acid metabolism in nutrition and health: Thinking outside the box. Proceedings of Gentech's First International Nutrition Forum.* Shanghai China: Gentech Industries Group; 2011:2–32.

21

Changes in various metabolic parameters in blood and milk during experimental *Escherichia coli* mastitis for primiparous Holstein dairy cows during early lactation

Kasey M Moyes[1]*, Torben Larsen[2], Peter Sørensen[2] and Klaus L Ingvartsen[2]

Abstract

Background: The objective of this study was to characterize the changes in various metabolic parameters in blood and milk during IMI challenge with *Escherichia coli* (**E. coli**) for dairy cows during early lactation. Thirty, healthy primiparous Holstein cows were infused (h = 0) with ~20-40 cfu of live *E. coli* into one front mammary quarter at ~4-6 wk in lactation. Daily feed intake and milk yield were recorded. At −12, 0, 3, 6, 12, 18, 24, 36, 48, 60, 72, 96, 108, 120, 132, 144, 156, 168, 180 and 192 h relative to challenge rectal temperatures were recorded and quarter foremilk was collected for analysis of shedding of *E. coli*. Composite milk samples were collected at -180, -132, -84, -36, -12, 12, 24, 36, 48, 60, 72, 84, 96, 132 and 180 h relative to challenge (h = 0) and analyzed for lactate dehydrogenase (**LDH**), somatic cell count, fat, protein, lactose, citrate, beta-hydroxybutyrate (**BHBA**), free glucose (**fglu**), and glucose-6-phosphate (**G6P**). Blood was collected at -12, 0, 3, 6, 12, 18, 24, 36, 60, 72, 84, 132 and 180 h relative to challenge and analyzed for plasma non-esterified fatty acids (**NEFA**), BHBA and glucose concentration. A generalized linear mixed model was used to determine the effect of IMI challenge on metabolic responses of cows during early lactation.

Results: By 12 h, *E. coli* was recovered from challenged quarters and shedding continued through 72 h. Rectal temperature peaked by 12 h post-challenge and returned to pre-challenge values by 36 h post-IMI challenge. Daily feed intake and milk yield decreased ($P < 0.05$) by 1 and 2 d, respectively, after mastitis challenge. Plasma BHBA decreased (12 h; $P < 0.05$) from 0.96 ± 1.1 at 0 h to 0.57 ± 0.64 mmol/L by 18 h whereas concentration of plasma NEFA (18 h) and glucose (24 h) were significantly greater, 11 and 27%, respectively, after challenge. In milk, fglu, lactose, citrate, fat and protein yield were lower whereas yield of BHBA and G6P were higher after challenge when compared to pre-challenge values.

Conclusions: Changes in metabolites in blood and milk were most likely associated with drops in feed intake and milk yield. However, the early rise in plasma NEFA may also signify enhanced adipose tissue lipolysis. Lower concentrations of plasma BHBA may be attributed to an increase transfer into milk after IMI. Decreases in both milk lactose yield and % after challenge may be partly attributed to reduced conversion of fglu to lactose. Rises in G6P yield and concentration in milk after challenge (24 h) may signify increased conversion of fglu to G6P. Results identify changes in various metabolic parameters in blood and milk after IMI challenge with *E. coli* in dairy cows that may partly explain the partitioning of nutrients and changes in milk components after IMI for cows during early lactation.

Keywords: Cow, Early lactation, *Escherichia coli*, Metabolism

* Correspondence: kmoyes@umd.edu
[1]Department of Animal and Avian Sciences, University of Maryland, 142 Animal Sciences Building, MD 20742-2311, 20910 College Park, MD, USA
Full list of author information is available at the end of the article

Background

During early lactation (i.e. 0-8 wk in milk), the homeorhetic mechanisms associated with hormonal changes, as well as changes in the nervous system and immune system, shift the partitioning of nutrients from peripheral tissues towards the synthesis of milk. This massive re-partition has been identified as a major contributor to the high risk of disease at this time [1]. Mastitis, an inflammation of the mammary gland, is the most costly of all diseases and occurs more frequently after parturition [2,3]. The innate immune response patterns to major mastitis-causing pathogens (e.g. *E. coli*, *Streptococcus uberis* and *Staphylococcus aureus*) have been well-documented [4-6] but the characterization of the metabolic responses in dairy cows during an IMI are not fully understood.

Most studies have focused on the effect of metabolic status on immune response for dairy cows [1,7,8]. During mastitis, the immunometabolic responses primarily focus on the transcription-level responses in liver and mammary tissue [9-11]. Previous work indicates that the ability of the liver to metabolize fatty acids is reduced and key genes associated with metabolic processes are down-regulated after intramammary *E. coli* challenge [12] as well as after intramammary endotoxin challenge [13]. Furthermore, changes in circulating non-esterified fatty acids (**NEFA**), beta-hydroxybutyrate (**BHBA**) and glucose, prior to decreases in feed intake and milk production, during an IMI in dairy cows have been reported [14-16]. To our knowledge, changes in free glucose (**fglu**) and glucose-6-phosphate (**G6P**) in milk during mastitis in relation to changes in circulating metabolites and other milk components has not been elucidated. The mammary gland primarily relies on circulating glucose for the synthesis of lactose, a disaccharide composed of the monosaccharides D-glucose and D-galactose [17]. Other fates of glucose in the mammary gland include the conversion to G6P for the synthesis of galactose [18]. Characterizing the metabolic responses of fglu and G6P in relation to other metabolic components during inflammation may further elucidate the partitioning of nutrients and changes in milk composition that occur during mastitis. The objective of this study was to characterize the changes in various metabolic parameters in blood and milk during IMI with *E. coli* for dairy cows during early lactation.

Methods

The experiment was carried out at the cattle research facilities at Department of Animal Science, Aarhus University. Experimental procedures involving animals were approved by the Danish Animal Experiments Inspectorate and complied with the Danish Ministry of Justice Laws concerning animal experimentation and care of experimental animals.

Animals, experimental design and sample collection

Thirty primiparous Holstein cows at ~4-6 wk in lactation were used for this study. Only healthy cows not treated for any clinical signs of disease before the study period were included. Details on animal housing, total mixed ration fed and refused, treatment, preparation and infusion of *E. coli* and clinical examinations have been previously described [19,20]. Briefly, eligible cows were considered healthy and free of mastitis-causing pathogens based on body temperature, white blood cell count, glutaraldehyde test, California Mastitis Test (Kruuse, Marslev, Denmark) and bacteriological examinations of aseptic quarter foremilk samples prior to the start of the study period. Using the portable DeLaval Cell Counter (DeLaval, Tumba, Sweden), the front quarter with the lowest somatic cell count (**SCC**; <27,000 cells/mL) was used for *E. coli* infusion.

Cows were housed and fed in individual straw-bedded tie-stalls, had free access to water, and were milked twice at 0600 and 1700 h. Cows averaged 27.5 ± 5.5 kg milk/d at the start of the trial. Cows were fed a standard total mixed ration for lactating cows ad libitum twice at 0800 and 1530 h. Daily feed intake and milk yield were recorded throughout the study period. Orts were collected in the mornings (~0800 h). To clarify, IMI challenge occurred after the afternoon milking, and therefore, d = 0 was calculated from milk yield and feed intakes from -36 to -12 h prior to IMI challenge where -12 h represents the morning relative to challenge.

All cows were infused with ~20-40 cfu of live *E. coli* (Danish field isolate k2bh2) into one front mammary quarter immediately following the afternoon milking (h = 0). The IMI challenge was imposed in the same year (i.e. 2007) and stage of lactation but in 4 different blocks: May (n = 8), June (n = 7), August (n = 8) and September (n = 8). Rectal temperature was recorded at -12, 0, 3, 6, 12, 18, 24, 48, 60, 72, 84, 96, 108, 120, 132, 144, 156, 168, 180 and 192 h relative to IMI challenge. Composite milk samples were collected relative to IMI challenge (h = 0) during the morning milking period at -180, -132, -84, -36, and -12 h, at each milking after challenge at 12, 24, 36, 48, 60, 72, 84, 96 h, and during the morning milking period at 132 and 180 h. Aseptic quarter foremilk samples were collected from challenged quarters at -12, 0, 3, 6, 12, 18, 24, 48, 60, 84, 96, 108, 120, 132, 144, 156, 168, 180 and 192 h relative to IMI challenge. One day prior to IMI challenge, sterile Micro-Renathane polyvinyl catheters were inserted into the jugular vein and flushed with a sterile 0.9% NaCl solution containing 50 IU Na-heparin (Loevens Kemiske Fabrik, Ballerup, Denmark). Blood was collected at -12, 0, 3, 6, 12, 18, 24, 36, 60, 84, 132 and 180 h relative to IMI challenge. For a subset of cows (n = 16), liver biopsies were collected at -144, 12, 24 and 192 h relative to IMI challenge for gene expression profiling and results are reported elsewhere [12]; and a mammary biopsy was collected at 24 and 192 h relative to

IMI challenge for gene expression analysis using a minimally invasive biopsy technique [19]. After the mammary biopsies had been collected (i.e. 24 and 192 h post-biopsy), cows were administered a prophylactic antibiotic treatment against infection with Gram-positive bacteria by intramuscular injection of 30 mL of Penovet® vet (300,000 IE benzylpenicillinprocain/ml; Boehringer Ingelheim Danmark A/S, Copenhagen, Denmark). No other antibiotic therapy was administered after IMI challenge for all cows, regardless of biopsy.

Sample analysis

Composite milk samples were analyzed for fat, protein, lactose, citrate and SCC (cells/mL) using a CombiFoss 4000 (Foss Electric A/S, Hillerød, Denmark) and BHBA, lactate dehydrogenase (LDH), N-acetyl-β-D-glucosaminidase (NAGase) and alkaline phosphatase activity (ALP) were analyzed according to methods previously described [21-23]. Free glucose and G6P were analyzed by an

enzymatic-fluorometric method as described by Larsen [24]. Quarter foremilk was analyzed for SCC and quantification of E. coli (cfu/mL) as previously described [19].

Plasma was harvested following centrifugation at 2,000 × g for 20 min at 4°C and stored at -18°C until further analysis. All plasma components were analyzed for NEFA, BHBA and glucose using an autoanalyzer, ADVIA 1650® Chemistry System (Siemens Medical Solution, Tarrytown, NY, USA) according to methods described by Bjerre-Harpøth et al. [21].

Statistical analysis

Plasma NEFA and BHBA were normalized by natural log (ln) transformation and SCC and shedding of E. coli were $\log_{10}$ transformed for statistical analyses. Yields for all milk components were calculated at each milking. Data were analyzed via a generalized linear mixed model using the MIXED procedure of SAS, version 9.3 [25] with the repeated measure of time (i.e. hour) relative to

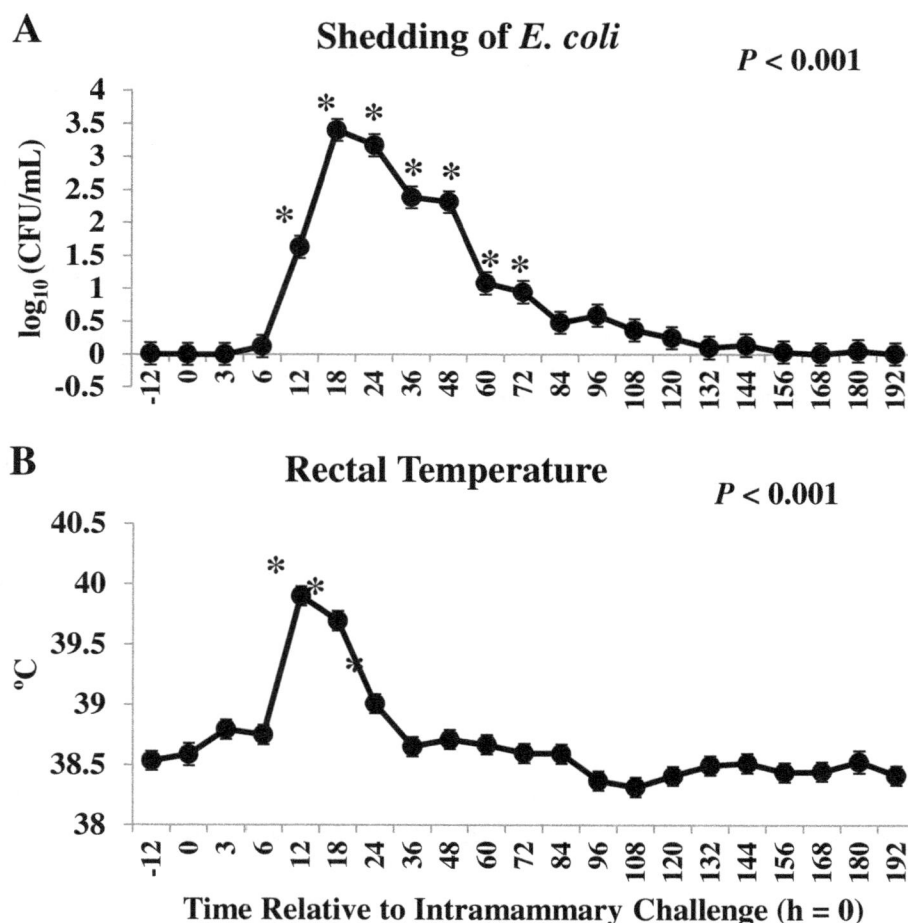

Figure 1 Colony forming units (cfu) of *Escherichia coli (E. coli)*/mL of mammary secretion (A) and rectal temperature (B) after intramammary challenge with *Escherichia coli* (h = 0) for 30 primiparous Holstein cows during early lactation. *Differences (P <0.05) when compared to h = 0.

IMI challenge (h = 0). Using the MIXED procedure of SAS, combined biopsy had no effect (P <0.05) on any of the metabolic parameters in blood and milk for this study and was, therefore, left out of the final model. The random effect of cow within block was used as the error term in the REPEATED statement with compound symmetry (CS) as the covariance structure. The model was used to determine the effect of IMI challenge on metabolic and immune responses in blood and milk for cows in early lactation. The class variables included cow, block and time relative to IMI challenge. Degrees of freedom were estimated with the Kenward-Roger specification in the model statements. Separation of least square means (LSM) for significant effects was accomplished using the Tukey's option within the MIXED procedure of SAS. Statistical differences were declared as significant and highly significant at P <0.05 and P <0.01, respectively. Trends towards significance are discussed at P <0.10. Plasma NEFA and BHBA were back-transformed for presentation in figures.

Results and discussion
Indicators of infection and immune response
The bacterial counts of E. coli (A) and rectal temperature (B) relative to IMI challenge are shown in Figure 1. By 12 h, E. coli was recovered from challenged quarters and shedding continued through 72 h similar to those of others [26,27]. Rectal temperature returned to pre-challenge values by 36 h post-IMI challenge as observed by Scaletti

and Harmon [27]. These findings are consistent with signs of experimental E. coli mastitis and confirmed the model system.

Composite milk LDH (A), SCC (B), NAGase (C) and ALP (D) concentrations were greater after IMI challenge with E. coli (Figure 2). During an IMI challenge, a cascade of changes occur including increased LDH, ALP and NAGase activity in milk associated with infiltrating neutrophils and resident macrophages [28,29]. These indigenous enzymes are accurate real-time indicators for detecting mastitis on-farm when compared to composite SCC or bacterial culture [28-30].

Cow-level and metabolic responses
Changes in daily feed intake (as fed; A) and milk yield (B) relative to IMI challenge are shown in Figure 3. To clarify, d = 0 reflects the daily feed intake and milk yield from the 24 h period prior to IMI challenge (i.e. -36 to -12 h relative to IMI challenge). Day = 1 reflects -12 to 12 h post-IMI challenge. Daily milk yield decreased 23% from d = 0 to d = 1 and 36% by d = 2 whereas feed intake was not significantly reduced until 2 d post-IMI challenge from 29.7 kg at d = 0 to 21.7 kg by d = 2. Feed intake and milk yield returned to pre-challenge values by 3 and 4 d post-IMI challenge, respectively. As lactation progressed, feed intake continued to increase as normally observed during early lactation [31]. Decreases in feed intake and milk yield have been previously shown

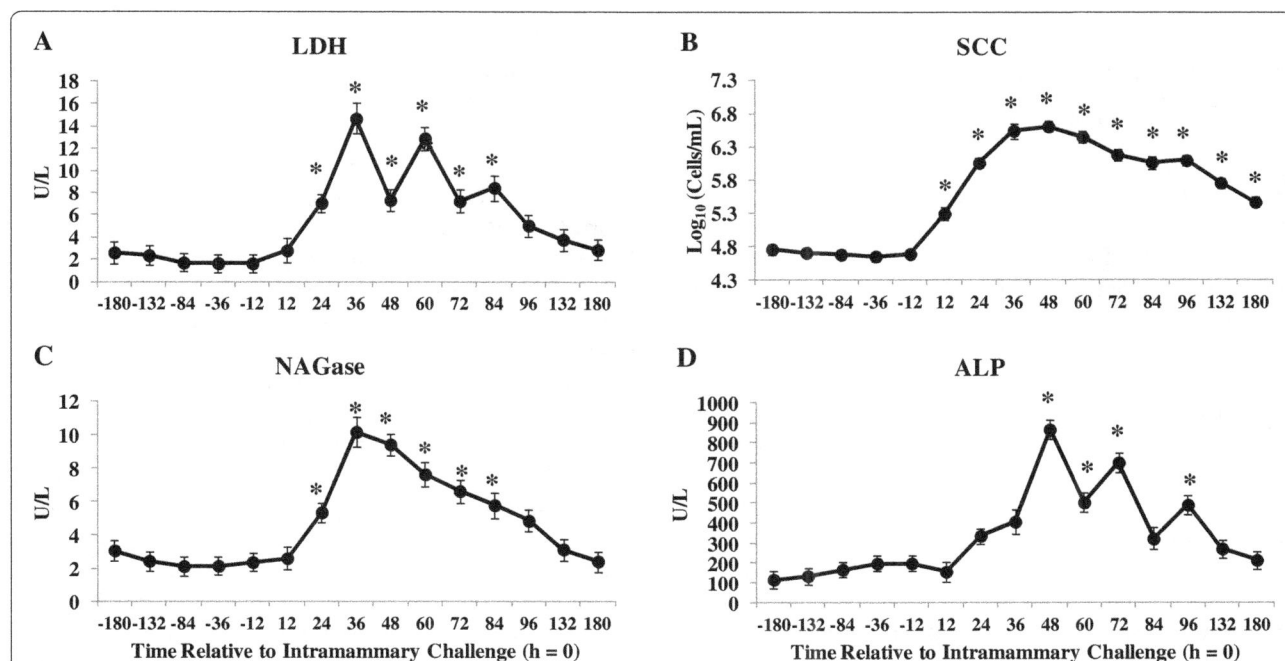

Figure 2 Composite milk concentrations of lactate dehydrogenase (LDH; A), somatic cell count (SCC; B), N-acetyl-β-D-glucosaminidase (NAGase; C) and alkaline phosphatase (ALP; D) at time points relative to intramammary challenge with *Escherichia coli* (h = 0) in 30 primiparous Holstein cows during early lactation. *Differences (P <0.05) when compared to h = 0.

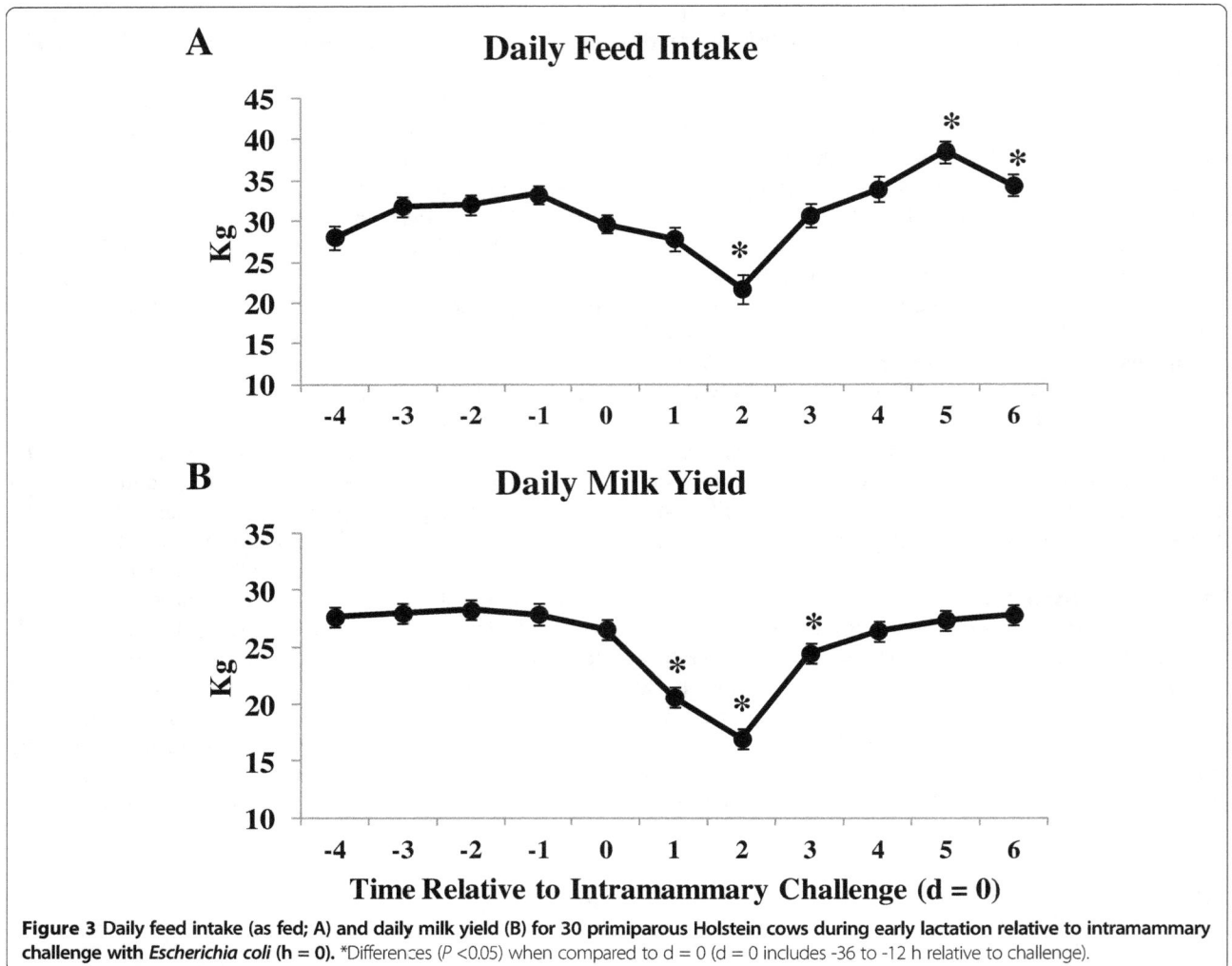

Figure 3 Daily feed intake (as fed; A) and daily milk yield (B) for 30 primiparous Holstein cows during early lactation relative to intramammary challenge with *Escherichia coli* (h = 0). *Differences (*P* <0.05) when compared to d = 0 (d = 0 includes -36 to -12 h relative to challenge).

for cows experimentally challenged with *E. coli* during early lactation [27]. Multiple local and systemic factors, i.e. production of cytokines and the changing hormonal environment, contribute to reduced feed intake and milk production observed during an IMI [32] and most likely explain the majority of variation in blood and milk metabolites for this study.

Changes in concentration of plasma NEFA (A), glucose (AB) and BHBA (C) relative to IMI challenge are shown in Figure 4. Blood samples collected prior to the morning feeding included -12, 12, 36, 60, 84, 132 and 180 h relative to IMI challenge. Plasma NEFA increased 11% by 18 h, after the morning feeding, but no other time points differed from pre-challenge values throughout the study period. Although changes in plasma NEFA may be primarily attributed to changes in feed intake, increases in plasma NEFA may also be associated with enhanced lipolysis in adipose tissue, regardless of changes in feed intake, which has been proposed to be the primary source of

elevated NEFA in blood during inflammation [33]. Steiger et al. [34] showed increased NEFA in blood following a prolonged low-dose intravenous (**IV**) lipopolysaccharide (**LPS**) infusion in non-lactating heifers. Similar results were reported after intramammary challenge with LPS for both primiparous and multiparous cows at 7 days in milk [14] and after IV infusion of LPS for multiparous cows in mid-lactation [16]. These results contradict those of Waldron et al. [15] where plasma NEFA decreased after intramammary LPS infusion for multiparous cows in early lactation as well as Moyes et al. [7] where no change in plasma NEFA was observed after IMI challenge with *Streptococcus uberis* for multiparous cows in mid-lactation. Our results, and those of others, indicate that plasma NEFA increase, regardless of stage of lactation or parity, in response to both live bacteria reported here (i.e. *E. coli*) and LPS administered either intramammary [16] or IV [34].

Plasma glucose was greater at 24 and 36 h when compared to -12 h (Figure 4B) where plasma glucose was

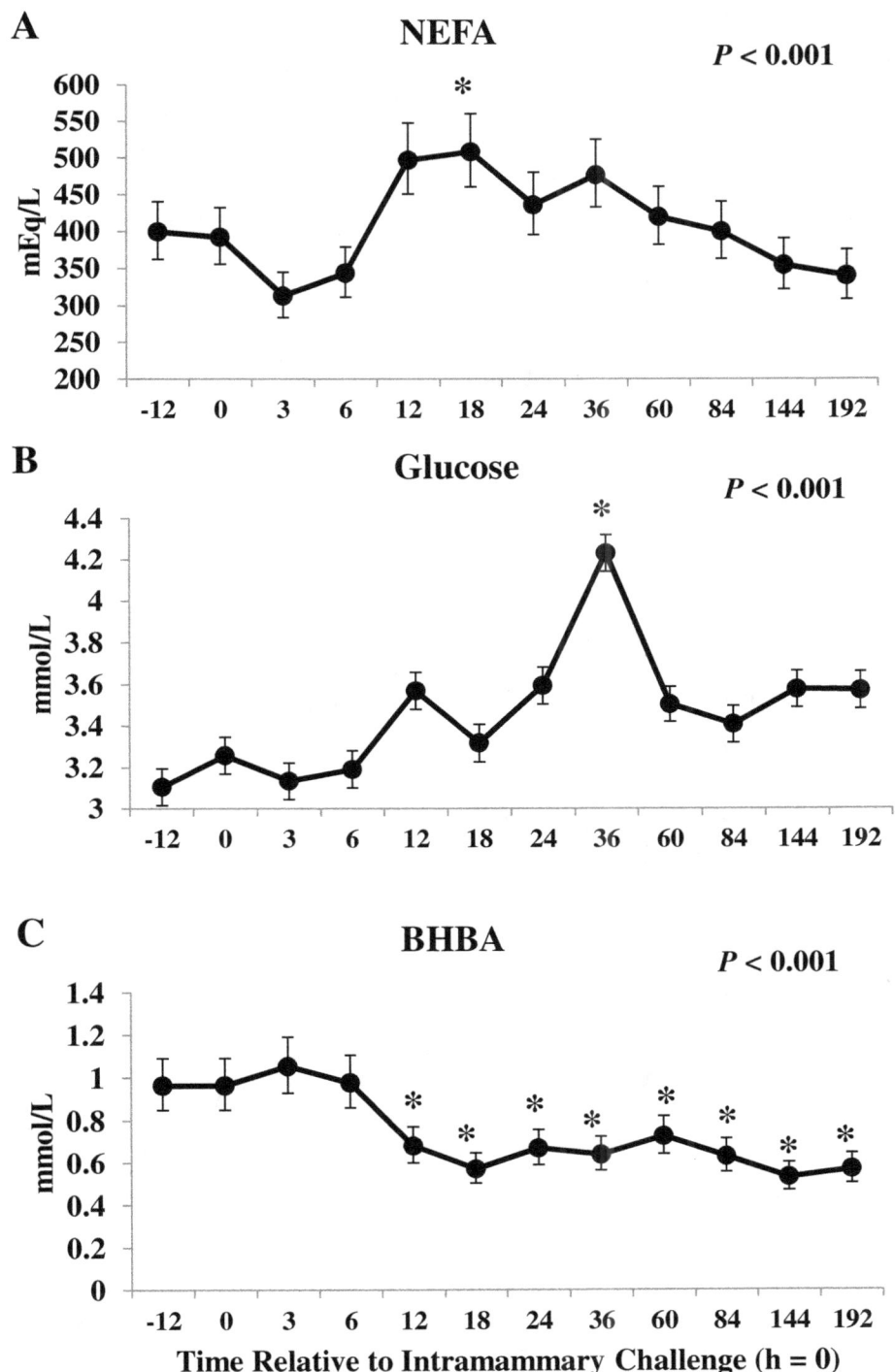

Figure 4 The effect of intramammary challenge with *Escherichia coli* (h = 0) on concentration of plasma non-esterified fatty acids (NEFA; A), glucose (B) and beta-hydroxybutyrate (BHBA; C) in 30 primiparous Holstein cows during early lactation. Samples collected at -12, 12, 36, 60, 84, 132 and 180 h were collected prior to morning feeding. *Differences ($P \le 0.05$) when compared to h = -12.

26.6% greater at 36 when compared to -12 h relative to IMI challenge. Increases in plasma glucose are primarily attributed to changes in feed intake and reduced demand for lactose synthesis in the mammary gland in response to IMI challenge. However, Steiger et al. [34] observed increases in plasma glucose after IV LPS infusion in

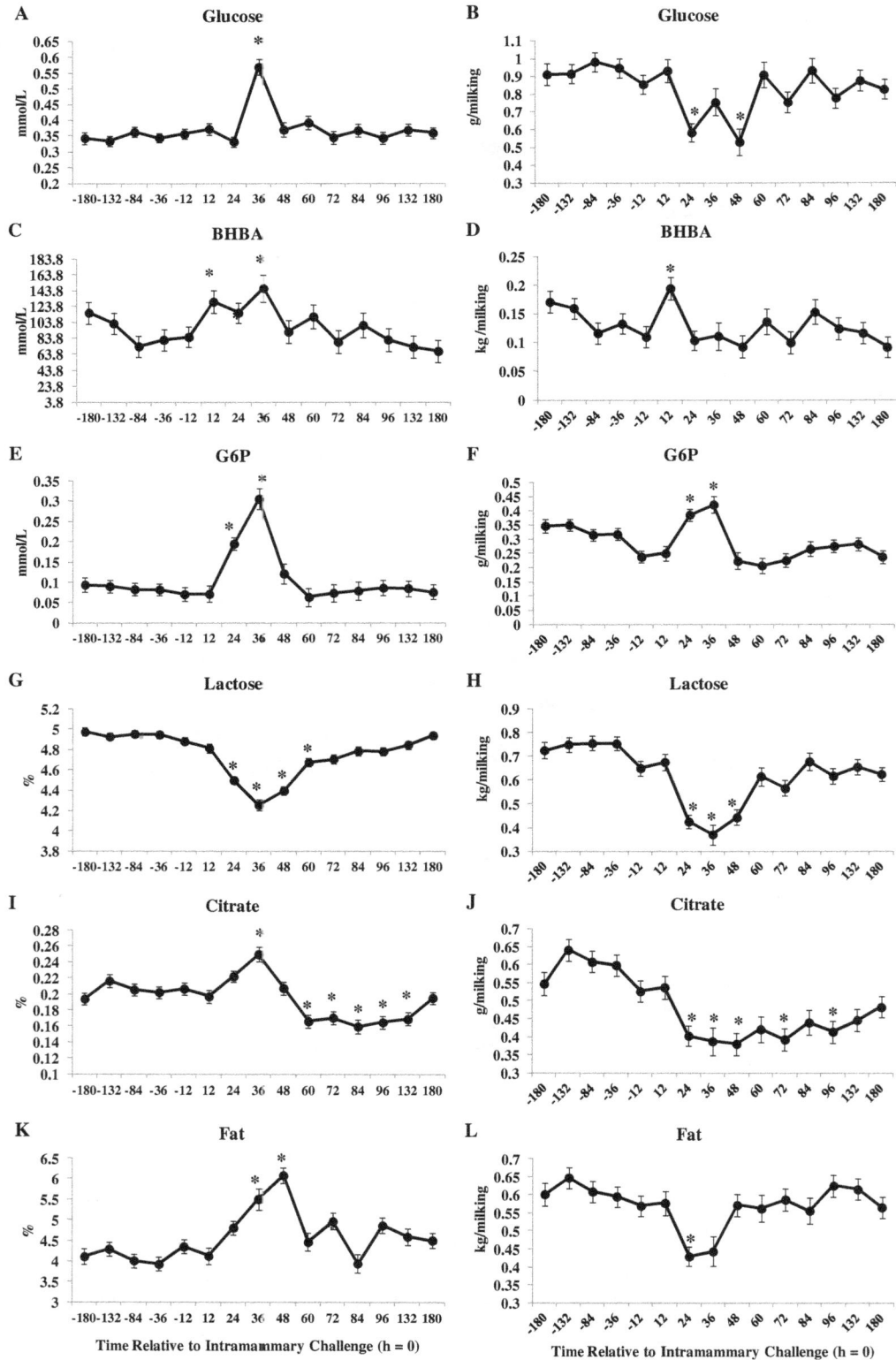

Figure 5 (See legend on next page.)

Changes in various metabolic parameters in blood and milk during experimental Escherichia coli...

201

(See figure on previous page.)
Figure 5 The effect of intramammary challenge with *Escherichia coli* (h = 0) on concentration and yield of milk glucose (A and B, respectively), BHBA (C and D, respectively), glucose-6-phosphate (G6P; E and F, respectively), lactose (G and H, respectively), citrate (I and J, respectively) and fat (K and L, respectively) in 30 primiparous Holstein cows during early lactation. Samples collected at -12, 12, 36, 60, 84, 132 and 180 h were collected prior to morning feeding. *Differences ($P \leq 0.05$) when compared to h = -12.

non-lactating heifers indicating that hyperglycemia is independent of changes in milk production.

The mechanisms regulating glucose homeostasis during IMI are unclear and the primary theories are 1) changes in feed intake, 2) increased circulating glucocorticoids observed during infection [16,35], 3) decreased lactose synthesis in the mammary gland, 4) increased hepatic lactate recycling via the Cori cycle [16,36], 5) increased glycogenolysis in peripheral tissues [34,37] and/or 6) increased hepatic gluconeogenesis [36]. Increases in glucocorticoids observed after infection are associated with increased adipose tissue lipolysis, increased hepatic gluconeogenesis and inhibition of insulin sensitivity in skeletal muscle [38]. Changes in plasma glucocorticoid concentrations were not assessed for this study and the contribution to changes in plasma glucose are unknown. Hyperglycemia has been reported in sheep [36] and in non-lactating dairy heifers [34] after inflammation and therefore decreases in milk lactose synthesis may not be the only factor explaining increases in plasma glucose at this time. Furthermore, glycogen stores are largely depleted during early lactation [39] and increased glycogenolysis unlikely explains changes in glucose supply based on transcriptional responses in liver for this study [12]. A down-regulation of key genes associated with hepatic gluconeogenesis was observed in liver tissue for this study by 24 h post-IMI challenge [12] including phosphoenolpyruvate carboxykinase 1 (*PCK1*; -8.2-fold change versus pre-IMI challenge) and glucose-6-phosphatase (*G6PC*; -1.7-fold change). However, an up-regulation of both lactate dehydrogenase A (*LDHA*; 1.2-fold change) and B (*LDHB*; 1.1-fold-change) were observed in liver at 24 h post-IMI challenge. Both *LDHA* and *LDHB* code for functional LDH, the enzyme responsible for the reversible conversion of lactate to pyruvate, and supports the theory of a potential increase in lactate recycling via the Cori cycle during IMI challenge [16,36].

Plasma BHBA decreased by 12 h post-IMI challenge when compared to pre-challenge levels (Figure 4C). Other studies have reported decreased circulating concentrations of BHBA during an intramammary challenge with LPS for lactating dairy cows [14,15,34]. During inflammation, decreases in BHBA in blood are a consequence of either 1) increased blood glucose; 2) changes in BHBA supply via reduced rumen motility [37,40]; 3) impairment of hepatic ketogenesis [34,41]; or 4) a combination of the above. Transcriptional profiling of liver tissue indicated a down-regulation of genes associated with hepatic ketogenesis [12] including 3-hydroxy-3-methylglutaryl-CoA synthase

1 (*HMGCS1*; -2.5-fold change) and *HMGCS2* (-28.0-fold change) and may partly explain lower plasma BHBA after IMI challenge for this study. Concentration of milk BHBA were elevated at 12 and 36 h post-IMI challenge and returned to pre-challenge levels by 48 h (Figure 5C). Furthermore, yield of milk BHBA increased at 12 h post-IMI challenge (Figure 5D) and indicates that the lower BHBA observed in blood is due to an increase transfer into milk after IMI challenge.

Changes in fglu, BHBA, G6P, lactose, citrate and fat in milk relative to IMI challenge are shown in Figure 5. In addition, milk protein % decreased during IMI challenge (data not shown). Yield of fglu, lactose, fat, protein (data not shown) and citrate were lower whereas yield of G6P and BHBA were higher after IMI challenge. Decreases in milk component yield are mostly explained by lower milk yield (-36% by d = 2 post-challenge) observed when compared to the pre-challenge period. Transcriptomic-level profiling of mammary quarters at 24 h post-IMI challenge revealed no changes in key genes associated with glucose metabolism and utilization between challenged and unchallenged quarters [19] and therefore cannot support changes in major milk components discussed in this study. Concentration of citrate was greater by 36 h post-IMI challenge followed by a decrease in citrate to concentrations below those observed at h = -12 (Figure 5I) whereas citrate yield was lower by 24 h post-IMI challenge (Figure 5J). Concentration of milk citrate was also shown to decrease in LPS challenged quarters from lactating dairy cows [42]. Milk citrate, a marker of mitochondrial metabolism in the mammary gland [43], induces the ferric citrate transport system and is competing with lactoferrin for iron [42]. Lactoferrin contributes to host defense by binding iron thereby reducing availability of iron to invading bacteria [44] and lower yield of citrate may indicate an increase in the iron-binding capacity of bovine lactoferrin [45].

Concentration and yield of milk lactose decreased by 24 h after IMI challenge and returned to pre-challenge levels by 72 h (Figures 5G and H, respectively). Lactose is the major osmole in milk and decreases in milk yield and the synthesis milk components, such as lactose and protein, most likely explain the majority of changes in lactose during IMI challenge. However, lower milk lactose yield may be attributed to lower yield of fglu (Figure 5B where the yield of fglu (Figure 5B) was lower at 24 and 48 h when compared to -12 h relative to IMI challenge. Both concentration and yield of G6P (Figure 5E and F, respectively) rose after IMI challenge. Elevated levels of G6P by 24 h post-IMI challenge may signify increased conversion of fglu from lactose synthesis and towards the synthesis of G6P. However, this is not supported by the transcription-level profiling in mammary tissue at 24-h after IMI challenge [19]. Glucose-6-phosphate may serve as a substrate for the pentose phosphate pathway for the production of reducing equivalents used for several anabolic processes [24].

Conclusions

Although drops in feed intake and milk yield are major contributors to changes in the metabolic response in blood and milk, the early rise in plasma NEFA during IMI challenge with *E. coli* may be partly attributed to increased adipose tissue lipolysis. Lower plasma BHBA may be associated with increase transfer into milk. We are the first to characterize changes in fglu and G6P in milk during IMI challenge. Lower yield of milk lactose may be attributed to lower yield of fglu. Higher G6P yield after IMI challenge may signify increased conversion of fglu to G6P. Results identify the metabolic response of various parameters in blood and milk and characterize the changes in fglu and G6P after IMI challenge with *E. coli* for cows in early lactation that may partly explain the partitioning of nutrients and changes in milk components in dairy cows with mastitis during early lactation. Future research is needed to determine how i.e. stage of lactation, parity, bacteria alter these metabolic changes that may help identify risk factors for the development, severity and duration of mastitis for dairy cows during lactation.

Competing interests
The authors declare that they have no competing interests.

Authors' contributions
KMM performed statistical analysis and interpretation and writing of the manuscript. TL provided financial support, analyzed all blood and milk parameters and contributed to the interpretation and writing of the manuscript. PS contributed financial support, design of the experimental model, acquisition of data and contributed to the interpretation and writing of the manuscript. KLI contributed financial support and contributed to the interpretation and writing of the manuscript. All authors read and approved the final manuscript.

Acknowledgements
The authors would like to thank and acknowledge the staff at the Department of Animal Science's Dairy Cattle facility, Aarhus University, and Martin Bjerring, Jens Clausen, and Hanne Møller Purup for their excellent technical assistance and analytical skills. This study was partly funded by the European Commission, within the 6th Framework Program (contract No. FOOD-CT-2006-016250) and the BIOSENS project granted by the Danish Ministry of Food, Agriculture and Fisheries (Innovations Law), Lattec I/S, the Danish Cattle Association, and the Faculty of Science and Technology, Aarhus University.

Author details
[1]Department of Animal and Avian Sciences, University of Maryland, 142 Animal Sciences Building, MD 20742-2311, 20910 College Park, MD, USA. [2]Department of Animal Science, Faculty of Science and Technology, Aarhus University, Tjele 8830, Denmark.

References
1. Ingvartsen KL, Moyes KM: Nutrition, immune function and health of dairy cattle. *Animal* 2013, 7(Suppl 1):112–122.
2. Bar D, Tauer LW, Bennett G, Gonzalez RN, Hertl JA, Schukken YH, Schulte HF, Welcome FL, Grohn YT: The cost of generic clinical mastitis in dairy cows as estimated by using dynamic programming. *J Dairy Sci* 2008, 91:2205–2214.
3. Green MJ, Green LE, Medley GF, Schukken YH, Bradley AJ: Influence of dry period bacterial intramammary infection on clinical mastitis in dairy cows. *J Dairy Sci* 2002, 85:2589–2599.
4. Schukken YH, Gunther J, Fitzpatrick J, Fontaine MC, Goetze L, Holst O, Leigh J, Petzl W, Schuberth HJ, Sipka A, Smith DG, Quesnell R, Watts J, Yancey R, Zerbe H, Gurjar A, Zadoks RN, Seyfert HM, members of the Pfizer mastitis research consortium: Host-response patterns of intramammary infections in dairy cows. *Vet Immunol Immunopathol* 2011, 144:270–289.
5. Ballou MA: Inflammation: Role in the etiology and pathophysiology of clinical mastitis in dairy cows. *J Anim Sci* 2011, 10:1466–1478.
6. Bannerman DD, Paape MJ, Goff JP, Kimura K, Lippolis JD, Hope JC: Innate immune response to intramammary infection with *Serratia marcescens* and *Streptococcus uberis*. *Vet Res* 2004, 35:681–700.
7. Moyes KM, Drackley JK, Salak-Johnson JL, Morin DE, Hope JC, Loor JJ: Dietary-induced negative energy balance has minimal effects on innate immunity during a *Streptococcus uberis* mastitis challenge in dairy cows during mid-lactation. *J Dairy Sci* 2009, 92:4301–4316.
8. Zarrin M, Wellnitz O, Van Dorland HA, Bruckmaier RM: Induced hyperketonemia affects the mammary immune response during lipopolysaccharide challenge in dairy cows. *J Dairy Sci* 2014, 97:330–339.
9. Loor JJ, Moyes KM, Bionaz M: Functional adaptations of the transcriptome to mastitis-causing pathogens: the mammary gland and beyond. *J Mammary Gland Biol Neoplasia* 2011, 16:305–322.
10. Moyes KM, Drackley JK, Morin DE, Bionaz M, Rodriguez-Zas SL, Everts RE, Lewin HA, Loor JJ: Gene network and pathway analysis of bovine mammary tissue challenged with Streptococcus uberis reveals induction of cell proliferation and inhibition of PPARgamma signaling as potential mechanism for the negative relationships between immune response and lipid metabolism. *BMC Genomics* 2009, 10:542–571.
11. Vels L, Rontved CM, Bjerring M, Ingvartsen KL: Cytokine and acute phase protein gene expression in repeated liver biopsies of dairy cows with a lipopolysaccharide-induced mastitis. *J Dairy Sci* 2009, 92:922–934.
12. Jorgensen HB, Buitenhuis B, Rontved CM, Jiang L, Ingvartsen KL, Sorensen P: Transcriptional profiling of the bovine hepatic response to experimentally induced *E. coli* mastitis. *Physiol Genomics* 2012, 44:595–606.
13. Jiang L, Sorensen P, Rontved C, Vels L, Ingvartsen KL: Gene expression profiling of liver from dairy cows treated intra-mammary with lipopolysaccharide. *BMC Genomics* 2008, 9:443.
14. Graugnard DE, Moyes KM, Trevisi E, Khan MJ, Keisler D, Drackley JK, Bertoni G, Loor JJ: Liver lipid content and inflammometabolic indices in peripartal dairy cows are altered in response to prepartal energy intake and postpartal intramammary inflammatory challenge. *J Dairy Sci* 2013, 96:918–935.
15. Waldron MR, Kulick AE, Bell AW, Overton TR: Acute experimental mastitis is not causal toward the development of energy-related metabolic disorders in early postpartum dairy cows. *J Dairy Sci* 2006, 89:596–610.
16. Waldron MR, Nishida T, Nonnecke BJ, Overton TR: Effect of lipopolysaccharide on indices of peripheral and hepatic metabolism in lactating cows. *J Dairy Sci* 2003, 86:3447–3459.
17. Annison EF, Linzell JL, West CE: Mammary and whole animal metabolism of glucose and fatty acids in fasting lactating goats. *J Physiol* 1968, 197:445–459.
18. Scott RA, Bauman DE, Clark JH: Cellular gluconeogenesis by lactating bovine mammary tissue. *J Dairy Sci* 1976, 50:50–56.
19. Buitenhuis B, Rontved CM, Edwards SM, Ingvartsen KL, Sorensen P: In depth analysis of genes and pathways of the mammary gland involved in the pathogenesis of bovine *Escherichia coli*-mastitis. *BMC Genomics* 2011, 12:130–140.
20. Fogsgaard KK, Rontved CM, Sorensen P, Herskin MS: Sickness behavior in dairy cows during *Escherichia coli* mastitis. *J Dairy Sci* 2012, 95:630–638.
21. Bjerre-Harpøth V, Friggens NC, Thorup VM, Larsen T, Damgaard BM, Ingvartsen KL, Moyes KM: Metabolic and production profiles of dairy cows

in response to decreased nutrient density to increase physiological imbalance at different stages of lactation. *J Dairy Sci* 2012, **95**:2362–2380.

22. Larsen T: **Determination of lactate dehydrogenase (LDH) activity in milk by a fluorometric assay.** *J Dairy Res* 2005, **72**:209–216.

23. Larsen T, Nielsen NI: **Fluorometric determination of beta-hydroxybutyrate in milk and blood plasma.** *J Dairy Sci* 2005, **88**:2004–2009.

24. Larsen T: **Fluorometric determination of free glucose and glucose-6-phosphate in cow milk and other opaque matrices.** *J Food Chem* 2014, Accepted.

25. SAS User's Guide: *Statistics V93E.* Cary, NC: SAS Inst., Inc; 2012.

26. Ma JL, Zhu YH, Zhang L, Zhuge ZY, Liu PQ, Yan XD, Gao HS, Wang JF: **Serum concentration and mRNA expression in milk somatic cells of toll-like receptor 2, toll-like receptor 4, and cytokines in dairy cows following intramammary inoculation with *Escherichia coli*.** *J Dairy Sci* 2011, **94**:5903–5912.

27. Scaletti RW, Harmon RJ: **Effect of dietary copper source on response to coliform mastitis in dairy cows.** *J Dairy Sci* 2012, **95**:654–662.

28. Babaei H, Mansouri-Najand L, Molaei MM, Kheradmand A, Sharifan M: **Assessment of lactate dehydrogenase, alkaline phosphatase and aspartate aminotransferase activities in cow's milk as an indicator of subclinical mastitis.** *Vet Res Commun* 2007, **31**:419–425.

29. Chagunda MG, Larsen T, Bjerring M, Ingvartsen KL: **L-lactate dehydrogenase and N-acetyl-beta-D-glucosaminidase activities in bovine milk as indicators of non-specific mastitis.** *J Dairy Res* 2006, **73**:431–440.

30. Friggens NC, Chagunda MG, Bjerring M, Ridder C, Hojsgaard S, Larsen T: **Estimating degree of mastitis from time-series measurements in milk: a test of a model based on lactate dehydrogenase measurements.** *J Dairy Sci* 2007, **90**:5415–5427.

31. Janovick NA, Drackley JK: **Prepartum dietary management of energy intake affects postpartum intake and lactation performance by primiparous and multiparous Holstein cows.** *J Dairy Sci* 2010, **93**:3086–3102.

32. Ingvartsen KL, Andersen JB: **Integration of metabolism and intake regulation: a review focusing on periparturient animals.** *J Dairy Sci* 2000, **83**:1573–1597.

33. Zu L, He J, Jiang H, Xu C, Pu S, Xu G: **Bacterial endotoxin stimulates adipose lipolysis via toll-like receptor 4 and extracellular signal-regulated kinase pathway.** *J Biol Chem* 2009, **284**:5915–5926.

34. Steiger M, Senn M, Altreuther G, Werling D, Sutter F, Kreuzer M, Langhans W: **Effect of a prolonged low-dose lipopolysaccharide infusion on feed intake and metabolism in heifers.** *J Anim Sci* 1999, **77**:2523–2532.

35. Jamieson AM, Yu S, Annicelli CH, Medzhitov R: **Influenza virus-induced glucocorticoids compromise innate host defense against a secondary bacterial infection.** *Cell Host Microbe* 2010, **7**:103–114.

36. Naylor JM, Kronfeld DS: **In vivo studies of hypoglycemia and lactic acidosis in endotoxic shock.** *Am J Physiol* 1985, **248**:E309–E316.

37. Lohuis JA, Verheijden JH, Burvenich C, Van Miert AS: **Pathophysiological effects of endotoxins in ruminants. 2. Metabolic aspects.** *Vet Q* 1988, **10**:117–125.

38. Park SY, Bae JH, Cho YS: **Cortisone induces insulin resistance in C2C12 myotubes through activation of 11beta-hydroxysteroid dehydrogenase 1 and autocrinal regulation.** *Cell Biochem Funct* 2014, **32**:249–257.

39. Drackley JK: **Biology of dairy cows during the transition period: the final frontier?** *J Dairy Sci* 1999, **82**:2259–2273.

40. Huhtanen P, Miettinen H, Ylinen M: **Effect of increasing ruminal butyrate on milk yield and blood constituents in dairy cows fed a grass silage-based diet.** *J Dairy Sci* 1993, **76**:1114–1124.

41. Kaminski MV Jr, Neufeld HA, Pace JG: **Effect of inflammatory and noninflammatory stress on plasma ketone bodies and free fatty acids and on glucagon and insulin in peripheral and portal blood.** *Inflammation* 1979, **3**:289–294.

42. Hyvönen P, Haarahiltunen T, Lehtolainen T, Heikkinen J, Isomäki R, Pyörälä S: **Concentrations of bovine lactoferrin and citrate in milk during experimental endotoxin mastitis in early- versus late-lactating dairy cows.** *J Dairy Res* 2010, **77**:474–480.

43. Faulkner A, Peaker M: **Reviews of the progress of dairy science: secretion of citrate into milk.** *J Dairy Res* 1982, **49**:159–169.

44. Brock JH: **The physiology of lactoferrin.** *Biochem Cell Biol* 2002, **80**:1–6.

45. Bishop JG, Schanbacher FL, Ferguson LC, Smith KL: **In vitro growth inhibition of mastitis-causing coliform bacteria by bovine apo-lactoferrin and reversal of inhibition by citrate and high concentrations of apo-lactoferin.** *Infect Immun* 1976, **14**:911–918.

22

Transforming growth factor β signaling in uterine development and function

Qinglei Li

Abstract

Transforming growth factor β (TGFβ) superfamily is evolutionarily conserved and plays fundamental roles in cell growth and differentiation. Mounting evidence supports its important role in female reproduction and development. TGFBs1-3 are founding members of this growth factor family, however, the *in vivo* function of TGFβ signaling in the uterus remains poorly defined. By drawing on mouse and human studies as a main source, this review focuses on the recent progress on understanding TGFβ signaling in the uterus. The review also considers the involvement of dysregulated TGFβ signaling in pathological conditions that cause pregnancy loss and fertility problems in women.

Keywords: Decidualization, Development, Embryonic development, Implantation, Myometrium, Pregnancy, Transforming growth factor β, Uterus

Introduction

Transforming growth factor β (TGFβ) superfamily proteins are versatile and fundamental regulators in metazoans. The TGFβ signal transduction pathway has been extensively studied. The application of mouse genetic approaches has catalyzed the identification of the roles of core signaling components of TGFβ superfamily members in reproductive processes. Recent studies using tissue/cell-specific knockout approaches represent a milestone towards understanding the *in vivo* function of TGFβ superfamily signaling in reproduction and development. These studies have yielded new insights into this growth factor superfamily in uterine development, function, and diseases. This review will focus on TGFβ signaling in the uterus, primarily using results from studies with mice and humans.

TGFβ superfamily
Core components of the TGFβ signaling pathway
Core components of the TGFβ signaling pathway consist of ligands, receptors, and SMA and MAD (mother against decapentaplegic)-related proteins (SMAD). TGFβ ligands

bind to their receptors and impinge on SMADs to activate gene transcription. TGFβ superfamily ligands include TGFβs, activins, inhibins, bone morphogenetic proteins (BMPs), growth differentiation factors (GDFs), anti-Müllerian hormone (AMH), and nodal growth differentiation factor (NODAL). Seven type I (i.e., ACVRL1, ACVR1, BMPR1A, ACVR1B, TGFBR1, BMPR1B, and ACVR1C) and five type II receptors (i.e., TGFBR2, ACVR2, ACVR2B, BMPR2, and AMHR2) have been identified [1-4]. SMADs are intracellular transducers. In mammalian species, eight SMAD proteins have been identified and are classified into receptor-regulated SMADs (R-SMADs; SMAD1, 2, 3, 5, and 8), common SMAD (Co-SMAD), and inhibitory SMADs (I-SMADs; SMAD6 and SMAD7). R-SMADs are tethered by SMAD anchor for receptor activation (SARA) [5]. In general, SMAD1/5/8 mediate BMP signaling, whereas SMAD2/3 mediate TGFβ and activin signaling. SMAD6 and SMAD7 can bind type I receptors and inhibit TGFβ and/or BMP signaling [6,7]. A plethora of ligands versus a fixed number of receptors and SMADs suggests the usage of shared receptor(s) and SMAD cell signaling molecules in this system.

TGFβ signaling paradigm: canonical versus non-canonical pathway
To initiate signal transduction, a ligand forms a hetero-meric type II and type I receptor complex, where the

Correspondence: qli@cvm.tamu.edu
Department of Veterinary Integrative Biosciences, College of Veterinary Medicine and Biomedical Sciences, Texas A&M University, College Station, TX 77843, USA

constitutively active type II receptor phosphorylates type I receptor at the glycine and serine (GS) domain. Subsequent phosphorylation of R-SMADs by the type I receptor and formation and translocation of R-SMAD-SMAD4 complex to the nucleus are critical steps for gene regulation [2,8-10]. Activation of transcription is achieved by SMAD binding to the consensus DNA binding sequence (AGAC) termed SMAD binding element (SBE) [11,12], in concert with co-activators and co-repressors. Of note, SMADs can promote chromatin remodeling and histone modification, which facilitates gene transcription by recruiting co-regulators to the promoters of genes of preference [13].

TGFβ signals through both SMAD-dependent (i.e., canonical) and SMAD-independent (i.e., non-canonical) pathways in a contextually dependent manner [2,8,14-16] (Figure 1). The non-canonical pathways serve to integrate signaling from other signaling cascades, resulting in a quantitative output in a given context. Davis and colleagues [17] have recently suggested the presence of microRNA (miRNA)-mediated non-canonical pathway, where TGFβ signaling promotes the biosynthesis of a subset of miRNAs via interactions between R-SMADs and a consensus RNA sequence of miRNAs within the DROSHA (drosha, ribonuclease type III) complex [17-19]. Thus, this type of non-canonical signaling requires R-SMADs but not SMAD4. Multiple regulatory layers including ligand traps (e.g., follistatin), inhibitory SMADs, and interactive pathways exist to determine the signaling output and precisely control TGFβ signaling activity [4,8,20-23]. For instance, the linker region of R-SMADs is subject to the phosphorylation modification by mitogen-activated protein kinases (MAPKs) [24].

Therefore, the variable responses triggered by this growth factor superfamily and the complex signaling circuitries within a given cell population underscore the importance of a fine-tuned TGFβ signaling system at both the cellular and systemic levels.

TGFβ superfamily signaling regulates female reproduction

TGFβ superfamily is evolutionarily conserved and plays fundamental roles in cell growth and differentiation. The signal transduction and biological functions of this signaling pathway have been extensively investigated [2,4,8,9,25]. TGFβ superfamily signaling is essential for female reproduction (Figure 2), and dysregulation of TGFβ signaling may cause catastrophic consequences, leading to reproductive diseases and cancers [26-33].

Recent studies have uncovered the roles of key receptors and intracellular SMADs of this pathway in female reproduction. *Smad1* and *Smad5* null mice are embryonically lethal, but *Smad8* null mice are viable and fertile [34,35]. SMAD1/5 and ALK3/6 act as tumor suppressors with functional redundancy in the ovary [27,29]. *Smad3*[Δex8] mice demonstrate impaired follicular growth and atresia, altered ovarian cell differentiation, and defective granulosa cell response to follicle-stimulating hormone (FSH) [36,37]. We have shown that SMAD2 and SMAD3 are redundantly required to maintain normal fertility and ovarian function [38]. Disruption of *Smad4* signaling in ovarian granulosa cells leads to premature luteinization [39]. However, oocyte-specific knockout of *Smad4* causes minimal fertility defects in mice [40]. SMAD7 mediates TGFβ-induced apoptosis [41] and antagonizes key TGFβ signaling in ovarian granulosa cells [42], suggesting inhibitory

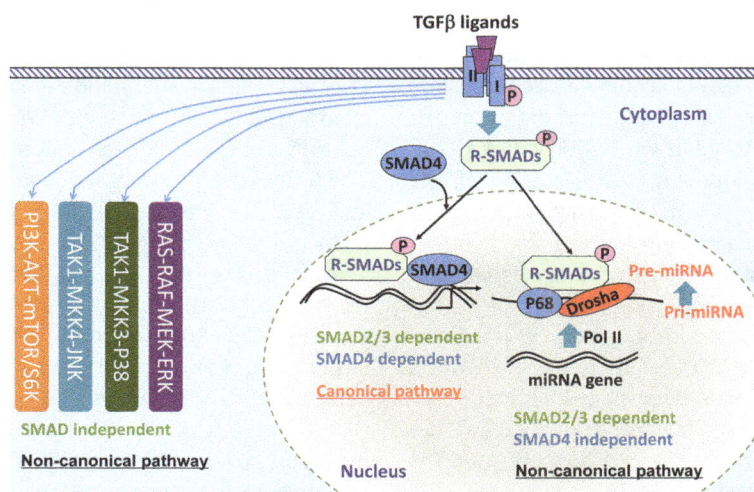

Figure 1 Canonical and non-canonical TGFβ signaling. In the canonical pathway, TGFβ ligands bind to serine/threonine kinase type II and type I receptors and phosphorylate R-SMADs, which form heteromeric complexes with SMAD4 and translocate into the nucleus to regulate gene transcription. The non-canonical pathway generally refers to the SMAD-independent pathway such as PI3K-AKT, ERK1/2, p38, and JNK pathways. Recent studies have identified an "R-SMAD-dependent but SMAD4-independent" non-canonical pathway that regulates miRNA maturation.

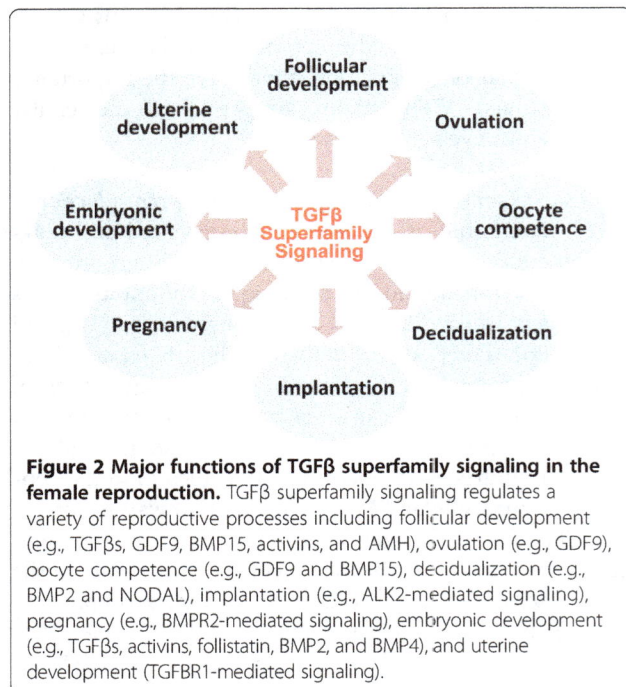

Figure 2 Major functions of TGFβ superfamily signaling in the female reproduction. TGFβ superfamily signaling regulates a variety of reproductive processes including follicular development (e.g., TGFβs, GDF9, BMP15, activins, and AMH), ovulation (e.g., GDF9), oocyte competence (e.g., GDF9 and BMP15), decidualization (e.g., BMP2 and NODAL), implantation (e.g., ALK2-mediated signaling), pregnancy (e.g., BMPR2-mediated signaling), embryonic development (e.g., TGFβs, activins, follistatin, BMP2, and BMP4), and uterine development (TGFBR1-mediated signaling).

SMADs are potentially novel regulators of ovarian function. Recent studies show that TGFBR1 is indispensable for female reproductive tract development [43,44], while ALK2 and BMPR2 are required for uterine decidualization and/or pregnancy maintenance [45,46].

TGFβ signaling in uterine development

The uterus develops from the Müllerian duct, which forms at embryonic day E11.75 in mice [47]. Uterine mesenchymal cells remain randomly oriented and undifferentiated until after birth. Between birth and postnatal day 3, circular and longitudinal myometrial layers are differentiated from the mesenchyme [48]. The uterus acquires basic layers and structures by postnatal day 15 [48,49]. Maturation of the myometrium continues into adulthood. Mechanisms controlling myometrial development are poorly defined. Wingless-type MMTV integration site family (Wnt)7a null females demonstrate defects in reproductive tract formation, suggesting a critical role of Wnt/β catenin signaling in myometrial development [50-53].

Myometrial contractility is critical for successful pregnancy and labor. The myometrial cells transform from a quiescent to a contractile phenotype trigged by the decline of progesterone levels during late pregnancy. What has long puzzled scientists is how this transformation occurs during pregnancy, and how myometrial development and function are coordinately regulated. Uterine contraction is controlled by hormonal, cellular, and molecular signals [54-65]. Recent studies have discovered that miRNAs are key regulators of contraction-associated genes and

suppressors including oxytocin receptor (*Oxtr*), cyclooxygenase 2 (*Cox2*), connexin 43 (*Cx43*), zinc finger E-box binding homeobox 1 (*Zeb1*), and *Zeb2* [65,66]. However, signaling pathways that control the development of morphologically normal and functionally competent myometrium are poorly understood.

TGFβ signaling plays a pleiotropic role in fundamental cellular and developmental events [2,3,8]. Using a *Tgfbr1* conditional knockout (cKO) mouse model created using anti-Müllerian hormone receptor type 2 (*Amhr2*)-Cre, we have shown that TGFβ signaling is essential for smooth muscle development in the female reproductive tract [43,44]. The female mice develop a striking oviductal phenotype that includes a diverticulum. The *Tgfbr1* cKO mice are infertile and embryos are unable to be transported to the uterus due to the presence of the physical barrier of oviductal diverticula [43]. Meanwhile, disrupted uterine smooth muscle formation is another prominent feature in these mice, which is associated with a developmental failure of the myometrium during early postnatal uterine development [44]. However, the expression of the majority of smooth muscle genes in the uterus of the conditional knockout mice does not significantly differ from that of controls, suggesting that the developmental abnormality might not be a direct result of intrinsic deficiency in smooth muscle cell differentiation. Our studies point to the contributions of reduced deposition of extracellular matrix proteins, derailed signaling of platelet-derived growth factors, and potentially altered migration of uterine cells during a critical time window of development [44]. The *Tgfbr1* cKO mouse model can be further exploited to understand the pathogenesis of myometrium-associated diseases, such as adenomyosis that is present in these mice [44].

TGFβ signaling and uterine function

Pre-implantation embryonic development refers to a period from fertilization to blastocyst implantation, which requires coordinated expression of maternal and embryonic genes. The fertilized egg undergoes dynamic genetic programming and divisions to reach the blastocyst stage. The pluripotent inner cell mass of the blastocyst will develop into the embryonic proper, while the trophectoderm and the primitive endoderm form extra-embryonic tissues during development [67]. Preimplantation embryonic development largely depends on maternal proteins and transcripts before zygotic genome activation (ZGA), which initiates the expression of genes that are needed for continued development of the embryos. ZGA occurs at the two-cell stage in the mouse [68].

Blastocyst implantation is a complex event that is controlled by both intrinsic embryonic programs and extrinsic cues including hormonal and uterine signals. Implantation in the mouse can be divided into three phases: apposition, attachment, and penetration. Following attachment, uterine stromal cells extensively proliferate and differentiate into decidual cells (i.e., decidualization) [69]. The roles of steroid hormones, cytokines, growth factors, integrins, and angiogenic factors have been explored, and more recently, a number of novel genes/pathways underlying implantation have been identified. Several elegant reviews are available on these topics [70-72]. The important roles of embryonic TGFβ superfamily signaling in embryo development have been reviewed [3]. This article will focus on the role of maternal TGFβ signaling in implantation and embryonic development.

TGFBs1-3 are founding members of the TGFβ superfamily. The majority of currently available studies are confined to the identification of tissue/cell-specific expression of TGFBs and *in vitro* analysis of the ligand function. In the uterus, the *in vivo* role of TGFβ signaling remains elusive, partially because of the redundancy of the ligands [73,74] and the lack of appropriate animal models as a result of the embryonic lethality in mice lacking TGFβ ligands. TGFB1 is involved in preimplantation development and yolk sac vasculogenesis/hematopoiesis [75]. To allow the *Tgfb1* null mice survive to reproductive age, they were bred onto the severe combined immunodeficiency (*SCID*) background [76]. Although the uterus of *Tgfb1* mutant mice appears to be morphologically normal [76], embryos are arrested in the morula stage.

An *in vitro* model has been used to determine the effect of growth factors on preimplantation development, and the results showed that TGFB1 or epidermal growth factor (EGF) dramatically improves the inferior development of singly cultured embryos between eight-cell/morula and blastocyst stages. This study suggests that embryo and/or reproductive tract-derived growth factors are involved in the development of preimplantation embryos [77]. *In vitro* treatment of preimplantation stage embryos with TGFB1 increases total numbers of cells in expanded and hatching blastocysts [78]. Furthermore, TGFB1-promoted *in vitro* blastocyst outgrowth is blocked by an antibody directed to parathyroid hormone-related protein [79], which suggests the involvement of parathyroid hormone-related protein in mediating the effect of TGFB1 on blastocyst outgrowth. In addition, TGFB1 increases the *in vitro* expression of oncofetal fibronectin, an anchoring trophoblast marker, indicating a potential role of TGFβ in trophoblast adhesion during implantation [80]. TGFB1 also inhibits human trophoblast cell invasion, at least partially, by promoting the production of tissue inhibitor of metalloproteinases (TIMP) [81]. An elegant study showed that maternal TGFB1 can cross the placenta and rescue the developmental defects of *Tgfb1* null embryos, leading to perinatal survival of these mice [82]. As further evidence, both maternal and fetal TGFB1 may act to maintain pregnancy [83].

TGFβ signaling and uterine diseases
Uterine fibroids
Leiomyoma, generally known as uterine fibroid, is a benign tumor arising from the myometrium (i.e., smooth muscle layers). Although leiomyoma is commonly benign, it could be the cause of fertility disorders and morbidity and mortality in women [84].

Increasing lines of evidence point to the involvement of TGFβ signaling in the development of leiomyoma. It has been shown that the expression of TGFBs and receptors is elevated in leiomyomata versus unaffected myometrium [85]. Among all the three TGFβ isoforms, TGFB3 seems to play a major role in leiomyoma development by promoting cell growth and fibrogenic process [86]. *Tgfb3* transcript and protein levels are elevated in human leiomyoma cells, compared with myometrial cells in two-dimensional (2D) and 3D cultures [87-90]. In a 3D culture system, a higher level of TGFB3 and SMAD2/3 activation is present in the leiomyoma cells versus myometrial cells [87,89]. However, it does not support that connective tissue growth factor 2 (CCN2/CTGF) is a major mediator of TGFβ action in leiomyoma tissues [91].

Although a link between overexpression of TGFBs and leiomyoma has been recognized, the precise mechanisms of TGFβ signaling in leiomyoma are largely unknown. It has been demonstrated that TGFB1-stimulated expression of fibromodulin may contribute to the fibrotic properties of leiomyoma [92]. Moreover, treatment of myometrial cells with TGFB3 promotes the expression of ECM components such as collagen 1A1 (COL1A1), fibronectin 1 (FN1), and versican, but reduces the expression of those associated with ECM degradation [88,93]. Thus, TGFβ signaling induces molecular changes that facilitate leiomyoma formation. Consistent with the enhanced TGFβ signaling in the etiology of leiomyoma, a number of substances or drugs, such as genistein [94], relaxin [95], halofuginone [96], asoprisnil [97], gonadotropin-releasing hormone-analogs (GnRH-a), and tibolone [98] may influence leiomyoma development via affecting TGFβ signaling. For the therapeutic purpose, an ideal drug is one that only targets TGFβ signaling in the leiomyoma cells but not normal myometrial cells. In this vein, asoprisnil, a steroidal 11β-benzaldoxime-substituted selective progesterone receptor modulator (SPRM), targets TGFB3 and TGFBR2 in leiomyoma cells but not normal myometrial cells [97], providing a potentially effective treatment option for leiomyoma. The high levels of leiomyoma-secreted TGFBs, in turn, may compromise uterine function of the patients. For example, by producing excessive amount of TGFB3, leiomyoma antagonizes decidualization mediated by BMP2 [99].

Preeclampsia
Preeclampsia often occurs in pregnant women after the 20th week of gestation, characterized by hypertension and

proteinuria. The causes of preeclampsia are complex and beyond the scope of this review. It has been shown that plasma TGFB1 [100-104] and TGFB2 [105] levels are elevated in patients with preeclampsia. Experimental evidence also suggests that failure to downregulate the expression of TGFB3 during early gestation may cause trophoblast hypoinvasion and preeclampsia [106]. Interestingly, the levels of soluble endoglin, a transmembrane TGFβ co-receptor, are elevated in sera of women with preeclampsia, which may be associated with vascular complications and hypertension in these patients [107,108]. Based on these findings, TGFB proteins may serve as potential biomarkers for preeclampsia [105]. It is thus plausible that optimal TGFβ signaling activity is required to keep preeclampsia in check by maintaining normal trophoblast invasion during implantation and placentation. However, another study showed that TGFBs1-3 are not expressed in villous trophoblasts, and TGFB1 and TGFB3 are not expressed in the extravillous trophoblast either. The expression of TGFBs1-3 in the placenta is not altered in patients with preeclampsia [109]. Moreover, there are also reports indicating that concentrations of TGFB1 in serum are indistinguishable between patients with preeclampsia and normal controls [110-112]. In addition, the levels of activin A and inhibin A, but not inhibin B, are increased in patients with preeclampsia [113-116]. Thus, the role of TGFβ signaling in the pathophysiological events of preeclampsia awaits further elucidation.

Intrauterine growth restriction

Intrauterine growth restriction (IUGR), also called fetal growth restriction (FGR), refers to a complication of fetal growth during pregnancy. The estimated weight of the fetus with IUGR is often less than 90% of other fetuses at the same stage of pregnancy [117]. Circumstantial evidence indicates that TGFβ signaling is involved in the development of IUGR. Serum levels of TGFB1 in the IUGR fetus are lower [118]. TGFB2 is required for normal embryo growth, as supported by the fact that *Tgfb2* mutant fetuses weigh less than littermate controls [119]. Soluble endoglin levels are elevated in IUGR pregnancies [108], although it is debatable [120]. It has been shown that the higher expression of endoglin in IUGR pregnancies may be caused by placental hypoxia involving TGFB3 [121]. Mouse models for IUGR are valuable to study the mechanism of this pathological condition, which may have devastating effects on the pregnancy and newborns. Notably, *Nodal* knockout mice show diminished decidua basalis due to reduced proliferation and enhanced apoptosis as well as defects in placental development, resulting in IUGR and preterm fetal loss [122]. Conditional ablation of *Bmpr2* in the uterus causes defects in decidualization, trophoblast invasion, and vascularization, which are causes of IUGR in the pregnant females [46].

Endometrial hyperplasia

Endometrial hyperplasia is a pathological condition where endometrial cells undergo excessive proliferation [123]. Categories of endometrial hyperplasia include simple hyperplasia, simple atypical hyperplasia, complex hyperplasia, and complex atypical hyperplasia [124]. Endometrial hyperplasia is recognized as a premalignant lesion of endometrial carcinoma [125] and a potential cause of abnormal uterine bleeding and fertility disorders. The high prevalence of endometrial carcinoma is associated with atypical hyperplasia in women [126-128]. It has been reported that up to 29% of untreated complex atypical hyperplasia progresses to carcinoma [124]. Endometrial hyperplasia is generally caused by excessive or chronic estrogen stimulation that is unopposed by progesterone, as in patients with chronic anovulation and polycystic ovary syndrome. Although progestin treatment is commonly effective for this disease [129], approximately 30% of patients with complex hyperplasia are progestin resistant [130]. Genetic alterations including mutations of *Pten* tumor suppressor have been shown to be associated with endometrial hyperplasia [131,132]. Elegant work has shown that inactivation of TGFβ signaling and loss of growth inhibition are associated with human endometrial carcinogenesis [133,134]. The role of TGFβ signaling in endometrial cancer has been reviewed and will not be covered in this article [135]. Our recent study shows that loss of TGFBR1 in the mouse uterus using *Amhr2*-Cre enhances epithelial cell proliferation. The aberration culminates in endometrial hyperplasia. Further studies have uncovered potential TGFBR1-mediated paracrine signaling in the regulation of uterine epithelial cell proliferation, and provided genetic evidence supporting the role of uterine epithelial cell proliferation in the pathogenesis of endometrial hyperplasia [136]. Further elucidating the role and the underlying mechanisms of TGFβ signaling in the pathogenesis of endometrial hyperplasia and/or cancer will benefit the design of new therapies.

Conclusions and future directions

A precisely controlled endogenous TGFβ signaling system is of critical importance for the development and function of female reproductive tract. Mouse genetics has proven to be a powerful tool to address many of the fundamental questions posed in the field of TGFβ and reproduction. Conditional knockout approaches have been utilized over the last two decades to decipher the reproductive function of TGFβ superfamily in female reproduction. These studies are at an exciting stage and are advancing at a rapid pace. The functional role of TGFβ signaling in the uterus is beginning to be unveiled. We anticipate that the genetic approach will continue to have large impacts and lead to new breakthroughs in this field. However, understanding how the hormonal, cellular, and molecular signals induce a specific biological response and functional outcome in the context of the uterine microenvironment *in vivo* represents a challenging task. It remains unclear how specific or integrated signals act on the chromatin to shape the epigenetic landscape in physiological and/or pathological conditions of the uterus. Therefore, the interaction between TGFβ signaling and other regulatory pathways (e.g., small RNA pathways) and potential epigenetic mechanisms underlying specific reproductive processes and/or diseases in the uterus need to be clarified. This knowledge will help to design new treatment options for uterine diseases and fertility disorders.

Competing interests
The author declares that he has no competing interests.

Author's contributions
The author reviewed and analyzed the literature and wrote this paper.

Acknowledgements
The author thanks the great support and collaboration from colleagues at Texas A&M University, especially Drs. Kayla Bayless, Gregory Johnson, Robert Burghardt, and Fuller Bazer. Several trainees (Yang Gao, Samantha Duran, Chao Wang, and Haixia Wen) in the author's lab have contributed to the related work. Yang Gao is also acknowledged for the assistance with literature review. Research in this area is supported by the National Institutes of Health grant R21HD073756 from the Eunice Kennedy Shriver National Institute of Child Health & Human Development and the Ralph E. Powe Junior Faculty Enhancement Awards from Oak Ridge Associated Universities.

References

1. Massague J: **Receptors for the TGF-beta family.** *Cell* 1992, **69:**1067–1070.
2. Massague J: **TGF-beta signal transduction.** *Annu Rev Biochem* 1998, **67:**753–791.
3. Chang H, Brown CW, Matzuk MM: **Genetic analysis of the mammalian transforming growth factor-β superfamily.** *Endocr Rev* 2002, **23:**787–823.
4. Schmierer B, Hill CS: **TGFbeta-SMAD signal transduction: molecular specificity and functional flexibility.** *Nat Rev Mol Cell Biol* 2007, **8:**970–982.
5. Tsukazaki T, Chiang TA, Davison AF, Attisano L, Wrana JL: **SARA, a FYVE domain protein that recruits Smad2 to the TGF beta receptor.** *Cell* 1998, **95:**779–791.
6. Imamura T, Takase M, Nishihara A, Oeda E, Hanai J, Kawabata M, Miyazono K: **Smad6 inhibits signalling by the TGF-beta superfamily.** *Nature* 1997, **389:**622–626.
7. Nakao A, Afrakhte M, Moren A, Nakayama T, Christian JL, Heuchel R, Itoh S, Kawabata M, Heldin NE, Heldin CH, ten Dijke P: **Identification of Smad7, a TGFbeta-inducible antagonist of TGF-beta signalling.** *Nature* 1997, **389:**631–635.
8. Massague J: **How cells read TGF-beta signals.** *Nat Rev Mol Cell Biol* 2000, **1:**169–178.
9. Massague J: **TGFbeta signalling in context.** *Nat Rev Mol Cell Biol* 2012, **13:**616–630.
10. Akhurst RJ, Hata A: **Targeting the TGFbeta signalling pathway in disease.** *Nat Rev Drug Discov* 2012, **11:**790–811.
11. Jonk LJC, Itoh S, Heldin CH, ten Dijke P, Kruijer W: **Identification and functional characterization of a Smad binding element (SBE) in the JunB promoter that acts as a transforming growth factor-beta, activin, and bone morphogenetic protein-inducible enhancer.** *J Biol Chem* 1998, **273:**21145–21152.
12. Shi Y, Wang YF, Jayaraman L, Yang H, Massague J, Pavletich NP: **Crystal structure of a Smad MH1 domain bound to DNA: insights on DNA binding in TGF-beta signaling.** *Cell* 1998, **94:**585–594.
13. Ross S, Cheung E, Petrakis TG, Howell M, Kraus WL, Hill CS: **Smads orchestrate specific histone modifications and chromatin remodeling to activate transcription.** *Embo J* 2006, **25:**4490–4502.
14. Moustakas A, Heldin CH: **Non-Smad TGF-beta signals.** *J Cell Sci* 2005, **118:**3573–3584.
15. Zhang YE: **Non-Smad pathways in TGF-beta signaling.** *Cell Res* 2009, **19:**128–139.
16. Guo X, Wang XF: **Signaling cross-talk between TGF-beta/BMP and other pathways.** *Cell Res* 2009, **19:**71–88.
17. Davis BN, Hilyard AC, Lagna G, Hata A: **SMAD proteins control DROSHA-mediated microRNA maturation.** *Nature* 2008, **454:**56–61.
18. Davis BN, Hilyard AC, Nguyen PH, Lagna G, Hata A: **Smad proteins bind a conserved RNA sequence to promote microRNA maturation by Drosha.** *Mol Cell* 2010, **39:**373–384.
19. Davis-Dusenbery BN, Hata A: **Smad-mediated miRNA processing: A critical role for a conserved RNA sequence.** *RNA Biol* 2011, **8:**71–76.
20. Attisano L, Wrana JL: **Signal transduction by the TGF-beta superfamily.** *Science* 2002, **296:**1646–1647.
21. Derynck R, Zhang YE: **Smad-dependent and Smad-independent pathways in TGF-beta family signalling.** *Nature* 2003, **425:**577–584.
22. Yan X, Liu Z, Chen Y: **Regulation of TGF-beta signaling by Smad7.** *Acta Biochim Biophys Sin (Shanghai)* 2009, **41:**263–272.
23. Yan XH, Chen YG: **Smad7: not only a regulator, but also a cross-talk mediator of TGF-beta signalling.** *Biochem J* 2011, **434:**1–10.
24. Pera EM, Ikeda A, Eivers E, De Robertis EM: **Integration of IGF, FGF, and anti-BMP signals via Smad1 phosphorylation in neural induction.** *Genes Dev* 2003, **17:**3023–3028.
25. Wakefield LM, Hill CS: **Beyond TGFbeta: roles of other TGFbeta superfamily members in cancer.** *Nat Rev Cancer* 2013, **13:**328–341.
26. Li Q, Graff JM, O'Connor AE, Loveland KL, Matzuk MM: **SMAD3 regulates gonadal tumorigenesis.** *Mol Endocrinol* 2007, **21:**2472–2486.
27. Pangas SA, Li X, Umans L, Zwijsen A, Huylebroeck D, Gutierrez C, Wang D, Martin JF, Jamin SP, Behringer RR, Robertson EJ, Matzuk MM: **Conditional deletion of Smad1 and Smad5 in somatic cells of male and female gonads leads to metastatic tumor development in mice.** *Mol Cell Biol* 2008, **28:**248–257.
28. Matzuk MM, Finegold MJ, Su JG, Hsueh AJ, Bradley A: **Alpha-inhibin is a tumour-suppressor gene with gonadal specificity in mice.** *Nature* 1992, **360:**313–319.
29. Edson MA, Nalam RL, Clementi C, Franco HL, Demayo FJ, Lyons KM, Pangas SA, Matzuk MM: **Granulosa cell-expressed BMPR1A and BMPR1B have unique functions in regulating fertility but act redundantly to suppress ovarian tumor development.** *Mol Endocrinol* 2010, **24:**1251–1266.
30. Middlebrook BS, Eldin K, Li X, Shivasankaran S, Pangas SA: **Smad1-Smad5 ovarian conditional knockout mice develop a disease profile similar to the juvenile form of human granulosa cell tumors.** *Endocrinology* 2009, **150:**5208–5217.
31. Neptune ER, Frischmeyer PA, Arking DE, Myers L, Bunton TE, Gayraud B, Ramirez F, Sakai LY, Dietz HC: **Dysregulation of TGF-beta activation contributes to pathogenesis in Marfan syndrome.** *Nat Genet* 2003, **33:**407–411.
32. Huang XR, Chung AC, Wang XJ, Lai KN, Lan HY: **Mice overexpressing latent TGF-beta1 are protected against renal fibrosis in obstructive kidney disease.** *Am J Physiol Renal Physiol* 2008, **295:**F118–F127.
33. Massague J: **TGFbeta in Cancer.** *Cell* 2008, **134:**215–230.
34. Arnold SJ, Maretto S, Islam A, Bikoff EK, Robertson EJ: **Dose-dependent Smad1, Smad5 and Smad8 signaling in the early mouse embryo.** *Dev Biol* 2006, **296:**104–118.
35. Huang Z, Wang DG, Ihida-Stansbury K, Jones PL, Martin JF: **Defective pulmonary vascular remodeling in Smad8 mutant mice.** *Hum Mol Genet* 2009, **18:**2791–2801.
36. Tomic D, Miller KP, Kenny HA, Woodruff TK, Hoyer P, Flaws JA: **Ovarian follicle development requires Smad3.** *Mol Endocrinol* 2004, **18:**2224–2240.
37. Gong X, McGee EA: **Smad3 is required for normal follicular follicle-stimulating hormone responsiveness in the mouse.** *Biol Reprod* 2009, **81:**730–738.
38. Li Q, Pangas SA, Jorgez CJ, Graff JM, Weinstein M, Matzuk MM: **Redundant roles of SMAD2 and SMAD3 in ovarian granulosa cells in vivo.** *Mol Cell Biol* 2008, **28:**7001–7011.
39. Pangas SA, Li X, Robertson EJ, Matzuk MM: **Premature luteinization and cumulus cell defects in ovarian-specific Smad4 knockout mice.** *Mol Endocrinol* 2006, **20:**1406–1422.
40. Li X, Tripurani SK, James R, Pangas SA: **Minimal fertility defects in mice deficient in oocyte-expressed Smad4.** *Biol Reprod* 2012, **86:**1–6.

41. Quezada M, Wang J, Hoang V, McGee EA: Smad7 is a transforming growth factor-beta-inducible mediator of apoptosis in granulosa cells. *Fertil Steril* 2012, 97:1452–1459. e1451-1456.

42. Gao Y, Wen H, Wang C, Li Q: SMAD7 antagonizes key TGFbeta superfamily signaling in mouse granulosa cells in vitro. *Reproduction* 2013, 146:1–11.

43. Li Q, Agno JE, Edson MA, Nagaraja AK, Nagashima T, Matzuk MM: Transforming growth factor beta receptor type 1 is essential for female reproductive tract integrity and function. *PLoS Genet* 2011, 7:e1002320.

44. Gao Y, Bayless KJ, Li Q: TGFBR1 is required for mouse myometrial development. *Mol Endocrinol* 2014, 28:380–394.

45. Clementi C, Tripurani SK, Large MJ, Edson MA, Creighton CJ, Hawkins SM, Kovanci E, Kaartinen V, Lydon JP, Pangas SA, DeMayo FJ, Matzuk MM: Activin-like kinase 2 functions in peri-implantation uterine signaling in mice and humans. *PLoS Genet* 2013, 9:e1003863.

46. Nagashima T, Li Q, Clementi C, Lydon JP, Demayo FJ, Matzuk MM: BMPR2 is required for postimplantation uterine function and pregnancy maintenance. *J Clin Invest* 2013, 123:2539–2550.

47. Orvis GD, Behringer RR: Cellular mechanisms of Mullerian duct formation in the mouse. *Dev Biol* 2007, 306:493–504.

48. Brody JR, Cunha GR: Histologic, morphometric, and immunocytochemical analysis of myometrial development in rats and mice: I. Normal development. *Am J Anat* 1989, 186:1–20.

49. Brody JR, Cunha GR: Histologic, morphometric, and immunocytochemical analysis of myometrial development in rats and mice: II. Effects of DES on development. *Am J Anat* 1989, 186:21–42.

50. Miller C, Sassoon DA: Wnt-7a maintains appropriate uterine patterning during the development of the mouse female reproductive tract. *Development* 1998, 125:3201–3211.

51. Wang Y, Jia Y, Franken P, Smits R, Ewing PC, Lydon JP, Demayo FJ, Burger CW, Anton Grootegoed J, Fodde R, Blok LJ: Loss of APC function in mesenchymal cells surrounding the Mullerian duct leads to myometrial defects in adult mice. *Mol Cell Endocrinol* 2011, 341:48–54.

52. Arango NA, Szotek PP, Manganaro TF, Oliva E, Donahoe PK, Teixeira J: Conditional deletion of beta-catenin in the mesenchyme of the developing mouse uterus results in a switch to adipogenesis in the myometrium. *Dev Biol* 2005, 288:276–283.

53. Parr BA, McMahon AP: Sexually dimorphic development of the mammalian reproductive tract requires Wnt-7a. *Nature* 1998, 395:707–710.

54. Mesiano S, Chan EC, Fitter JT, Kwek K, Yeo G, Smith R: Progesterone withdrawal and estrogen activation in human parturition are coordinated by progesterone receptor A expression in the myometrium. *J Clin Endocrinol Metab* 2002, 87:2924–2930.

55. Condon JC, Jeyasuria P, Faust JM, Wilson JW, Mendelson CR: A decline in the levels of progesterone receptor coactivators in the pregnant uterus at term may antagonize progesterone receptor function and contribute to the initiation of parturition. *Proc Natl Acad Sci U S A* 2003, 100:9518–9523.

56. Brainard AM, Miller AJ, Martens JR, England SK: Maxi-K channels localize to caveolae in human myometrium: a role for an actin-channel-caveolin complex in the regulation of myometrial smooth muscle K + current. *Am J Physiol Cell Physiol* 2005, 289:C49–C57.

57. Brainard AM, Korovkina VP, England SK: Potassium channels and uterine function. *Semin Cell Dev Biol* 2007, 18:332–339.

58. Pierce SL, Kresowik JD, Lamping KG, England SK: Overexpression of SK3 channels dampens uterine contractility to prevent preterm labor in mice. *Bio Reprod* 2008, 78:1058–1063.

59. Pierce SL, England SK: SK3 channel expression during pregnancy is regulated through estrogen and Sp factor-mediated transcriptional control of the KCNN3 gene. *Am J Physiol Endocrinol Metab* 2010, 299:E640–E646.

60. Yallampalli C, Dong YL: Estradiol-17beta inhibits nitric oxide synthase (NOS)-II and stimulates NOS-III gene expression in the rat uterus. *Bio Reprod* 2000, 63:34–41.

61. Yallampalli C, Garfield RE, Byam-Smith M: Nitric oxide inhibits uterine contractility during pregnancy but not during delivery. *Endocrinology* 1993, 133:1899–1902.

62. Yallampalli C, Izumi H, Byam-Smith M, Garfield RE: An L-arginine-nitric oxide-cyclic guanosine monophosphate system exists in the uterus and inhibits contractility during pregnancy. *Am J Obstet Gynecol* 1994, 170:175–185.

63. Dong YL, Yallampalli C: Interaction between nitric oxide and prostaglandin E2 pathways in pregnant rat uteri. *Am J Physiol* 1996, 270:E471–E476.

64. Tong D, Lu X, Wang HX, Plante I, Lui E, Laird DW, Bai D, Kidder GM: A dominant loss-of-function GJA1 (Cx43) mutant impairs parturition in the mouse. *Biol Reprod* 2009, 80:1099–1106.

65. Renthal NE, Chen CC, Williams KC, Gerard RD, Prange-Kiel J, Mendelson CR: miR-200 family and targets, ZEB1 and ZEB2, modulate uterine quiescence and contractility during pregnancy and labor. *Proc Natl Acad Sci U S A* 2010, 107:20828–20833.

66. Williams KC, Renthal NE, Gerard RD, Mendelson CR: The microRNA (miR)-199a/214 cluster mediates opposing effects of progesterone and estrogen on uterine contractility during pregnancy and labor. *Mol Endocrinol* 2012, 26:1857–1867.

67. Cockburn K, Rossant J: Making the blastocyst: lessons from the mouse. *J Clin Invest* 2010, 120:995–1003.

68. Flach G, Johnson MH, Braude PR, Taylor RA, Bolton VN: The transition from maternal to embryonic control in the 2-cell mouse embryo. *Embo J* 1982, 1:681–686.

69. Salamonsen LA, Dimitriadis E, Jones RL, Nie G: Complex regulation of decidualization: a role for cytokines and proteases–a review. *Placenta* 2003, 24(Suppl A):S76–S85.

70. Wang H, Dey SK: Roadmap to embryo implantation: clues from mouse models. *Nat Rev Genet* 2006, 7:185–199.

71. Cha J, Sun X, Dey SK: Mechanisms of implantation: strategies for successful pregnancy. *Nat Med* 2012, 18:1754–1767.

72. Guzelogiu-Kayisli Z, Kayisli UA, Taylor HS: The role of growth factors and cytokines during implantation: endocrine and paracrine interactions. *Semin Reprod Med* 2009, 27:62–79.

73. Memon MA, Anway MD, Covert TR, Uzumcu M, Skinner MK: Transforming growth factor beta (TGF beta 1, TGF beta 2 and TGF beta 3) null-mutant phenotypes in embryonic gonadal development. *Mol Cell Endocrinol* 2008, 294:70–80.

74. Mu Z, Yang Z, Yu D, Zhao Z, Munger JS: TGFbeta1 and TGFbeta3 are partially redundant effectors in brain vascular morphogenesis. *Mech Dev* 2008, 125:508–516.

75. Kallapur S, Ormsby I, Doetschman T: Strain dependency of TGFbeta1 function during embryogenesis. *Mol Reprod Dev* 1999, 52:341–349.

76. Ingman WV, Robker RL, Woittiez K, Robertson SA: Null mutation in transforming growth factor beta1 disrupts ovarian function and causes oocyte incompetence and early embryo arrest. *Endocrinology* 2006, 147:835–845.

77. Paria BC, Dey SK: Preimplantation embryo development in vitro - cooperative interactions among embryos and role of growth-factors. *Proc Natl Acad Sci U S A* 1990, 87:4756–4760.

78. Lim J, Bongso A, Ratnam S: Mitogenic and cytogenetic evaluation of transforming growth-factor-beta on murine preimplantation embryonic-development in-vitro. *Mol Reprod Dev* 1993, 36:482–487.

79. Nowak RA, Haimovici F, Biggers JD, Erbach GT: Transforming growth factor-beta stimulates mouse blastocyst outgrowth through a mechanism involving parathyroid hormone-related protein. *Biol Reprod* 1999, 60:85–93.

80. Feinberg RF, Kliman HJ, Wang CL: Transforming growth factor-beta stimulates trophoblast oncofetal fibronectin synthesis in vitro: implications for trophoblast implantation in vivo. *J Clin Endocrinol Metab* 1994, 78:1241–1248.

81. Graham CH, Connelly I, Macdougall JR, Kerbel RS, Stetlerstevenson WG, Lala PK: Resistance of malignant trophoblast cells to both the antiproliferative and anti-invasive effects of transforming growth-factor-beta. *Exp Cell Res* 1994, 214:93–99.

82. Letterio JJ, Geiser AG, Kulkarni AB, Roche NS, Sporn MB, Roberts AB: Maternal rescue of transforming growth factor-beta 1 null mice. *Science* 1994, 264:1936–1938.

83. McLennan IS, Koishi K: Fetal and maternal transforming growth factor-beta 1 may combine to maintain pregnancy in mice. *Biol Reprod* 2004, 70:1614–1618.

84. Akinyemi BO, Adewoye BR, Fakoya TA: Uterine fibroid: a review. *Niger J Med* 2004, 13:318–329.

85. Dou Q, Zhao Y, Tarnuzzer RW, Rong H, Williams RS, Schultz GS, Chegini N: Suppression of transforming growth factor-beta (TGF beta) and TGF beta receptor messenger ribonucleic acid and protein expression in leiomyomata in women receiving gonadotropin-releasing hormone agonist therapy. *J Clin Endocrinol Metab* 1996, 81:3222–3230.

86. Arici A, Sozen I: Transforming growth factor-beta3 is expressed at high levels in leiomyoma where it stimulates fibronectin expression and cell proliferation. *Fertil Steril* 2000, 73:1006–1011.

87. Levy G, Malik M, Britten J, Gilden M, Segars J, Catherino WH: **Liarozole inhibits transforming growth factor-beta3-mediated extracellular matrix formation in human three-dimensional leiomyoma cultures.** *Fertil Steril* 2014, **102**:272–281.

88. Joseph DS, Malik M, Nurudeen S, Catherino WH: **Myometrial cells undergo fibrotic transformation under the influence of transforming growth factor beta-3.** *Fertil Steril* 2010, **93**:1500–1508.

89. Malik M, Catherino WH: **Development and validation of a three-dimensional in vitro model for uterine leiomyoma and patient-matched myometrium.** *Fertil Steril* 2012, **97**:1287–1293.

90. Malik M, Catherino WH: **Novel method to characterize primary cultures of leiomyoma and myometrium with the use of confirmatory biomarker gene arrays.** *Fertil Steril* 2007, **87**:1166–1172.

91. Luo X, Ding L, Chegini N: **CCNs, fibulin-1C and S100A4 expression in leiomyoma and myometrium: inverse association with TGF-beta and regulation by TGF-beta in leiomyoma and myometrial smooth muscle cells.** *Mol Hum Reprod* 2006, **12**:245–256.

92. Levens E, Luo X, Ding L, Williams RS, Chegini N: **Fibromodulin is expressed in leiomyoma and myometrium and regulated by gonadotropin-releasing hormone analogue therapy and TGF-beta through Smad and MAPK-mediated signalling.** *Mol Hum Reprod* 2005, **11**:489–494.

93. Norian JM, Malik M, Parker CY, Joseph D, Leppert PC, Segars JH, Catherino WH: **Transforming growth factor beta3 regulates the versican variants in the extracellular matrix-rich uterine leiomyomas.** *Reprod Sci* 2009, **16**:1153–1164.

94. Di X, Andrews DM, Tucker CJ, Yu L, Moore AB, Zheng X, Castro L, Hermon T, Xiao H, Dixon D: **A high concentration of genistein down-regulates activin A, Smad3 and other TGF-beta pathway genes in human uterine leiomyoma cells.** *Exp Mol Med* 2012, **44**:281–292.

95. Li Z, Burzawa JK, Troung A, Feng S, Agoulnik IU, Tong X, Anderson ML, Kovanci E, Rajkovic A, Agoulnik AI: **Relaxin signaling in uterine fibroids.** *Ann N Y Acad Sci* 2009, **1160**:374–378.

96. Grudzien MM, Low PS, Manning PC, Arredondo M, Belton RJ Jr, Nowak RA: **The antifibrotic drug halofuginone inhibits proliferation and collagen production by human leiomyoma and myometrial smooth muscle cells.** *Fertil Steril* 2010, **93**:1290–1298.

97. Ohara N, Morikawa A, Chen W, Wang J, DeManno DA, Chwalisz K, Maruo T: **Comparative effects of SPRM asoprisnil (J867) on proliferation, apoptosis, and the expression of growth factors in cultured uterine leiomyoma cells and normal myometrial cells.** *Reprod Sci* 2007, **14**:20–27.

98. De Falco M, Staibano S, D'Armiento FP, Mascolo M, Salvatore G, Busiello A, Carbone IF, Pollio F, Di Lieto A: **Preoperative treatment of uterine leiomyomas: Clinical findings and expression of transforming growth factor-beta 3 and connective tissue growth factor.** *J Soc Gynecol Investig* 2006, **13**:297–303.

99. Sinclair DC, Mastroyannis A, Taylor HS: **Leiomyoma simultaneously impair endometrial BMP-2-mediated decidualization and anticoagulant expression through secretion of TGF-beta 3.** *J Clin Endocr Metab* 2011, **96**:412–421.

100. Peracoli MT, Menegon FT, Borges VT, de Araujo Costa RA, Thomazini-Santos IA, Peracoli JC: **Platelet aggregation and TGF-beta(1) plasma levels in pregnant women with preeclampsia.** *J Reprod Immunol* 2008, **79**:79–84.

101. Djurovic S, Schjetlein R, Wisloff F, Haugen G, Husby H, Berg K: **Plasma concentrations of Lp(a) lipoprotein and TGF-beta1 are altered in preeclampsia.** *Clin Genet* 1997, **52**:371–376.

102. Enquobahrie DA, Williams MA, Qiu C, Woelk GB, Mahomed K: **Maternal plasma transforming growth factor-beta1 concentrations in preeclamptic and normotensive pregnant Zimbabwean women.** *J Matern Fetal Neona* 2005, **17**:343–348.

103. Wang XJ, Zhou ZY, Xu YJ: **Changes of plasma uPA and TGF-beta1 in patients with preeclampsia.** *Sichuan Da Xue Xue Bao Yi Xue Ban* 2010, **41**:118–120.

104. Feizollahzadeh S, Taheripanah R, Khani M, Farokhi B, Amani D: **Promoter region polymorphisms in the transforming growth factor beta-1 (TGFbeta1) gene and serum TGFbeta1 concentration in preeclamptic and control Iranian women.** *J Reprod Immunol* 2012, **94**:216–221.

105. Shaarawy M, El Meleigy M, Rasheed K: **Maternal serum transforming growth factor beta-2 in preeclampsia and eclampsia, a potential biomarker for the assessment of disease severity and fetal outcome.** *J Soc Gynecol Investig* 2001, **8**:27–31.

106. Caniggia I, Grisaru-Gravnosky S, Kuliszewsky M, Post M, Lye SJ: **Inhibition of TGF-beta 3 restores the invasive capability of extravillous trophoblasts in preeclamptic pregnancies.** *J Clin Invest* 1999, **103**:1641–1650.

107. Venkatesha S, Toporsian M, Lam C, Hanai J, Mammoto T, Kim YM, Bdolah Y, Lim KH, Yuan HT, Libermann TA, Stillman IE, Roberts D, D'Amore PA, Epstein FH, Sellke FW, Romero R, Sukhatme VP, Letarte M, Karumanchi SA: **Soluble endoglin contributes to the pathogenesis of preeclampsia.** *Nat Med* 2006, **12**:642–649.

108. Stepan H, Kramer T, Faber R: **Maternal plasma concentrations of soluble endoglin in pregnancies with intrauterine growth restriction.** *J Clin Endocrinol Metab* 2007, **92**:2831–2834.

109. Lyall F, Simpson H, Bulmer JN, Barber A, Robson SC: **Transforming growth factor-beta expression in human placenta and placental bed in third trimester normal pregnancy, preeclampsia, and fetal growth restriction.** *Am J Pathol* 2001, **159**:1827–1838.

110. Szarka A, Rigo J Jr, Lazar L, Beko G, Molvarec A: **Circulating cytokines, chemokines and adhesion molecules in normal pregnancy and preeclampsia determined by multiplex suspension array.** *BMC Immunol* 2010, **11**:59.

111. Perucci LO, Gomes KB, Freitas LG, Godoi LC, Alpoim PN, Pinheiro MB, Miranda AS, Teixeira AL, Dusse LM, Sousa LP: **Soluble endoglin, transforming growth factor-Beta 1 and soluble tumor necrosis factor alpha receptors in different clinical manifestations of preeclampsia.** *PLoS One* 2014, **9**:e97632.

112. Huber A, Hefler L, Tempfer C, Zeisler H, Lebrecht A, Husslein P: **Transforming growth factor-beta 1 serum levels in pregnancy and pre-eclampsia.** *Acta Obstet Gynecol Scand* 2002, **81**:168–171.

113. Bersinger NA, Smarason AK, Muttukrishna S, Groome NP, Redman CW: **Women with preeclampsia have increased serum levels of pregnancy-associated plasma protein a (PAPP-A), inhibin A, activin A, and soluble E-selectin.** *Hypertens Pregnancy* 2003, **22**:45–55.

114. Silver HM, Lambert-Messerlian GM, Reis FM, Diblasio AM, Petraglia F, Canick JA: **Mechanism of increased maternal serum total activin A and inhibin A in preeclampsia.** *J Soc Gynecol Investig* 2002, **9**:308–312.

115. Yair D, Eshed-Englender T, Kupferminc MJ, Geva E, Frenkel J, Sherman D: **Serum levels of inhibin B, unlike inhibin A and activin A, are not altered in women with preeclampsia.** *Am J Reprod Immunol* 2001, **45**:180–187.

116. Laivuori H, Kaaja R, Turpeinen U, Stenman UH, Ylikorkala O: **Serum activin A and inhibin A elevated in pre-eclampsia: no relation to insulin sensitivity.** *BJOG* 1999, **106**:1298–1303.

117. Figueras F, Gardosi J: **Intrauterine growth restriction: new concepts in antenatal surveillance, diagnosis, and management.** *Am J Obstet Gynecol* 2011, **204**:288–300.

118. Ostlund E, Tally M, Fried G: **Transforming growth factor-beta1 in fetal serum correlates with insulin-like growth factor-I and fetal growth.** *Obstet Gynecol* 2002, **100**:567–573.

119. Sanford LP, Ormsby I, GittenbergerdeGroot AC, Sariola H, Friedman R, Boivin GP, Cardell EL, Doetschman T: **TGF beta 2 knockout mice have multiple developmental defects that are nonoverlapping with other TGF beta knockout phenotypes.** *Development* 1997, **124**:2659–2670.

120. Jeyabalan A, McGonigal S, Gilmour C, Hubel CA, Rajakumar A: **Circulating and placental endoglin concentrations in pregnancies complicated by intrauterine growth restriction and preeclampsia.** *Placenta* 2008, **29**:555–563.

121. Yinon Y, Nevo O, Xu J, Many A, Rolfo A, Todros T, Post M, Caniggia I: **Severe intrauterine growth restriction pregnancies have increased placental endoglin levels: hypoxic regulation via transforming growth factor-beta 3.** *Am J Pathol* 2008, **172**:77–85.

122. Park CB, DeMayo FJ, Lydon JP, Dufort D: **NODAL in the uterus is necessary for proper placental development and maintenance of pregnancy.** *Biol Reprod* 2012, **86**:194.

123. Mills AM, Longacre TA: **Endometrial hyperplasia.** *Semin Diagn Pathol* 2010, **27**:199–214.

124. Kurman RJ, Kaminski PF, Norris HJ: **The behavior of endometrial hyperplasia. A long-term study of "untreated" hyperplasia in 170 patients.** *Cancer* 1985, **56**:403–412.

125. Montgomery BE, Daum GS, Dunton CJ: **Endometrial hyperplasia: a review.** *Obstet Gynecol Surv* 2004, **59**:368–378.

126. Shutter J, Wright TC: **Prevalence of underlying adenocarcinoma in women with atypical endometrial hyperplasia.** *Int J Gynecol Pathol* 2005, **24**:313–318.

127. Lacey JV, Chia VM: **Endometrial hyperplasia and the risk of progression to carcinoma.** *Maturitas* 2009, **63**:39–44.

128. Hahn HS, Chun YK, Kwon YI, Kim TJ, Lee KH, Shim JU, Mok JE, Lim KT: **Concurrent endometrial carcinoma following hysterectomy for atypical endometrial hyperplasia.** *Eur J Obstet Gynecol Reprod Biol* 2010, **150**:80–83.

129. Gambrell RD: **Progestogens in estrogen-replacement therapy.** *Clin Obstet Gynecol* 1995, **38:**890–901.

130. Reed SD, Voigt LF, Newton KM, Garcia RH, Allison HK, Epplein M, Jordan D, Swisher E, Weiss NS: **Progestin therapy of complex endometrial hyperplasia with and without atypia.** *Obstet Gynecol* 2009, **113:**655–662.

131. Stambolic V, Tsao MS, Macpherson D, Suzuki A, Chapman WR, Mak TW: **High incidence of breast and endometrial neoplasia resembling human Cowden syndrome in pten(+/−) mice.** *Cancer Res* 2000, **60:**3605–3611.

132. Milam MR, Soliman PT, Chung LH, Schmeler KM, Bassett RL, Broaddus RR, Lu KH: **Loss of phosphatase and tensin homologue deleted on chromosome 10 and phosphorylation of mammalian target of rapamycin are associated with progesterone refractory endometrial hyperplasia.** *Int J Gynecol Cancer* 2008, **18:**146–151.

133. Parekh TV, Gama P, Wen X, Demopoulos R, Munger S, Carcangiu ML, Reiss M, Gold LI: **Transforming growth factor beta signaling is disabled early in human endometrial carcinogenesis concomitant with loss of growth inhibition.** *Cancer Res* 2002, **62:**2778–2790.

134. Lecanda J, Parekh TV, Gama P, Lin K, Liarski V, Uretsky S, Mittal K, Gold LI: **Transforming growth factor-beta, estrogen, and progesterone converge on the regulation of p27Kip1 in the normal and malignant endometrium.** *Cancer Res* 2007, **67:**1007–1018.

135. Piestrzeniewicz-Ulanska D, McGuinness D, Yeaman G: **TGF-β Signaling in Endometrial Cancer.** In *Transforming Growth Factor-β in Cancer Therapy, Volume II*. Edited by Jakowlew S. Totowa, NJ: Humana Press; 2008:63–78.

136. Gao Y, Li S, Li Q: **Uterine epithelial cell proliferation and endometrial hyperplasia: evidence from a mouse model.** *Mol Hum Reprod* 2014, **20:**776–786.

Permissions

The contributors of this book come from diverse backgrounds, making this book a truly international effort. This book will bring forth new frontiers with its revolutionizing research information and detailed analysis of the nascent developments around the world.

We would like to thank all the contributing authors for lending their expertise to make the book truly unique. They have played a crucial role in the development of this book. Without their invaluable contributions this book wouldn't have been possible. They have made vital efforts to compile up to date information on the varied aspects of this subject to make this book a valuable addition to the collection of many professionals and students.

This book was conceptualized with the vision of imparting up-to-date information and advanced data in this field. To ensure the same, a matchless editorial board was set up. Every individual on the board went through rigorous rounds of assessment to prove their worth. After which they invested a large part of their time researching and compiling the most relevant data for our readers.

The editorial board has been involved in producing this book since its inception. They have spent rigorous hours researching and exploring the diverse topics which have resulted in the successful publishing of this book. They have passed on their knowledge of decades through this book. To expedite this challenging task, the publisher supported the team at every step. A small team of assistant editors was also appointed to further simplify the editing procedure and attain best results for the readers.

Apart from the editorial board, the designing team has also invested a significant amount of their time in understanding the subject and creating the most relevant covers. They scrutinized every image to scout for the most suitable representation of the subject and create an appropriate cover for the book.

The publishing team has been an ardent support to the editorial, designing and production team. Their endless efforts to recruit the best for this project, has resulted in the accomplishment of this book. They are a veteran in the field of academics and their pool of knowledge is as vast as their experience in printing. Their expertise and guidance has proved useful at every step. Their uncompromising quality standards have made this book an exceptional effort. Their encouragement from time to time has been an inspiration for everyone.

The publisher and the editorial board hope that this book will prove to be a valuable piece of knowledge for researchers, students, practitioners and scholars across the globe.

List of Contributors

Megan R Ruth
Department of Agricultural, Food and Nutritional Science, 4-126A Li Ka Shing Health Research Innovation Centre, University of Alberta, Edmonton, AB T6G 2E1, Canada

Catherine J Field
Department of Agricultural, Food and Nutritional Science, 4-126A Li Ka Shing Health Research Innovation Centre, University of Alberta, Edmonton, AB T6G 2E1, Canada

Andrea D Stapp
Department of Animal Science, Oklahoma State University, Stillwater, OK 74078, USA

Craig A Gifford
Department of Animal Science, Oklahoma State University, Stillwater, OK 74078, USA

Dennis M Hallford
Department of Animal and Range Science, New Mexico State University, Las Cruces, NM 88003, USA

Jennifer A Hernandez Gifford
Department of Animal Science, Oklahoma State University, Stillwater, OK 74078, USA
Department of Animal Science, 114B Animal Science Building, Oklahoma State University, Stillwater, OK 74078, USA

Claudia Maria Bertan Membrive
São Paulo State University, Rod. Comandante João Ribeiro de Barros (SP 294) Km 651, Dracena, SP 17900-000, Brazil

Pauline Martins da Cunha
Department of Animal Reproduction, School of Veterinary Medicine and Animal Science, University of São Paulo, São Paulo, Brazil

Flávio Vieira Meirelles
Department of Veterinary Medicine, School of Animal Sciences and Food Engineering, University of São Paulo, Pirassununga, Brazil

Mario Binelli
Department of Animal Reproduction, School of Veterinary Medicine and Animal Science, University of São Paulo, São Paulo, Brazil

Juliana Abranches Soares Almeida
Department of Animal Sciences, University of Illinois, Urbana 61801, USA

Yanhong Liu
Department of Animal Sciences, University of Illinois, Urbana 61801, USA

Minho Song
Department of Animal Sciences, University of Illinois, Urbana 61801, USA
Department of Animal Science and Biotechnology, Chungnam National University, Daejeon, South Korea

Jeong Jae Lee
Department of Animal Sciences, University of Illinois, Urbana 61801, USA

H Rex Gaskins
Department of Animal Sciences, University of Illinois, Urbana 61801, USA

Carol Wolfgang Maddox
Department of Pathobiology, University of Illinois, Urbana 61801, USA

Orlando Osuna
Milwhite, Inc., Brownsville, TX, USA

James Eugene Pettigrew
Department of Animal Sciences, University of Illinois, Urbana 61801, USA

Govind Kannan
Agricultural Research Station, Fort Valley State University, 1005 State University Drive, Fort Valley, GA 31030, USA

Venkat R Gutta
Agricultural Research Station, Fort Valley State University, 1005 State University Drive, Fort Valley, GA 31030, USA

Jung Hoon Lee
Agricultural Research Station, Fort Valley State University, 1005 State University Drive, Fort Valley, GA 31030, USA

Brou Kouakou
Agricultural Research Station, Fort Valley State University, 1005 State University Drive, Fort Valley, GA 31030, USA

Will R Getz
Agricultural Research Station, Fort Valley State University, 1005 State University Drive, Fort Valley, GA 31030, USA

George W McCommon
Agricultural Research Station, Fort Valley State University, 1005 State University Drive, Fort Valley, GA 31030, USA

Michael E McCormick
Southeast Region LSU Agricultural Center, 21549 Old Covington Hwy, Hammond, LA 70403, USA

Kun Jun Han
Louisiana State University School of Plant, Soil and Environmental Sciences, Baton Rouge 70803, USA

Vinicius R Moreira
Louisiana State University Agricultural Center, Southeast Research Station, P.O. Drawer 569, Franklinton 70438, USA

David C Blouin
Department of Experimental Statistics, Louisiana State University, Baton Rouge 70803, USA

Laura D Brown
Perinatal Research Center, Division of Neonatology, Department of Pediatrics, University of Colorado Denver school of Medicine, Aurora, CO, USA
Center for Women's Health Research, University of Colorado Denver School of Medicine, Aurora, CO, USA

Stephanie R Thorn
Perinatal Research Center, Division of Neonatology, Department of Pediatrics, University of Colorado Denver School of Medicine, Aurora, CO, USA

Alex Cheung
Perinatal Research Center, Division of Neonatology, Department of Pediatrics, University of Colorado Denver School of Medicine, Aurora, CO, USA

Jinny R Lavezzi
Perinatal Research Center, Division of Neonatology, Department of Pediatrics, University of Colorado Denver School of Medicine, Aurora, CO, USA

Frederick C Battaglia
Perinatal Research Center, Division of Neonatology, Department of Pediatrics, University of Colorado Denver School of Medicine, Aurora, CO, USA

Paul J Rozance
Perinatal Research Center, Division of Neonatology, Department of Pediatrics, University of Colorado Denver School of Medicine, Aurora, CO, USA
Center for Women's Health Research, University of Colorado Denver School of Medicine, Aurora, CO, USA

Kathrin A Dunlap
Department of Animal Science, Texas A&M University, 2471 TAMU, College Station, Texas 77843, USA

Jacob D Brown
Department of Animal Science, Texas A&M University, 2471 TAMU, College Station, Texas 77843, USA

Ashley B Keith
Department of Animal Science, Texas A&M University, 2471 TAMU, College Station, Texas 77843, USA

M Carey Satterfield
Department of Animal Science, Texas A&M University, 2471 TAMU, College Station, Texas 77843, USA

Rodney D Geisert
Animal Sciences Research Center, University of Missouri, 920 East Campus Drive, Columbia, MO 65211, USA

Matthew C Lucy
Animal Sciences Research Center, University of Missouri, 920 East Campus Drive, Columbia, MO 65211, USA

Jeffrey J Whyte
Animal Sciences Research Center, University of Missouri, 920 East Campus Drive, Columbia, MO 65211, USA

Jason W Ross
Department of Animal Science, Iowa State University, 2356 Kildee Hall, Ames, IA 50011, USA

Daniel J Mathew
Animal Sciences Research Center, University of Missouri, 920 East Campus Drive, Columbia, MO 65211, USA

Jianjun Zang
State Key Laboratory of Animal Nutrition, Ministry of Agriculture Feed Industry Centre, China Agricultural University, Beijing 100193, China

Jingshu Chen
State Key Laboratory of Animal Nutrition, Ministry of Agriculture Feed Industry Centre, China Agricultural University, Beijing 100193, China

Ji Tian
State Key Laboratory of Animal Nutrition, Ministry of Agriculture Feed Industry Centre, China Agricultural University, Beijing 100193, China

Aina Wang
Weifang Business Vocational College, Zhucheng, Shandong 262234, China

Hong Liu
State Key Laboratory of Animal Nutrition, Ministry of Agriculture Feed Industry Centre, China Agricultural University, Beijing 100193, China

Shengdi Hu
State Key Laboratory of Animal Nutrition, Ministry of Agriculture Feed Industry Centre, China Agricultural University, Beijing 100193, China

Xiangrong Che
College of Animal Science and Veterinary Medicine, Shanxi Agricultural University, Taigu, Shanxi 030801, China

Yongxi Ma
State Key Laboratory of Animal Nutrition, Ministry of Agriculture Feed Industry Centre, China Agricultural University, Beijing 100193, China

Junjun Wang
State Key Laboratory of Animal Nutrition, Ministry of Agriculture Feed Industry Centre, China Agricultural University, Beijing 100193, China

hunlin Wang
State Key Laboratory of Animal Nutrition, Ministry of Agriculture Feed Industry Centre, China Agricultural University, Beijing 100193, China

Guanghua Du
College of Animal Science and Veterinary Medicine, Shanxi Agricultural University, Taigu, Shanxi 030801, China

Xi Ma
State Key Laboratory of Animal Nutrition, Ministry of Agriculture Feed Industry Centre, China Agricultural University, Beijing 100193, China

Lingyan Li
National Beef Cattle Industry and Technology System, College of Animal Science and Technology, China Agricultural University, Beijing 100193, China

Yuankui Zhu
Tin Wah Industrial Co.Ltd, Shimen Industrial Area of Lian Yuan 417100, Hunan Province, China

Xianyou Wang
National Beef Cattle Industry and Technology System, College of Animal Science and Technology, China Agricultural University, Beijing 100193, China

Yang He
National Beef Cattle Industry and Technology System, College of Animal Science and Technology, China Agricultural University, Beijing 100193, China

Binghai Cao
National Beef Cattle Industry and Technology System, College of Animal Science and Technology, China Agricultural University, Beijing 100193, China

Da Chuan Piao
Laboratory of Animal Cell Biotechnology, Department of Agricultural Biotechnology, Seoul National University, Shinlim-dong, Kwanak-gu, Seoul 151-742, South Korea

Tao Wan
College of Animal Science and Technology, Jilin Agricultural University, 2888 Xincheng Street, Nan-guan District, Changchun 130118, People's Republic of China Key Laboratory of Animal Nutrition and Feed Science, Jilin Province, Jilin Agricultural University, 2888 Xincheng Street, Nan-guan District, Changchun 130118, People' Republic of China

Jae Sung Lee
Department of Animal Science and Technology, College of Animal Bioscience & Technology, Konkuk University, 120 Neungdong-ro, Gwangjin-gu, Seoul 143-701, South Korea

Renato SA Vega
Animal & Dairy Sciences Cluster, College of Agriculture, University of the Philippines Los Bañose, Los Baños 4031, Laguna, Philippines

Sang Ki Kang
Laboratory of Animal Cell Biotechnology, Department of Agricultural Biotechnology, Seoul National University, Shinlim-dong, Kwanak-gu, Seoul 151-742, South Korea

Yun Jaie Choi
Laboratory of Animal Cell Biotechnology, Department of Agricultural Biotechnology, Seoul National University, Shinlim-dong, Kwanak-gu, Seoul 151-742, South Korea

Hong Gu Lee
Department of Animal Science and Technology, College of Animal Bioscience & Technology, Konkuk University, 120 Neungdong-ro, Gwangjin-gu, Seoul 143-701, South Korea

Robert R Kraeling
L&R Research Associates, Watkinsville, GA, USA

Stephen K Webel
JBS United Animal Health, Sheridan, IN, USA

Greg A Johnson
Department of Veterinary Integrative Biosciences, Texas A&M University, College Station, TX 77843-4458, USA

Robert C Burghardt
Department of Veterinary Integrative Biosciences, Texas A&M University, College Station, TX 77843-4458, USA

Fuller W Bazer
Department of Animal Science, Texas A&M University, College Station, TX 77843, USA

Xi Lin
Laboratory of Developmental Nutrition, Department of Animal Sciences, North Carolina State University, Box 7621, Raleigh, NC 27695, USA

Sheila Jacobi
Laboratory of Developmental Nutrition, Department of Animal Sciences, North Carolina State University, Box 7621, Raleigh, NC 27695, USA

Jack Odle
Laboratory of Developmental Nutrition, Department of Animal Sciences, North Carolina State University, Box 7621, Raleigh, NC 27695, USA

Yang Wang
Ministry of Agriculture Key Laboratory of Animal Genetics, Breeding and Reproduction, National engineering laboratory for animal breeding, College of Animal Sciences and Technology, China Agricultural University, No.2 Yuanmingyuan Xi Lu, Haidian, Beijing 100193, China

Chao Wang
Ministry of Agriculture Key Laboratory of Animal Genetics, Breeding and Reproduction, National engineering laboratory for animal breeding, College of Animal Sciences and Technology, China Agricultural University, No.2Yuanmingyuan Xi Lu, Haidian, Beijing 100193, China

Zhuocheng Hou
National Engineering Laboratory for Animal Breeding and MOA Key Laboratory of Animal Genetics and Breeding, China Agricultural University, Beijing 100193, China

Kai Miao
Ministry of Agriculture Key Laboratory of Animal Genetics, Breeding and Reproduction, National engineering laboratory for animal breeding, College of Animal Sciences and Technology, China Agricultural University, No.2 Yuanmingyuan Xi Lu, Haidian, Beijing 100193, China

Haichao Zhao
Ministry of Agriculture Key Laboratory of Animal Genetics, Breeding and Reproduction, National engineering laboratory for animal breeding, College of Animal Sciences and Technology, China Agricultural University, No.2 Yuanmingyuan Xi Lu, Haidian, Beijing 100193, China

Rui Wang
Ministry of Agriculture Key Laboratory of Animal Genetics, Breeding and Reproduction, National engineering laboratory for animal breeding, College of Animal Sciences and Technology, China Agricultural University, No.2 Yuanmingyuan Xi Lu, Haidian, Beijing 100193, China

Min Guo
Ministry of Agriculture Key Laboratory of Animal Genetics, Breeding and Reproduction, National engineering laboratory for animal breeding, College of Animal Sciences and Technology, China Agricultural University, No.2 Yuanmingyuan Xi Lu, Haidian, Beijing 100193, China

Zhonghong Wu
State Key Laboratory of Animal Nutrition, College of Animal Sciences and Technology, China Agricultural University, No.2 Yuanmingyuan Xi Lu, Haidian, Beijing 100193, China

Jianhui Tian
Ministry of Agriculture Key Laboratory of Animal Genetics, Breeding and Reproduction, National engineering laboratory for animal breeding, College of Animal Sciences and Technology, China Agricultural University, No.2Yuanmingyuan Xi Lu, Haidian, Beijing 100193, China

Lei An
Ministry of Agriculture Key Laboratory of Animal Genetics, Breeding and Reproduction, National engineering laboratory for animal breeding, College of Animal Sciences and Technology, China Agricultural University, No.2 Yuanmingyuan Xi Lu, Haidian, Beijing 100193, China

Yong Shao and Feng-Qi Zhao
Laboratory of Lactation and Metabolic Physiology, Department of Animal Science, University of Vermont, Burlington, Vermont 05405, USA

Yanhua Zhou
National Engineering Laboratory for Animal Breeding, Key Laboratory of Animal genetics, Breeding and Reproduction, Ministry of Agriculture, College of Animal Science and Technology, China Agricultural University, Beijing 100193, P.R. China

Xiangwei Fu
National Engineering Laboratory for Animal Breeding, Key Laboratory of Animal genetics, Breeding and Reproduction, Ministry of Agriculture, College of Animal Science and Technology, China Agricultural University, Beijing 100193, P.R. China

Guangbin Zhou
Institute of Animal Genetics and Breeding, College of Animal Science and Technology, Sichuan Agricultural University (Chengdu Campus), Wenjiang 611130, P.R. China

Baoyu Jia
National Engineering Laboratory for Animal Breeding, Key Laboratory of Animal genetics, Breeding and Reproduction, Ministry of Agriculture, College of Animal Science and Technology, China Agricultural University, Beijing 100193, P.R. China

Yi Fang
National Engineering Laboratory for Animal Breeding, Key Laboratory of Animal genetics, Breeding and Reproduction, Ministry of Agriculture, College of Animal Science and Technology, China Agricultural University, Beijing 100193, P.R. China

Yunpeng Hou
State Key Laboratory for Agrobiotechnology, College of Biological Sciences, China Agricultural University, Beijing 100193, P.R. China

Shien Zhu
National Engineering Laboratory for Animal Breeding, Key Laboratory of Animal genetics, Breeding and Reproduction, Ministry of Agriculture, College of Animal Science and Technology, China Agricultural University, Beijing 100193, P.R. China

Ratan K Choudhary
School of Animal Biotechnology, Guru Angad Dev Veterinary and Animal Science University, Ludhiana, Punjab 141004, India

Bob Goodband
Department of Animal Sciences and Industry, Kansas State University, Manhattan, KS 66506-0201, USA

Mike Tokac
Department of Animal Sciences and Industry, Kansas State University, Manhattan, KS 66506-0201, USA

Steve Dritz
Department of Diagnostic Medicine/ Pathobiology, College of Veterinary Medicine, Kansas State University, Manhattan, KS 66506-0201, USA

Joel DeRouchey
Department of Animal Sciences and Industry, Kansas State University, Manhattan, KS 66506-0201, USA

Jason Woodworth
Department of Animal Sciences and Industry, Kansas State University, Manhattan, KS 66506-0201, USA
Department of Diagnostic Medicine/ Pathobiology, College of Veterinary Medicine, Kansas State University, Manhattan, KS 66506-0201, USA

Kasey M Moyes
Department of Animal and Avian Sciences, University of Maryland, 142 Animal Sciences Building, MD 20742-2311, 20910 College Park, MD, USA

Torben Larsen
Department of Animal Science, Faculty of Science and Technology, Aarhus University, Tjele 8830, Denmark

Peter Sørensen
Department of Animal Science, Faculty of Science and Technology, Aarhus University, Tjele 8830, Denmark

Klaus L Ingvartsen
Department of Animal Science, Faculty of Science and Technology, Aarhus University, Tjele 8830, Denmark

Qinglei Li
Department of Veterinary Integrative Biosciences, College of Veterinary Medicine and Biomedical Sciences, Texas A&M University, College Station, TX 77843, USA

www.ingramcontent.com/pod-product-compliance
Lightning Source LLC
Chambersburg PA
CBHW080626200326
41458CB00013B/4523